KU-282-082

Rock Weathering and Landform Evolution

CHESTER COLLEGE

ACC. No.
0100307 9

CLASS No.
551·302 ROB

LIBRARY

British Geomorphological Research Group Symposia Series

Rock Weathering and Landform Evolution

Edited by

D. A. Robinson

and

R. B. G. Williams

Geography Laboratory
University of Sussex, Brighton, UK

JOHN WILEY & SONS
Chichester · New York · Brisbane · Toronto · Singapore

Copyright © 1994 by John Wiley & Sons Ltd,
                 Baffins Lane, Chichester,
                 West Sussex PO19 1UD, England
                 Telephone National Chichester (0243) 779777
                 International (+44) (243) 779777

All rights reserved.

No part of this book may be reproduced by any means,
or transmitted, or translated into a machine language
without the written permission of the publisher.

Other Wiley Editorial Offices

John Wiley & Sons, Inc., 605 Third Avenue,
New York, NY 10158-0012, USA

Jacaranda Wiley Ltd, 33 Park Road, Milton,
Queensland 4064, Australia

John Wiley & Sons (Canada) Ltd, 22 Worcester Road,
Rexdale, Ontario M9W 1L1, Canada

John Wiley & Sons (SEA) Pte Ltd, 37 Jalan Pemimpin #05-04,
Block B, Union Industrial Building, Singapore 2057

Library of Congress Cataloging-in-Publication Data

Rock weathering and landform evolution / edited by D. A. Robinson and
R. B. G. Williams.
       p.   cm.—(British Geomorphological Research Group symposia
series)
   Includes bibliographical references and index.
   ISBN 0-471-95119-6
   1. Weathering.   2. Landscape changes.   I. Robinson, D. A.
II. Williams, R. B. G.   III. Series.
QE570.R63   1994
551.3′02—dc20   94-4636
                                                  CIP

British Library Cataloguing in Publication Data

A catalogue record for this book is available from the British Library

ISBN 0-471-95119-6

Typeset in 10/12pt Times by Acorn Bookwork, Salisbury, Wilts
Printed and bound in Great Britain by Bookcraft (Bath) Ltd

Contents

Section 5   WEATHERING AND LANDFORM DEVELOPMENT IN TEMPERATE ENVIRONMENTS

List of Contributors

R. J. Allison Department of Geography, University of Durham, Durham DH1 3LE, UK

D. J. Bowden Newman College, University of Birmingham, Genners Lane, Bartley Green, Birmingham B32 3NT, UK

R. U. Cooke Vice-Chancellor, University of York, Heslington, York YO1 5DD, UK

G. R. Douglas Mentaskólinn vid Hamrahlid, Reykjavik, Iceland

J. Ehlen US Army Topographic Engineering Center, Alexandria, Virginia 22060 5546, USA

I. S. Evans Earth Surface Systems Research Group, Department of Geography, University of Durham, South Road, Durham DH1 3LE, UK

C-H. Fan Department of Geological Sciences, University College London, Gower Street, London WC1E 6BT, UK

R. A. M. Gardner Department of Geography, Queen Mary and Westfield College, Mile End Road, London E1 4NS, UK

A. J. Gerrard School of Geography, The University of Birmingham, Edgbaston, Birmingham B15 2TT, UK

L. Guisti Division of Environmental Sciences, University of Lancaster, University House, Lancaster LA1 4YW, UK

A. S. Goudie School of Geography, University of Oxford, Mansfield Road, Oxford OX1 3TB, UK

R. J. Inkpen Department of Geography, University of Portsmouth, Buckingham Building, Lion Terrace, Portsmouth PO1 3HE, UK

M. E. Jones Department of Geological Sciences, University College London, Gower Street, London WC1E 6BT, UK

M. Kimata Institute of Geoscience, University of Tsukuba, Ibaraki 305, Japan

J. Lewin Institute of Earth Studies, UCW Aberystwyth, Aberystwyth, Dyfed SY23 3DB, UK

M. G. Macklin Department of Geography, University of Leeds, Leeds, LS2 9JT, UK

R. W. Magee School of Geosciences, The Queen's University of Belfast, Belfast BT7 1NN, UK

Y. Matsukura Institute of Geoscience, University of Tsukuba, Ibaraki 305, Japan

M. J. McFarlane   Department of Environmental Science, University of Botswana, Gaborone, Botswana

J. P. McGreevy   Conservation Laboratory, Ulster Museum, Belfast BT9 5AB, UK

C. A. Moses   School of Geosciences, The Queen's University of Belfast, Belfast BT7 1NN, UK

D. N. Mottershead   Department of Geography, Edge Hill College of Higher Education, St Helen's Road, Ormskirk, Lancashire L39 4QP, UK

R. Parish   Geography Laboratory, University of Sussex, Falmer, Brighton BN1 9QN, UK

A. Pentecost   Division of Biosphere Sciences, King's College London, The Strand, London WC2R 2LS, UK

E. T. Power   School of Geosciences, The Queen's University of Belfast, Belfast BT7 1NN, UK

B. R. Rea   School of Geosciences, The Queen's University of Belfast, Belfast BT7 1NN, UK

D. A. Robinson   Geography Laboratory, University of Sussex, Falmer, Brighton, BN1 9QN, UK

K-H. Schmidt   Geomorphologisches Laboratorium, Freie Universität Berlin, Altensteinstrasse 19, D-1000 Berlin 33, Germany

R. Sjöberg   Centre for Arctic Research, Umeå University, S-901 87 Umeå, Sweden

B. J. Smith   School of Geosciences, The Queen's University of Belfast, BT7 1NN, UK

T. Suzuki   Institute of Geosciences, Chuo University, Tokyo 112, Japan

J. O. H. Swantesson   Department of Geography, University of Karlstad, Box 9501, S-650 09 Karlstad, Sweden

K. Takahashi   Faculty of Literature, Chuo University, Tokyo 192-03, Japan

R. M. Teeuw   Environmental Science Division, University of Hertfordshire, College Lane, Hatfield, Herts AL10 9AB, UK

M. F. Thomas   Department of Environmental Science, University of Stirling, Stirling FK9 4LA, UK

M. B. Thorp   Department of Geography, University College Dublin, Dublin 4, Eire

H. A. Viles   St Catherine's College, University of Oxford, Oxford OX1 3UJ, UK

P. A. Warke   School of Geosciences, The Queen's University of Belfast, Belfast BT7 1NN, UK

W. B. Whalley   School of Geosciences, The Queen's University of Belfast, Belfast BT7 1NN, UK

R. B. G. Williams   Geography Laboratory, University of Sussex, Falmer, Brighton BN1 9QN, UK

J. C. Woodward   Department of Environmental and Geographical Sciences, Manchester Metropolitan University, Manchester M1 5GO, UK

S. Yokoyama   Department of Geography, Faculty of Education, Kumamoto University, Kumamoto 806, Japan

Series Preface

The British Geomorphological Research Group (BGRG) is a national multidisciplinary Society whose object is 'the advancement of research and education in geomorphology'. Today the BGRG enjoys an international reputation and has a strong membership from both Britain and overseas. Indeed, the Group has been actively involved in stimulating the development of geomorphology and geomorphological societies in several countries. The BGRG was constituted in 1961 but its beginnings lie in a meeting held in Sheffield under the chairmanship of Professor D. L. Linton in 1958. Throughout its development the Group has sustained important links with both the Institute of British Geographers and the Geological Society of London.

Over the past three decades the BGRG has been highly successful and productive. This is reflected not least by BGRG publications. Following its launch in 1976 the Group's journal, *Earth Surface Processes* (since 1981 *Earth Surface Processes and Landforms*) has become acclaimed internationally as a leader in its field, and to a large extent the journal has been responsible for advancing the reputation of the BGRG. In addition to an impressive list of other publications on technical and educational issues, including 30 *Technical Bulletins* and the influential *Geomorphological Techniques* edited by A. Goudie, BGRG symposia have led to the production of a number of important works. These have included *Nearshore Sediment Dynamics and Sedimentation* edited by J. R. Hails and A. P. Carr: *Geomorphology and Climate*, edited by E. Derbyshire; *Geomorphology, Present Problems and Future Prospects*, edited by C. Embleton, D. Brunsden and D. K. C. Jones; *Megageomorphology*, edited by R. Gardner and H. Scoging; *River Channel Changes*, edited by K. J. Gregory; and *Timescales in Geomorphology*, edited by R. Cullingford, D. Davison and J. Lewin. This sequence of books culminated in 1987 with a publication, in two volumes, of the *Proceedings of the First International Geomorphology Conference*, edited by Vince Gardiner. This international meeting, arguably the most important in the history of geomorphology, provided the foundation for the development of geomorphology into the next century.

This current BGRG Symposia Series has been founded and is now being fostered to help maintain the research momentum generated during the past three decades, as well as to further the widening of knowledge in component fields of geomorphological endeavour. The series consists of authoritative volumes based on the themes of BGRG meetings, incorporating, where appropriate, invited contributions to complement chapters selected from presentations at these meetings under the guidance and editorship of one or more suitable specialists. Although maintaining a strong emphasis on pure geomorphological research, BGRG meetings are diversifying, in a very positive way, to consider links between geomorphology *per se* and other disciplines such as ecology, agriculture, engineering and planning.

The first volume in the series was published in 1988. *Geomorphology in Environmental Planning*, edited by Janet Hooke, reflected the trend towards applied studies. The second volume, edited by Keith Beven and Paul Carling, *Floods—Hydrological, Sedimentological and Geomorphological Implications*, focused on a traditional research theme. *Soil Erosion in Agricultural Land* reflected the international importance of the topic for researchers during the 1980s. This volume, edited by John Boardman, John Dearing and Ian Foster formed the third in the series. The role of vegetation in geomorphology is a traditional research theme, recently revitalised with the move towards interdisciplinary studies. The fourth in the series, *Vegetation and Erosion—Processes and Environments*, edited by John Thornes, reflected this development in geomorphological endeavour, and raised several research issues for the next decade. The fifth volume, *Lowland Floodplain Rivers—Geomorphological Perspectives*, edited by Paul Carling and Geoff Petts, reflects recent research into river channel adjustments, especially those consequent to engineering works and land use change. The sixth volume *Landscape Sensitivity*, edited by David Thomas and Robert Allison addressed a vital geomorphological topic. This concerns the way in which landscape and landforms respond to external changes; important concepts for understanding landform development and crucial to an appreciation of human-induced response in the landscape. The seventh in the series, *Geomorphology and Sedimentology of Lakes and Reservoirs* by John McManus and Robert Duck, provided a stimulating mixture of pure and applied research which appealed to a wide audience, and the eighth in the series, *Process Models and Theoretical Geomorphology*, edited by Mike Kirkby, reported some important new numerical and field results from a variety of environments. The ninth in the series, *Environmental Change in Drylands*, edited by Andrew Millington and Ken Pye, continued the current trend of analysing recent palaeoenvironmental change in order to understand and to control present-day systems and included examples from Europe, Africa, North America and Australia.

The present volume (the tenth in the series) *Rock Weathering and Landform Evolution*, edited by D. A. Robinson and R. B. G. Williams, forms both an extensive review of rock weathering and also presents some of the most recent research findings. The book is divided into six sections which progress logically from descriptions of work on the rock weathering processes through geochronology to the application of the research results to landforms in tropical, temperate and high latitudes. *Rock Weathering and Landform Evolution* represents a very valuable addition to the Symposia Series which will appeal to a wide readership.

Jack Hardisty
BGRG Publications Committee

Introduction: Advances in Rock Weathering Studies

The aim of this book is to illustrate the advances currently being made by geomorphologists in the study of weathering processes and landforms, and to examine the variety of approaches and techniques that have been developed. The book has its origin in a series of papers presented at the Annual Conference of the British Geomorphological Research Group at the University of Sussex in September 1992, but includes additional material.

The first section of the book examines the nature of the various weathering processes and the types of debris that they produce. Weathering processes are of crucial importance in geomorphology because most material removed in the development of landforms has first to be weakened or loosened by one or more weathering processes before erosion can commence. Despite this importance, weathering processes have often been curiously neglected or taken for granted in landform studies.

The principal variables controlling the nature and rate of weathering have long been recognised to be the composition and structure of the parent rock, the nature of the climate and the length of time over which weathering has operated. However, there are still many uncertainties about precisely how the different processes achieve rock breakdown, and why weathering varies greatly on different rocks and in different climates. Many weathering processes would appear to be much more complex than has generally been thought to be the case.

The study of weathering processes has always posed difficulties because most act only very slowly over very long time spans. The processes can be exasperatingly difficult to observe in action in the field, and their effects are often impossible to monitor accurately except over a period of many years or decades. Because of the problems associated with fieldwork, some researchers have resorted to laboratory experiments, but unfortunately it is hard to simulate weathering mechanisms satisfactorily in the laboratory without so speeding them up as to cast serious doubt on the validity of the results.

In the field, unweathered and weathered rock tend to occur in close juxtaposition, but it is often impossible to be certain which weathering processes have transformed the one into the other, because different processes can produce identical or very similar results. The problems of interpretation are further complicated by past climactic changes. Many regoliths contain weathered debris produced under more than one climatic regime, particularly in the temperate regions of the world where climatic oscillations in recent millennia have been especially marked, but also in hot deserts and other areas. As a result, as the contributions by Gerrard (Chapter 1) and Power and Smith (Chapter 2) demonstrate, the precise nature of weathering products produced in different environments remains uncertain, even for a common rock type such as granite that has been intensively studied.

Despite its importance to geomorphology, experimental research into the fundamental mechanisms of weathering has until quite recently been badly neglected. Many geologists and geomorphologists in the past apparently thought that rock weathering was such an obvious phenomenon, and the nature of the different processes so self-evident, that detailed research was either unnecessary or at any rate not a major priority. Before 1950 only a very few geologists and geomorphologists conducted experiments to try to discover precisely how the different weathering processes actually worked, and, if they worked, how effective they were. The most famous of the experimenters was undoubtedly Griggs (1936). His experiments were seriously flawed, and yet for many decades his results were frequently quoted as proof that temperature changes (above 0°C) are almost totally ineffective in causing rock breakdown except in the presence of water. The fact that his results were entirely ambiguous owing to poor experimental design seemed to escape most commentators.

In the absence of a sustained research effort by geologists and geomorphologists, the impetus for developing experimental weathering studies came largely from architects and engineers who were concerned about the strength and durability of building stones, concrete and other constructional materials (see, for example, Schaffer, 1932; Honeyborne and Harris, 1958; Winkler, 1973). Thomas (1938) deserves special mention for his major contribution to the understanding of weathering processes.

After 1950, geomorphologists began to take a greater interest in rock weathering studies. Pioneering experimental work was undertaken by Battle (Battle, 1960; Whalley and McGreevy, 1991), but after his untimely death, there was a lull until the increasing interest in geomorphological processes that characterised the 1960s led to a resumption of experimental work designed to test the efficacy of weathering processes (Wiman, 1963; Williams, 1968; Potts, 1970). Recent years have seen a steady stream of papers reporting the results of experimental rock weathering, and major progress has been made in elucidating both the processes and products of weathering. Of particular note is the extensive programme of experiments on frost shattering and other weathering processes, which commenced in the late 1960s at the Centre de Geomorphologie du CNRS at Caen (Lautridou and Ozouf, 1982), and has greatly increased levels of understanding. The experimental approach is continued in this book with the contributions by Allison and Goudie (Chapter 3) and Warke and Smith (Chapter 4).

A major problem common to many experimental weathering studies is how to measure the weakening of rock samples prior to breakdown without further weakening them or causing premature breakdown. An innovative feature of Allison and Goudie's work, which is concerned with the effects of fire on rock breakdown, is their use of a 'Grindsonic' instrument that provides a measure of changes in rock strength during the course of weathering experiments, yet does not damage the rocks.

Although researchers generally record the changes that they make to the air temperatures during their weathering simulations, and the initial degree of saturation of their specimens, they often fail to monitor the fluctuating temperatures and humidity levels within their rock samples, which makes the results of the experiments difficult to interpret. It is essential to know how the temperatures and moisture levels vary inside the samples if the weathering processes at work are to be interpreted correctly. Rock materials differ in their specific heat and conductivity, and consequently the temperature and humidity fluctuations in rocks never correspond precisely with the changes recorded in the surrounding air space, and vary from rock to rock. In addition, there may be major lag effects due to heat given out as a result of changes of state from liquid to solid or absorbed as a result of changes from solid to liquid (Jerwood *et al.*, 1987). Thus, there is a real need to monitor rock, rather than atmospheric, conditions during simulated weathering experiments, and the paper by Warke and Smith (Chapter 4) is a useful addition to our knowledge in this relatively unexplored field.

One of the most fundamental weathering processes that is still inadequately understood is crack propagation. Although there are several theories as to how and why cracks develop in different materials (Barber and Meredith, 1990) there have been few studies that have tried to link crack development to the macro-failure of rock in the field. In Chapter 5, Douglas, McGreevy and Whalley suggest that crack development in the basalts that they have studied in Northern Ireland can be explained, in part at least, by the mineralogy of the rocks, and that this, in turn, controls the freeface activity on the cliffs.

A common misconception about weathering processes is that they are invariably destructive, whereas in reality they may strengthen rocks, particularly sandstones, through the formation of surface rinds or crusts. Some weathering rinds result from mineral depletion and weakening of the surface layers, but the majority involve reprecipitation and enrichment of the surface layers, which normally helps to strengthen and waterproof the rock. In Chapter 6, Matsukura, Kimata and Yokoyama discuss an unusual example of crust formation on andesite blocks around the Aso volcano in Japan that they ascribe to the fumes given off by the volcano.

Their work is a good illustration of the successful use of thin sections and the conventional petrographic microscope to observe and interpret internal changes within a rock. However, in weathering studies, the preparation of thin sections often causes difficulties because the weathered rock tends to disintegrate when cut. Impregnation to hold the samples together does not always provide a satisfactory solution since the resulting slides may be so thick as to make identification of minerals problematic. The arrival of the scanning electron microscope (SEM), which allows the examination of undisturbed pieces of rock under very high magnification, has offered new opportunities and led to a burst of renewed interest and activity in weathering studies. The technique has proved particularly valuable for studying alteration products, and for interpreting the weathering history of rocks and sediments, as well as for the study of weathering crusts. The varied use of this type of analysis is demonstrated in a number of contributions to this volume. The ability to marry SEM with elemental analysis of the mineral components visible on or within a weathered rock by X-ray diffraction analysis (XRD), or differential thermal analysis (DTA), has proved particularly valuable, and has led to major advances in the understanding of weathering processes and products, although the use of more traditional techniques of chemical analysis such as volumetric analysis, atomic absorption spectrophotometry and colorimetry remains important.

A particularly difficult form of weathering to study, where the application of SEM and other high technology equipment may help to resolve many outstanding issues, is the problematic field of the biochemical weathering of rocks by lichens. In Chapter 7, Viles and Pentecost provide a welcome collaboration between a geomorphologist and a biologist using a combination of field work and SEM observations to begin to unravel the interrelationships between the lichen cover and the weathering of the underlying sandstone in South Africa.

In the second section of the book, attention switches from laboratory experiments and the weathering of natural rock surfaces to the weathering of buildings constructed of cut or dressed rock. Following the lead given by architects and engineers, geomorphologists have become increasingly concerned to study the weathering of rocks used in construction and building work. Processes and rates of urban weathering are examined by Inkpen, Cooke and Viles in Chapter 8, who provide a useful progress report on current research. Good examples of this move into applied weathering studies are the work described in Chapter 9 by Smith, Magee and Whalley, on the weathering of sandstone buildings in Belfast, and in Chapter 12 by Cooke, on the destruction of buildings in desert environments by salt weathering.

Building stones, and also tombstones, can provide excellent conditions for studying long-term rates of weathering and the evolution of weathering features, provided it can be established when the stones were first exposed to weathering. This is discussed by Inkpen, Cooke and Viles in Chapter 8, in a contribution on urban limestone weathering, and forms the basis for a study by Mottershead of the alveolar weathering of sea walls at Weston-super-Mare, which is presented in Chapter 10, and the work by Takahashi, Matsukara and Suzuki on the erosion of a bridge pier by salt spray, described in Chapter 11. The idea of a cycle of erosion, historically popular as a model of landform development in other fields of geomorphology, is adapted by Mottershead to provide an attractive explanation of the development and subsequent destruction of the weathering features observed on the sea walls at Weston-super-Mare.

Papers in the third section of the book are concerned with the application of weathering studies and weathering rates to the dating of deposits or rock surfaces. Since the introduction of a successful design by High and Hanna (1970), it has been possible to obtain estimates of rates of downwearing of rock surfaces by the use of micro-erosion meters (see, for example, Trudgill et al., 1989). Recent developments in microchip technology now offer researchers exciting opportunities to obtain much more detailed and accurate measurements of the downwearing of rock surfaces. In Chapter 13, Swantesson describes a sophisticated device that utilises modern laser equipment linked to microprocessor recording and computer analysis to measure and record micro-changes in the downwearing of rock surfaces with very high levels of accuracy.

Researchers have used both qualitative observations and quantitative measurements to record changes in the physical properties of rocks that are induced by weathering. The Schmidt hammer is a tool that has been widely used to obtain a non-destructive quantitative measure of changes in the surface strength of rocks. In Chapter 14, Sjöberg gives an illustration of this technique, in which he uses Schmidt hammer rebound values to compare weathering rates on rock surfaces in different environments.

In Chapter 15, Parish shows that weathering of minerals in some sediments can alter physical properties that were previously assumed to be unaffected by weathering, such as the luminescence signals emitted by feldspars. As these signals are used for the luminescence dating of many sediments, her findings have important implications in geochronology.

The fact that sediments deposited on the Earth's surface and exposed to atmospheric conditions undergo progressive chemical weathering and pedogenesis means that, under favourable circumstances, they can be assigned relative ages by comparing the different stages of weathering that they have reached. In Chapter 16, Woodward, Macklin and Lewin apply this technique to date Quaternary alluvial sediments in the Pindus Mountains of Greece.

The final three sections of the book examine the relationship between weathering and landform development, which is the focus of much current research. Geomorphologists concerned with the tropics have been particularly concerned to study subsurface weathering profiles and their relationships to past and present geomorphological processes. This tradition is well represented in contributions concerned with aspects of tropical deep weathering by Thomas (Chapter 17), Teeuw, Thomas and Thorp (Chapter 18), and McFarlane, Bowden and Giusti (Chapter 19).

Changes that occur to the mechanical properties of rocks as they weather, particularly changes in the strength properties of building materials, and of weathered mantles into which foundations are placed, are of great concern to engineers. The contribution by Fan, Allison and Jones on the effects of tropical weathering on the characteristics of argillaceous rocks (Chapter 20) shows that geomorphologists can contribute to this field, and is a good example of the interdisciplinary nature of many weathering studies.

It has long been recognised that rock composition and structure are major factors influencing the relationship between weathering and landform evolution, but there are many details as to how they interact in different environments that remain unanswered. The interactions can be studied at various scales. In Chapter 27, Rea, for example, discusses the influence of joint control on the formation of rock steps in the subglacial environment, on the basis of detailed field observations on a recently deglaciated foreland in Norway. In contrast, Schmidt, in a study of the groundplan of cuesta scarps in dry regions (Chapter 21), and Evans on the form of cirques and lake basins in glaciated regions (Chapter 26) both seek significant statistical relationships between lithology and structure at a macro-scale. At a meso-scale, Ehlen (Chapter 23) provides detailed analysis on the much discussed topic of the relationship between the form and distribution of tors on Dartmoor, and the jointing patterns and detailed mineralogy of the granite in which they are developed.

Even in temperate regions, where geomorphologists have been most active, the current state of knowledge about the relationships between weathering and landforms is still regrettably incomplete, especially for some rock types. Whilst there are extensive literatures on the weathering and evolution of landforms on limestones and granites, sandstone landforms in particular have been largely ignored. Yet, as Robinson and Williams explain in Chapter 22, sandstones exhibit a very characteristic suite of macro- and micro-landform features, whose origin is only poorly understood. Many of the features have been very little studied and yet are not uncommon, such as polygonal cracking and fluting (which is discussed in detail in Chapter 24).

The weathering of coastal platforms in a variety of temperate environments has been a major focus of research in recent years (see Trenhaile, 1987), and the paper by Moses and Smith (Chapter 25) is a useful addition to this growing literature.

The final contribution to the book, by Gardner (Chapter 28) highlights the fact that identical, or near-identical, weathered debris can be produced by more than one process. Her studies in the Himalayas suggest that some loess may be the product of chemical weathering in a humid environment, rather than a product of the mechanical processes of abrasion or salt weathering, to which it is more commonly attributed (see Pye, 1987, for discussion).

Overall, the contributions to this book demonstrate that geomorphological research into weathering is active on many fronts, using a wide array of approaches and techniques. Some investigators employ sophisticated, state-of-the-art technology, others the traditional geomorphological skills of observation and deductive reasoning, backed up where appropriate by statistical analysis. The availability of more sophisticated equipment is enabling researchers to answer questions that defeated earlier geomorphologists, but, as in all realms of knowledge, uncovering more facts generates still more questions that need answering. Much work remains to be done before we can hope to understand all the basic weathering processes and their varied contributions to landform development.

It is a curious paradox of modern weathering research that experimenters have tended to concentrate their efforts on physical processes, notably frost weathering, and have been seemingly reluctant to study chemical weathering, yet field workers have carried out intensive research on chemical weathering, particularly in the tropics, but have made relatively few studies of physical weathering. There would seem to be much that could be learnt from a well-organised programme of experiments into the chemical weathering of different rock types under controlled environmental conditions, and equally there is a need for more detailed field investigation of physical weathering processes and their weathering products.

A criticism that can be levelled at many weathering experiments is that they test the effects of individual processes, but not the processes in combination (Smith, 1994). Until more work is done to explore the possible ways in which processes interact, understanding of rock weathering will inevitably be incomplete. That the interactions can sometimes be unexpectedly complex is shown by Williams and Robinson (1991).

Most of the contributions to the field of weathering research have been made by geologists and geomorphologists, but it is to be hoped that chemists and physicists will play a greater role in future research. Progress in research would undoubtedly be faster if interdisciplinary teams could be more easily organised. The investigation of such pressing issues as stress corrosion and fatigue effects in rocks requires the cooperation of specialists across a wide range of disciplines.

It is also to be hoped that more biologists will join with geomorphologists and geologists in weathering research. Until recently, biological weathering was assumed to have a relatively minor role in rock breakdown, but evidence is accumulating to suggest that it may be of great importance (Schwartzman and Volk, 1989), and that even such lowly animals as molluscs can greatly enhance rates of weathering. Thus Schachak et al. (1987) have found that two species of snail grazing on lichen-covered limestone in the Negev desert increase the annual weathering rate by 0.7 to 1.1 tonnes per hectare. As Viles and Pentecost point out in Chapter 7, certain lichens are major agents of biodeterioration in their own right, capable even of dissolving quartz (Hallbauer and Jahns, 1977). There is good reason to believe that bacteria also are very destructive, promoting the weathering of rock minerals both at the surface and deep in the interior of rocks (Schwartzman et al., 1991). Weathering effects that until now have been seen as abiotic may prove on further investigation to be the work of bacteria. It would not be surprising if microbiologists were to contribute the greatest advances in our understanding of rock weathering processes over the next decade.

David Robinson
Rendel Williams

REFERENCES

Barber, D. J. and Meredith, P. G. (eds) (1990). *Deformation Processes in Minerals, Ceramics and Rocks*. Unwin Hyman, London.

Battle, W. R. B. (1960). Temperature observations in bergschrunds and their relationship to frost shattering. In Lewis, W. V. (ed.), *Norwegian Cirque Glaciers*, Royal Geographical Society Research Series 4, London.

Griggs, D. T. (1936). The factor of fatigue in rock weathering. *Journal of Geology*, **44**, 781–96.

Hallbauer, D. K. and Jahns, H. M. (1977). Attack of lichens on quartzitic rock surfaces. *Lichenologist*, **9**, 119–122.

High, C. and Hanna, F. K. (1970). A method for the direct measurement of erosion on rock surfaces. *Technical Bulletin 5, British Geomorphological Research Group*.

Honeyborne, D. B. and Harris, P. B. (1958). The structure of porous building stone and its relation to weathering behaviour. *Colston Papers*, **10**, 343–365.

Jerwood, L. C., Robinson, D. A. and Williams, R. B. G. (1987). Frost and salt weathering as periglacial processes: the results and implications of some laboratory experiments. In Boardman, J. (ed.), *Periglacial Processes and Landforms in Britain and Ireland*, Cambridge University Press, Cambridge, pp. 135–144.

Lautridou, J.P. and Ozouf, J.C. (1982). Experimental frost shattering: 15 years of research at the Centre de Geomorphologie du CNRS. *Progress in Physical Geography*, **6**, 215–232.

Potts, A. S. (1970). Frost action in rocks: some experimental data. *Transactions of the Institute of British Geographers*, **49**, 109–124.

Pye, K. (1987) *Aeolian Dust and Dust Deposits*. Academic Press, London.

Schaffer, R. T. (1932). *The Weathering of Natural Building Stones*. Building Research Special Report 18, DSIR, London.

Schwartzman, D., Evans, J., Okrend, H. and Aung, S. (1991). Microbial weathering and Gaia. In Schneider, S. H. and Boston, P. J. (eds), *Scientists on Gaia*, MIT Press, Cambridge, Massachusetts.

Schwartzman, D. W. and Volk, T. (1989). Biotic enhancement of weathering and the habitability of Earth. *Nature*, **340**, 457–460.

Shachak, M., Jones, C.G. and Granot, Y. (1987). Herbivory in rocks and the weathering of a desert. *Science*, **236**, 1098–1099.

Smith, B. J. (1994). Weathering processes and forms. In Abrahams, A. D. and Parsons, A. J. (eds), *Geomorphology of Desert Environments*, Chapman and Hall, London.

Thomas, W. N. (1938). Experiments on the freezing of certain building stones. *Building Research Technical Paper 17*, DSIR, London.

Trenhaile, A. S. (1987). *The Geomorphology of Rock Coasts*. Clarendon Press, Oxford.

Trudgill, S. T., Crabtree, R. W., Viles, H. and Cooke, R. U. (1989). Remeasurement of weathering rates, St Paul's Cathedral, London. *Earth Surface Processes and Landforms*, **14**, 175–196.

Whalley, W. B. and McGreevy, J. P. (1991). The contribution of W. R. B. Battle to mechanical weathering studies. *Permafrost and Periglacial Processes*, **2**, 341–346.

Williams, R. B. G. (1968). Periglacial climate and its relation to landforms. Unpublished Ph.D. Thesis, University of Cambridge.

Williams, R. B. G. and Robinson, D. A. (1991) Frost weathering of rocks in the presence of salts—a review. *Permafrost and Periglacial Processes*, **2**, 347–353.

Wiman, S. (1963). A preliminary study of frost weathering. *Geografiska Annaler*, **45**, 113–121.

Winkler, E. M. (1973). *Stone: Properties, Durability in Man's Environment*. Springer-Verlag, New York.

Section 1

WEATHERING PROCESSES

1 Weathering of Granitic Rocks: Environment and Clay Mineral Formation

JOHN GERRARD
The University of Birmingham, UK

ABSTRACT

There has been much speculation concerning relationships between clay mineral formation and climatic conditions. Clay minerals have often been used to indicate the palaeoclimates under which some weathered residues were produced. But, as this review of granite weathering indicates, clay mineral formation is a highly complex process and many factors have to be considered. However, with care and the complete analysis of the weathering system, a number of general relationships may be established. Montmorillonite production seems to be related to arid and semi-arid conditions. An abundance of kaolinite might indicate weathering under humid tropical conditions, but kaolinite can be produced in a variety of environments. The same appears to be the case with gibbsite. Illite appears not to be a reliable indicator of environmental conditions.

INTRODUCTION

Much has been written about the role of climate in chemical weathering and there has been great speculation concerning the type of clay minerals produced under specific climate conditions. Many regions of the world have experienced marked changes of climate in the comparatively recent geological past and there have been attempts to reconstruct former environments using the clay mineralogy of weathered residues. Thus, it has been stressed that swelling clays, such as montmorillonite, are associated with arid and semi-arid climates, while other clays, such as kaolinite and gibbsite, are associated with weathering under humid tropical conditions. There has been more disagreement concerning clays formed by weathering under temperate climatic conditions. The question that needs to be asked is how reliable are these assumptions?

Such, apparently straightforward, relationships have been demonstrated in the clay mineral suites of alluvial soils in many environments (Gerrard, 1987 and 1992). One such area is Sri Lanka which can be divided into wet and dry zones with a narrow intermediate band between (Figure 1.1). The components in the alluvial soils are derived essentially from the weathering of largely Precambrian igneous and metamorphic rocks (Herath and Grimshaw, 1971). Gibbsite is common in the wet zone

Rock Weathering and Landform Evolution. Edited by D. A. Robinson and R. B. G. Williams
© 1994 John Wiley & Sons Ltd.

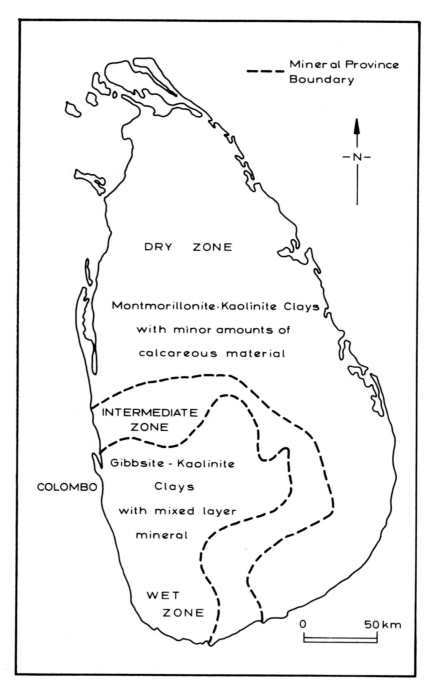

Figure 1.1 Clay mineralogy of alluvial soils in different climatic regions of Sri Lanka (after Herath and Grimshaw, 1971)

whereas in the drier zone kaolinite and montmorillonite are present but gibbsite is absent. In the intermediate zone, minor amounts of gibbsite may be present together with montmorillonite.

Such 'models' have often been used to interpret clays in weathering profiles in areas which have experienced major climatic fluctuations in the past, the idea being that specific clay minerals will act as fingerprints for the climatic conditions. However, unless it is possible to establish relations between clay minerals and climate, beyond all reasonable doubt, there is a great danger of circular argument. The assessment, presented here, of clay minerals produced by the weathering of granitic rocks is offered in the hope that it will clarify some of the major uncertainties that exist. It is also hoped that such a review will allow a more rational interpretation of previous work. Granite has been chosen because it is an extremely widespread rock. Intrusive igneous rocks are exposed over approximately 15% of the continental land surface and granite is the dominant igneous rock occupying large areas of the United States, much of Canada, and found extensively in South America, Africa, Asia and Australia as well as Scandinavia and other parts of Europe. Also, compared to many rocks, its mineralogy is relatively simple and its weathering has been extensively studied (Ehlen and Gerrard, 1988). Before the review is presented two specific examples are used to illustrate the nature of the problems encountered in relating clay mineral suites to climatic conditions. The first of these concerns clay mineralogy in arid and semi-arid areas, and the second the interpretation of 'sandy weathering' profiles in temperate regions.

CLAY MINERALOGY IN ARID AND SEMI-ARID AREAS

As noted earlier, a number of workers have stressed the dominance of montmorillonite in arid and semi-arid regions (e.g. Bakker, 1960; Barshad, 1966; Ismail, 1970). Barshad (1966) has also argued that smectite is dominant in areas with less than 380 mm, and similar observations have been noted in Iraq (Al Rawi et al., 1969; Abtahi, 1977). However, other results have cast doubt on this apparently simple relationship. Dredgne (1976) has shown that smectite is common except where soil potassium levels are high, when illite might be expected. Illite has also been noted in the dry areas of China (Hseung and Jackson, 1952) and illite and chlorite in central Asia (Minashina and Gradusov, 1973). In other dry areas palygorskite has been identified in some of the weathered products (Singer, 1982).

The most confusing results have been produced by Buol (1965), who noted that, contrary to popular belief, kaolinite may dominate in some arid areas with subsidiary amounts of illite. However, there may be special reasons for these results because, in general, most workers assume that it is very unusual for kaolinite to form in such environments. One such reason may be the specific mineral which is being weathered. Thus, Singer (1984) has noted how a microgranite was weathered to smectite, chlorite and some subsidiary kaolinite, but it was only the mafic components of the rock which were altered. The potassium feldspars remained unaltered and illite was only produced when sericitisation of feldspars had occurred. This stresses the need to understand the specific details of the weathering process and the microenvironment within which weathering takes place. The use of scanning electron microscopes and electron probes now makes this possible and allows a more complete understanding of how specific clay minerals may have been produced. Such procedures were unavailable to earlier workers; consequently early results are not as precise as later work and are possibly misleading.

SANDY WEATHERING PROFILES

The second example concerns the problems faced by geomorphologists over the interpretation of the 'sandy weathering' profiles found on many granite slopes in temperate regions (Jahn, 1962; Bakker, 1967). In the Massif Central, France, Cheshworth and Dejou (1980) have identified goethite, gibbsite, kaolinite, illite and vermiculite in a 3 m weathered profile. On Dartmoor, England, much stress has been placed on the lack of clay, high feldspar content and lack of alteration of the feldspars (Green and Eden, 1971; Doornkamp, 1974; Gerrard, 1990). The Dartmoor materials have been variously attributed to chemical decomposition by acid surface waters, to frost shattering and to hydrothermal alteration (Linton, 1955; Palmer and Nielson, 1962; Brunsden, 1964 and 1968; Waters, 1964). The possibility of hydrothermal alteration must always be considered when examining weathered products in granite areas. Te Punga (1957) attributed similar material on Bodmin Moor, Cornwall, England, to Pleistocene frost shattering, and in the Central Massif, France, Collier (1961) has suggested an origin related to the present cool climate. Similar differences of opinion have been expressed about the incoherent granite in the Sierra Nevada of California. Prokopovich (1965) suggested frost action, but Wahrhaftig (1965) has shown how the alteration of biotite to chlorite results in expansion which shatters the rock. Such an effect has also been stressed by Eggler *et al.* (1969), Nettleton *et al.* (1970) and Bustin and Matthews (1978). However, such a 'grussification' process has not been accepted by all workers.

The discovery of gibbsite in the Dartmoor growan has further complicated interpretations (Green and Eden, 1971). Gibbsite has been frequently reported in weathered granite in the humid tropics, but less frequently in temperate latitudes. Does its presence in the Dartmoor weathered granite imply that humid tropical conditions formerly existed or is our understanding of the factors favouring gibbsite formation at fault? Are there special circumstances which have led to its production? This confusion is clear from two French examples. The presence of gibbsite at Tarn has been attributed to an earlier, hot humid climate (Maurel, 1968) while in the Central Limousin a temperate origin was suggested (Dejou *et al.*, 1968). Bakker (1967) also related its presence in the Harz Mountains, Germany, to a former sub-tropical environment. However, Green and Eden (1971) conclude that gibbsite in the Dartmoor growan is an initial product of weathering, and they agree with the view of Erhart (1968) that the presence of gibbsite does not necessarily imply a humid tropical environment. The production of gibbsite is one example where it is difficult to relate a specific clay mineral to specific climatic characteristics. It seems to be more related to the stage of the weathering process and the particular leaching conditions. The production of gibbsite is examined in greater detail later. These examples suggest that the association between clay minerals and climate is not as precise as many workers have suggested. It also suggests that a number of spurious geomorphological conclusions may have been made.

GRANITE WEATHERING—GENERAL OBSERVATIONS

A few specific examples of clay minerals produced by the weathering of granitic rocks illustrate the potential variability of such weathering. Smectite has been noted as the dominant clay mineral in weathered granite at Holmsbu, south of Oslo, Norway (Bergseth et al., 1980). In addition to those noted earlier a number of studies have examined the chemical weathering of granite at various localities in France. Seddoh (1973), in Morvan, noted that biotite weathers to hydrobiotite, interstratified mica–vermiculite, vermiculite, vermiculite–chlorite intergrade and kaolinite, whereas plagioclase feldspar alters to kaolinite, gibbsite, sericite and montmorillonite. In the granite massif of Parthenay (Deux-Sevres) Meunier and Velde (1976) noted that, in the early stages of weathering, illite was produced, together with vermiculite and kaolinite in small amounts. Dejou et al. (1977), summarising many French studies, concluded that various secondary minerals, such as illite, vermiculite, montmorillonite, kaolinite, gibbsite and amorphous minerals, occur in varying percentages depending on local situations. Granite weathering in Galicia, northwest Spain, under a humid temperate climate, produced gibbsite and kaolinite (Calvo et al., 1983).

A number of studies in the USA, in areas such as Missouri (Blaxland, 1974), Maryland (Wolff, 1967), Oklahoma (Harriss and Adams, 1966), Georgia (Grant, 1963 and 1964; Van Tassell and Grant, 1980), have observed kaolinite, illite, halloysite and hydrated oxides of aluminium and iron. Fritz (1988) has argued for the production of gibbsite and kaolinite from plagioclase in the Piedmont of the southeast USA.

In the following examples from Japan great use is made of Yatsu's (1988) review. Many papers on the weathering of granitic rock have been produced (e.g. Oyagi et al., 1969; Miura, 1976; Kitagawa and Kakitani, 1977; Suzuki et al., 1977; Nagatsuka, 1979; Egashira and Tsuda, 1983; Matsukura et al., 1983). Nagatsuka (1979) observed illite–vermiculite interchange minerals and gibbsite in weathered crusts of granite, and Al-vermiculite, metahalloysite, illite and a small amount of goethite in the yellow-brown forest soils derived from granite weathering. Halloysite was the principal mineral accompanied by vermiculite in weathered granite rocks of the Kyushu regions (Egashira and Tsuda, 1983). Matsukura et al. (1983) noted how biotite had been altered into vermiculite and that subsequently plagioclase had been changed to kaolinite.

One of the most intensive studies has been undertaken on granitic rocks in the southeast coasts of Brazil under semi-arid, sub-humid, humid sub-tropical and humid temperate climates (Melfi et al., 1983). The most abundant secondary mineral was kaolinite, principally as a result of feldspar breakdown. In regions with a pronounced dry season, leaching was advanced, profiles shallow and small quantities of smectite and illite occurred as mica weathering products. Where the weathering profile was deeper, in more humid conditions, biotite alteration products were interlayered mica

–vermiculite, vermiculite, kaolinite and feldspar. Weathering sometimes produced gibbsite, which was more prevalent in the soil horizons or zone of maximum leaching. Montmorillonite, halloysite, kaolinite, gibbsite and goethite have all been found in weathered granite in Vietnam (Samotoin *et al.*, 1989).

This very brief review has demonstrated a number of crucial issues. Most workers would agree that the type of clay mineral produced is the result of the particular leaching conditions at the specific location being analysed. The nature, amount and timing of leaching processes is to a great extent governed by climate. Hence the link between climate and clay mineralogy. However, superimposed on this general relationship there are specific site characteristics which will influence leaching. These can loosely be called the geomorphological factors such as slope angle and position on the slope. Many studies have demonstrated a catena-like arrangement of weathering profiles on slopes (e.g. Ruxton and Berry, 1957; Thomas, 1974; Gerrard, 1988) largely as a function of water distribution. Position within the weathering profile, the microenvironment, will also influence leaching conditions.

It is the specific conditions at the site of weathering that are important and not necessarily the general climatic environment. Thus, vermiculite, mixed layer vermiculite–phlogopite and smectite are presently forming in the alpine zone of the northern Cascades, Washington (Reynolds, 1971). Weathering was mostly occurring on south-facing exposures, above the snow line, where temperatures of 40°C were measured on rock faces. The results are significant for climatic interpretation based on palaeosols or palaeo-weathering profiles. Thus, any assessment of clay mineral formation must take account of such factors. Also, the minerals being weathered and the order in which they are weathered will govern which clay minerals are formed. It is important to remember the cautionary note of Trudgill (1976) that the multifactorial nature of weathering systems means that they must be analysed in terms of the isolation of control or noise variables before the dynamics can be understood. Failure to take account of such control and noise variables has clearly been responsible for some of the conflicting results. Some of the more important of these variables relate to rock mineral variability, and the specific leaching conditions under which weathering has taken place. These points are now considered in greater detail.

MINERAL VARIABILITY

Granitic rocks include the alkali-feldspar granites, the granites *sensu stricto*, the granodiorites and the tonalites. Feldspars make up the greater part of most of the granitic rocks. The alkali feldspars are normally orthoclase, orthoclase microperthite, microcline, microcline microperthite or microcline perthite, and the plagioclases range in composition from oligoclase to andesine. Alkali feldspar is usually the only feldspar in the alkali-feldspar granites; however, according to Streckheisen (1974), up to 10% of their total feldspar content may be plagioclase. In granites *sensu stricto*, 10–65% of the total feldspar is plagioclase, in the granodiorites, 65–90%, whereas in the tonalites 90–100% is plagioclase. The quartz content is usually between 20 and 40%. Mafic mineral content is highest in the granodiorites and some of the tonalites. The most common mafic minerals are biotite, muscovite and/or amphibole. Mineral composition has been stressed because it governs the course of weathering. Also, considerable variability can exist within granitic rocks and this variability is often overlooked when assessing clay mineral formation. Studies on seemingly similar rocks may not be comparing like with like. Some large bodies of granite exhibit remarkable homogeneity, such as the fine-grained calc-alkali granites of New England, USA (Chayes, 1952), whereas other areas, such as Dartmoor, exhibit systematic variations (Ehlen, 1991 and 1992) which might be expected to be reflected in the spatial variation of weathering products.

WEATHERING ENVIRONMENT

It is important to stress that the studies reviewed earlier have been conducted in a variety of ways. Most results have been obtained from the analysis of weathering profiles, but some results have been derived from the analysis of clay minerals in granitic soils. The pedogenic environment is different to that in the weathering profile proper and processes of clay formation might differ. Pedro (1968 and 1983) has argued that the principal types of chemical weathering are basic pedological processes, such as podzolisation, bisiallitisation, monosiallitisation and allitisation. Figure 1.2 illustrates some possible pathways for clay formation during weathering and during pedogenesis respectively. This is essentially the two-cycle concept of soil clay genesis as proposed by Matsui (1966 and 1969). The equivalent processes have been termed geochemical weathering and pedogenetic transformation by Millot (1982).

The important point is that additional factors are involved in clay transformation within the soil, the most important being the biotic component. It is widely accepted that vegetation, together with microorganisms, influence the transformation of minerals in the pedosphere. Spyridakis *et al.* (1967) have shown how the rhizospheric

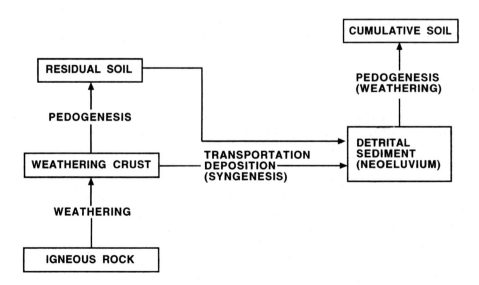

Figure 1.2   Two-cycle concept of soil clay genesis (after Matsui, 1966)

activity of coniferous and deciduous trees affected the transformation of biotite to kaolinite. In New Zealand, McIntosh (1980) found halloysite in more profiles under radiata pine than under manuka native scrub. Thus, one must agree with Eswarin and Bin (1978) and urge caution in the interpretation of mineralogical trends in soil profiles as indicators of mineral alteration sequences in deep profiles. In thick profiles conditions may also have differed at different stages of soil development. Similar caution must be exercised in the analysis of clay mineralogy in transported alluvial soils. Also, it must be assumed that clay minerallogy is in some sort of equilibrium with the conditions. In the West Carpathians, Vrana (1981) established that the geochemistry of the groundwater in granite areas implied that the unstable minerals in the weathering zone were illite and gibbsite, indicating a tendency to monosiallitisation and the formation of kaolinite, whereas the constituents of the clay fractions reflected the tendency of bisiallitisation (the formation of illite, montmorillonite and chlorite). This evidence was used to suggest that existing clay minerals were not in equilibrium with present environments.

MINERAL WEATHERING

Some of the problems involved in clay mineral interpretation are caused because it is not clear which primary minerals are being weathered. Also, some clay minerals may be formed as temporary initial products of weathering and might not be an indication of the environmental conditions. The major minerals that are transformed to secondary clays during the weathering of granitic rocks are feldspars and micas. The order in which the main minerals are weathered has been generally well established and mostly follows the sequence enumerated by Goldich (1938), namely:

plagioclase > biotite > orthoclase > muscovite > quartz

However, specific exceptions have been noted. Kato (1964a and b, 1965a,b and c) was one of the first to question the general validity of Goldich's stability series, and Wilson et al. (1971) suggested, from their study in northeast Scotland, an order of:

plagioclase = orthoclase > muscovite > biotite > microcline > quartz

Thus, it is clear that the relative stabilities of minerals cannot be considered invariant but must always be assessed in terms of the prevailing weathering conditions (Henin et al., 1968). In the Wilson et al. (1971) study, closed alkaline conditions were prevalent and not the more open acid environment in which most studies have been conducted. This is further reinforcement of the importance of microenvironmental factors.

The breakdown of feldspars can be extremely variable. Thus, feldspars have been noted as weathering to a non-crystalline phase (Eswaran and Bin, 1978; Rimsaite, 1979; Eggleton and Buseck, 1980); to illite (Eggleton and Buseck, 1980); to kaolinite and illite (Loughnan, 1969); to smectite (Wilson et al., 1971); a non-crystalline phase and smectite (Guilbert and Sloane, 1968); to kaolinite and smectite (Loughnan, 1969); to smectite, kaolinite and gibbsite (Carroll, 1970; Tardy et al., 1973); to kaolinite and gibbsite (Anand et al., 1985); to kaolinite and several forms of halloysite (Keller, 1978; Wilke et al., 1978) and to gibbsite (Parham, 1969a; Lodding, 1972).

Kaolinite is the dominant clay mineral produced by the weathering of feldspars but other clay minerals can be produced. Parham (1969b) has argued against the general statement that kaolinite is the clay mineral formed from the chemical weathering of potassium feldspar. He suggested that the weathering of feldspar in the early stages should follow the sequence:

feldspar → allophane → halloysite and boehmite

However, his views are probably in the minority. The alteration sequence suggested by Kitagawa and Kakitani (1977) is:

plagioclase → allophane → halloysite → metahalloysite → kaolinite

It is clear from the last example that specific clay minerals can be produced directly from the weathering of feldspars as well as going through intermediate phases. Alexanian (1951 and 1960) noted how feldspars had apparently weathered directly to montmorillonite whereas Guilbert and Sloan (1968) observed an intermediate stage. In weathered granite, in an Old Red Sandstone conglomerate in northeast Scotland, the observed sequence was orthoclase to microperthite, plagioclase direct to montmorillonite and muscovite to kaolinite (Wilson *et al.*, 1971). In highly leached tropical soils, feldspars may be converted to gibbsite, either through an intermediate halloysite stage (Bates, 1962), or directly (Stephen, 1963). To complicate matters, such a transformation may also occur in cool temperate (Wilson, 1969), or even alpine environments (Reynolds, 1971). Helgeson (1971) has postulated that feldspar hydrolysis involves the sequential appearance of intermediate products such as gibbsite, kaolinite and mica, whereas Meilhac and Tardy (1970) have argued that vermiculite and montmorillonite will be formed directly from feldspar or mica. Similarly it is always assumed that the halloysite comes from the weathering of feldspar.

The weathering of biotite has been extensively studied and it is almost unanimously accepted that it alters through vermiculite or mixed layer clay minerals to smectite and eventually kaolinite in temperate climates. In humid tropical environments, biotite weathers to halloysite and further alteration produces gibbsite, goethite and hematite depending upon weathering conditions. Some of the more recent studies on the weathering of biotite are those by Eswarin and Heng (1976), Weiss (1980), Penven *et al.* (1981), Bisdom *et al.* (1982), Crawford *et al.* (1983) and Idefonse *et al.* (1986).

MICROENVIRONMENT

Much of the previous discussion has shown that the chemical weathering of granitic rocks can be extremely variable. It is the specific leaching conditions at the mineral surface that determine the pathway of weathering. Such leaching conditions, although partially governed by broad climatic parameters such as temperature and moisture availability, are more likely to be governed by specific site characteristics. The microenvironment of mineral weathering is the key to an understanding of clay mineral formation.

The scale of the possible differences between almost adjacent weathering profiles has been demonstrated by Gilkes *et al.* (1973) in western Australia. In the saprolite, feldspars have been weathered to kaolinite and halloysite. In the pallid zone halloysite dominated and in the mottled zone gibbsite. However, the major differences were between profiles. Such variation may reflect variations in intensity of leaching since water movement in deeply weathered granite tends to follow cracks and other fissures rather than pervading the bulk material. Differences in the original composition of feldspars may also be important.

A number of detailed studies have shown the value of examining the weathering environment at the grain level within weathering profiles. Such examinations have often been able to explain results seemingly at variance with accepted doctrine. It is generally assumed that kaolinite is produced under open acid conditions with continual removal of silica and metal ions (Keller, 1957). This is in contrast with montmorillonite which is favoured by relatively closed alkaline environments with retention of alkaline earths and silica, hence the apparent relationship between montmorillonite and arid and semi-arid climates. However, Wilson *et al.* (1971) discovered kaolinites being produced in a weathering environment clearly dominated by alkaline conditions. Thus, under certain circumstances kaolinite might be able to attain chemical equilibrium in relatively closed, alkaline conditions, a point stressed by Zen (1959). Weathering needs to be assessed at the sub-micrometre scale, as the changing stability of weathering products may be due to increased access of dilute oxygenated solutions as etch pits become enlarged and form networks (Banfield and Eggleton, 1990).

Glassman and Simonson (1985) have characterised the weathering process in the following terms. The initial zone of alteration consists of mineral dissolution leading to the formation of etch pits and hairline cracks on mineral surfaces. The second stage is characterised by advanced mineral dissolution and the development of isotropic domains which may completely replace primary igneous phases. The third stage is characterised by the formation of authigenic clays which occur as pseudomorphous replacements of original minerals. The mineralogy of the authigenic clay is dependent upon the microchemistry of the weathering environment and may change significantly over a few tens of micrometres in response to such factors as increased permeability and leaching associated with the development of microcracks. Similarly, Baynes and Dearman (1978) have stated that the initial stages of weathering are dominated by the opening of grain boundaries, microfracturing and the development of an intragranular porosity in feldspars. Thus, Rodgers and Holland (1979) found feldspars weathered to clay along microcracks in tonalite cobbles from moraines in the Tobacco Rout Mountains, Montana, USA. Microcracks in oligoclase contained a central zone of kaolinite surrounded by a zone of smectite. They concluded that the development of microcracks and clay fillings involved diffusion of proton-rich species, probably $H_2ca^{2+}CO_3$, the dissolution of feldspar, the precipitation of clay minerals and the diffusion of Na^+, K^+, Ca^{2+}, H_4SiO_4 and HCO_3 out of cracks.

The presence of quartz and gibbsite together would normally indicate a lack of equilibrium. However, the presence of gibbsite in a quartz-bearing soil may be explained in terms of local equilibrium at the level of individual grains. This may be part of the explanation for the presence of gibbsite in temperate 'sandy weathering' profiles rich in quartz noted earlier. Similarly gibbsite, which is common in tropical moist climates, is generally unstable relative to kaolinite except in extremely silica-poor solutions. Sites of lateritic weathering may be the only natural environments where free alumina solids are in equilibrium (Curtis and Spears, 1971). In a similar way halloysite formation seems to need slightly higher base saturations and higher soil moisture conditions.

Microenvironments will of course change as weathering progresses. Thus, when muscovite alters to kaolinite, the collapse of the muscovite structure should produce a volume decrease of about 30% which will allow ingress of weathering fluids. Robertson and Eggleton (1991) have shown how such a process allows the hydration of kaolinite to halloysite. Unaltered potassium feldspar seems to have resisted collapse and to provide strength for voids to remain open. Within a single horizon feldspar grains may alter to different secondary minerals as a result of the presence of voids, and the association of gibbsite with halloysite and gibbsite with kaolinite may reflect the range of microenvironments occurring in a single horizon (Eswarin and Bin, 1978).

Microenvironments and leaching conditions often vary systematically with depth within weathering profiles. This means that different clay minerals may be being produced at different levels in the profile, and emphasises the need to sample the weathering profile very carefully before making general statements on the factors influencing weathering. In well-drained environments in the humid tropics, biotite weathers to kaolinite (Ojanuga, 1973). However, at the base of a deep weathering profile, where moisture conditions and cationic composition are different from surface horizons, vermiculite and montmorillonite may be found (Furtado, 1968). Gilkes and Suddhiprakarn (1979) have charted the transformation of biotite in a weathering profile as shown in Figure 1.3. Even more elaborately, Churchman and Gilkes (1989) have noted not only systematic changes with depth but have shown two contrasting pathways whereby the same end product can be created (Figure 1.4).

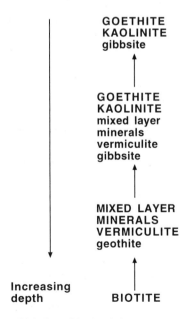

Figure 1.3 Transformation of biotite with depth in the weathering profile (after Gilkes and Suddhiprakarn, 1979)

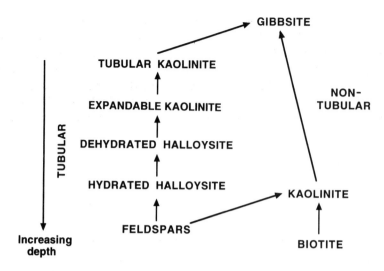

Figure 1.4 Two possible pathways for the production of gibbsite within the weathering profile (after Churchman and Gilkes, 1989)

Major variations also exist within soil profiles reinforcing the point about pedogenesis stressed earlier. Wilson (1970), examining the weathering of biotite in soil profiles developed on biotite–hornblende rock in Inverness-shire, Scotland, noted how in the C horizon a direct transformation to hydrobiotite occurred. This involved the oxidation of all the ferrous iron and the subsequent movement of ferric ions from the octahedral sheet. In the B horizon weathering produced an interstratified vermiculite–chloride mineral. In the more acid surface horizons the interlayers broke down producing vermiculite.

DISCUSSION

This review has demonstrated a certain amount of conflicting evidence concerning the production of clay minerals from the weathering of granitic rocks in different environments. The main problem is establishing a standard against which clay formation can be judged. Experimental studies are often inadequate because it is not possible to replicate exactly natural conditions. Field studies are also beset with numerous problems and it is often difficult to establish the conditions under which weathering took place. There are very few areas of the world that have not experienced major changes in climate over the last million years and clay minerals formed under the impact of palaeoclimatic events may persist to the present time. Studies should be conducted, if possible, on the weathering catena and not on single weathered profiles. Also, weathered profiles should be thoroughly sampled with depth and general relationships established.

It also must be assumed that weathering has reached the point where there is some sort of equilibrium between clay mineral production and environmental conditions.

A variety of clay minerals may be produced in the early stages of weathering that do not necessarily reflect the environmental conditions. It is clearly also important to establish which mineral is being weathered. It is often tacitly assumed that it is feldspar weathering that is producing most of the clay. If the rock contains appreciable amounts of biotite this may be a false assumption. If analysis of clay mineralogy is part of a wider geomorphological investigation it should be possible to use such evidence in reconstructing the environment in which weathering took place.

CONCLUSIONS

On the basis of this analysis it is possible to make a number of general statements. These are:

1. Montmorillonite is probably the best climatic indicator. It is very rarely found in areas that have not experienced arid or semi-arid climates.
2. Although kaolinite would be expected to be the dominant clay mineral in deeply weathered zones under humid tropical conditions, kaolinite can be produced under a variety of conditions. It is also necessary to establish that it is a weathering product and not hydrothermal in origin.
3. Halloysite may be a better indicator of weathering under humid tropical conditions.
4. Gibbsite, in appreciable amounts, probably indicates lateritic-type weathering, but small amounts are possible under a variety of conditions.
5. If mafic minerals are being weathered chlorite and smectite might be expected, especially under arid and semi-arid conditions.
6. Illite is a poor indicator of environmental conditions.
7. The mix of clay minerals may be more important than individual minerals. A single clay mineral type might not be diagnostic.
8. The absence of a clay mineral may be more significant than the presence of others.

Finally, it is important to remember that knowledge is best increased by gathering data and ideas from all sources. 'The specialist following but one line of research lives in grave danger of misinterpreting observations' (Curtis, 1976, p. 57). This is especially true in the examination of weathering products.

REFERENCES

Abtahi, A. (1977). Effect of a saline and alkaline groundwater on soil genesis in semi-arid southern Iran. *Soil Science Society of America, Journal*, **41**, 583–588.

Al Rawi, A. H., Jackson, M. L. and Hole, F. D. (1969). Mineralogy of some arid and semiarid land soils of Iraq. *Soil Science*, **107**, 480–486.

Alexanian, C. (1951). Sur la presence de la montmorillonite dans le feldspar altered de la France. *Comptes Rendus Academic des Sciences, Paris*, **233**, 1203.

Alexanian, C. (1960). Constations faits sur la genese et la formations des argiles. *Bulletin Groupe Francaise Argiles*, **12**, 69.

Anand, R. R., Gilkes, R. J., Armitage, T. M. and Hillyer, J. W. (1985). Feldspar weathering in lateritic saprolite. *Clays and Clay Minerals*, **33**, 31–43.

Bakker, J. P. (1960). Some observations in connection with recent Dutch investigations about granite weathering and slope development in different climates and climate changes. *Zeitschrift für Geomorphologie, Supplementband*, **1**, 69–92.

Bakker, J. P. (1967). Weathering of granites in different climates, particularly in Europe. In Macar, P. (ed.), *L'Evolution des Versants*, Université de Liège et Academie Royale de Belgique, Liège, pp. 51–68.

Banfield, J. F. and Eggleton, R. A. (1990). Analytical transmission electron microscope studies of plagioclase, muscovite and k-feldspar weathering. *Clays and Clay Minerals*, **38**, 77–89.

Barshad, I. (1966). The effects of variation of precipitation on the nature of clay minerals formed in soils from acid and basic igneous rocks. *Proceedings International Clay Conference, Jerusalem*, **1**, 167–173.

Bates, T. F. (1962). Halloysite and gibbsite formation in Hawaii. *Clays and Clay Minerals, Ninth National Conference on Clays and Clay Minerals*, Pergamon, pp. 315–238.

Baynes, J. and Dearman, W. R. (1978). The microfabric of a chemically weathered granite. *Bulletin International Association of Engineering Geology*, **18**, 91–100.

Bergseth, H., Lag, J. and Tunsvik, K. (1980). Smectite formed as a weathering of granite at Holmsbu, southern Norway. *Norsk Geologisk Tidsskrift*, **60**, 270–281.

Bisdom, E. B. A., Stoops, G., Delvigne, J., Curmi, P. and Altemuller, H. J. (1982). Micromorphology of weathering biotite and its secondary products. *Pedologie*, **32**, 225–252.

Blaxland, A. B. (1974). Geochemistry and geochronology of chemical weathering, Butler Hill granite, Missouri. *Geochemica et Cosmochemica Acta*, **38**, 843–852.

Brunsden, D. (1964). The origin of decomposed granite. In Simmons I. G. (ed.), *Dartmoor Essays*, Devonshire Association, Exeter, pp. 97–116.

Brunsden, D. (1968). *Dartmoor*. Geographical Association, Sheffield.

Buol, S. W. (1965). Present soil forming factors and processes in arid and semi-arid regions. *Soil Science*, **99**, 45–49.

Bustin, R. M. and Matthews, W. H. (1978). Selective weathering of granitic clasts. *Canadian Journal of Earth Sciences*, **16**, 215–223.

Calvo, R. M., Garcia-Rodeja, E. and Macias, F. (1983). Mineralogical variability in weathering microsystems of a granitic outcrop of Galicia (Spain). *Catena*, **10**, 225–236.

Carroll, D. (1970). *Rock Weathering*. Plenum Press, New York.

Chayes, F. (1952). The finer-grained calcalkaline granites of New England. *Journal of Geology*, **60**, 207–254.

Chesworth, W. and Dejou, J. (1980). Are considerations of mineralogical equilibrium relevant to pedology? Evidence from a weathered granite in central France. *Soil Science*, **130**, 290–292.

Churchman, G. J. and Gilkes, R. J. (1989). Recognition of intermediates in the possible transformation of halloysite to kaolinite in weathering profiles. *Clay Minerals*, **24**, 579–590.

Collier, D. (1961). Mise au point sur les processus de l'alteration des granites en pays tempere. *American Agronomist*, **12**, 273–332.

Crawford, T. W. Jr, Whittig, L. D., Begg, E. L. and Huntingdon, G. L. (1983). Eolian influence on development and weathering of some soils of Point Reyes Peninsula, California. *Soil Science Society of America, Journal*, **57**, 1179–1185.

Curtis, C. D. (1976). Chemistry of rock weathering: fundamental reactions and controls. In Derbyshire, E. (ed.), *Geomorphology and Climate*, Wiley, Chichester, pp. 25–57.

Curtis, C. D. and Spears, D. A. (1971). Diagenetic development of kaolinite. *Clays and clay Minerals*, **19**, 219–227.

Dejou, J., Guyot, J., Pedro, G., Chaumont, C. and Huguette, A. (1968). Nouvelles donnees concernant la presence de gibbsite dans les formations d'alternation superficielle des massifs granitique (Cas du Cantal et du Limousin). *Comptes Rendus Academie des Sciences*, Paris, **266D**, 1825–1827.

Dejou, J., Guyot, J. and Robert, M. (1977). *Evolution superficielle des roches cristallines et cristallophylliennes dans les regions tempérées*. Institut national de la recherche agronomique, Paris.

Doornkamp, J. C. (1974). Tropical weathering and the ultra-microscopic characteristics of regolith quartz on Dartmoor. *Geografiska Annaler*, **56**, 73–82.

Dregne, H. E. (1976). *Soils of Arid Regions*. Elsevier, New York.

Egashira, K. and Tsuda, S. (1983). High-charge smectite found in weathered granitic rocks of Kyushu. *Clay Science*, **6**, 67–71.

Eggler, D. H., Larson, E. E. and Bradley, W. C. (1969). Granites, grusses and the Sherman erosion surface, southern Laramie Range, Colorado–Wyoming. *American Journal of Science*, **267**, 510–522.

Eggleton, R. A. and Buseck, P. R. (1980). High resolution electron microscopy of feldspar weathering. *Clays and Clay Minerals*, **28**, 173–178.

Ehlen, J. (1991). Significant geomorphic and petrographic relations with joint spacing in the Dartmoor granites, Southwest England. *Zeitschrift für Geomorphologie*, NF **35**, 425–438.

Ehlen, J. (1992). Analysis of spatial relationships among geomorphic, petrographic and structural characteristics of the Dartmoor tors. *Earth Surface Processes and Landforms*, **17**, 53–67.

Ehlen, J. and Gerrard, A. J. (1988). *A Bibliography on the Chemical Weathering of Granitic rocks*. US Army Engineer Topographic Laboratories Report ETL-0505l, Fort Belvoir, Virginia, USA.

Erhart, H. (1968). Sur les trois modes geochemiques d'accumulation des hydroxides d'alumine dans la nature. *Comptes Rendus Academie des Sciences, Paris*, **267D**, 2081–2083.

Eswaran, H. and Bin, W. C. (1978). A study of a deep weathering profile on granite in peninsular Malaysia: III, Alteration of feldspars. *Soil Science Society of America*, Journal, **42**, 154–158.

Eswarin, H. and Heng, Y. Y. (1976). The weathering of biotite in a profile on gneiss in Malaysia, *Geoderma*, **16**, 9–20.

Fritz, S. J. (1988). A comparative study of gabbro and granite weathering. *Chemical Geology*, **68**, 275–290.

Furtado, A. F. A. (1968). Alteration des granites dans les regions intertropicales sous differents climats. *Proceedings 9th International Congress Soil Science, Adelaide*, **4**, 403–409.

Gerrard, A. J. (1987). *Alluvial Soils*. Van Nostrand Rheinhold, New York.

Gerrard, A. J. (1988). *Rocks and Landforms*. Unwin and Hyman, London.

Gerrard, A. J. (1990). Variations within and between weathered granite and head on Dartmoor. *Proceedings of the Ussher Society*, **7**, 285–288.

Gerrard, A. J. (1992). *Soil Geomorphology*. Chapman and Hall, London.

Gilkes, R. J. and Suddhiprakarn, A. (1979). Biotite alteration in deeply weathered granite, I. Morphological, mineralogical and chemical properties. *Clays and Clay Minerals*, **27**, 349–360.

Gilkes, R. J., Scholz, G. and Dimmock, G. M. (1973). Lateritic deep weathering of granite. *Journal of Soil Science*, **24**, 523–536.

Glassmann, J. R. and Simonson, G. H. (1985). Alteration of basalt in soils of western Oregon. *Soil Science Society of America, Journal*, **49**, 262–273.

Goldich, S. S. (1938). A study of rock weathering. *Journal of Geology*, **46**, 17–58.

Grant, W. H. (1963). Chemical weathering of biotite–plagioclase gneiss. *Clays and Clay Minerals, Twelfth National Conference on Clays and Clay Minerals*, pp. 455–463.

Grant, W. H. (1964). Kaolinite stability in the Central Piedmont of Georgia. *Clays and Clay Minerals, Thirteenth National Conference on Clays and Clay Minerals*, pp. 131–140.

Green, C. P. and Eden, M. J. (1971). Gibbsite in the Dartmoor weathered granite. *Geoderma*, **6**, 315–317.

Guilbert, J. M. and Sloane, R. L. (1968). Electron-optical study of hydrothermal fringe alteration of plagioclase in quartz monzonite, Butte District, Montana. *Clays and Clay Minerals*, **16**, 215–221.

Harriss, R. C. and Adams, J. A. S. (1966). Geochemical and mineralogical studies on the weathering of granitic rocks. *American Journal of Science*, **264**, 146–173.

Helgeson, H. C. (1971). Kinetics of mass transfer among silicates and aqueous solutions, *Geochemica et Cosmochimica Acta*, **35**, 41321–4168.

Henin, S., Pedro, G. and Robert, M. (1968). Considerations sur les notions de stabilité et d'instabilites des mineraux en fonction des conditions du milieu: Essai de classification des "systemès d'aggression". *Transactions 9th International Congress of Soil Science*, **19**, 135–149.

Herath, J. w. and Grimshaw, R. W. (1971). A general evaluation of the frequency distribution of clay and associated minerals in the alluvial soils of Ceylon. *Geoderma*, **5**, 119–130.

Hseung, Y. and Jackson, M. L. (1952). Mineral composition of the clay fraction: 3. Some main soil groups of China. *Proceedings of the Soil Science Society of America*, **16**, 294–297.

Idefonse, P., Manceau, A., Prost, D. and Groke, M. C. T. (1986). Hydroxy-Cu-vermiculite formed by the weathering of Fe-biotites at Salobo, Carajas, Brazil. *Clays and Clay Minerals*, **34**, 338–345.

Ismail, F. T. (1970). Biotite weathering and clay formation in arid and humid regions, California. *Soil Science*, **109**, 257–261.

Jahn, A. (1962). Geneza skalel granitowych. *Geograficky Casopis*, **33**, 19–44.

Kato, Y. (1964a). Mineralogical study of weathering products of granodiorite at Shinshiro city, (I) Occurrence and environmental conditions. *Soil Science and Plant Nutrition*, **10**, 258–263.

Kato, Y. (1964b). Mineralogical study of weathering products of granodiorite at Shinshiro city, (II) Weathering of primary minerals—stability of primary minerals. *Soil Science and Plant Nutrition*, **109**, 264–269.

Kato, Y. (1965a). Mineralogical study of weathering products of granodiorite at Shinshiro city, (III) Weathering of primary minerals—mineralogical characteristics of weathered primary mineral grains, *Soil Science and Plant Nutrition*, **11**, 30–40.

Kato, Y. (1965b). Mineralogical study of weathering products of granodiorite at Shinshiro city, (IV) Mineralogical compositions of silt and clay fractions. *Soil Science and Plant Nutrition*, **11**, 62–73.

Kato, Y. (1965c). Mineralogical study of weathering products of granodiorite at Shinshiro city, (V) Trioctahedral aluminium–vermiculite as a weathering product of biotite. *Soil Science and Plant Nutrition*, **11**, 114–122.

Keller, W. D. (1957). *The Principles of Chemical Weathering*. Lucas Bros, Columbia, Missouri.

Keller, W. D. (1978). Kaolinization of feldspars as displayed in scanning electron micrographs. *Geology*, **6**, 184–188.

Kitagawa, R. and Kakitani, S. (1977). Alteration of plagioclase in granite during weathering. *Journal of Science, Hiroshima University*, Ser. C, **7**, 183–197.

Linton, D. L. (1955). The problem of tors. *Geographical Journal*, **121**, 470–487.

Lodding, W. (1972). Conditions for the direct formation of gibbsite from K-feldspar. Discussion. *American Mineralogist*, **57**, 292–294.

Loughnan, F. c. (1969). *Chemical Weathering of Silicate Minerals*. Elsevier, Amsterdam.

Matsui, T. (1966). A two-cycle concept on the genesis of soil clay material. *Clay Science (Nendo Kagaku, Journal of Clay Science Society of Japan)*, **5**, 2–13.

Matsui, T. (1969). Two-cycle concept of soil clay genesis and its application to the study of the polygenetic red soils in Japan. *Proceedings of the International Clay Conference, Tokyo, 1969*, Israel Universities Press, **1**, 533–540.

Matsukura, Y., Maekado, A., Hatta, T. and Yatsu, E. (1983). Vertical changes in mineral, physical, chemical and mechanical properties due to deep weathering of Inada granitic rocks. *Transactions Japanese Geomorphological Union*, **4**, 65–80.

Maurel, P. (1968). Sur la presence de gibbsite dans les arènes du massif du Sidobre (Tarn) et de la Montagne Noire. *Comptes Rendus Academie des Sciences, Paris*, **26D**, 652–653.

McIntosh, P. D. (1980). Weathering products in Vitrandept profiles under pine and manuka, New Zealand. *Geoderma*, **24**, 255–239.

Meilhac, A. and Tardy, Y. (1970). Genese et évolution des sericites, vermiculites et montmorillonites au cours de l'alteration des plagioclases un pays témperè. *Bulletin Servis Carte Geologique, Alsace Lorraine*, **23**, 145–161.

Melfi, A. J., Cerri, C. C., Kronberg, B. I., Fyfe, W. S. and McKinnon, B. (1983). Granitic weathering: a Brazilian study. *Journal of Soil Science*, **34**, 841–851.

Meunier, A. and Velde, B. (1976). Mineral reactions at grain contacts in early stages of granite weathering. *Clay Minerals*, **11**, 235–240.

Millot, G. (1982). Weathering sequence, 'climatic' planations, levelled surface and palaeosurface. International Clay Conference, Bologna and Pavia, Italy, September 6–12, 1981, *Development in Sedimentology*, **35**, Elsevier, Amsterdam, 585–593.

Minashina, N. G. and Gradusov, B. P. (1973). The mineralogy of clay in some desert soils. *Soviet Soil Science*, **5**, 459–472.

Miura, K. (1976). The characteristic altered biotites in the weathered Tottori granite and their significances to applied biology. *Journal of the Japanese Society of Engineering Geology*, **17**, 169–175.

Nagatsuka, S. (1979). Genesis of yellow-brown forest soils from granite under natural Lucidophyllus forest (part 2), Characteristics of clay mineral composition. *Journal of the Science of Soil and Manure, Japan*, **50**, 91–97.

Nettleton, W. D., Flach, K. W. and Nelson, R. E. (1970). Pedogenic weathering of tonalite in southern California. *Geoderma*, **4**, 387–402.

Ojanuga, A. G. (1973). Weathering of biotite in soils of a humid tropical climate. *Proceedings of the Soil Science Society of America*, **37**, 644–646.

Oyagi, N., Uchida, T. and Suzuki, H. (1969). Clay minerals of weathered-zone in Kamo-district granodiorite region. Report 1, Clay minerals and their formation sequence in weathering. *Report of the National Research Centre for Disaster Prevention*, No. 2, 21–44.

Palmer, J. and Nielson, R. A. (1962). The origin of granite tors on Dartmoor, Devonshire. *Proceedings of the Yorkshire Geological Society*, **33**, 315–340.

Parham, W. E. (1969a). Formation of halloysite from feldspar: Low temperature artificial weathering versus natural weathering. *Clays and Clay Minerals*, **17**, 13–22.

Parham, W. E. (1969b). Halloysite rich tropical weathering products of Hong Kong. *Proceedings of the International Clay Conference, Tokyo*, **1**, 463.

Pedro, G. (1968). Distribution des principaux types d'alteration chimique a la surface du globe, presentation d'une esquisse geographique. *Revue de Geographie Physique et de Geologie Dynamique*, **10**, 457–470.

Pedro, G. (1983). Structuring of some basic pedological processes. *Geoderma*, **31**, 289–299.

Penven, M.-J., Federoff, N. and Robert, M. (1981). Alteration meteorique des biotites en Algerie. *Geoderma*, **26**, 287–309.

Prokopovich, N. P. (1965). Pleistocene periglacial weathering in the Sierra Nevada, California. *Geological Society of America, Special Paper*, **82**, 271.

Reynolds, R. C. Jr (1971). Clay mineral formation in an alpine environment. *Clays and Clay Minerals*, **19**, 361–374.

Rimsaite, J. (1979). Natural amorphous materials, their origin and identification procedures. In Mortland, M. M. and Farmer, V. C. (eds), *Proceedings International Clay Conference, Oxford, 1978*, Elsevier, Amsterdam, pp. 567–577.

Robertson, I. D. M. and Eggleton, R. A. (1991). Weathering of granitic muscovite to kaolinite and halloysite and of plagioclase-derived kaolinite to halloysite. *Clays and Clay Minerals*, **39**, 113–126.

Rodgers, G. P. and Holland, H. D. (1979). Weathering products within microcracks in feldspars. *Geology*, **7**, 278–280.

Ruxton, B. P. and Berry, L. (1957). Weathering of granite and associated features in Hong Kong. *Bulletin Geological Society of America*, **68**, 1263–1292.

Samotoin, N. D., Novikov, V. M. and Magazine, L. O. (1989). Ontogenesis of minerals in a bauxite bearing crust of weathering of granites. *Izvestiya-Akademiya Nauk SSSR Seriya Geologicheskaya*, **8**, 1987, 77–91 (in Russian).

Seddoh, F. (1973). Alteration des roches cristallines du Morvan. *Memoire geologiques de l'universite de Dijon*, **1**.

Singer, A. (1982). Pedogeneic palygorskite in the arid environment. *Transactions 12th International Congress Soil Science, Abstracts*, p. 205.

Singer, A. (1984). Clay formation in saprolites of igneous rocks under semiarid to arid conditions, Negev, Southern Israel. *Soil Science*, **137**, 332–340.

Spyridakis, D. E., Chesters, G. and Wilde, S. A. (1967). Kaolinization of biotite as a result of coniferous and deciduous seedling growth. *Proceedings Soil Science Society of America*, **31**, 203–210.

Stephen, I. (1963). Bauxite weathering at Mount Zamba, Nyasaland. *Clay Mineral Bulletin*, **5**, 203–208.

Streckheisen, A. L. (1974). Classification and nomenclature of plutonic rocks. *Geologische Rundschau*, **63**, 773–786.

Suzuki, T., Hirano, M., Takahashi, K. and Yatsu, E. (1977). The interaction between the weathering processes of granites and the formation of landforms in the Rokko mountains, Japan. *Bulletin of the Faculty of Science and Engineering, Chuo University*, **20**, 343–389.

Tardy, Y., Bocquier, G., Parquet, H. and Millot, G. (1973). Formation of clay from granite and its distribution in relation to climate and topography. *Geoderma*, **10**, 271–284.

Te Punga, M. T. (1957). Periglaciation in southern England. *Tijdschrift kon Nederlands Aard Genoot*, **74**, 401–412.

Thomas, M. F. (1974). *Tropical Geomorphology*. Macmillan, London.

Trudgill, S. T. (1976). Rock weathering and climate: quantitative and experimental aspects. In Derbyshire, E. (ed.), *Geomorphology and Climate*, Wiley, Chichester, pp. 59–99.

Van Tassell, J. and Grant, W. H. (1980). Granite disintegration, Panola Mountain, Georgia. *Journal of Geology*, **88**, 360–364.

Vrana, K. (1981). To the question of origin and stability of secondary minerals in the weathering zone of Male Karpaty, Mts. *Geological Carpatica*, **32**, 353–364.

Wahrhaftig, C. (1965). Stepped topography of the southern Sierra Nevada, California. *Bulletin Geological Society of America*, **76**, 1165–1190.

Waters, R. S. (1964). The Pleistocene legacy to the geomorphology of Dartmoor. In Simmons, I. G. (ed.), *Dartmoor Essays*, Devonshire Association, Exeter, pp. 73–96.

Weiss, Z. (1980). Single-crystal X-ray study of mixed structures of vermiculite and biotite (hydrobiotites). *Clay Minerals*, **15**, 275–281.

Wilke, B. S., Schwertmann, U. and Murad, E. (1978). An occurrence of polymorphic halloysite in granite saprolite of the Beyerischer Wald, Germany. *Clay Minerals*, **13**, 67–77.

Wilson, M. J. (1969). A gibbsite soil derived from the weathering of an ultrabasic rock on the island of Rhum. *Scottish Journal of Geology*, **5**, 81–89.

Wilson, M. J. (1970). A study of weathering in a soil derived from a biotite–hornblende rock, I. Weathering of biotite. *Clay Minerals*, **8**, 291–303.

Wilson, M. J., Bain, D. C. and McHardy, W. J. (1971). Clay mineral formation in deeply weathered boulder conglomerate in north-east Scotland. *Clays and Clay Minerals*, **19**, 345–252.

Wolff, R. G. (1967). Weathering of Woodstock granite near Baltimore, Maryland. *American Journal of Science*, **265**, 106–117.

Yatsu, E. (1988). *The Nature of Weathering*. Sozosha, Tokyo.

Zen, E. (1959). Clay mineral–carbonate relations in sedimentary rocks. *American Journal of Science*, **257**, 29.

2 A Comparative Study of Deep Weathering and Weathering Products: Case Studies from Ireland, Corsica and Southeast Brazil

E. T. POWER and B. J. SMITH
The Queen's University of Belfast, UK

ABSTRACT

Various climate-related characteristics have previously been sought to exemplify climatic control on the weathering of granitoid rocks. These have included textural differences and clay and sesquioxide mineralogy. Use of mineralogy has been restricted, however, because data have been drawn from numerous studies employing different sampling and analytical procedures. This study addresses this problem through a systematic and consistent analysis of weathering products from Ireland, Corsica and sub-tropical Brazil. Within each area there is local variability related in particular to relief and drainage but a number of profile characteristics attributable to regional climatic control also emerge. These include: the presence of illite and vermiculite as well as kaolinite and gibbsite in Brazil; the occurrence in Ireland of kaolinite together with vermiculite and illite as intermediate products of biotite and feldspar weathering respectively; and of vermiculite (from biotite) and illite–smectite (from muscovite) in Corsica.

INTRODUCTION

Deeply weathered profiles on granitoid rocks have traditionally been associated with humid tropic and sub-tropical environments. When deep weathering is found under other climatic regimes it has frequently been explained as relict from former times when the climate was wetter and/or warmer. The simple equation of depth with wet tropical conditions has, however, been shown to be spurious. Depth may represent length of weathering rather than intensity, and other profiles may have been truncated by subsequent erosion. Investigators have, therefore, sought other criteria for distinguishing weathering under different climates. Some have used clay content to distinguish between 'temperate arenaceous profiles' and 'tropical argillaceous profiles'. Others have examined clay and sesquioxide mineralogy in an attempt to identify minerals or suites of minerals which can be used to fingerprint environments. In support of the latter there are numerous analyses available from specific sites (Table 2.1). Comparison is, however, difficult because materials have been collected and analysed using different techniques and for many purposes. In this study samples

Rock Weathering and Landform Evolution. Edited by D. A. Robinson and R. B. G. Williams
© 1994 John Wiley & Sons Ltd.

of deeply weathered granitoid rocks have been systematically collected and analysed from three different climatic regimes to facilitate some initial comments on climate control on the mineralogy of weathering products.

BACKGROUND

The Influence of Climate on the Weathering of Granitoid Rocks

'Climate' is frequently invoked as the overriding influence upon the characteristics of weathered material from region to region. Climatic factors control the geochemical weathering environment (Nagell, 1962) whereby the degree of weathering is seen to increase with mean annual temperature (Jenny, 1941) and annual average precipitation (Sherman, 1952). Temperature controls the rate of chemical reactions, while rainfall provides moisture for those reactions as well as a medium for ion removal (Loughnan, 1969). Many writers have favoured the influence of rainfall over that of temperature in inducing weathering variations (e.g. Mulcahy, 1960; Ollier, 1976). It must be borne in mind, however, that as temperature increases evapotranspiration also increases, and 'effective' rainfall is reduced. These climatic influences are manifested in both the specific clay mineral types within a profile as well as the more general characteristics of the profile itself. James *et al.* (1981, p. 161) have stated, for example, that 'the composition of the alteration product is sensitive to climate'. Clay mineral assemblages are particularly seen to differ between arid, temperate and tropical areas, due to climatic influences upon the degree of leaching (ion removal) and hydrolysis (ion availability) (Keller, 1964). Table 2.1 provides a summary of clay minerals reported from different climatic regimes or annual rainfall averages. Variations in the definitions of climatic regimes make exact comparisons difficult, but overall similarities in observations can be noted. The consensus is that in arid regions, where leaching is not strong, smectites (e.g. montmorillonite) and chlorite are the dominant alteration products. Illite is dominant in dry-temperate regions, with vermiculite in humid-temperate areas or montmorillonite if poorly drained. Very wet temperate areas can be dominated by kaolinite, which is in turn the most commonly reported product of tropical weathering. In addition to kaolinite, montmorillonite can occur in less humid tropical regions while gibbsite is present in the wettest. Seasonality is important, and weathering profiles which experience monsoons or concentrated rainfall can contain some arid/dry temperate-type minerals such as illite and montmorillonite (Bakker, 1967). Oklahoma, for example, experiences a seasonally wet climate so that a mixture of illite and kaolinite is produced which reflects the dry and wet months respectively (Harriss and Adams, 1966).

Clay minerals are important components in any assessment of climatic impact on the weathering regime, but other profile characteristics, such as particle size distributions, have also been used as indicators of climatic effect. Ruxton and Berry (1961, p. 16) have stated that 'weathering profile patterns . . . are thought to be members of an homologous series whose contrasts are dependent on climate'. Two types of weathered profile on granitoid rocks have been distinguished and it remains to be seen whether these are endpoints on a continuum as suggested above, or whether they are distinct and separate types which can be differentiated on the basis

Table 2.1 Documented observations on clay mineral assemblages in different climates on acid igneous rocks

Climate/rainfall (location)	Dominant minerals	Author
1. Arid (Negev Desert)	Smectite–chlorite intergrades	Singer, 1984
2. Arid	Chlorite, illite, montmorillonite	Loughnan, 1969
3. 'Dry' (China)	Illite	Hseung and Jackson, 1952
4. Arid/semi-arid	Montmorillonite	Bakker, 1960
5. Arid/semi-arid (California)	Montmorillonite	Ismail, 1970
6. <500 mm (Sierra Nevada)	Montmorillonite	Barshad, 1966
7. Semi-arid	Smectite	James et al., 1981
8. 'Temperate'	Illite, montmorillonite	Tan and Troth, 1982
9. Mediterranean	Illite, montmorillonite	Bakker, 1960
10. Dry sub-humid (Oklahoma)	Illite, kaolinite	Harriss and Adams, 1966
11. Cool humid temperate	Montmorillonite, vermiculite	Novikoff et al., 1972
12. Humid temperate	Vermiculite	Ismail, 1970
13. Humid temperate (1250 m, Georgia, USA)	Kaolinite	Harriss and Adams, 1966
14. Humid temperate (W. Australia)	Kaolinite	Butt, 1985
15. Seasonally humid tropical	Kaolinite, illite, smectite	Bakker, 1960
16. 800–1000 mm	Kaolinite, montmorillonite	Sanches Furtado, 1968
17. 1000–1200 mm	Kaolinite	Sanches Furtado, 1968
18. 'Humid'	Kaolinite	James et al., 1981
19. >750 mm (Hong Kong)	Kaolinite, gibbsite	Ruxton and Berry, 1961
20. >1000 mm	Kaolinite, gibbsite	Barshad, 1966
21. 1200–2000 mm	Kaolinite, gibbsite	Sanches Furtado, 1968
22. 'Humid tropical'	Kaolinite, gibbsite	Bakker, 1960 and 1967
23. 'Tropical'	Kaolinite, gibbsite	Tan and Troth, 1982
24. 'Humid'	Kaolinite, gibbsite	Loughnan, 1969

of weathering mechanisms rather than weathering rate. These two profile types are defined as 'arenaceous', or sandy profiles, and 'argillaceous', or clayey profiles.

Sandy regoliths have a wide occurrence in temperate areas (Thomas, 1974). They have been observed in New Zealand (Thomas, 1976), Cornwall (Rice, 1973), Dartmoor (Brunsden, 1964), Missouri (Humbert and Marshall, 1943) and Corsica (Klaer, 1956). The particle size distribution within this type of profile has been the dominant diagnostic and classificatory tool. Definitions of this type of profile based on particle size limits are presented in Table 2.2. The process of arenisation is seen to be one in which there is some alteration of plagioclase feldspars to clay along with mechanical fracture of quartz and orthoclase to sand-sized material (Curmi and Maurice, 1980). Quartz sand or gruss is the residual deposit of granite weathering on Dartmoor (Brunsden, 1964) and has a high percentage of weatherable minerals exemplified by a feldspar:quartz ratio of nearly 1 (Eden and Green, 1971).

In contrast, argillaceous regoliths have few weatherable minerals, perhaps 4% of the original amount (Buringh, 1970). These profiles have a high clay content and a low silt content due to the rapid transformation of feldspars directly to clay

Table 2.2 Published particle size distributions of arenaceous regoliths

Author	Location	% clay	% silt	% sand	Other
Bakker (1967)	Western Europe	7	15–20	75–85	
Flageollet (1977)	NW France	<6	15–25	70–85	median <1 mm
Collier (1961)	France	1.6–4.6			
Peulvast (1986)	Norway	<2		<5	
Eden and Green (1971)	Dartmoor, England	2–10.5			silt + clay = 13.5–28%
Hall and Mellor (1988)	Grampian Highlands, Scotland	0–5			

(Radwanski and Ollier, 1959), with a silt:clay ratio of less than 0.25 (Buringh, 1970). In Surinam, the amount of silt is, for example, negligible especially when the profile contains more than 30% clay-sized material (Wensink, 1968). Silt:clay ratios are, however, variable with depth in the profile. In Malaysia, Eswaran and Bin (1978) described clay contents that are high in the saprolite, corresponding to the zone of maximum halloysite. Clay contents then fall in relation to silt until surface layers are reached where intense weathering has produced kaolinite, goethite and gibbsite. Work in Angola suggests that the silt:clay ratio is particularly sensitive to changes in clay, since silt remains constant with depth while clay decreases and sand increases (Sousa and Eswaran, 1975). These considerations have led to formalised conceptions of the 'tropical' weathering profile.

Ruxton and Berry's (1957) generalised profile, derived from work in Hong Kong, is frequently referred to and is particularly concerned with changes in particle size distributions through the section. Tardy and Nahon (1985) are more concerned with mineralogical changes, while Vargas (1953) concentrated on variations in engineering properties with depth in profiles from Southern Brazil. These three profile descriptions are basically similar and indicate an increase in weathering intensity towards the surface since the aggressivity of the percolating water decreases as it passes down the profile (Rice, 1973). This gradation differentiates chemically weathered profiles from those which have been hydrothermically altered, usually to a consistent degree throughout (Dearman and Baynes, 1978).

Aside from these idealised tropical profiles, the one supposedly universal characteristic of these regoliths is their 'bimodal' particle size distribution. Ruxton and Berry (1961, p. 19) state that it is 'well known that granitic debris consists of boulders, chips and dust without fragments of intermediate size'. That is to say, the quartz grains remain unaltered while micas and feldspars are chemically altered to clays (Ruxton, 1958). Such distributions have been noted by Brock (1949) and Lumb (1962) in Hong Kong, Ahn (1970) in West Africa, and Verheye and Stoops (1975) in the Ivory Coast. In Singapore, Nossin and Levelt (1967) observed that sand fractions contain only primary quartz so that 'the relative sandiness of the regolith is at least partly controlled by the quartz content of the parent rock' (p. 20). The assumption is that quartz is resistant to chemical breakdown. It must be noted, however, that quartz does appear in the silt fraction, and that the amount of this silt-sized quartz increases with rainfall and temperature (Vanderford, 1942).

There is a general consensus, therefore, that profiles found in temperate areas are 'arenaceous' in character while those from the tropics are 'argillaceous'. Not all observations, however, substantiate this hypothesis. Bakker (1967) states that 6 m deep regoliths containing 15–30% clay are to be found in Europe, while Watson (1962) noted arenaceous material overlying argillaceous material in sub-tropical Rhodesia. In Scotland, Fitzpatrick (1963) found argillaceous profiles with corestones. The usual explanation in such cases, is to assume that the profiles are relict and were formed under climates different from those of the present day. Bakker (1967, p. 64) states that 'relics of different types of deep weathering on granites . . . can be explained only by assuming totally different climates, and periodic climate changes'. On Dartmoor, chemically weathered granite was seen to be relict from the Tertiary by some (e.g. Brunsden, 1964) while others contended that chemical weathering is active and continuous (Ternan and Williams, 1979; Williams et al., 1986). Nearby, on the Cornubian batholith, Sheppard (1977) found kaolinite which he considers to have formed in the tropical/warm temperate climate of the Cretaceous–Tertiary. Others consider arenaceous profiles to be relict from preglacial and interglacial periods; for example, Goldthwait and Kruger (1938) in New Hampshire and Wilson (1967) in Aberdeenshire. On the other hand, Bakker (1967) and Hall and Mellor (1988) consider such profiles to be Pliocene relicts. Many other examples of such considerations could be quoted, but the problem is that such palaeoclimatic interpretations may not provide a total explanation of why profiles in certain climatic zones do not conform to expectations. Both palaeoclimates and contemporary climates are seen by some to be overrated. Climatic variations are said to affect only the rate of weathering, rather than the more fundamental mechanisms of weathering (e.g. Pickering, 1962). Others maintain that atmospheric climate 'has little or nothing to do with the weathering at depth' (e.g. Ollier, 1988, p. 287). By this reasoning, there is a 'climatic convergence' of regolith due to, for example, textural factors (Kostrzewski, 1985) or the 'ubiquity of groundwater' (Twidale, 1986).

LOCATION AND SAMPLING

Three locations were studied that reflect a range of climatic conditions from cool temperate Ireland to warm temperate Corsica and humid sub-tropical southeast Brazil. Basic climatic and geological information is given in Tables 2.3 and 2.4. Variations in age and texture of the sampled rocks inevitably make the isolation of climatic controls on weathering difficult. Because of this, observed weathering patterns are related, where possible, to specific minerals to enhance comparability.

Profile depths at the three locations were variable and ranged from the order of 4 m in Ireland to greater than 30 m at some sites in Brazil. Profiles were sampled in quarries and road cuttings and, where possible, were selected to reflect a range of topographical conditions. At each profile horizons with overt evidence of soil formation were avoided and samples were taken instead from the underlying saprolite (corresponding to weathering zones IIa, IIb and III of Ruxton and Berry (1957)) and unweathered parent rock. Samples were generally collected from the top, middle and bottom of each profile, although the Mourne Mountains profiles were sampled systematically at 30 cm intervals. Saprolite at the Irish and Corsican sites is universally

Table 2.3 Climatic characteristics of the field areas. Leaching factor = mean annual precipitation (cm) minus 3.3 × mean annual temperature (°C) (Crowther, 1930)

Location	Annual average temperature (°C)	Annual average rainfall (mm)	Climate type	Leaching factor
Ireland:				
Mourne Mtns	9	1361	Cool temperate	106.4
Wicklow Mtns	10	1859	Cool temperate	152.9
Corsica:				
Ajaccio	16.1	698	Warm temperate (mediterranean)	16.67
Brazil:				
Rio de Janeiro	24	975	Sub-tropical	18.3

arenaceous, with never more than 15% clay and with more than 60% of samples with a less than 5% clay content (Power, 1989). Sites in Brazil can be separated into two sub-sets. The first consists of arenaceous saprolites which are similar to those from Ireland and Corsica described in this study and elsewhere from a variety of environments (Table 2.2). These are coherent and retain many visual characteristics of the parent rock and have high proportions of unweathered feldspar crystal. The second

Table 2.4 Geological characteristics of the field areas, and profile and sample numbers

Location	Age	Principal rock types	Number of profiles sampled	Number of samples
Ireland:				
Mourne Mountains	Early Tertiary *c.* 55 my BP	Older, coarse-grained granite and younger microgranite	4	27
Wicklow Mountains	Late silurian/ Early Devonian *c.* 392 my BP	Porphyritic microcline granite	3	9
Corsica:				
Numeorus intrusions along western coast	Variscan *c.* 280 my BP	Muscovite granite, biotite granite and granodiorite	10	31
Southeast Brazil:				
Fluminense Complex of coastal range near Niteroi	Early Precambrian, metamorphosed in late Precambrian	Coarse-grained porphyritic granite, biotite–orthoclase granite, fine-grained granite and augen gneiss	25	56

sub-set consists of argillaceous profiles that characteristically have greater than 45% clay and fine silt and retain little of the original rock structure and few weatherable minerals (Power, 1989). Weathering characteristics may be further complicated by complex climatic histories that include Pleistocene glacial and periglacial conditions at the Irish sites, periodic cold pluvial conditions in Corsica and episodes of aridity in southeast Brazil (Power, 1989).

ANALYTICAL PROCEDURES

Analysis was carried out on the <63 μm fraction of samples obtained by gentle grinding with a rubber pestle and mortar and sieving through a 63 μm stainless steel mesh. Approximately 0.5 g of fines from each sample was taken and shaken overnight in ammonium acetate solution (pH 4.4) to remove major cations and calcium from mineral lattices. Samples were then washed of extractant and disaggregated ultrasonically. Immediately after dispersion a little of the suspension was pipetted onto a clean, dry glass slide. The slides were then dried beneath infra-red lamps to provide thin layers of material suitable for mineralogical analysis by X-ray diffraction (XRD) using a Phillips Diffractometer with a copper target. Further subsamples of the <63 μm fraction were mixed in a 1:1 ratio with calcined aluminium oxide for mineralogical analysis by differential thermal analysis (DTA) using a Stanton Redcroft 673-4 DTA instrument. Twenty-six samples were also prepared for thin section analysis by optical microscope. All samples were analysed for elemental composition by atomic absorption spectrophotometry (Perkin Elmer 306) using an Aqua Regia/Hydrofluoric acid digest in teflon bombs.

RESULTS

Results of mineralogical analyses are given in Table 2.5, where, to aid comparison, they are presented as the frequency of occurrence of different minerals in each profile type and are subdivided into primary minerals, secondary minerals and metal oxides. The dominant clay mineral species associated with each field location are further summarised in Table 2.6, where they are compared to climatic associations derived from published literature—see Table 2.1.

DISCUSSION

Primary Minerals

The appearance of primary minerals in the <63 μm fraction indicates that they have been physically comminuted while retaining their crystal structure for XRD identification. If bedrock minerals only occur in a small percentage of samples, this implies that either the mineral has been chemically altered to secondary mineral form, or that it remains as less weathered grains >63 μm. Plagioclase feldspars are, for example, rare in the samples examined from Brazil. These minerals are readily weathered to

28

Table 2.5 Percentages of samples from each field area/profile type (numbers of samples are given in parentheses after each location) containing specific minerals in the <63 μm fraction (identified by XRD and DTA)

	Mourne Mountains	Wicklow Mountains	Corsica			Brazil		
	BG (27)	MG (9)	BG (18)	MG (10)	GN (3)	ARG (34)	ARN (18)	GN (ARG) (4)
Plag.	100	100	89	100	100	3	39	25
K-feld	100	100	100	100	100	45	79	100
Biotite	38	100	42	83	100	48	23	50
Musc.	62	100	5	–	–	48	83	–
Quartz	92	100	–	–	–	76	28	100
Apatite	–	–	–	–	67	64	–	–
Pyrophyl.	–	13	5	–	–	12	–	–
Palygors.	–	–	–	–	–	9	10	–
Imogolite	8	–	–	–	–	–	–	–
Ill.-smect.	15	–	100	100	100	–	–	–
Illite	31	100	–	–	–	85	89	50
Hyd.-biot	23	–	26	25	–	3	–	25
Vermic.	100	88	100	83	–	88	32	–
Chlorite	15	50	–	58	100	–	–	–
Kaolinite	77	100	84	100	–	97	44	100
Halloysite	–	–	–	–	–	–	72	50
Endellite	38	–	–	–	–	–	–	–
Gibbsite	8	–	–	–	–	82	22	75
Boehmite	–	–	11	–	–	15	–	–
Diaspore	8	–	5	17	33	3	–	25
Anatase	15	–	5	17	–	–	–	–
Goethite	–	–	–	–	33	39	50	–
Haematite	–	–	–	–	–	18	6	–
Lepidoc.	–	–	–	–	–	9	–	–
Berthier.	–	–	–	–	–	3	–	–

BG = biotite granite; MG = muscovite granite; GN = granodiorite; ARG = argillaceous profiles; ARN = arenaceous profiles.

Table 2.6 Climatic associations of key clay minerals species identified in this study (a) and in other published literature (b)

(a)

Climate type	Dominant clay minerals
Cool temperate (Ireland	Illite, vermiculite, kaolinite
Warm temperate (Corsica)	Illite–smectite, vermiculite, kaolinite
Sub-tropical (Brazil)	Illite, vermiculite, kaolinite, gibbsite, halloysite

(b)

Climate type	Dominant clay minerals
Arid–semi-arid	Illite, montmorillonite, chlorite
Cool temperate	Vermiculite, montmorillonite, kaolinite
Warm temperature	Illite, montmorillonite
Sub-tropical	Kaolinite, gibbsite

clay in a chemically aggressive environment, so that few have been physically broken to this size. Their disappearance is linked to high kaolinite and gibbsite levels in Brazil. More of the Brazilian profiles have potassium feldspars in their $<63\mu m$ fractions indicating the relative resistance of this feldspar group to chemical weathering and its tendency to break down by physical means. The other field areas have plagioclase and potassium feldspars in all of their samples except for the Corsican biotite granite, where plagioclase feldspars are absent from some samples.

The micas are differentially weathered in all areas except Wicklow. Biotite is more chemically weathered than muscovite in the Mournes which corresponds with high vermiculite and hydrobiotite percentages. Weathered biotite in Brazil, especially in the arenaceous profiles, is linked to high goethite contents where iron oxides have formed from iron released during biotite weathering. In Corsica, muscovite rather than biotite appears to be preferentially weathered to kaolinite and illite–smectite. Corsica is a special case with regard to muscovite breakdown. This mica is found in only 5% of the weathered biotite granite samples and in none of the others, including those from muscovite granite. In Corsica, muscovite appears to be more susceptible to chemical weathering than biotite yet less susceptible to physical breakdown. Thin section analysis indicates that microcracks in Corsican granite are neither wide nor excessive. The environment is sufficiently conducive to chemical activity in that once a crack exploits muscovite cleavage the mineral is weathered to illite–smectite rather than persisting to be mechanically disintegrated. Illite–smectite has been noted by Wilson and Nadeau (1985) as an intermediate product of the chemical weathering of muscovites and Table 2.5 shows that as a corollary to the lack of muscovite, this secondary form is present throughout the Corsican profiles. Five per cent of the biotite granite samples do contain physically comminuted muscovite, and thin sections show that some cracking is biotite-induced. However, biotite-induced cracking is not of sufficient extent or magnitude to effect large-scale cracking in the samples examined from Corsica. Biotite is also weathered to a hydrobiotite intermediate form and it has been assumed (e.g. Wahrhaftig, 1965) that the conversion of biotite to an expandable hydrobiotite form will effect cracking (especially of quartz) and grussification through wetting and drying. However, quartz is not found in <63 μm fractions from Corsica and although thin sections show some tensional disruption of quartz it is rare and does not appear to be induced by biotite expansion as hydrobiotite grains can be seen without any surrounding cracking.

Clearly a re-evaluation of how the chemical weathering of biotite can cause microfracturing is required. The essential factor here is the relative activity of hydrolysis versus hydration. The rate of hydrolysis depends on temperature, the size of particle (the large surface areas of clays accelerate the process) and the concentration of ions in solution (Keller, 1964). Hydration depends more on water volume. In Corsica, hydrolysis appears to convert the mineral gradually to hydrobiotite, then to vermiculite. At the intermediate stage, the warm environment probably allows hydrolysis to continue at the expense of hydration, so that expansion is negligible. Also in Corsica, water for hydration does not appear to be as readily available as in the other areas (leaching factor 16.67, Table 2.3). In sub-tropical Brazil, hydrolysis is dominant due to the high temperatures and the inputs of organic acid, so that biotite is converted directly to vermiculite. Sorenson (1988) noted a volume increase of 40% following biotite alteration to vermiculite, and perhaps the more dramatic direct conversion and the concomitantly greater replacement of potassium by water is a more important expansion/cracking mechanism. Alternatively, the conversion of biotite to hydrobiotite in the Mournes may aid cracking in that hydration is favoured over hydrolysis in this cool, wet climate, especially in the presence of organic acids from overlying peaty soils.

Secondary Minerals

Palygorskite, or attapulgite, is a magnesium mineral which is formed early in the weathering of silicate minerals. It is delicate and becomes unstable with further weathering due to the solubility of the principal magnesium cation (Loughnan, 1969). Surprisingly, therefore, it is found in both arenaceous and argillaceous samples from Brazil. Elemental analysis shows that the Brazilian profiles are indeed enriched in magnesium with respect to the parent rock (Table 2.7), so that the formation of palygorskite does seem to correspond with local relative ion mobilities rather than apparent weathering degree.

Illite minerals and/or interlayered illite–smectite clays are common in samples from all areas. Illite, most usually an early alteration product of feldspars, is found in Ireland and Brazil. In warm temperate Corsica an illite–smectite mix is found as a degradation product of muscovite. Hydrobiotite, an intermediate product between biotite and vermiculite, has been noted in the Mournes, Corsica and Brazil. Vermiculite is common in the cool temperate field areas, but it is also found in Corsica and Brazil as an intermediate product of biotite weathering which may later alter to kaolinite. In the Brazilian case it may also be derived from feldspars where potassium removal has occurred (see Table 2.7). It is not found in either the Corsican or Brazilian granodiorite samples, where chlorite is instead preferentially formed from biotite.

Table 2.7 Mean elemental ratios in weathered profiles with respect to parent rocks, <1 = depletion, >1 = enrichment

	Mournes	Wicklow	Corsica (biot. granite)	Brazil Aren.	Brazil Arg
Ca	0.89	0.35	1.32	0.49	0.44
Mg	2.72	0.53	1.92	1.23	1.15
Na	0.78	2.40	1.25	1.65	2.56
K	1.08	2.51	1.18	0.69	0.70
Fe	0.92	0.71	3.20	1.24	0.30
Al	1.08	1.68	1.65	0.89	0.69

The results show that kaolin minerals (kaolinite, halloysite and endellite) are the most common alteration products of granitoid breakdown, derived from the weathering of both micas and feldspars. Kaolinite is found in all areas and its occurrence most probably depends on local site conditions where it is favoured by free-draining conditions. However, the endellite versus halloysite differentiation appears to be climactically determined. Endellite peaks at 10.16Å and 7.37Å on XRD traces and is found in the wet, cool Mournes. Alternatively, halloysite (7.25Å) is found in the Brazilian arenaceous profiles, where annual average precipitation is lower. Rainfall seasonality might be significant in this area in that samples were taken during the dry season and halloysite might hydrate to endellite during the wet season.

Residual Oxides

Gibbsite is formed in conjunction with kaolinite in Brazil. This is thought to be the 'secondary' gibbsite of Macias-Mazquez' (1981) that forms from the desilicification of kaolinite. Of the Mournes samples, 8% contain gibbsite, possibly as a 'primary' weathering product precipitated from an alumina-rich solution transported by cheluviation in this podzolic environment (pH = 5).

Other oxides (boehmite, anatase and diaspore) become prevalent in warm temperate Corsica indicating the enrichment of aluminium by the weathering of aluminosilicates (see Table 2.7), and titanium by immobility. In this sense, weathering in Corsica is more advanced than in Ireland. However, anatase is also found in the Mournes indicating a relative enrichment of titanium. Table 2.7 also shows iron depletion in Ireland (particularly in Wicklow) and in argillaceous profiles from Brazil and aluminium depletion in Brazilian arenaceous profiles. These data indicate possible dangers in the assumption that Fe_2O_3 or Al_2O_3 is not lost during weathering when calculating quantitative weathering indices.

The Brazilian argillaceous profiles, despite iron depletion, contain all four of the major iron oxides, namely goethite, hematite, lepidocrite and berthierine. The Corsican granodiorite and the Brazilian arenaceous profiles contain predominantly goethite. Goethite typically forms by the precipitation of iron from solution, whereas haematite forms from the dehydration of either amorphous material or goethite (Schwertmann, 1985). Goethite may dehydrate to give hematite, but the inorganic rehydration of hematite to goethite is not feasible (Langmuir, 1971) so that this particular process is unidirectional. Goethite can, however, form from hematite if organic compounds enter the regolith, complex the hematite iron and subsequently reprecipitate it as goethite (Schwertmann, 1971). The inference then arises that hematite is formed in a climate that satisfies two conditions: first, the climate must be wet and warm to allow the release of iron from primary minerals and secondly, the climate must also incorporate dry periods which allow hematite to form via dehydration. Walker (1967) suggested that seasonally wet arid climates satisfy the above criteria, while Schmalz (1968) advocated formation under a savanna-type climate. Schwertmann (1971) proposed that a later change to a cooler, humid climate will encourage the slower surface decomposition of organic matter, so that organic compounds survive to enter the subsurface material and cause the precipitation of goethite from hematite. In Brazil, it is possible that the climatic change from arid interpluvials to present-day humid sub-tropical conditions has allowed goethite to become the stable iron oxide. Consequently, in the argillaceous profiles, red hematite layers typically occur below yellow goethite-stained material as also described by Schwertmann (1971). The red hematite horizons are possibly relict from the Pleistocene, whereas goethite is forming today through the degradation of hematite by organic compounds. In Corsica, goethite appears to be forming directly from the weathering of the primary ferromagnesium minerals, most of which have already been weathered in the Brazilian argillaceous profiles.

In these discussions there is an assumption that 'redness' of regolith is synonymous with the presence of hematite. Torrent *et al.* (1980), working in Spain, suggested that Munsell colours can be translated into a redness rating, which is positively correlated with soil age and hematite content and where:

$$\text{Redness rating} = (H \times C)/V$$

where H = soil hue weighting (7.5 R = 12.5; 10 R = 10; 2.5 YR = 7.5; 5 YR = 5; 7.5 YR = 2.5; 10 YR = 0), C = soil chroma and V = soil value.

The redness ratings of the argillaceous Brazilian profiles are given in Table 2.8 together with their iron and iron oxide contents. Iron oxides may be present in those layers labelled as having 'none', but they are of insufficient quantity and crystallinity to register on the XRD trace.

The assertion that redness is linked with hematite content is clearly not supported by the Brazilian results. Goethite and hematite often occur together in yellow, brown and red horizons indicating the neoformation of goethite from hematite. 'Redness' may be correlated with hematite content in areas where this oxide is forming today, but in Brazil, where hematite-rich layers are essentially relict, the contemporary formation of goethite has skewed the results. Alternatively, 'yellowness' is apparently linked with goethite content, providing further evidence for the hypothesis that the red layers are centres of hematite destruction, and the yellow layers sites of geothite formation. Redness is not therefore linked directly with iron content and has a low positive Pearson's product–moment correlation coefficient of 0.43. Watson (1900) in Georgia and Ahn (1970) in West Africa also found that redness is not necessarily related to a high iron content, and, with iron present in appreciable quantities throughout these deeply weathered profiles, differentiation on the basis of colour alone becomes very tentative.

Table 2.8 Iron oxide mineralogies and 'redness ratings' of argillaceous Brazilian profiles

Profile no.	Redness rating	Fe (%)	Fe oxides present
10	0	4.75	GT, HM
9	0	2.65	GT, HM
8	0	N/D	None
21	0	2.15	GT
25	0	3.3	None
10	0.63	3.8	GT, HM
25	0.63	3.15	HM
8	1.25	1.6	GT
21	2.5	2.2	>GT, L
24	2.5	2.1	None
14	2.86	2.9	None
18	2.86	2.65	None
19	2.86	5.25	GT
19	3.21	6.35	None
23	3.33	2.35	None
25	3.33	4.35	None
10	4.0	N/D	GT, HM
7	4.0	2.4	None
10	4.29	3.1	GT, HM
18	4.29	5.15	BR
24	4.29	3.35	None
18	5.0	4.9	None
25	6.67	3.8	None
7	10.0	7.35	None
14	12.0	5.1	None
21	12.0	2.55	GT

N/D = not determined
GT = goethite
HM = hematite
BR = berthierine
L = lepidocrocite

NON-CLIMATIC INFLUENCES

As indicated earlier, there are a number of additional factors that can influence weathering patterns, the effects of which are superimposed upon any exerted by climate. Certain distinct patterns were, for example, observed in the chemical weathering of feldspars that can be best explained by local factors and feldspar type. Illite is derived from feldspars in confining weathering environments as noted by Stephen (1952) and Wilson (1975). In the Mourne and Wicklow mountains this is a function of temporary poor drainage under peaty conditions. Conversely, in Brazil, illite is possibly relict from arid interpluvials when leaching was weaker. There was, however, no conclusive evidence to support the contention that gibbsite is more likely to derive from calcite plagioclases than alkali feldspars, as suggested by Alexander *et al.* (1942), except perhaps in Brazil where argillaceous and granodiorite profiles contain more gibbsite than the arenaceous profiles (Table 2.5). However, the occurrence of secondary gibbsite could be related to more advanced desilicification rather than feldspar type. Feldspar type does appear to have an impact upon the degree of weathering so that in an indirect sense gibbsite may more often be derived from calcic plagioclases, but this is a function of weatherability rather than chemical affinity with gibbsite. It has also been proposed that kaolinite is most commonly derived from plagioclase feldspars (Stoch and Sikora, 1976), with halloysite more commonly associated with K-feldspars (Parham, 1969). Halloysite was differentiated from kaolinite in the Mournes by XRD and SEM, and by XRD in Brazilian arenaceous profiles. Soil moisture conditions linked to topographical-drainage systems were, however, seen to be more likely causes of this differentiation, with halloysite occurring where water retention permits the formation of an extra water layer within kaolin minerals. Such environmental conditions have also been favoured over mineralogical effects by Cady (1950) and Heng and Eswaran (1976), and similar occurrences were observed

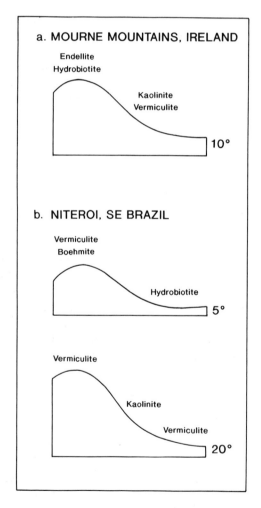

Figure 2.1 Topographical differentiation of secondary weathering products

in the Mournes, where halloysite forms in wet upper profile sites overlain by moisture-retaining peaty topsoils.

The weathering of biotite to vermiculite can also be influenced by local factors. Typically this proceeds either via a chloride–vermiculite intergrade mineral (corrensite) or via hydrobiotite intermediates. The former is more likely to occur where local aluminium levels are high and organic input low. This pattern was observed in Wicklow and in Corsican muscovite granite. The complete conversion of biotite to vermiculite most often occurs where water supply is sufficient to allow rapid hydrolysis of biotite, and where drainage is efficient enough to ensure the removal of magnesium and possibly potassium cations. Conversion to kaolinite occurs if leaching is excessive, whereas 'vermiculitisation' occurs at sites with intermediate drainage conditions. In the Mourne Mountains, conversion of biotite to vermiculite occurred in midslopes where drainage is more efficient than upslope beneath peat. In the Brazilian argillaceous profiles, vermiculite was found at the top of gentle slopes (of the order of 5°), whereas in midslope locations intermediate clay minerals occurred. Leaching is strong in upslope zones and slower downslope, but when the slope angle increases beyond approximately 20° in Brazil, drainage is more rapid in the middle sections so that kaolinisation occurs there. Vermiculite is found at the top of slopes, where water supply is limited and at the bottom where drainage is slower. These topographical influences on the presence of vermiculite are represented in Figure 2.1, where the slope angles are representative of the steepest section of the transect (i.e. maximum slope angle).

CLIMATIC IMPLICATIONS

In cool temperate Iceland, where leaching potential is strong, kaolinite is formed from the weathering of both feldspars and micas, with vermiculite as an intermediate product of biotite weathering. Illite rather than montmorillonite is the intermediate product of the weathering of feldspars, and is indicative of the preferential removal of sodium relative to potassium (Stephen, 1952). Table 2.9 shows the potassium and sodium concentration ratios with respect to the parent granite of the eight Mourne samples containing illite. Potassium is indeed generally enriched during weathering as compared to sodium which is depleted, thus favouring the formation of illite.

Illite also occurs in the Brazilian profiles, but may be relict from previous arid interpluvials. Vermiculite and kaolinite also occur, which may reflect oscillating leaching potential through the year, with kaolinite formed by 'monosiallitisation'

Table 2.9 Ratios of K and Na in those Mourne Mountain sample which contain illite calculated with respect to parent granite

Granite		Weathered samples							
	79	71	72	75	76	77	80	81	82
K	1.0	1.11	1.11	1.13	1.08	1.06	1.07	1.05	0.91
Na	1.0	0.71	0.69	0.92	0.79	0.75	0.83	0.58	0.73

when some silica is removed and some remains to combine with aluminium during the wettest season. Vermiculite is probably formed by 'bisiallitisation' just before and after this wettest time when silica is not removed from the profiles.

The only major contrast between these secondary minerals found in Brazil as compared with Ireland is the dominance of gibbsite (along with other residual oxides) in Brazil. Mineralogical convergence occurs since both areas experience high rainfall, although the leaching factor in the Mournes is much higher. A kaolinite and gibbsite combination, as found in Brazil, is documented in Table 2.6 as traditionally indicative of sub-tropical conditions. The presence of vermiculite in Brazil cannot, however, be similarly explained by contemporary climatic factors. Instead, local drainage effectiveness could be a more important control on vermiculitisation. The most obvious climatically linked characteristic remains, however, the dominance of residual oxides in the Brazilian profiles, indicative of more complete alteration and leaching.

In warm temperate Corsica, vermiculite and kaolinite are somewhat unexpected occurrences. Vermiculite has previously been seen to occur in humid temperate areas rather than Mediterranean environments, so that it may be a relict product from Pleistocene cold humid conditions. Kaolinite has only rarely been noted in warm temperate environments (e.g. Harriss and Adams, 1966, in Oklahoma). However, Corsica's leaching factor of 16.67 may in fact be sufficient to allow kaolinite to form under present-day conditions with vermiculite as an intermediate biotite product and illite–smectite as the corresponding muscovite derivative. Indeed, Tardy (1971) has also found kaolinite and montmorillonite in Corsica. He attributes kaolinite formation to the removal of silica during wet periods (September to March), and the formation of the intermediate minerals to the drier periods of the year (April to August) when less silica is removed.

In the three study areas, considerable explanation of the mineralogy found is afforded by current climate modified by local topographic controls on drainage. Interestingly, seasonal ratios rather than mean annual climate conditions seems frequently responsible for the variety of weathering products (especially clay minerals). One consequence of this is a degree of convergence between the clay minerals identified in each area. Correlation between mineralogy and climate, particularly in extra-tropical locations, also suggests that alteration is continuing under present-day conditions, where drainage permits removal of decomposition products. If so, it highlights the dangers of assuming that all arenaceous profiles encountered in high latitudes are necessarily relict. Equally, the occurrence of deep, arenaceous profiles in Brazil emphasises the range of palaeoclimates that might be invoked if other factors do point to climatic inheritance. Work is continuing to identify possible local factors which control the formation of argillaceous and arenaceous profiles.

ACKNOWLEDGEMENTS

The writers are grateful to Professors Braz Sanchez and Brian Whalley and Dr John McAlister for their encouragement and enlightenment. Financial support was provided by a grant from the Department of Education for Northern Ireland.

REFERENCES

Ahn, P. M. (1970). *West African Soils*. Oxford University Press, London.

Alexander, L. T., Hendricks, S. B. and Faust, G. T. (1942). Occurrence of gibbsite in some soil-forming materials. *Soil Science Society America Proceedings*, **6**, 52–57.

Bakker, J. P. (1960). Some observations in connection with recent Dutch investigations about granite weathering and slope development in different climates and climate changes. *Zeitschrift für Geomorphologie*, Supplementband **1**, 69–92.

Bakker, J. P. (1967). Weathering of granites in different climates, particularly in Europe. *Union Geographie International Symposium, Liège, Evolution des Versants*, **1**, 51–68.

Barshad, I. (1966). The effect of precipitation on the nature of clay mineral formation in soils from acid and basic igneous rocks. *International Clay Conference Proceedings, Jerusalem*, 1966, pp. 167–173.

Brock, R. W. (1949). Weathering of igneous rocks near Hong Kong. *Geological Society America Bulletin*, **54**, 717–738.

Brunsden, D. (1964). The origin of decomposed granite on Dartmoor. In Simmons, I. G. (ed.), *Dartmoor Essays*, Devonshire Association, Exeter, pp. 97–116.

Buringh, P. (1970). *Introduction to the Study of Soils in Tropical and Subtropical Regions*. Pudoc, Wageningen.

Butt, C. R. M. (1985). Granite weathering and silcrete formation on the Yilgarn Block, Western Australia. *Australian Journal Earth Science*, **32**, 415–432.

Cady, J. G. (1950). Rock weathering and soil formation in the North Carolina Region. *Soil Science Society America Proceedings*, **15**, 337–342.

Collier, D. (1961). Mise au point sur les processus de l'alteration des granties en pays tempere. *Annales Agronomie*, **12**, 273–331.

Crowther, E. M. (193). The relation of climate and geological factors to the composition of the clay and the distribution of soil type. *Proceedings Royal Society*, **107**, 10–30.

Curmi, P. and Maurce, F. (1980). Caractisation microscopique de l'alteration dans une arene granitique a structure conservee. In *Submicroscopy of Soils and Weathered Rocks*, Wageningen, Netherlands, **1**, 249–270.

Dearman, W. R. and Baynes, F. J. (1978). A field study of the basic controls of weathering patterns in the Dartmoor granite. *Ussher Society Proceedings*, **4**, 192–203.

Eden, M. J. and Green, C. P. (1971). Some aspects of granite weathering and tor formation on Dartmoor, England. *Geografiska Annaler*, **53A**, 92–99.

Eswaran, H. and Bin, W. C. (1978). A study of a deep weathering profile on granite in peninsular Mayalsia: 1. Physico-chemical and micromorphological properties. *Soil Science Society America Journal*, **42**, 144–149.

Eswaran, H. and Heng, Y. Y. (1976). The weathering of biotite in a profile on gneiss in Malaya. *Geoderma*, **16**, 9–20.

Fitzpatrick, E. A. (1963). Deeply weathered rock in Scotland, its occurrence, age, and contribution to the soils. *Journal of Soil Science*, **14**, 33–43.

Flageollet, J. C. (1977). Origine des reliefs, alterations et formations superficielle; contribution a l'étude geomorphologique des massifs anciens cristallines; l'exemple du Limousin et de la Vendee du Nord-Ouest. *Science de la Terre Memoire*, **35**.

Goldthwait, J. W. and Kruger, F. C. (1938). Weathered rock in and under the drift in New Hampshire. *Geological Society America Bulletin*, **49**, 1183–1198.

Hall, A. M. and Mellor, A. (1988). The characteristics and significance of deep weathering in the Gaick area, Grampian Highlands, Scotland. *Geografiska Annaler*, **70A**, 309–314.

Harriss, R. C. and Adams, J. A. S. (1966). Geochemical and mineralogical studies on the weathering of granitic rocks. *American Journal Science*, **264**, 146–173.

Hseung, Y. and Jackson, M. L. (1952). Mineral composition of the clay fraction: 3. Some main soil groups of China. *Soil Science Society America Proceedings*, **16**, 294–297.

Humbert, R. P. and Marshall, C. E. (1943). Mineralogical and chemical studies of soil formation from acid and basic igneous rocks in Missouri. *University of Missouri Research Bulletin*, **359**.

Ismail, F. T. (1970). Biotite weathering and clay formation in arid and humid regions, California. *Soil Science*, **109**, 257–261.

James, W. C., Mack, G. H. and Suttner, L. J. (1981). Relative alteration of microcline and sodic plagioclase in semi-arid and humid climates. *Journal of Sedimentary Petrology*, **51**, 151–164.

Jenny, H. (1941). *Factors of Soil Formation*. McGraw-Hill, New York.

Keller, W. D. (1964). Processes and alteration of clay minerals. In Rich, C. I. and Kunze, G. W. (eds), *Soil Clay Mineralogy*, University of North Carolina Press, pp. 3–76.

Klaer, W. (1956). Vermitterungsformen im granit auf Korsica. *Pettermanns Geographische Mitteilungen*, **261**.

Kostrzewski, A. (1985). Granulometric features of crystalline massif regolith in extreme climatic zones (Spitsbergen, Australia): Interpretation of textural properties. *Questiones Geographie*, **9**, 77–98.

Langmuir, D. (1971). Particle size effect on the reaction goethite = haematite + water. *American Journal of Science*, **271**, 147–156.

Loughnan, F. G. (1969). *Chemical Weathering of the Silicate Minerals*. Elsevier, New York.

Lumb, P. (1962). The properties of decomposed granite. *Geotechnique*, **12**, 226–243.

Macias-Vazquez, F. (1981). Formation of gibbsite in soil and saprolites of temperate-humid zones. *Clays and Clay Minerals*, **16**, 43–52.

Mulcahy, M. J. (1960). Laterites and lateritic soils in Southwestern Australia. *Journal of Soil Science*, **11**, 206–225.

Nagell, R. H. (1962). Geology of the Serro do Navio manganese district, Brazil. *Economic Geology*, **57**, 481–497.

Nossin, J. J. and Levelt, W. M. (1967). Igneous rock weathering on Singapore Island. *Zeitschrift für Geomorphologie*, NF11, 14–35.

Novikoff, A., Tsawlasson, G., Gac, T.Y., Bourgeut, F. and Tardy, Y. (1972). Alteration des biotites dans les arenes des pays temperes, tropicaux et equatoriaux. *Science Géologie Bulletin*, **25**, 287–305.

Ollier, C. D. (1976). Catenas in different climates. In Derbyshire, E. (ed.), *Geomorphology and Climate*, Wiley, Chichester, pp. 137–169.

Ollier, C. D. (1988). Deep weathering, groundwater and climate. *Geografiska Annaler*, **70A**, 285–290.

Parham, W. E. (1969). Formation of halloysite from feldspar: low temperature artificial weathering versus natural weathering. *Clays and Clay Minerals*, **17**, 13–22.

Peulvast, J. P. (1986). Structural geomorphology and morphological development in the Lofoten–Vestralen area, Norway. *Norsk Geografische Tidsskrift*, **40**, 135–161.

Pickering, R. J. (1962). Some leaching experiments on three quartz-free silicate rocks and their contribution to an understanding of laterization. *Economic Geology*, **57**, 1185–1206.

Power, E. T. (1989) *Subsurface weathering of granitoid rocks in different climates*. Unpublished Ph.D. Thesis, Queen's University, Belfast.

Radwanski, S.A. and Ollier, C.D. (1959). A study of an East African catena. *Journal of Soil Science*, **10**, 149–170.

Rice, C.M. (1973). Chemical weathering on the Carnmenellis granite. *Mineralogical Magazine*, **39**, 429–447.

Ruxton, B. P. (1958). Weathering and subsurface erosion in granite at the piedmont angle, Balos, Sudan. *Geological Magazine*, **95**, 29–31.

Ruxton, B. P. and Berry, L. (1957). Weathering of granite and associated erosional features in Hong Kong. *Geological Society America Bulletin*, **68**, 1263–1292.

Ruxton, B. P. and Berry, L. (1961). Weathering profiles and geomorphic position on granite in two tropical regions. *Revue Géomorphologie Dynamique*, **12**, 16–31.

Sanches Furtado, A. F. S. (1968). Alteration des granites intertropicales sous differents climats. *9th International Congress, Soil Science Transactions, Adelaide*, **4**, 403–409.

Schmalz, R. F. (1968). Formation of red beds in modern and ancient deserts. Discussion. *Geological Society America Bulletin*, **79**, 277–280.

Schwertman, U. (1971). Transformation of haematite to geothite in soils. *Nature*, **232**, 624–625.

Schwertmann, U. (1985). Formation of secondary iron oxin oxides in various environments. In Drever, J. I. (ed.), *The Chemistry of Weathering*, D. Reidel, Dordrecht, pp. 119–120.

Sheppard, S. M. F. (1977). The Cornubian Batholith, S.W. England: D/H and $^{18}O/^{16}O$ studies of kaolinite and other alteration products. *Journal Geological Society London*, **33**, 573–591.

Sherman, G. D. (1952). The genesis and morphology of the alumina rich laterite clays in clay and laterite genesis. *American Institute Mining Metallurgy*, **16**, 154–161.

Singer, A. (1984). Clay formation in saprolites of igneous rocks under semi-arid to arid conditions, Negev, Southern Israel. *Soil Science*, **137**, 332–340.

Sorenson, R. (1988). In-situ weathering in Vestfiord, Southeastern Norway. *Geografiska Annaler*, **70A**, 299–307.

Sousa, E.C. and Eswaran, H. (1975). Alteration of micas in the saprolite of a profile from Angola: a morphological study. *Pedologie*, **25**, 71–79.

Stephen, I. (1952). A study of rock weathering with reference to the soils of the Malvern Hills. Part 1. Weathering of biotite and granite. *Journal Soil Science*, **3**, 20–33.

Stoch, L. and Sikora, N. (1976). Transformation of micas in the process of kaolinization of granites and gneisses. *Clay Minerals*, **24**, 156–162.

Tan, K. H. and Troth, P. S. (1982). Silica-sesquioxide ratios as aids in characterization of some temperate regions and tropical soil clays. *Soil Science Society America Journal*, **46**, 1109–1114.

Tardy, Y. (1971). Characterisation of the principal weathering types by the geochemistry of waters from some European and African crystalline massifs. *Chemical Geology*, **7**, 253–271.

Tardy, Y. and Nahon, D. (1985). Geochemistry of laterites, stability of Al-goethite, Al-hematite, and Fe^{3+}-kaolinite in bauxites and ferricrites: an approach to the mechanism of concretion formation. *American Journal Science*, **285**, 865–903.

Ternan, J. L. and Williams, A. G. (1979). Hydrological pathways and granite weathering on Dartmoor. In Pitty, A. F. (ed.), *Geographical Approaches to Fluvial Processes*, Geo Abstracts, Norwich, pp. 5–30.

Thomas, M. F. (1974). *Tropical Geomorphology—A Study of Weathering and Landford Development in Warm Climates*. Macmillan, London.

Thomas, M. F. (1976). Criteria for the recognition of climatically-induced variations in granite landforms. In Derbyshire, E. (ed.), *Geomorphology and Climate*, Wiley, Chichester, 411–447.

Torrent, J., Schwertmann, U. and Schulze, D. G. (1980). Iron oxide mineralogy of some soils of two river terrace sequences in Spain. *Geoderma*, **23**, 191–208.

Twidale, C.R. (1986). Granite landform evolution: factors and implications. *Geologische Rundschau*, **75**, 769–779.

Vanderford, H. B. (1942). Variations in the silt and clay fractions of loessial soils caused by climatic differences. *Soil Science Society America Proceedings*, **6**, 83–85.

Vargas, M. (1953). Some engineering properties of residual clay soils occurring in Southern Brazil. *3rd International Conference Soil Mechanics Foundations Engineering, Zurich*, **1**, 67–71.

Verheye, W. and Stoops, G. (1975). Nature and evolution of soils developed on the granite complex in the subhumid tropics (Ivory coast): 2, Micromorphology and mineralogy. *Pedologie*, **25**, 40–55.

Wahrhaftig, C. (1965). Stepped topography of the Southern Sierra Nevada, California. *Geological Society America Bulletin*, **76**, 1165–1190.

Walker, T. R. (1967). Formation of red beds in modern and ancient deserts. *Geological Society America Bulletin*, **78**, 353–368.

Watson, J. P. (1962). Formation of gibbsite as a primary weathering product of acid igneous rocks. *Nature*, **196**, 1123–1124.

Watson, T. L. (1990). Weathering of granitic rocks of Georgia. *Geological Society America Bulletin*, **12**, 93–108.

Wensink, J. J. (1968). The Emma Range in Surinam. *Fysisch-Geografie Bodemkund Laboratory Publication, Amsterdam*, **13**.

Williams, A. G., Ternan, L. and Kent, M. (1986). Some observations on the chemical weathering of the Dartmoor Granite. *Earth Surface Processes Landforms*, **11**, 557–574.

Wilson, M. F. and Nadeau, P. H. (1985). Interstratified clay minerals and weathering processes. In Drever, J. I. (ed.), *The Chemistry of Weathering*, D. Reidel, Dordrecht, pp. 97–118.

Wilson, M. J. (1967). The clay mineralogy of some soils derived from a biotite-rich quartz–gabbro in the Strathdon area, Aberdeenshire. *Clays and clay Minerals*, **7**, 91–100.

Wilson, M. J. (1975). Chemical weathering of some primary rock-forming minerals. *Soil Science*, **119**, 349–355.

3 The Effects of Fire on Rock Weathering: An Experimental Study

ROBERT J. ALLISON
University of Durham, UK

and

ANDREW S. GOUDIE
University of Oxford, UK

ABSTRACT

Laboratory simulation studies have been undertaken to examine the effects of fire on rock weathering. The results presented here develop work previously reported by Goudie *et al.* (1992), in which the effect of fire on seven rock materials was simulated in the laboratory using a furnace. This study focuses on the effects of water on rates of rock material disintegration. Thin sections prepared from samples subjected to a range of temperatures from 50°C to 900°C show that the process of material disintegration varies, depending on important rock physical characteristics such as texture, mineral composition and the cementing agent between individual mineral grains. The modulus of elasticity has been recorded before and after fire simulation experiments for both dry and wet samples using non-destructive apparatus. The results for wet rock samples suggest that the effects of moisture on weathering processes at elevated temperatures is closely dependent on those material physical characteristics which control water absorption.

INTRODUCTION

The results reported here follow a hypothesis generated by Francis (1992, p. 172, para. 3) that wet rocks crack when placed in a fire, and develop work previously presented (Goudie *et al.*, 1992) to examine the role of fire in promoting weathering in the natural environment and the significance of fire as an agent of rock disintegration. There are many environments in which both the frequency of and the temperatures attained in fires are important in the overall processes of weathering, erosion and landform development. Temperatures at the ground surface rise rapidly as fire sweeps across a landscape and the thermal gradient set up in rock within the top few centimetres of the ground surface will be significant, resulting in large changes in stress regime.

Two new sets of data are presented. The first is based on thin section analysis of materials used in simulation studies designed to assess the effects of fire on dry rock

---

Rock Weathering and Landform Evolution. Edited by D. A. Robinson and R. B. G. Williams
© 1994 John Wiley & Sons Ltd.

materials. Comparisons made between thin sections prepared from samples subjected to heating at a temperature of 900°C, and control samples, can be used to assess differences in the precise mechanism of mineral matrix disintegration for different rock types. The second new data set presents results to elucidate the effects of fire on rock materials which are wet prior to experiencing rapid changes in temperature. Since rock outcrops in the natural environment are frequently damp prior to fire events, it seems logical to develop previous studies in this manner.

BACKGROUND

Fires are a characteristic of many natural environments (Davis, 1959; Kozlowski and Ahlgren, 1974; Pyne, 1982 and 1991; Wein and Maclean, 1983; Booysen and Tainton, 1984; Goldammer, 1990, for example). Once a fire has ignited, it can spread rapidly and engulf extensive tracts of terrain. Maximum temperatures can exceed 1000°C and values of 500°C are quite normal. The vegetation cover of an area can act both to speed up the spread of a fire and increase temperatures (Table 3.1). Eucalypts act in this way, for example (Adamson et al., 1983).

One characteristic of many natural fires is that the thermal conditions at the ground surface change very rapidly, having important geomorphological implications. The time lapse from natural or equilibrium conditions to the most intense part of a blaze can be as little as a few minutes. At the same time, the duration at which temperatures maintain their maximum value may be short. As a consequence, the stresses set up in rock by the rapidly changing thermal conditions at the ground surface may be significant and the resulting thermal shock extreme. There are implications here for both physical and chemical weathering. Physically, rock materials subjected to large, rapid changes in temperature will have to respond to the resulting sudden changes in stress regime. Indeed, in a geomechanical sense, physical weathering is a response to changes in the stress conditions imposed on a rock material. Chemically, the mineral matrix may alter; alternatively, cementing agents within a rock mass may start to disintegrate.

Table 3.1 Maximum temperatures attained in fires and the associated vegetation types (after Goudie et al., 1992)

Vegetation type	Temperature (°C)	Source
Minnesota Jack Pine	800	Ahlgren (1974)
Chaparral scrub	538	Ahlgren (1974)
British heath	840	Whittaker (1961)
Senegal savanna	715	Daubenmire (1968)
Japanese grassland	887	Daubenmire (1968)
Chaparral scrub	1100	Mooney and Parsons (1973)
Nigerian savanna	538–640	Hopkins (1965)
Sudanese savanna	850	Hopkins (1965)
Heathland	550	Kruger and Bigalke (1984)
Fynbos	770	Trollope (1984)
Grass fires (High Plains)	388	Wright and Bailey (1982)
Chaparral scrub	685	Wright and Bailey (1982)

MATERIALS

Previous work (Goudie et al., 1992), which forms the basis for this study, investigated seven different rock types selected to permit comparison between materials of different physical and chemical composition. Numerous studies exist which discuss the significance of a particular rock material characteristic to disintegration at elevated temperatures (Aires-Barros et al., 1975; Aires-Barros, 1977). Three examples can be cited. Quartz expands approximately four times more than feldspar and twice as much as hornblende as temperature increases (Winkler, 1973), with a 3.76% volume expansion between room temperature and 570°C. Very heterogenous materials are less responsive to thermal changes (Homand-Etienne and Troalen, 1984). On the other hand, limestone may be susceptible to the effects of fire because of the thermal degradation of organics, the expulsion of fluid inclusions and the conversion of aragonite to calcite (Gaffey et al., 1991).

Of the seven original study materials, namely gabbro, slate, white marble, Shap Granite, white granite, York Stone and Portland Limestone, four were selected for further study. The four materials include igneous, metamorphic and sedimentary rocks which display a variety of physical and chemical characteristics.

Their main differences are as follows:

1. Portland Limestone: a white, Jurassic, oolitic limestone, with a micritic mass. Portland Limestone has a reasonably constant grain size distribution with a matrix cement of less than 5%. The water absorption capacity averages 6.8%.
2. York Stone: a weakly foliated, arkosic sandstone, with a high proportion of feldspar and some muscovite. York Stone possesses a strongly interlocking texture with medium size quartz grains. The water absorption capacity averages 4.3%.
3. Shap Granite: a coarse-grained, pink granite, with an interlocking texture and some alteration to alkali feldspar minerals present within the rock. Shap Granite has a quartz content of around 40% and a 5% biotite content, which is marginally

Table 3.2 Percentage change in the modulus of elasticity (decrease unless otherwise stated) at selected temperatures between 50°C and 900°C (after Goudie et al., 1992)

Rock type	Number of cycles	Starting temperature (°C)									
		50	100	200	300	400	500	600	700	800	900
Portland	1	+1.0	+0.9	7.5	17.5	19.8	32.3	49.5	61.2	58.2	63.7
Limestone	5	+1.9	+2.2	7.2	19.2	32.8	42.2	56.0	66.6	72.7	78.3
York Stone	1	+7.7	+8.2	+4.4	+6.7	+4.6	1.6	13.9	12.5	26.9	22.6
	5	+29.8	+12.3	+6.5	+8.9	+6.8	2.7	16.7	16.2	37.0	48.8
Shap Granite	1	+1.1	+0.8	7.9	21.7	27.4	43.3	43.5	63.9	65.4	77.2
	5	+0.7	+0.3	12.0	24.4	33.2	52.2	59.8	70.4	76.1	93.1
Slate	1	1.2	0.5	0.6	0.6	0.6	0.6	0.6	Split		
	5	0.0	0.0	0.0	0.0	0.0	0.0	0.0	Split		

altered to chlorite and displays some radiation damage. The water absorption capacity averages 0.18%.

4. Slate: in this instance a grey slate from North Wales, which displays a very strong foliation pattern and small, angular fragments. The smaller grains within the rock are mostly of angular quartz, surrounded by a strongly foliated matrix of chloride. The water absorption capacity averages 0.08%.

The results of earlier studies undertaken as part of this research programme (Table 3.2) show that Portland Limestone, York Stone, Shap Granite and slate respond in different ways to sudden changes in temperature.

METHODS

Previous studies of temperature effects on rocks, rock weathering and landforms, have been qualitative in approach. There are numerous records of rock cracking and boulder spalling, for example. An important aspect of this study is the attempt to provide geotechnical data to quantify the effects of fire and the associated implications for weathering processes.

The modulus of elasticity was selected as the measure of rock material response to fire for a number of reasons. First, as the elastic behaviour of rock changes so does its durability. Secondly, changes in elasticity are closely controlled by the composition of rock and parameters such as mineralogy, texture, density, porosity, water content, the nature and composition of any cementing material and anisotropy (Richter and Simmons, 1974). The modulus of elasticity can thus be used as an aggregate index of such properties. Thirdly, there is a close relationship between modulus of elasticity and compressive strength (Judd and Huber, 1962; D'Andrea et al., 1965; Deere and Miller, 1966). Fourthly, in the context of fire, the variables noted above are important in determining the significance of processes such as spalling and mineral matrix disintegration. They will, therefore, exert a control on rates of weathering and the precise mechanisms by which rocks disintegrate. Fifthly, higher elasticity values indicate less weathered, more competent materials, and lower elasticity values generally indicate more highly weathered materials. A number of earlier studies have examined these associations (see Attewell and Farmer, 1976; Dearman et al., 1978; and Cooks, 1983, for example). Finally, the modulus of elasticity can be measured non-destructively using ultrasonic techniques. Thus, in the context of geomorphological studies, it is possible to repeat laboratory simulations on the same specimens, eliminating between-sample variability.

The modulus of elasticity was measured using the *Grindosonic* apparatus, a non-destructive technique which measures the ultrasonic response of a regular-shaped sample, which has been struck to set up a mechanical vibration pattern within it. Full details of the technique have been presented previously (Allison, 1988 and 1989). Tests were conducted with samples resting on a foam mat to damp spurious harmonics. Each test piece was struck in the centre of its upper face using a glass rod and a piezo-electric detector was held in contact with and at the centre of one of the side faces. The apparatus measures the time of eight wave passes and the resulting value is used with details of sample length, width, thickness and mass to determine the modulus of elasticity.

Table 3.3 Starting temperatures and temperature change during simulation tests

Starting temperature (°C)	Minimum temp. dry cycle (°C)	Minimum temp. wet cycle (°C)	Difference	Wet temp. as % of dry temp.
50	50	50	0	100
100	100	99	1	99
200	200	195	5	98
300	300	279	21	93
400	375	358	17	95
500	471	448	23	93
600	560	520	40	93
700	634	597	37	94
800	707	670	37	95
900	761	753	8	99

A full discussion of a set of tests conducted to determine the effects of fire on dry samples has been presented previously (Goudie *et al.*, 1992), and a summary of the results which are relevant here is presented in Table 3.2. The experimental procedure adopted to examine the effect of water involved submerging the samples for a set period of time before placing them in the furnace. The method used for the test programme was designed, as far as possible, to replicate the conditions of the earlier study.

Samples were cut using a diamond-bladed saw and a flat-bed grinder into bars measuring approximately 150 mm × 30 mm × 20 mm. The blocks were placed in a climatic cabinet and left to reach constant weight at a temperature of 50°C and relative humidity of 20%. The weight and modulus of elasticity of each sample were recorded. The samples were immersed in a bath of distilled water for 24 hours. The blocks were placed in a furnace for five minutes at one of a range of starting temperatures between 50°C and 900°C (Table 3.3). Temperatures dropped during each cycle due to the thermal transfer of energy to the rock samples. The temperature drop was greater for the wet treatments than for the dry. The maximum temperature difference (40°C) between the wet and dry treatments occurs at a starting temperature of 600°C. At low (<200°C) and at high (900°C) temperatures the differences are small. This range of temperatures and time duration were considered to be reasonable in the light of published data on natural fires. Following five minutes in the furnace, the samples were returned to the climatic cabinet and left to reach equilibrium at 50°C and 20% relative humidity. The weight and modulus of elasticity of each sample was then remeasured. This procedure was repeated five or ten times at intervals of 24 hours in an attempt to simulate the effects of a number of fires on the rock materials. Seven materials and therefore seven rock specimens were used in each of the original simulation tests. To ensure comparability between the wet and dry treatments, additional 'dummy' test pieces were employed during the wet cycles.

As well as undertaking a new set of furnace tests to examine the water variable, thin sections were prepared for samples tested using the above procedure but without submerging specimens in water prior to firing. Two types of thin section were prepared. The first were from samples which had been placed in the furnace at a temperature of 900°C. The samples were sectioned across the middle of the specimen so as to ascertain variations between the exterior and the interior of the sample. Secondly, thin sections were prepared in exactly the same way for control rock samples which had not been subjected to any fire simulation.

RESULTS

Simulations Using Water

The results of the fire simulation tests for wet samples are presented in Table 3.4 and Figures 3.1 and 3.2. The results need to be considered in two contexts. First, by comparing the data for different rock types within the same simulation programme and, secondly, by comparing the wet and dry simulation data for the same rock types.

A number of interesting trends emerge by comparing the data for different rock types within the same simulation programme using fire and water:

1. For the Portland Limestone there is no significant change in the modulus of elasticity at temperatures below 200°C after either one, five or ten cycles. A significant change occurs between 500°C and 600°C and again between 700°C and 800°C for both cycle durations. After one cycle, the drop in the modulus of elasticity is approximately 10% at both temperatures. After five cycles, the drops are approximately 15% and 25% respectively. In other words, as the number of cycles increases, the percentage drop in elasticity gets larger at higher temperatures. In addition, after five cycles at 900°C and after eight cycles at 800°C, the sample disintegrated. For a given increase in temperature, the change in elasticity

Table 3.4 Results of fire and water simulation, showing the percentage change (decrease unless otherwise stated) in the modulus of elasticity at selected temperatures between 50°C and 900°C

Rock type	Number of cycles	Starting temperature (°C)									
		50	100	200	300	400	500	600	700	800	900
Portland Limestone	1	0.36	0.67	1.77	2.99	4.28	5.88	13.88	13.83	23.11	29.42
	5	0.36	0.67	2.09	3.95	5.24	7.11	21.18	20.88	45.92	Split
	10	+1.11	+0.35	3.12	2.35	5.04	11.81	21.66	21.25	Split	
York Stone	1	1.49	2.06	1.91	2.36	3.45	8.69	11.34	11.54	15.16	18.91
	5	0.00	+0.27	+0.46	0.77	2.22	5.60	14.66	17.71	22.21	28.95
	10	4.26	5.42	4.69	5.37	7.42	9.74	18.67	20.66	27.32	32.45
Shap Granite	1	+0.38	0.00	1.53	12.46	21.30	36.23	41.77	41.79	60.34	78.40
	5	0.00	0.73	2.63	14.73	29.70	42.58	60.92	51.36	69.31	84.22
	10	2.94	1.09	5.17	18.13	38.22	45.70	61.01	53.41	80.27	86.39
Slate	1	0.00	+1.08	0.57	0.55	0.54	1.52	Split			
	5	0.54	+1.62	0.00	0.00	0.54	Split				
	10	1.60	1.32	4.27	1.02	1.09	Split				

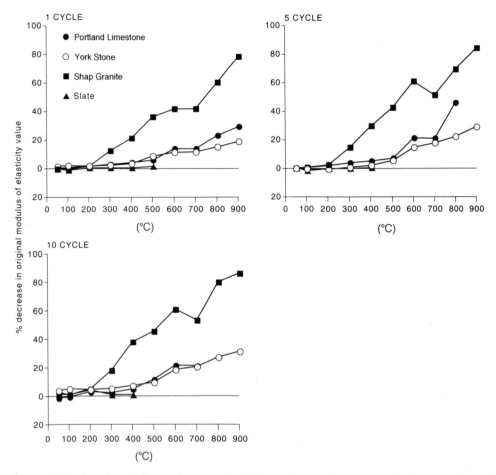

Figure 3.1 Results of fire and water simulation, showing the change in the modulus of elasticity after five minutes of heating at various starting temperatures: (a) one cycle of heating; (b) five cycles of heating; (c) ten cycles of heating

is always greater after five cycles than after one. The difference is not as marked, however, between five and ten cycles.

2. The behaviour of York Stone is somewhat different. First, the drop in elasticity both at increasing temperatures and as the number of cycles increases is not as marked as it is for the Portland Limestone. There are also no big step drops as temperature increases but a more gradual transition towards lower modulus of elasticity values. At temperatures of 100°C and 200°C there is a slight increase in the modulus of elasticity after five cycles, but the change is not large enough to be of significance.

3. Shap Granite displays no significant drop in the modulus of elasticity below temperatures of 300°C, but the decline increases in magnitude for temperature increments of 100°C thereafter. Again, the percentage decrease is greatest first after ten cycles and then after five in every case. There is some variation to an otherwise smoothly decreasing trend at temperatures around 500°C to 600°C

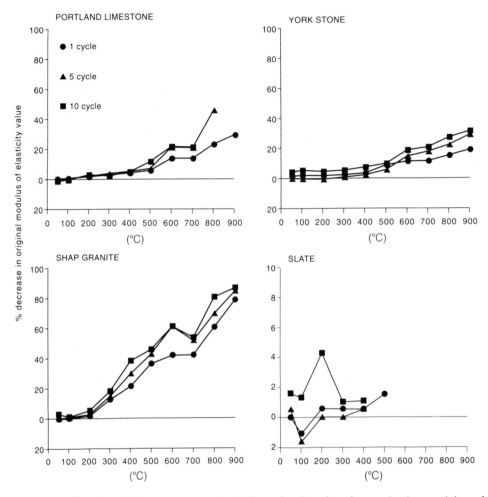

Figure 3.2 Results of fire and water simulation, showing the change in the modulus of elasticity: (a) Portland Limestone; (b) York Stone; (c) Shap Granite; (d) Slate

where, after five and ten cycles, the percentage drop in elasticity doubles. This may be because some mineral alteration occurs around these temperatures, but further work is required to examine this in greater detail.

4. None of the tests conducted on the slate material demonstrate a major decrease in the modulus of elasticity. However, at 600°C and above the test pieces split during the first cycle, and the sample tested at 500°C disintegrated during the fourth cycle.

Some tentative suggestions can be made for the differences in material behaviour. Portland Limestone and York Stone are both relatively porous sedimentary rocks with high water absorption capacities. Shap Granite and slate, on the other hand, have a much denser mineral matrix and much lower porosity. When submerged in water, the sedimentary materials will take up significant amounts of fluid. As a consequence, when placed within a furnace at elevated temperatures, the thermal shock to which the four materials are subjected will be dissipated in somewhat different ways. For the Portland Limestone and York Stone heat transfer will take place most rapidly through the water held within the pore spaces due to the higher thermal conductivity of the fluid. The initial result will be the rapid evaporation of water within the rock. Only after this has occurred will the mineral matrix itself be significantly affected by the increase in temperature. Indeed, at temperatures of 300°C or less, the heating effect on the rock mineral matter appears to be minimal. Further tests are required to examine whether this continues to be the case for durations of more than five minutes.

In the case of the igneous and metamorphic rocks the amount of water permeating the test specimens during immersion in water is small in comparison to that for the sedimentary materials. The result is that the amount of energy initially used in the latent heat of water vaporisation is small. The mineral matrix is, therefore, much more immediately affected by elevated temperatures.

It therefore appears that the influence of water will vary depending on the physical characteristics of the rock, in particular those which control the extent to which a rock can absorb significant amounts of fluid during immersion. The conclusions are confirmed if the data collected during the water and fire simulations are compared with those collected for the dry tests (Table 3.5).

Drops in the modulus of elasticity for both simulations are similar for wet and dry Shap Granite and slate samples. For Shap Granite, particularly at lower temperature regimes, the drop in elasticity is smaller for the wet tests but there is a close similarity between both sets of results. The overall trend for the slate samples is similar, although the wet samples disintegrate at lower temperatures than do the dry ones.

The two sedimentary rocks show somewhat different variations. In both cases, the wet tests show considerably smaller drops in the modulus of elasticity than do the dry ones, highlighting the effect of water in ameliorating the thermal shock and associated stress gradient set up in Portland Limestone and York Stone. Indeed, as Table 3.5

Table 3.5 Comparison of results for fire simulations on dry and wet samples. The data shows differences between % change in modulus of elasticity values for dry and wet samples

Rock type	Number of cycles	Starting temperature (°C)									
		50	100	200	300	400	500	600	700	800	900
Portland	1	1.36	1.57	5.73	14.51	15.52	26.42	35.62	47.37	35.09	34.28
Limestone	5	2.26	2.87	5.11	15.25	27.56	35.09	34.82	45.72	26.78	Split
York Stone	1	9.19	10.26	6.31	9.06	8.05	7.09	2.56	0.96	11.74	3.69
	5	29.8	12.03	6.04	11.12	9.02	2.9	2.04	1.51	14.79	19.85
Shap Granite	1	0.72	0.8	6.37	9.24	6.1	7.07	1.73	22.11	5.06	1.2
	5	0.7	0.76	9.37	9.67	3.5	9.62	1.12	19.0	6.79	8.88
Slate	1	1.2	1.58	0.03	0.05	0.06	0.92	Split			
	5	0.54	1.62	0.00	0.00	0.54	Split				

shows, the difference in the elasticity decrease displays the same ranking as the water absorption capacity. For the Portland Limestone, after one cycle at 900°C for example, the difference in modulus decrease is greater than 30%. After five cycles, the wet sample has disintegrated, suggesting that under repeated fires at high temperatures, Portland Limestone begins to disintegrate at a faster rate when wet than when dry. A similar trend is evident for the York Stone. In addition, the increase in the modulus of elasticity recorded for samples tested when dry is not present for the wet materials. It has been suggested previously (Goudie *et al.*, 1992) that the recorded modulus increase at temperatures below 400°C for dry samples may reflect some sort of case hardening process. If this is the case, the hardening effect is lost in the presence of water.

Thin Section Examination

Further details of the mechanism of rock breakdown for the different materials were provided by the thin sections prepared from the samples tested at 900°C under dry conditions. The different rock types demonstrate different types of mineral matrix alteration.

York Stone (Figure 3.3a and b) shows little visible evidence of change to the mineral matrix. The drop in the modulus of elasticity is most likely to be due to a general weakening of between-grain contacts, particularly around the perimeter of the sample. There is some evidence of discoloration, particularly towards the edge of the fired test piece (Figure 3.3b) but it is difficult to confirm whether this is a temperature effect.

A major change can be detected for the Portland Limestone (Figure 3.4). In its unweathered state, complete spherical ooliths are clearly visible, with a pervasive cementing material between much of the mineral matrix (Figure 3.4a). The fired sample shows three major changes (Figure 3.4b). First, the entire sample has become discoloured. Secondly, the matrix cement has broken down and largely disappeared throughout much of the specimen. Thirdly, many of the individual ooliths have started to disintegrate, some have disappeared completely and others remain only in part.

Shap Granite (Figure 3.5) demonstrates a different kind of alteration. By comparing fired and unfired samples an increase in fracture density and fracture width can be seen across larger grains. This is similar to the thermal cracking reported by Bauer and Handin (1983). There also appears to be some mineral alteration taking place, particularly towards the edge of the fired test specimen. In other words, the drop in the modulus is due to cracks appearing across crystal grains, rather than the destruction of matrix cement and the complete disintegration of the mineral matter.

The slate displays yet another mechanism of physical alteration (Figure 3.6). The sample fired at 900°C shows the development of cracks around the edge of the test piece. The cracks are absent from the unweathered specimen. What appears to be happening here is that cracks are opening up along cleavage, representing a form of stress relief and a response to the thermal gradient set up in the sample while it is in the furnace. The foliation pattern and the direction of cleavage planes relative to the exposed surface of any outcrop will, therefore, be important in the natural environment.

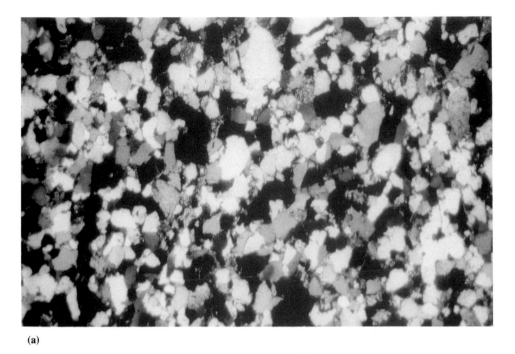

(a)

(b)

Figure 3.3 Photo-micrographs for York Stone: (a) unweathered material; (b) material subjected to 900°C temperature simulation

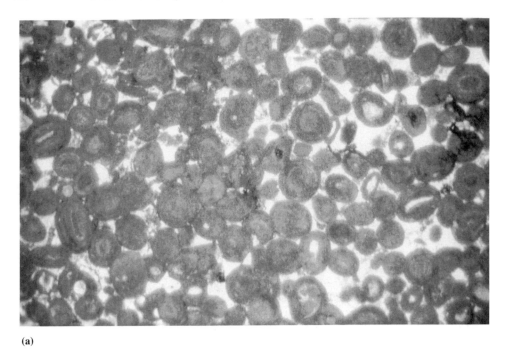

(a)

(b)

Figure 3.4   Photo-micrographs for Portland Limestone: (a) unweathered material; (b) material subjected to 900°C temperature simulation

(a)

(b)

Figure 3.5 Photo-micrographs for Shap Granite: (a) unweathered material; (b) material subjected to 900°C temperature simulation

(a)

(b)

Figure 3.6 Photo-micrographs for slate: (a) unweathered material; (b) material subjected to 900°C temperature simulation

The general conclusion to be drawn from the examination of the thin sections is that changes in the modulus of elasticity under simulated fires are a consequence of a variety of mechanisms of alteration. The process of breakdown not only varies between igneous, metamorphic and sedimentary rocks but also within each group, the detailed physical and chemical characteristics of the material having an important part to play in the rock disintegration process.

CONCLUSIONS

The main conclusions to be drawn from this study are as follows:

1. The geotechnical response of rock materials to fire is variable, depending on the rock itself, the intensity of the fire and other environmental conditions prevailing at the site, such as the amount of water in the rock.
2. Substantial changes in rock properties can occur at temperatures as low as 200°C, although the presence of moisture will raise the temperature at which significant effects can be recorded.
3. The duration of fire required to affect materials is not great, and in fact is sometimes less than one minute.
4. Although rock materials generally show a decrease in the modulus of elasticity as temperature increases, there are exceptions to this trend and the precise mechanism by which this drop is effected will vary from rock type to rock type.
5. Rock outcrops subjected to natural fires will respond in different ways, and the elastic properties of materials are an indication both of these changes and of subsequent susceptibility to weathering and erosion.

ACKNOWLEDGEMENTS

The simulation work was undertaken in the School of Geography, University of Oxford. The rock samples and thin sections were prepared in the Department of Geological Sciences, University College London by Arthur Beer, Sean Houlding and John Huggett. The diagrams were kindly prepared by Janet Baker and Mike Gray, University College London.

REFERENCES

Adamson, D., Selkirk, P. M. and Mitchell, P. (1983). The role of fire and lyre birds in the sandstone landscape of the Sydney Basin. In Young, R. W. and Nanson, C. G. (eds), *Aspects of Australian Sandstone Landscapes*, Australia and New Zealand Geomorphology Group, Wollongong, pp. 81–93.
Ahlgren, I. F. (1974). The effect of fire on soil organisms. In Kozlowski, T. T. and Ahlgren, C. E. (eds), *Fire and Ecosystems*, Academic Press, New York, pp. 47–72.
Aires-Barros, L. (1977). Experiments on thermal fatigue of non-igneous rocks. *Engineering Geology*, **11**, 227–238.
Aires-Barros, L., Graca, R. C. and Velez, A. (1975). Dry and wet laboratory tests and thermal fatigue of rocks. *Engineering Geology*, **9**, 249–265.
Allison, R. J. (1988). A non-destructive method of determining rock strength. *Earth Surface Processes and Landforms*, **13**, 729–736.

Allison, R. J. (1989). Developments in a technique for measuring rock strength. *Earth Surface Processes and Landforms*, **14**, 571–578.

Attewell, P. B. and Farmer, I. W. (1977). *Principles of Engineering Geology*. Chapman & Hall, London.

Bauer, S. J. and Handin, J. (1983). Thermal expansion and cracking of three confined, water saturated igneous rocks at 800°C. *Rock Mechanics and Rock Engineering*, **16**, 181–198.

Booysen, P. de V. and Tainton, N. M. (eds) (1984). *Ecological Effects of Fire in South African Ecosystems*. Springer-Verlag, Berlin.

Cooks, J. C. (1983). Geomorphic response to rock strength and elasticity. *Zeitschrift für Geomorphologie*, NF **27**, 483–493.

D'Andrea, D. V., Fischer, R. L. and Fogelson, D. E. (1965). Prediction of compressive strength from other rock properties. *US Bureau of Mines Report Investigation*, **6702**.

Daubenmire, R. (1968). Ecology of fire in grassland. *Advances in Ecological Research*, **5**, 209–266.

Davis, K. P. (1959). *Forest Fire, Control and Use*. McGraw-Hill, New York.

Dearman, W. R., Baynes, F. J. and Irfan, T. Y. (1978). Engineering grading of weathered granite. *Engineering Geology*, **12**, 345–374.

Deere, D. U. and Miller, F. D. (1966). Engineering classification and index properties for intact rock. *University of Illinois Technical Report*, **AFWL-TR-65-116**, Urbana Champagne.

Francis, D. (1992). *Longshot*. Pan, London.

Gaffey, S. J., Kulak, J. J. and Bronnimann, C. E. (1991). Effects of drying, heating, annealing and roasting on carbonate skeletal material with geochemical and diagenetic implications. *Geochimica et Cosmochimica Acta*, **55**, 1627–1640.

Goldammer, J. G. (ed.) (1990). *Fire in the Tropical Biota*. Springer-Verlag, Berlin.

Goudie, A. S., Allison, R. J. and McLaren, S. J. (1992). The relations between Modulus of Elasticity and temperature in the context of the experimental simulation of rock weathering by fire. *Earth Surface Processes and Landforms*, **17**, 605–615.

Homand-Etienne, F. and Troalen, J. P (1984). Behaviour of granites and limestones subjected to slow and homogeneous temperature changes. *Engineering Geology*, **20**, 219–233.

Hopkins, B. (1965). Observations on savanna burning in the Olikemeji forest reserve, Nigeria. *Journal of Applied Ecology*, **2**, 367–381.

Judd, W. R. and Huber, C. (1962). Correlation of rock properties by statistical methods. *International Symposium Mining Research*, **2**, 621–648.

Kozlowski, T. T. and Ahlgren, C. E. (eds) (1974). *Fire and Ecosystems*. Academic Press, New York.

Kruger, F. J. and Bigalke, R. C. (1984). Fire in fynbos. In Booysen, P. de V. and Tainton, N. M. (eds), *Ecological Effects of Fire in South African Ecosystems*, Springer-Verlag, Berlin, pp. 67–114.

Mooney, H. A. and Parsons, D. J. (1973). Structure and function of the California Chaparral —an example from San Dimas. *Ecological Studies*, **7**, 83–112.

Pyne, S. J. (1982). Fire in America—A Cultural History of Wildlife and Rural Fire. Princeton University Press, Princeton.

Pyne, S. J. (1991). *Burning Bush: A Fire History of Australia*. Holt, New York.

Richter, D. and Simmons, G. (1974). Thermal expansion behaviour of igneous rocks. *International Journal of Rock Mechanics and Mining Science & Geomechanical Abstracts*, **11**, 403–411.

Trollope, W. S. W. (1984). Fire behaviour. In Booysen, P. de V. and Tainton, N. M. (eds), *Ecological Effects of Fire in South African Ecosystems*, Springer-Verlag, Berlin, pp. 199–243.

Wein, R. W. and Maclean, D. A. (eds) (1983). *The Role of Fire in Northern Circumpolar Ecosystems*. Wiley, Chichester.

Whittaker, E. (1961). Temperatures in heath fires. *Journal of Ecology*, **49**, 709–715.

Winkler, E. M. (1973). *Stone Properties, Durability in Man's Environment*. Springer-Verlag, Berlin.

Wright, H. A. and Bailey, A. W. (1982). *Fire Ecology: United States and Southern Canada*. Wiley, New York.

4 Short-term Rock Temperature Fluctuations under Simulated Hot Desert Conditions: Some Preliminary Data

P. A. WARKE and B. J. SMITH
The Queen's University of Belfast, UK

ABSTRACT

Five centimetre cubes of basalt, granite, quartz sandstone and two types of limestone were subjected to 30 minute cycles of heating and cooling in a laboratory experiment. The simulation was designed to replicate short-term daytime rock temperature fluctuations experienced in hot deserts when insolation is interrupted, for example, by the passage of cloud. Results for the five rock types show markedly different surface and subsurface temperature responses related to albedo and thermal conductivity differences. All the rock types do, however, show initially rapid near-surface rates of temperature change on heating and cooling. The results suggest that the outer few millimetres of desert rock may be subject to numerous short-term thermally induced stress events. These are superimposed upon the longer-term cycles caused by diurnal temperature change, which have traditionally been invoked as the key factor in thermally related rock breakdown.

INTRODUCTION

Disagreement persists over the effectiveness of rock temperature fluctuations in causing rock breakdown. There is dispute between those influenced by early experiments which failed to achieve any noticeable insolation weathering (Blackwelder, 1933; Griggs, 1936), those swayed by field observations of shattered stones apparently unaffected by any chemical alteration (e.g. Ollier, 1963) and those who believe that insolation weathering *per se* is highly unlikely and that most breakdown occurs as a result of weathering mechanisms acting in combination or sequence. Because of this, debate has flourished concerning the possible effects of repeated stressing of rock (fatigue failure), the effects of extreme temperature changes (thermal shock), and the role of moisture in influencing not just the operation of specific weathering processes but also the thermal response characteristics of the rock.

Invariably, investigations into insolation-related weathering have concentrated upon two issues. First, there is the question of the control exerted on rock breakdown

Rock Weathering and Landform Evolution. Edited by D. A. Robinson and R. B. G. Williams
© 1994 John Wiley & Sons Ltd.

by external variables such as latitude, altitude, aspect, time of year, cloud cover and air temperature. Secondly, there is the issue of diurnal temperature variability whereby simulation experiments in particular have used 'smoothed' temperature regimes derived from data recorded at relatively long time intervals, i.e. >30 minutes.

Despite this concentration upon external controls and diurnal temperature change there is evidence that other factors play important roles in influencing the stresses to which rocks are subjected in hot desert environments. Intrinsic rock properties such as albedo, thermal conductivity and heat storage capacity should, as suggested by McGreevy (1985), have important effects upon the thermal response characteristics of different rock types. The effects of albedo were mentioned by Peel (1974) in field studies from Libya, and Kerr *et al.* (1984) presented some data on the influence of rock properties upon diurnal patterns of surface and subsurface heating and cooling from Morocco. The results of these studies suggest that there is a need for the systematic investigation of controls exerted by different thermal properties on rock temperature variability.

Another factor which needs further detailed investigation is short-term temperature change and its role in creating conditions of 'thermal shock' where the rate of stress development is extremely rapid. Recent technological advances have permitted almost continuous rock temperature monitoring and have indicated that rock surfaces are subject to numerous short-term fluctuations in temperature of less than 15 minutes duration. These fluctuations occur when insolation is interrupted by the passage of cloud or when rock surface temperatures are lowered by increases in windspeed or by the onset of rainfall. Such factors have been noted by Whalley *et al.* (1984), and Jenkins and Smith (1990), and are particularly evident when air temperatures are low and can cause rapid cooling of rock surfaces when placed in shadow.

The possible significance of these temperature changes was recently demonstrated by Hall and Hall (1991) who reported high rates of rock surface temperature change well in excess of $2°C min^{-1}$ for short periods in simulation experiments performed in subzero air conditions. Richter and Simmons (1974) also found that heating rates, when greater than $2°C min^{-1}$, produced microcracking in igneous rocks. Such temperature changes and the gradients established between the rock surface and substrate are restricted to the outer 1 to 2 cm of rocks, but this, as noted by Jenkins and Smith (1990), is precisely the zone in which rock breakdown by granular disintegration and flaking occurs.

It is the purpose of this chapter to present some preliminary observations of the combined effects of rock thermal properties and short-term temperature change on the thermal response characteristics of different rock types and their implications for subsequent weathering.

METHODOLOGY

Rock Types

Five rock types were used in this study, each possessing quite different structural, mineralogical and thermal properties (Table 4.1). The thermal conductivity value for Dunhouse Sandstone was not available, but its thermal response curve allows it to be

Table 4.1   Characteristics of rock types used in the study

Rock type	Age	Albedo[a]	Munsell colour notation	Thermal conductivity[b] (W m^{-1}K^{-1})
Basalt	Tertiary	12	N3/0	0.96
Portland Limestone	Jurassic	N.A.	2.5Y 8/1	1.53
Granite	Tertiary	18	5YR 6/1	1.65
Sandstone	Carboniferous	N.A.	10YR 7/2	N.A.
Hard white limestone (chalk)	Cretaceous	25	5YR 8/1	1.72

[a]Albedo is a measure of the reflectivity of a material. A low albedo results in increased absorption of solar radiation.
[b]Thermal conductivity is a measure of the ease of heat transmission from surface to subsurface.
Thermal conductivity and albedo values from McGreevy (1985), except for Portland Limestone thermal conductivity value which was provided by ARC Ltd.
N.A. = Not available.

placed relative to the other four rock types for which thermal conductivity values are known, as will be demonstrated in the following sections.

Equipment

The equipment used in this study was specially designed to allow simulation of short-term temperature fluctuations (Figure 4.1). It consists of a large drum 75 cm in diameter and 100 cm deep. Three ports located at regular intervals around the circumference of the drum allow access to the interior and three infrared heat lamps fixed over the drum provide the heat source. Interruptions to this heat source are achieved by a rotating semi-circular blade which can be set to revolve at different speeds ranging from a minimum of 1 rev h^{-1} to a maximum speed of 16 rev h^{-1}. When, for example, the blade is set to rotate at 2 rev h^{-1} any rock samples below will experience two shade episodes, each of 15 minutes duration, in every hour. The blade revolutions are controlled by a stepping motor incorporated into the base of the drum.

When all three infrared heat lamps are on full they create a relatively stable air temperature within the drum of 40°C. The attachment of a fan and cooling unit to one of the access ports reduces air temperatures within the drum from 40°C to around 20°C. It is hoped that further refinement of the cooling system will reduce air temperatures by a further 10–15°C to between 0 and 10°C giving an air temperature range within the drum of over 30°C. At present, however, the 20°C range available allows simulation of summer and winter conditions experienced by many hot desert regions and hereafter referred to as 'warm air' and 'cool air' conditions respectively.

Air and rock temperatures were recorded using a 12-bit Grant Instrument Squirrel Logger and bead thermistors. Rock surface and depth temperatures were recorded in small blocks with dimensions of 5 × 5 × 5 cm. The blocks were prepared by drilling a hollow approximately 1 mm deep in the block surface and by drilling 2.5 cm up from the centre of the block base. The surface bead thermistor was secured using a non-silicone heat transfer compound over which powder retrieved from the drilling was sprinkled in order to maintain a uniform surface albedo. Core temperatures (2.5 cm below the surface) were measured by inserting another bead thermistor into the hole drilled through the base of the block. This hole was then tightly packed with powdered rock and the last 0.5 cm plugged with cotton wool to hold the thermistor and powder securely in place. Each block was then embedded in a jacket of expanded polystyrene allowing heat to effectively move through the one exposed surface (Figure 4.2).

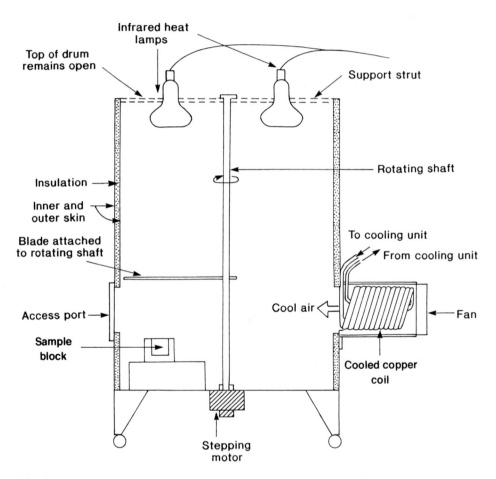

Figure 4.1 Detail of circular drum used in simulation experiments with the optional air cooling unit attached

A bead thermistor measuring air temperature was set just above the surface level of the blocks and shaded by a small foil canopy in order to prevent direct heating by the infrared lamps.

Air and rock temperatures were recorded at one minute intervals, and at the end of the recording period data were downloaded directly to computer.

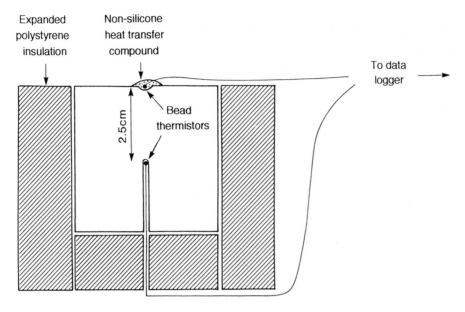

Figure 4.2   Insulated block and position of thermistors used in temperature measurements

Conditions of Exposure

Blocks were exposed over 72 hours for alternating 6 hour periods of warm air and cool air conditions. The rotating blade was set at a speed of 2 rev h^{-1}, and no additional moisture was introduced over and above the atmospheric humidity already present as this could have altered the thermal response characteristics of the rock samples including albedo and thermal conductivity.

RESULTS AND DISCUSSION

Surface Temperatures

The basalt block consistently produced the highest surface temperatures under both 'warm' and 'cool' air conditions. Under cool air conditions the sequence of surface temperature maxima in decreasing order was basalt > sandstone > granite > chalk > Portland Limestone. Under 'warm' air conditions this sequence changes slightly; basalt > sandstone > Portland Limestone > granite > chalk (Tables 4.2 and 4.3).

Under both sets of conditions the ranges of temperatures recorded for the basalt are almost twice those of the other rock types with the chalk consistently producing the lowest surface temperature range.

The rock temperatures achieved under these simulated conditions are comparable with temperatures reached by these rock types under actual hot desert conditions. For example, in Death Valley, California, surface maximum and minimum temperatures recorded during a 24 hour period in August 1992 on similarly insulated blocks

Table 4.2 Surface maximum and minimum temperatures under cool air conditions

Rock type	Surface max. (°C)	Surface min. (°C)	Range (°C)
Basalt	40.00	33.15	6.85
Sandstone	36.20	32.90	3.30
Granite	35.70	32.05	3.70
Hard white limestone (chalk)	32.70	29.95	2.75
Portland Limestone	31.85	28.45	3.40

Table 4.3 Surface maximum and minimum temperatures under warm air conditions

Rock type	Surface max. (°C)	Surface min. (°C)	Range (°C)
Basalt	53.00	45.85	7.15
Sandstone	40.60	37.30	3.30
Portland Limestone	39.65	35.60	4.05
Granite	39.55	35.80	3.75
Hard white limestone (chalk)	37.20	34.80	2.40

Table 4.4 Maximum and minimum rock surface temperatures recorded in August 1992 over a 24 hour period in Death Valley, California

Rock type	Surface max. (°C)	Surface min (°C)
Granite	61.85	28.35
Sandstone	61.40	28.95
Portland Limestone	57.65	28.10
Basalt	72.65	27.80
Air Temp. 5 cm above ground	65.80	35.10

were even higher (Table 4.4). The Death Valley measurements are amongst the most extreme that can be expected under desert conditions. However, the experimental values also accord with those of other rock types measured under summer and winter conditions in desert environments (McGreevy and Smith, 1982; Goudie, 1989).

Rates of Temperature Change with Depth

The effects of thermal conductivity are reflected in the surface/subsurface temperature gradients for the five rock types. Low thermal conductivity results in high surface temperatures when exposed to direct insolation, and because heat is only slowly conducted to subsurface layers a much steeper thermal gradient exists, resulting in correspondingly high thermal stresses. Comparison of temperature data for the basalt and chalk samples demonstrates this (Table 4.5). Under warm air conditions the difference between the basalt surface maximum temperature and the subsurface value recorded simultaneously was 4.1°C. The difference between these two values in the chalk block was only 1°C. The rate of temperature decrease in the basalt block is 1.64°C cm^{-1} from the surface to the centre of the block whereas the rate of temperature decrease in the chalk is only 0.4°C cm$^{-1.}$ The thermal conductivity of the chalk is greater than that of the basalt; therefore, heat is more readily conducted through the chalk but less so through the basalt allowing surface temperatures to rise much higher. The differences between the temperature gradients of these two rock types are also influenced by albedo. The basalt, which is much darker than the chalk, reflects much less incident radiation. When two rock types with similar albedo values are exposed to the same temperature fluctuations the resulting temperature curves reflect the differences in thermal conductivity. For example, Table 4.5 shows the subsurface temperature gradient of the Portland Limestone to be over twice that of the chalk. Even though the surface albedo of the two blocks is almost identical, because the thermal conductivity of Portland Limestone is lower than that of the chalk heat is conducted more readily through the chalk than it is through the Portland Limestone under warm air conditions.

Under cool air conditions surface heat is rapidly lost by the passage of cooler air over the rock surfaces when shaded. Those rock types with the highest unshaded surface temperatures should experience the greatest decrease in surface temperatures. The differences between surface maxima recorded under warm and cool air conditions are, in ascending order, 3.80°C, 4.35°C, 4.50°C, 7.80°C and 13.00°C for the granite, sandstone, chalk, Portland Limestone and basalt respectively.

Table 4.6 shows the rates of temperature decrease with depth to be most noticeably reduced for the basalt and Portland Limestone blocks, whereas the rates remain fairly constant in the granite, chalk and sandstone. This suggests that rock types with low thermal conductivities experience the greatest surface/subsurface temperature gradients, and hence will experience greater levels of thermally induced stress on sudden cooling or warming. This stress may be further enhanced if the rock also possesses a low albedo, as in the case of the basalt.

The rates of temperature decrease with depth recorded in this study are comparable to those recorded under natural conditions in hot desert environments over diurnal periods. For example, Roth (1965) found rock temperatures in quartz monzonite to decrease by 15°C in the outer 30 cm (0.5°C cm^{-1}), whereas Smith (1977) noted a decrease of 10.1°C in 10 cm (1°C cm^{-1}) in limestone while Peel (1974) found temperatures to decrease by 0.65°C cm^{-1} in sandstone. Although average rates of temperature decrease with depth are given, this most probably is not an accurate reflection of actual cooling patterns. Smith (1977) observed that the thermal gradient is not uniform. Greatest rates of temperature decrease occur in the outer few millimetres of rock where many weathering mechanisms operate. Actual thermal gradients in these areas may, therefore, be much greater, giving rise to more severe thermal stresses between the surface and subsurface than would otherwise be expected.

Table 4.5 Rates of temperature decrease with depth under warm air conditions

Rock type	Max surface temp. (°C)	Simultaneous subsurface temp. 2.5 cm deep)	Difference (°C)	Rate of decrease (°C cm^{-1})	Thermal conductivity[a] (W m^{-1}K^{-1})
Basalt	53.00	48.90	4.10	1.64	0.96
Portland Limestone	39.65	36.95	2.70	1.08	1.53
Granite	39.55	37.95	1.60	0.64	1.65
Sandstone	40.55	39.35	1.20	0.48	N.A.
Hard white limestone (chalk)	37.20	36.20	1.00	0.40	1.72

[a]Thermal conductivity values from McGreevy (1985), except for Portland Limestone value which was provided by ARC Ltd.
N.A. = Not available.

Table 4.6 Rates of temperature decrease with depth under cool air conditions

Rock type	Max surface temp. (°C)	Simultaneous subsurface temp. 2.5 cm deep)	Difference (°C)	Rate of decrease (°C cm^{-1})	Thermal conductivity[a] (W m^{-1}K^{-1})
Basalt	40.00	36.90	3.10	1.24	0.96
Portland Limestone	31.85	30.10	1.75	0.70	1.53
Granite	35.75	34.15	1.60	0.64	1.65
Sandstone	36.20	34.80	1.40	0.56	N.A.
Hard white limestone (chalk)	32.70	31.25	1.45	0.58	1.72

[a]Thermal conductivity values from McGreevy (1985), except for Portland Limestone value which was provided by ARC Ltd.
N.A. = Not available.

Rates of Temperature Change over Time

Rock surface temperature curves for all five rock samples under both warm and cool air conditions display a slight asymmetry (Figures 4.3 and 4.4) which results from variable rates of heating and cooling. Surface heating on coming out of shade is initially quite rapid, but gradually slows until re-entering shade. Once in the shade the rate of cooling initially proceeds relatively rapidly, but then slows until the block moves out of shade.

Under warm air conditions at the end of the first 5 minutes in shade, basalt surface temperatures fell by around 4°C, over half the total surface cooling occurring in only one third of the time spent in shade. After this period the rate of cooling gradually slows (Table 4.7). These results demonstrate how misleading average rates for surface temperature loss and gain can be with the basalt cooling almost twice as rapidly during the first 5 minutes in shade as the average rate for the full 15 minutes would suggest. The rates of heat gain are similar to those of heat loss with most of the increase in surface temperatures occurring during the first 5 to 10 minutes of exposure to the heat source.

Under cool air conditions a similar asymmetry exists in the pattern of rock surface temperature change (Table 4.8). The significance of these rapid temperature changes lies in the fact that the stresses established during rapid heating or cooling of rock surfaces may be sufficient to promote microfracturing. As Yatsu (1988, p. 132) suggests:

> If there is a mismatch in the thermoelastic behaviour of minerals across their grain boundaries, internal stresses may be generated when the rock is subjected to different temperature.

This is particularly relevant when considering the weathering behaviour of rocks such as granite and basalt, which are composed of a variety of minerals each with different thermal expansion characteristics, and which are prone to granular disintegration as a principal form of breakdown.

Under both warm and cool air conditions, the subsurface temperature curves reflect more even rates of heating and cooling, resulting from the slower response of rock at depth to changes in external temperature conditions. In this way, subsurface temperature responses are seen to lag behind those at the surface.

The 'Lag Effect'

When the heat supply is interrupted, rock surface temperatures immediately begin to fall at an initially rapid rate. Subsurface temperatures, however, continue to rise for a short period creating conditions in which the subsurface of the rock is warmer than the surface. During surface heating tensile stresses arise between the rock surface and the cooler subsurface layers. However, when this situation is reversed, compressive stresses will form as the rock surface contracts and the core tries to expand. Repeated reversals in temperature gradients and changes in the nature of the resulting thermally induced stresses will most readily affect the outer centimetre of rock where temperature gradients are steepest. Over a prolonged period such changes may possibly weaken the surface structure of the rock leading to the development of thermal stress fatigue. Rocks which will be most prone to this are those with low levels of thermal

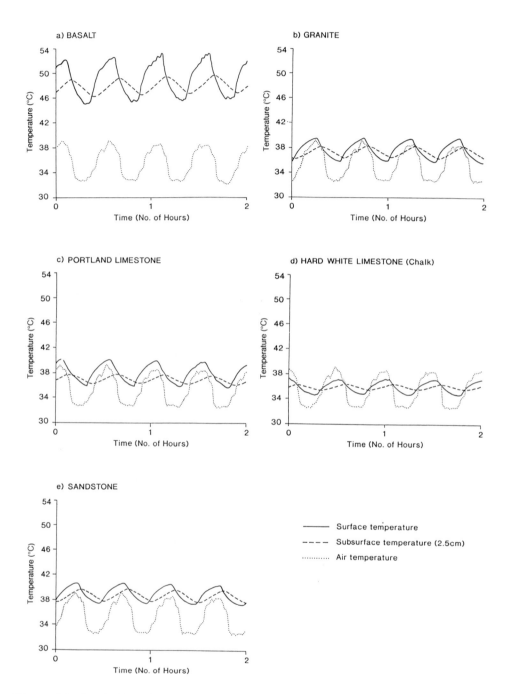

Figure 4.3 (a–e) Air and rock surface/subsurface (2.5 cm) temperatures recorded under warm air conditions

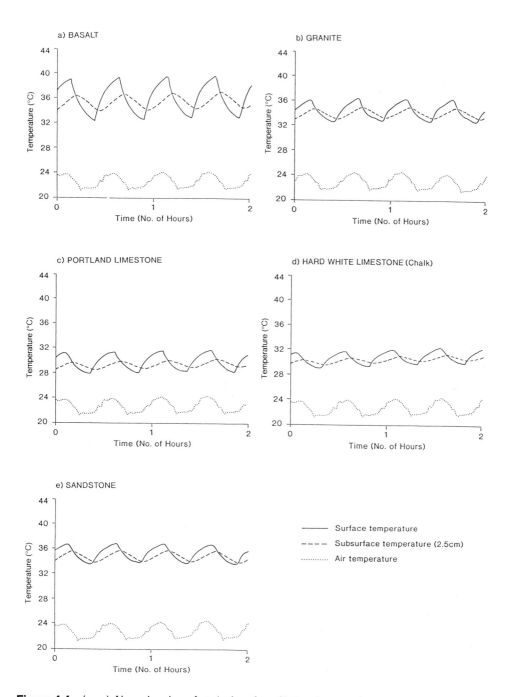

Figure 4.4   (a–e) Air and rock surface/subsurface (2.5 cm) temperatures recorded under cool air conditions

Table 4.7 Rates of rock surface cooling under warm air conditions

Rock type	Surface temp. (°C)	T.C. after 1st 5 mins (°C)	Rate of T.C. (°C min⁻¹)	T.C. after 2nd 5 mins (°C)	Rate of T.C. (°C min⁻¹)	T.C. after 3rd 5 mins (°C)	Rate of T.C. (°C min⁻¹)	Total T.C. (°C)	Average T.C during 15 mins (°C min⁻¹)
Basalt	53.00	4.25	0.85	2.05	0.41	0.95	0.19	7.25	0.48
Portland Limestone	39.65	2.25	0.45	1.40	0.28	0.45	0.09	4.10	0.27
Granite	39.55	1.85	0.37	0.70	0.26	0.60	0.12	3.75	0.25
Sandstone	40.55	1.75	0.35	1.30	0.20	0.50	0.10	3.30	0.22
Hard white limestone (chalk)	37.20	1.40	0.28	1.05	0.14	0.35	0.07	2.45	0.16

T.C. = Temperature change

Table 4.8 Rates of rock surface cooling under cool air conditions

Rock type	Surface temp. (°C)	T.C. after 1st 5 mins (°C)	Rate of T.C. (°C min⁻¹)	T.C. after 2nd 5 mins (°C)	Rate of T.C. (°C min⁻¹)	T.C. after 3rd 5 mins (°C)	Rate of T.C. (°C min⁻¹)	Total T.C. (°C)	Average T.C during 15 mins (°C min⁻¹)
Basalt	40.00	4.10	0.82	1.80	0.36	0.90	0.18	6.80	0.46
Portland Limestone	31.85	1.65	0.33	1.15	0.23	0.55	0.11	3.35	0.22
Granite	35.65	2.15	0.43	1.15	0.23	0.40	0.08	3.70	0.25
Sandstone	36.20	1.60	0.32	1.15	0.23	0.40	0.08	3.15	0.21
Hard white limestone (chalk)	32.50	1.15	0.23	0.95	0.19	0.45	0.09	2.55	0.17

T.C. = Temperature change

conductivity and low albedos, the combination of which results in the creation of steep thermal gradients between surface and depth.

CONCLUSIONS

These preliminary results show the variable nature of rock temperature changes with depth and over time. They show the asymmetry of surface heating and cooling patterns produced in all five rock types examined as well as highlighting the importance of thermal rock properties in determining response to short-term interruptions in insolation. A combination of low albedo and thermal conductivity leads to the development of steep thermal gradients and consequently much greater thermally induced stresses. Rock that naturally possesses such characteristics, or has its albedo changed, for example, by the development of rock varnish, will experience a high level of thermally induced stress, particularly in its outer layers. Perhaps most significantly, the data illustrate the different responses of five rock types to identical temperature regimes. These different patterns of response have implications for the operation of a range of weathering mechanisms, and ultimately for the rate and nature of rock breakdown.

In particular, identification of rapid, short-term temperature changes at or near rock surfaces has clear implications for the nature of 'fatigue failure' as it is believed to apply to rock breakdown. Previous studies have implied that if fatigue effects occur, they are a response to diurnal expansion and contraction of rocks to a depth of several centimetres. It seems likely, however, that there is an additional, superimposed fatigue effect perhaps operating several times each day. Although restricted to the outer few millimetres of the rock surface—depending upon internal rock properties as well as insolation changes—expansion and contraction nonetheless occurs in the zone where many weathering mechanisms operate, and with a much greater frequency and at a greater rate than that induced by diurnal thermal cycles.

Although insolation may not cause sudden and dramatic rock breakdown, it probably acts by enhancing the effectiveness of other weathering mechanisms through weakening outer layers of rock and facilitating the entrance of more widely recognised agents of weathering such as salt and moisture. Insolation like any other agent of weathering does not operate in isolation. Although each weathering mechanism proceeds according to its own laws, the effect on rock surfaces is much more than simply the sum of these individual parts. Such would be possibly only in the absence of relationships and interactions

ACKNOWLEDGEMENTS

We would like to thank Mr D. Wright and Mr D. Jamison for assistance in construction of the simulation equipment, Mrs G. Alexander for cartographic assistance and Mr R. Magee for his constructive comments. We would also like to acknowledge the assistance given by Mr J. Aardahl and Mr T. Coonan of the US Department of the Interior, National Park Service, Death Valley National Monument. P.A. Warke is in receipt of a post-graduate studentship from the Department of Education, Northern Ireland.

REFERENCES

Blackwelder, E. (1933). The insolation hypothesis of rock weathering. *American Journal of Science*, **226**, 97–113.

Goudie, A. S. (1989). Weathering processes. In Thomas, D. S. G. (ed.), *Arid Zone Geomorphology*, Chapter 2, Belhaven Press, London.

Griggs, D. T. (1936). The factor of fatigue in rock exfoliation. *Journal of Geology*, **44**, 783–796.

Hall, K. and Hall, A. (1991). Thermal gradients and rock weathering at low temperatures: some simulation data. *Permafrost and Periglacial Processes*, **2**, 103–112.

Jenkins, K. A. and Smith, B. J. (1990). Daytime rock surface temperature variability and its implications for mechanical rock weathering: Tenerife, Canary Islands. *Catena*, **17**, 449–459.

Kerr, A., Smith, B. J., Whalley, W. B. and McGreevy, J. P. (1984). Rock temperatures from southeast Morocco and their significance for experimental rock-weathering studies. *Geology*, **12**, 306–309.

McGreevy, J. P. (1985). Thermal properties as controls on rock surface temperature maxima, and possible implications for rock weathering. *Earth Surface Processes and Landforms*, **10**, 125–136.

McGreevy, J. P. and Smith, B. J. (1982). Salt weathering in hot deserts: observations on the design of simulation experiments. *Geografiska Annaler*, **64A**, 161–170.

Ollier, C. D. (1963). Insolation weathering: examples from central Australia. *American Journal Science*, **261**, 376–381.

Peel, R. F. (1974). Insolation weathering: some measurements of diurnal temperature changes in exposed rocks in the Tibesti region, central Sahara. *Zeitschrift für Geomorphologie*, Supplementband **21**, 19–28.

Richter, D. and Simmons, G. (1974). Thermal expansion behaviour of igneous rocks. *International Journal of Rock Mechanics and Mining Sciences*, **11**, 403–411.

Roth, E. S. (1965). Temperature and water content as factors in desert weathering. *Journal of Geology*, **73**, 454–468.

Smith, B. J. (1977). Rock temperature measurements from the northwest Sahara and their implications for rock weathering. *Catena*, **4**, 41–63.

Whalley, W. B., McGreevy, J.P. and Ferguson, R.I. (1984). Rock Temperature observations and chemical weathering in the Hunza region, Karakoram: preliminary data. In Miller, K. J. (ed.), *The International Karakoram Project*, vol. 2, Cambridge University Press, Cambridge, pp. 616–633.

Yatsu, E. (1988). *The Nature of Weathering*. Sozosha, Tokyo.

5 Mineralogical Aspects of Crack Development and Freeface Activity in some Basalt Cliffs, County Antrim, Northern Ireland

GEORGE R. DOUGLAS
Mentaskólinn vid Hamrahlid, Reykjavik, Iceland

JAMES P. McGREEVY
Ulster Museum, Belfast, UK

and

W. BRIAN WHALLEY
The Queen's University of Belfast, UK

ABSTRACT

A series of cliff faces developed in basalt lava flows (monitored over 17 months) have been shown to release widely differing amounts of material. The differences seem to be related primarily to the extent of the microcrack system and the physical properties of the rocks rather than the local environmental conditions. The flows contain a range of post-depositional minerals, some of which can be considered as typical 'weathering' products but others include hydrothermal alteration products such as zeolites. In some cases the latter may heal cracks but in other cases they may promote weakening. The rocks most susceptible to weathering appear to be those containing low temperature alteration minerals such as smectite which, although chemically more stable than the higher temperature products (such as chlorite), are physically less stable. This susceptibility is related to swelling, ion exchange and subaerial alteration to new clay minerals such as kaolinite. Within this range of alteration products local, internal factors may play a part; most significantly the presence of water in the cliff.

INTRODUCTION

There is now a reasonably large literature on cliff disintegration and rockfall from various parts of the world (e.g. Rapp, 1960; Gardner, 1980; Francou, 1982; André, 1986). However, despite acknowledgement of the potential role which cracks and microcracks play in massive rock breakdown (e.g. Whalley, 1984; Whalley *et al.*,

Rock Weathering and Landform Evolution. Edited by D. A. Robinson and R. B. G. Williams
© 1994 John Wiley & Sons Ltd.

1982), there has been little attempt to link the rock properties with geomorphology. Perhaps an exception is the work of Moon (1984) and Selby (1987), although the role of cracks is seen more in terms of a multi-component explanation of cliff profile rather than as an understanding of cliff decay.

In previous papers (Douglas, 1980 and 1981; Douglas *et al.*, 1987 and 1991) some of the basic controls on the weathering of a basalt freeface have been investigated. In the present chapter some basalt rock properties are examined in further detail as a way of establishing how the rock itself behaves and gives, in an indirect way, some idea of the likely mechanisms involved. This follows the basic reasoning of Yatsu's (1966) notion of 'rock control', where a knowledge of rock (or soil) properties is considered vital for a thorough understanding of geomorphological processes.

SITE

The site investigated was at Ardclinis, County Antrim, Northern Ireland (see Douglas, 1980, for a full description). A series of Tertiary lava flows (Figure 5.1) of varying properties have produced different amounts of rock debris (Table 5.1).

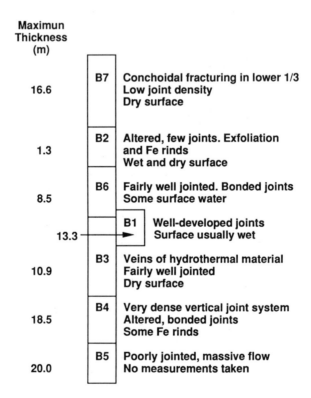

Figure 5.1 Schematic diagram of the cliff lava flows at Ardclinis with their main properties (see also Tables 5.1 and 5.2). The position of B1 appears to overlap B3 and B6 as it is separated from them by a fault

Table 5.1 Rock properties of the lava flows at Ardclinis

	Thickness of flow (m)	Main crack properties, etc.	Swelling coeff. (%)	Micro-fracture density (no. 500 mm⁻¹)	Micro-fracture width (mm)	Coeff. loosening (K)	Debris fall unit area⁻¹ (kg)	Moisture content (%)	Porosity density (%)	Bulk density (mg m⁻³)	Grain (mg m⁻³)
B1	13.3	Bonded and open fractures	0.013	66	0.02	1.008	2.09	5.76	14.42	2.51	3.09
B2a	1.3	Open fractures	0.035	89	0.17	1.100	5.90	6.64	15.82	2.38	2.87
B2b		Highly altered open fractures	0.077					16.46	32.30	1.96	2.72
B2c		Vesicular top of flow	0.0527					17.24	32.49	1.88	2.57
B3	4.0	Bonded (smectite) fractures	0.082	60	0.10	1.080	1.69	4.19	10.47	2.50	2.70
			0.064					10.16	22.22	2.19	
			0.850					31.16	44.20	1.42	
B4	18.5	Open fractures	0.081	15	0.08	1.010	1.01	4.51	10.18	2.48	2.87
B6			1.270					60.12	68.64	1.14	2.14
B7a	11.6	Bonded and open fractures	0.189	79	0.07	1.040	2.87	3.00	7.26	2.54	2.78
B7b		Highly altered, open fractures	0.731					21.66	31.96	2.09	2.19

The hydrothermal minerals in the Antrim basalts occur in approximately horizontal zones or facies (Walker, 1951 and 1960) and reflect the temperature distribution and associated depth of burial within the laval pile. The mineral range from lower temperature smectite minerals, through chlorite and low temperature zeolites such as chabazite, to high temperature zeolites, such as laumonite. It has also been shown, by comparison with Iceland, that at least 500 m of lavas are missing as a result of erosion and this implies an associated uplift of the rocks. The freeface is active today with constant material removal (Douglas, 1980), and extensive scree slopes all along the Antrim Plateau escarpment suggest that this has been so for some time.

ROCK CHARACTERISATION

Various properties were examined using both standard and improvised techniques. Definitions of properties are given in Douglas *et al.* (1991). The main aim has been to produce test results which come as close as possible to real field conditions. Values of field moisture content of the freeface rocks were obtained at weekly intervals.

Attention has been focused on three aspects of rock properties. First, an investigation has been made of the types of secondary mineralisation and mineral alteration. Many microfractures are occupied by hydrothermal minerals where changes may still be taking place. Similarly, the general degree of alteration of the rocks has an effect on fracture propagation as well as a direct effect on properties such as moisture absorption. For this reason we avoid a rigid separation of chemical and physical weathering effects.

Secondly, it has been thought important to define the size and nature of discontinuities which, especially the frequent microfractures, represent the most important single property of the rocks because: (i) they have a profound effect on other property determinations (Onadera *et al.*, 1974; Dearman and Irfan, 1978); (ii) they are likely sites for moisture absorption which is generally recognised as being an important factor in most types of rock weathering; and (iii) they represent local stress concentrations and changes both in the past and at present which are of fundamental importance in any consideration of rock breakdown (Whalley *et al.*, 1982).

Thirdly, the investigation has concentrated on obtaining data on void-dependent properties such as bulk density, moisture absorption, porosity and capillarity which are relevant to weathering susceptibility (cf. Lautridou and Ozouf, 1982).

Mineralogy

Thin sections, representative of both weathered and unweathered rock, were made of all the basalt flows at Ardclinis. Certain minerals are of importance because of their ease of weathering; these include olivine, which can be present in large phenocrysts and is often serpentinised. Serpentine replaces olivine in low temperature hydrothermal activity as well as by weathering, and serpentinisation takes place, significantly, along fracture systems in the olivines and may replace the entire crystal. Flow B2, which has a high debris field (Table 5.1), shows widespread olivine alteration. Flows B2, 3 and 4 also have high amounts of altered olivine.

The alteration of minerals is not due to their exposure on the freeface but is related to the deuteric history of the lava flows. Olivine is particularly prone to alteration, either to serpentine or on crystal rims to a mixture of clay minerals which includes goethite, hematite and smectite–chlorite. The latter is particularly important because it is a swelling clay mineral. Many of the rocks examined have a glassy phase, which is very susceptible to weathering. Again, this can be attributed to both present-day alteration as well as hydrothermal activity.

Microcracks

The terminology used here complies with the definitions of Simmons and Richter (1976). Two main crack types have been encountered: (i) multigrain cracks which involve several crystals by crossing them, usually along cleavage or crystal boundaries and (ii) single grain cracks which cross only single crystals, mainly as cleavage cracks or concordant grain boundary cracks. The multigrain cracks have a number of separate causes, have formed at different times, and can be either open or bonded by secondary minerals. However, as regards fabric, density and width they are treated as one group.

Common features which occur, such as 120° triple point intersections, crack bending and 90° interactions, are typical of continuously propagating fractures (Ernsberger, 1960). In addition to these common features, four groups of multigrain cracks have been recognised as being distinctive from a weathering viewpoint:

1. Bonded fractures which contain hydrothermally formed minerals ranging from smectite and chlorite to zeolite minerals and calcite or quartz minerals. These fractures are regarded as having formed in a late cooling and subsequent hydrothermal state (Walker, 1960; Douglas, 1981). An example is shown in Figure 5.2a.
2. Systems of open fractures which occur especially in the more altered rocks. These are common at sites B2 and B7. Examples are shown in Figure 5.2b.
3. Open fractures that occur entirely within the clay mineral material of some of the bonded microfractures. These are not as extensive as Groups 1 and 2. An example is shown in Figure 5.2c.
4. Open fractures associated with changes in the valency state of transition metals such as Mn and Fe. Typically, this is seen as Fe^{3+} red stained 'weathering rinds'. These include (a) a very extensive fracture, which sometimes forms just at or behind the limit of Fe^{3+} coloration, and (b) finer fractures which run normally to this outwards to the rock surface. Manganese has also been seen to occur in cracks (Douglas, 1981), and both Mn and Fe may have mixed valency materials in deposits because of their complex chemistry.

Group 4 fractures can often be seen in hand specimens, but the other types usually can only be seen by resorting to staining and/or microscopic detection methods. The widespread nature of microfractures of various types in basalt rocks is perhaps the most important feature relevant to weathering studies. The aggregate effect of the above fracture types can be gauged by examining density and width relationships, and their relationship to actual rock fall amounts.

76

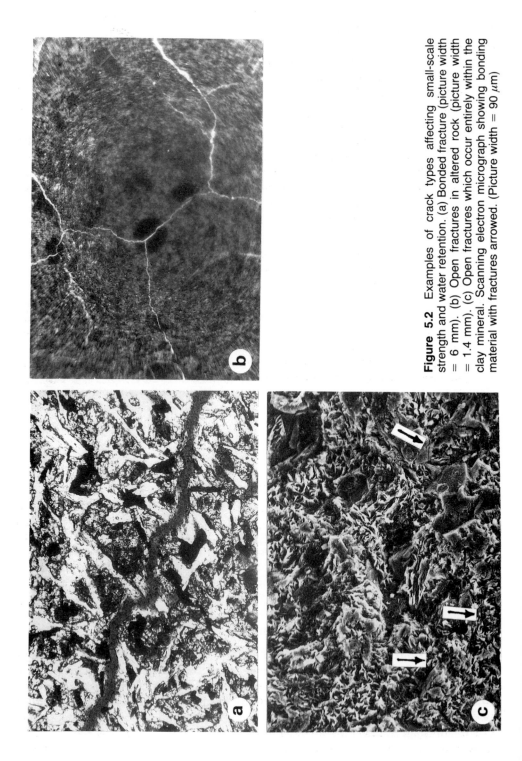

Figure 5.2 Examples of crack types affecting small-scale strength and water retention. (a) Bonded fracture (picture width = 6 mm). (b) Open fractures in altered rock (picture width = 1.4 mm). (c) Open fractures which occur entirely within the clay mineral. Scanning electron micrograph showing bonding material with fractures arrowed. (Picture width = 90 μm)

Void-dependent Properties

The void-related properties, density and porosity, reflect both the degree of alteration of the rocks and also the extent of development of microfracture systems. The density values (Table 5.1) were obtained by both direct measurement on cubes and by calculation. The grain density values also give a qualitative idea of the degree of alteration of the rocks, which in some cases is sufficient to produce densities approaching as low as $2.0 \, \text{Mg m}^{-3}$ (e.g. B2 and B3, Table 5.1). The porosity is defined as the ratio of the volume of voids to the total volume of the sample and the results are included here mainly to underline the great deviation which freeface rocks show as compared to typically quoted values for igneous rocks. Values of around 10% seem to be most common and are considerably higher than values often quoted (e.g. Whalley *et al.*, 1982). The high values are believed to be due to the multigrain microfractures on the one hand, and to the wide distribution of water-absorbing clay minerals on the other.

Of particular relevance to weathering studies are the field moisture content, saturation moisture content and swelling coefficient, all of which reflect closely the interaction of rock and water. The field moisture content is the ratio of the weight of water, including surface water (which is often a substantial proportion of the total), to the weight of dry solids in the sample. Field moisture was recorded by weekly random sampling of the freeface. The saturation moisture content is the same ratio for a sample fully saturated under vacuum in the laboratory. These values reach a maximum of around 20% in the most altered fractured rocks. The interrelationships between rock properties and the availability and movement of water in the rocks are now discussed in more detail.

ROCK PROPERTIES AND MOISTURE REGIMES

Determination of how and where water is held in rock is of fundamental importance to an understanding of rock breakdown mechanisms. The situation with the basalts is more complex than in the case of 'porous' rocks such as sandstone, for example, in which water is held in inter granular spaces. Humlum (1992) has recently presented some data on water uptake in a basalt under arctic conditions to show the variability of water content. In the present study, two areas appear to be of importance for basalts.

1. In the multigrain microfractures, especially when open or when bonded by clay minerals which absorb the water.
2. In a narrow 2–5 mm surface zone, particularly in the less altered rocks.

There are several findings which support this distinction. Observations on capillary movement show that water is absorbed along the microfractures at around 25 mm hr^{-1}, stabilizing within 2 hours. There is also a capillary rise of water along a narrow (2–5 mm) surface zone in the less altered samples. All of the rocks show signs of swelling on water uptake, but the most susceptible are those containing smectite-bonded microfractures where linear swelling of up to 0.8% (B3) has been recorded. This compares with values quoted elsewhere for a 'dense basalt' of 0.015% to 0.02%

(Nepper-Christensen, 1965). The importance of smectite expansion in cracks has been noted by Douglas *et al.* (1987).

Examination of thin sections shows that ferric iron zones occur widely in these rocks. In some cases this is deuteric alteration (e.g. of olivine phenocrysts) but there are also red-brown zones extended partly inwards along bonded microfractures (often up to 30 mm) as well as 'weathering rinds' which appear to coincide with the surface water zones mentioned earlier. These two areas are probably the result of fluctuating water content where the oxidation of the iron may be associated with dissolved oxygen in the water, ion exchange and solution, especially in the clay minerals. It is by no means established that all the weathering rinds are due to inward movement of a weathering front, indeed many of them may be exposed surfaces of multigrain bonded microfractures which undergo rapid colour change.

In some rocks, therefore, the main water areas are essentially on the rock surface or in the multigrain microfractures, and even here it may be largely surface water rather than capillary-held water in fractures. Water penetration into the rock along microfractures depends on factors such as capillary forces, spreading forces, the distribution of clay minerals on surfaces and the amount of water being supplied to the system. Further insight into the relationships between rock properties and the held water can be gained by considering Figure 5.3 comparing saturation moisture content with bulk density values (mean values). There is a rapid increase in saturation

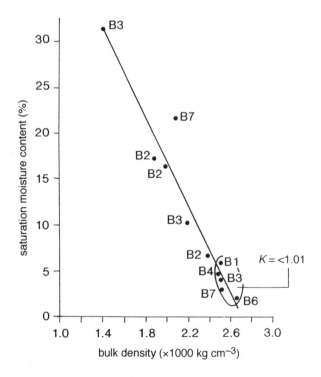

Figure 5.3 Saturation moisture content plotted against bulk density (Figure 2 from Douglas *et al.*, 1991. Reproduced by permission). The value of K is the 'coefficient of loosening' (Havir *et al.*, 1971)

moisture content from <5% in rocks of the group B1, B4 and B6 to values of up to 30% for the second main group of altered and fractured rocks (B2, B3, B7). The reasons for this rapid increase can be explained by the increase in loosening coefficient and associated increase in microfracture surface area and microfracture volume. However, added to these factors is the increase in clay minerals exposed on these surfaces in altered fractured rocks.

All of the water at Ardclinis moves from the rock surface zone inwards. There is no true groundwater, although some of the surface water may be supplied by semi-permanent rivulets from the cliff top. There is a general decrease in moisture content in the summer months at most sites (e.g. B1, B2, B3 and B7). In these periods moisture content can remain low for several weeks at a time (e.g. weeks 40–50 at site B4 and B3). High values of moisture content are of much shorter duration, consecutively high weekly values being the exception rather than the rule. Even on the weekly scale of recording this indicates that the rocks may lose water immediately the supply ceases. This effect is perhaps exaggerated by the surface moisture zone changes.

It is possible to identify typical or mean levels of saturation in the rocks in the simplest cases: sites B1, B4 and B6. Here microfractured rocks represent about 50% of the samples. The mean saturation moisture content of the laboratory samples is 5.76% and 4.51% respectively. This compares with a mean field moisture content of 5.42% and 3.60% for these rocks. On this basis it could be inferred that water occupies between 80% and 90% of the available space within these rocks. The high saturation coefficients are in part due to surface water included in the field values. On the other hand, we know from recordings (Douglas, 1972) that the moisture content remains low for long periods so that on many occasions when water is available, very high saturation coefficients must be attained to produce such high mean values. However, these occasions are likely to be short-lived, certainly of less than one week's duration and probably often only a matter of hours. This fits well with laboratory observation on moisture absorption patterns. It is more difficult to interpret results where there is some variation of rock properties. At sites B2, B3 and B7 there are two types of microfracture system in roughly equal proportions. By analogy with the simple cases just described, the situation at site B2 is similar. A mean laboratory value for saturation moisture of 11.40% and a mean field moisture content of 9.04% suggests that about 79% of available pores could be occupied. Again, there will be very brief periods of high saturation and longer periods of partly or poorly saturated rocks. At site B7 the situation is rather different. Allowing for even very large variations due to sampling techniques, the high porosity rocks at this site do not seem to reach saturation coefficients of anywhere near 100%. Comparison of mean laboratory values with mean field values gives a mean saturation coefficient of only 36%. Thus many of the rocks are never more than perhaps 50% saturated. The reasons for this may be that a limited water supply (by rainfall, for example) has to be distributed through a very dense microfracture system. In such a situation microfracture width is likely to become highly significant. If widths are very fine, water will move by capillary forces and we are dealing, in effect, with 100% saturated fractures even if the water does not reach all of the fractures. If widths are broader (>0.5 mm?) then we may be dealing instead with later movement by spreading forces on the microfracture surfaces. The consequences are likely to be very different in each case.

A safe conclusion seems to be that some of the rocks at Ardclinis can attain nearly 100% saturation coefficient for very brief periods. More often, rocks are partly or poorly saturated.

ROCK PROPERTIES AND WEATHERING PROCESSES

Diverse evidence suggests that rock weathering processes at this site are complex and that rock properties control them to a greater extent than anything else. The evidence includes generally poor correlation with simple climatic variables, all the year round rock breakdown, and variation in breakdown from one laval flow to another. Although there is no clear division between many of the processes of weathering involved, it is convenient to consider fracturing processes on the one hand and interactive processes between the loosened rock and the atmosphere on the other.

Fracturing Processes

Multigrain fractures of several types are the most important feature of these rocks and an understanding of their development is essential in working out just how they play a part in weathering processes. The bonded microfractures are certainly pre-freeface in age. Originally, they are brittle fractures, bonded by hydrothermal secondary minerals during post-eruptional hydrothermal phases. The same bonding materials are found in fractures in buried rocks from borehole samples, and this greatly reduces the possibility that the bonding materials could have formed by subaerial weathering to any great extent. Depending on the mineral phases occupying such fractures the effect can be either weakening or strengthening. The zeolites appear to have a strengthening effect but clay minerals are sensitive to further change on exposure at the freeface.

The dense systems of open multigrain fractures appear to be mainly associated with the most altered rocks. Typical rocks containing open microfractures are from sites B2 and B7 where pseudomorph mineralogy is general or where only feldspars remain relatively fresh in a highly altered glassy matrix. Just when such open fractures develop is harder to say than in the case of the bonded fractures, but the highly altered rocks are very weak and sensitive to changes in stress.

The freeface zone of largely unconfined rocks seems the most likely area for the formation of these fractures. They may form as a result of change in confining pressure on exposure at the surface or they may result partly from fluctuating water content in the rock in general, especially where clay minerals are widely dispersed in the rock. Porosity values of the most highly altered rocks can reach 30% (Table 5.1) and the field moisture content values described earlier for these rocks suggest a rather fluctuating pattern. With compressive strength values well below 10 MN m^{-2} (Douglas, 1972) for many of these rocks such conditions are likely to be conducive to further fracture propagation, in the freeface zone. Wetting and drying experiments on similar rocks at elevated temperatures have produced comparable fracture networks (McGreevy, 1982). A discoloration caused by an unidentified material occurs along some of these fractures which suggests that they remain in the freeface zone for some time, but in other cases this is not so.

Interactive Processes

These processes include subaerial chemical change on rock surfaces including micro-fracture surfaces, small-scale fracture propagation within bonding materials and further fracture formation or extension by processes such as freezing and hydration. Subaerial chemical change and associated fracturing seems to be quite active within the bonded microfractures where clay minerals, especially smectite, are the bonding materials (Douglas, 1981).

Changes within the clay mineral bonding include new open microfractures running longitudinally inwards from the rock prism edges. The surfaces of these new fractures are often coated with Mn^{4+} or red Fe^{3+} and this indicates that they are *in situ* fractures and not a result of handling methods (Figure 5.4). It also indicates that they could have formed as a result of fluctuating water content or that water has subsequently occupied these sites. It could mean a reduction in further water adsorption by the clay minerals and, therefore, a temporary stabilisation effect.

The exact cause of the oxidation of ferrous iron is difficult to ascertain especially whether it is oxidised while in solution or while in the silicate structure *in situ*. Experiments have not solved this problem (Siever and Woodford, 1979). However, it is known that such reactions are complex (Yatsu, 1988) with the formation of mixed compounds such as iddingsite (Smectite/chlorite/hydrous mica/saponite). The formation of new fractures within the clay minerals may be the result of stresses set up by their swelling or by their drying, perhaps associated with, for example, the smectite in these weathering products (McGreevy, 1982).

Figure 5.4 Scanning electron micrograph showing a curved manganese coating on an internal crack surface. Picture width = 20 μm

The rocks can be both confined and unconfined in the freeface situation so that induced local stress concentrations are likely. The fluctuating H_2O content of the rocks stressed earlier favours instability in the clay minerals, and the drying process in particular may be the most important aspect of this in producing small fractures in the clay mineral structure.

Atomic absorption analyses were made on freeface waters on 10 separate occasions, and, although the concentrations were extremely low (<4.0 ppm), there was a definite ranking, which in decreasing order was Ca^{2+}, Mg^{2+}, Na^+, Fe^{2+}, K^+, with occasional reversal of Ca^{2+} and Mg^{2+} (Douglas, 1972). The ranking possibly reflects the proportions in the whole rock or in the clay minerals. The iron is low in the ranking presumably because much of it is fixed as insoluble Fe^{3+} in the surface zones and near surface fractures. What effect such solution could have on ion exchange with the clay minerals has still to be investigated experimentally, but it is known, for instance, that octahedral substitution has an effect on swelling ability (Cole and Lancucki, 1976).

A fine fracture system running through clay minerals provides a large surface area with high water-absorbing capability and high loss rates by evaporation. However, thin sections suggest that such fractures are restricted to the outermost parts of the bonded fractures. This could be simply due to higher confining stress away from the freeface or joint surfaces.

The second group of interactive processes, such as freezing and hydration, has received much attention in the literature and will not be discussed in further detail. The open microfractures from rocks at site B2 or B7 or even B1 appear ideal for freezing processes to produce breakdown, at least in theory. The widths are mainly around 0.02 mm and 0.1 mm, and the outermost rocks are relatively unconfined. The question is rather whether moisture conditions on the one hand and freezing conditions on the other are suitable. Unfortunately, the equipment used to monitor temperatures does not permit calculation of rates of freezing (except on a weekly basis). However, long periods of frost were not common in the periods in question. On the other hand, air temperature fluctuations around 0°C were very common and are typical for many locations in Northern Ireland. This makes it clear that data on freezing rates should be obtained for the in-crack environment. More relevant, though, is the fact that the saturation coefficient of these rocks only reaches high values for short periods. Evaporation and free drainage is rapid. Thus frost shattering by simple volume expansion does not seem to be a very significant process under the available conditions.

A more likely process might be the growth of ice crystals on open fracture surfaces. Only slightly supercooled water would enable crystal growth on the variety of nuclei available on these surfaces. If crack temperatures were slightly lower than air temperatures, then an intermittent water supply spreading inwards from the rock surface would enable crystal growth in this way, but perhaps breakdown would only result where the rocks are unconfined, which they tend to be where open fractures dominate.

The situation may be quite different where the fractures are within clay mineral material, or where water is absorbed in now-fractured smectite material. Here, as is well known, the water is bonded to the clay mineral surface, and the area of this may be very large where, for instance, the clay has crystallized in a fibrous habit, which is evident in many cases. The adsorbed water on clay mineral surfaces can become ordered water and in clay-rich carbonates this has even been put forward as a disruptive force (Dunn and Hudec, 1972). However, this very property of water also causes a depression of freezing point as does simple capillary force. The question arises as to which is the more significant factor. In temperature conditions of near 0°C it may well be that water is not completely frozen in smectite-bonded fractures.

The importance of the narrow surface and near-surface water zone has already been emphasized in regard to the explanation of total moisture content. Under freezing conditions ice does form on the freeface in places and under the right conditions can reach thicknesses of several millimetres. Such conditions were considered by Powers (1945) in regard to frost damage of concrete, and he considered that in a porous material hydraulic pressure could arise towards the less saturated interior. The non-porous nature of many of the basalt rocks away from the surface zone means that the situation is not really comparable, and there is certainly not much evidence of active fracturing inwards from this zone. On the other hand, some small scale flaking is probably associated with freezing and thawing of this surface water.

Variations in density and widths of cracks are given in Tables 5.1 and 5.2 (see also Douglas et al., 1991), but these mean values sometimes reflect much higher local networks such as continuously bifurcating fractures in rocks from flow B2 (Figure 5.2). The rocks can be divided into two groups (Figure 1 of Douglas et al., 1991) when microfracture density is plotted against width. The group comprising rocks from B2, B7 and B3 have high density/high width fractures and those from B4 and B6 have low density/low width fractures. Such groups may be merely part of a continuous spectrum of rock fracture types, but at any rate the weathering behaviour of the two groups is likely to be different.

Table 5.2   The density of cracks in the basalt flows at Ardclinis

Lava flow	Intersections 50 mm (mean)	Standard deviation	Skewness	Number of values per sample
B7 Parallel to freeface	7.7	1.19	−0.08	25
B7 Normal to freeface	8.1	1.40	−0.04	25
B7 Parallel to joint	7.8	1.86	+0.02	5
B2 Parallel to freeface	8.8	2.50	+0.04	30
B2 Normal to freeface	9.2	2.00	+0.12	30
B2 Parallel to joint	8.8	2.20	+0.02	5
B1 Parallel to freeface	6.5	1.24	0.00	20
B1 Normal to freeface	6.7	1.10	0.10	20
B6 Parallel to freeface	0.8	0.55	+0.03	10
B3 Parallel to freeface	No microfractures			
B4 Parallel to freeface	1.5	1.40	+0.35	20
B4 Normal to freeface	1.5	1.70	+0.29	20

ROCK PROPERTIES AND FREEFACE DEVELOPMENT

Observations suggest a continuous evolution of rock properties and final destruction by rockfall, triggered by interaction with the atmosphere. Consideration of the geological history of the Antrim Plateau makes it possible to set up a general scheme to explain the weathering on cliffs subjected to previous hydrothermal activity. This is presented in Douglas et al. (1991) and reproduced here as Figure 5.5, which shows the main factors controlling cliff-face evolution. Despite the difficulties an attempt will be made to link the information from the six flows to this scheme.

The rocks considered here belong only to the chabazite and analcime zeolite zones which represent a rather limited temperature range (<150°C) in their hydrothermal history. Chemically, the laval flows can be very similar laterally, but there are important variations. For instance, glass content is variable so that the degree of alteration can depend on factors other than temperature. Very glassy rocks have tended to put rocks at site B7 in a somewhat anomalous position by favouring alteration in this way. A further difficulty is that the zeolite zones are not always simply belted. Local variations occur near fault zones, perhaps associated with heat loss there. These difficulties should not invalidate the general scheme proposed but it remains to develop it to satisfactorily account for all known events in the geological history and to take account of local rock variations. However, the data obtained from the six sites make it possible to define the main restrictions operating and to identify two main rock pathways involving different rock histories and response to near-surface weathering.

The more typical trend or pathway involves low temperature rock alterations producing smectite minerals, chlorite and low temperature zeolite minerals. These occur widely in bonded microfractures which become active in several ways in the freeface zone. Changes include swelling, oxidation of iron, development of new microfractures within the bonding material and change by freezing processes.

The other main trend or pathway suggested by the observations is one where more altered rocks reach the freeface zone, and develop dense opening microfracture networks, which then become sites for processes involving hydration and freezing, probably under highly fluctuating moisture conditions. Evidence for chemical change along such microfractures is very variable and suggests that their lifespan at the freeface can be short or long.

The evolution and behaviour of the first trend is suggested by rocks from sites B1 and B6 while the second trend is well demonstrated by rocks from sites B2 and B7. There are real differences in the properties of rocks from these sites and it is especially in the final development of properties of the freeface zone that the two pathways diverge and cause the rocks to break down in different ways. There are possible stabilizing effects which may operate on a local scale such as the fixation of insoluble Fe^{3+} in clay minerals.

The final stage in the evolution suggested is when a condition of instability results in fall by gravity. Local changes to freeface geometry undoubtedly have a strong effect on further instability and the actual size of the event. The patterns of activity have been discussed elsewhere (Douglas, 1980; Douglas et al., 1991).

85

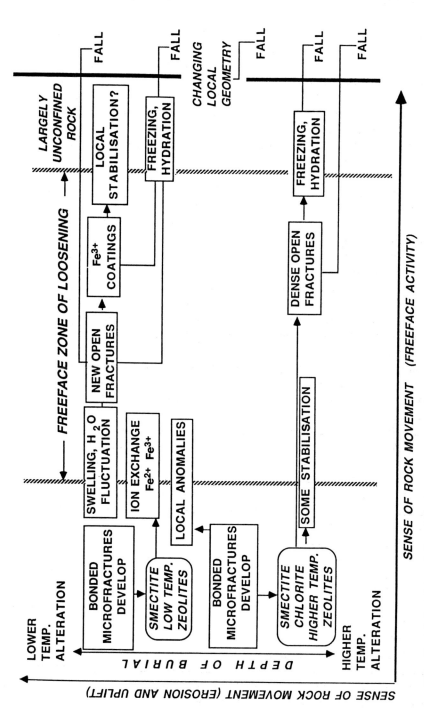

Figure 5.5 Postulated interrelationships between microfractures and mineralogy for a basalt cliff face (from Douglas *et al.*, 1991. Reproduced by permission)

DISCUSSION

As emphasised earlier, this chapter has been concerned with small-scale changes, but these have been shown to be active and varied. Their effects are also much further reaching since eventually they place larger-scale discontinuities in more effective instability conditions, or produce local overhangs, and such conditions favour periodic large rockfalls. The small-scale changes must, therefore, be regarded as the fundamental cause of all instability in these basalt freefaces. The exploitation of the crack system, in particular the microcrack system, is fundamental in the weathering of the rock and cliff face at all scales. The presence of clay minerals gives a weakening effect, especially if smectite is present. Zeolites may provide a stronger bonding in the cracks as they are less prone to weathering effects.

Hydrothermal activity in basaltic rocks is a frequent occurrence. Although cracks may sometimes be 'healed' by some alteration products, clay minerals especially are most likely to produce a weakening of the rock mass. At the large scale of joints, this may be superficially seen as 'weathering', i.e. subaerially produced weathering. In effect, however, hydrothermal alteration may be an intrinsic weakness in the rock, which operates even down to the small-scale (microcrack) as seen in Figure 5.2. Not only will (subaerial) weathering take place along such lines and allow further crack penetration, but the ingress of water will allow expansion of clays and freeze–thaw action. The combined effects of clay mineral expansion and frost action may be particularly effective. The precursor to such extrinsic activity is the exploitation of hydrothermally induced microcracking. The importance of crack-filling materials has been discussed at length by Barton (1973) and Samalikova (1985), and we would expect that the general principles they suggest would apply in the present case.

The cliff at Ardclinis represents the only freeface which has been examined from the standpoint of both geomorphology and mineralogy as well as overall structure. It is certainly not a unique instance as most cliffs represent situations that are far from equilibrium, whether mineralogical or tectonic. Therefore, there is a great need to extend this combined approach. We have investigated other sites of basalt cliffs in Northern Ireland and Iceland and see a similarity of decay features and mineralogical alteration. It is perhaps significant that the role of hydrothermal alteration has been insufficiently regarded from a geomorphological viewpoint (Dixon and Young, 1981; McGreevy, 1982; Ollier, 1983).

CONCLUSIONS

Although there have been many experiments investigating the basis mechanisms of rock breakdown and observations on the decay of cliffs there is still a large area of ignorance relating to the role of rock properties themselves. These provide the link between the laboratory simulations and field investigations. Examination of rock properties at the freeface has shown that there is considerable variation both in these properties on a cliff and in the geomorphic changes currently taking place. The essential characteristic features of the rocks examined in this study are: various stages of hydrothermal alteration, very extensive microfracturing which proves to be of different ages, and physico-chemical changes. The latter are due to interactions between the atmosphere and rock surfaces or fracture surfaces at scales between joints and microcracks.

REFERENCES

André, M. F. (1986). Dating slope deposits and estimating rates of rock wall retreat in northwestern Spitsbergen by lichenometry. *Geografiska Annaler*, **68A**, 65–75.

Barton, N. (1973). A review of the shear strength of filled discontinuities in rock. In Broch, E., Heltzen, A. and Joannessen, O. (eds), *Fjellsprengningsteknikk Bermekanikk*, Tapir, Oslo, pp. 19.1–19.38.

Cole, W. F. and Lancucki, C. J. (1976). Clay minerals developed by Deuteric alteration of basalt. In Bailey, S. W. (ed.), *International Clay Conference*, Mexico City, 1975, Applied Publishing, Wilmette, Illinois, pp. 35–43.

Dearman, W. R. and Irfan, T. F. (1978). Assessment of the degree of weathering in granite using petrographic and physical index tests. In International Colloquium, *Alteration et Protection des Monuments en Pierre*, RILEM/UNESCO, Paris, Report 2.3., 35pp.

Dixon, J. C. and Young, R. W. (1981). Character and origin of deep arenaceous weathering mantles on the Bega Batholith, southeastern Australia. *Catena*, **8**, 97–109.

Douglas, G. R. (1972) Processes of weathering and some properties of the Tertiary basalts of County Antrim, Northern Ireland. Unpublished Ph.D. Thesis, Queen's University, Belfast.

Douglas, G. R. (1980). Magnitude frequency study of rockfall in Co. Antrim, N. Ireland. *Earth Surface Processes and Landforms*, **5**, 123–129.

Douglas, G. R. (1981). The development of bonded discontinuities in basalt, and their significance to freeface weathering. *Jøkull*, **31**, 1–9.

Douglas, G. R., McGreevy, J. P. and Whalley, W. B. (1987). The use of strain gauges for experimental investigations on frost weathering. In Gardiner, V. (ed), *International Geomorphology*, Wiley, Chichester, pp. 605–621.

Douglas, G. R., Whalley, W. B. and McGreevy, J. P. (1991). Rock properties as controls on free-face debris fall activity. *Permafrost and Periglacial Processes*, **2**, 311–319.

Dunn, J. R. and Hudec, P. P. (1972). Frost and sorption effect in argillaceous rocks. *Highway Research Record*, **393**, 65–78.

Ernsberger, F. M. (1960). Detection of strength-impairing surface flaws in glass. *Proceedings of the Royal Society* A, **257**, 213–223.

Francou, B. (1982). Chutes de pierres et éboulisation dans les parois de l'étage périglaciaire. *Revue de Géographie Alpine*, **70**, 279–300.

Gardner, J. S. (1980). Frequency, magnitude and spatial distribution of mountain rockfalls and rockslides in the Highwood Pass area, Alberta Canada. In Coates, D. R. and Vitek, J. D. (eds), *Thresholds in Geomorphology*, Allen and Unwin, London, pp. 267–295.

Havir, J., Bordia, S. K. and Petros, V. (1971). A simple method of determining the coefficient of loosening of rocks. *International Journal of Rock Mechanics and Mining Science*, **8**, 97–103.

Humlum, O. (1992). Observations on rock moisture variability in gneiss and basalt under natural, arctic conditions. *Geografiska Annaler*, **74A**, 197–205.

Lautridou, J. P. and Ozouf, J. C. (1982) Experimental frost shattering: 15 years of research at the Centre de Geomorphologie du C.N.R.S. *Progress in Physical Geography*, **6**, 215–232.

McGreevy, J. P. (1982). Hydrothermal alteration and earth surface rock weathering: a basalt example. *Earth Surface Processes and Landforms*, **7**, 189–195.

Moon, C. B. P. (1984). Refinement of a technique for determining rock mass strength for geomorphological purposes. *Earth Surface Processes and Landforms*, **9**, 189–193.

Nepper-Christensen, P. (1965). Shrinkage and swelling or rocks due to moisture movement. *Meddeleser fra Dansk Geologisk Forening*, **15**, 548–555.

Ollier, C. D. (1983). Weathering or hydrothermal alteration. *Catena*, **10**, 57–59.

Onadera, T. F., Yoshinaka, R. and Oda, M. (1974). Weathering and its relation to mechanical properties of granite. *Proceedings of the 3rd Congress of the International Society of Rock Mechanics*, AIME New York, **2A**, 71–78.

Powers, T. C. (1945). A working hypothesis for further studies of frost resistance of concrete. *Journal of the American Concrete Institute*, **16**, 245–272.

Rapp, A. (1960). Recent development of mountain slopes in Kärkevagge and surroundings, northern Scandinavia. *Geografiska Annaler*, **42A**, 66–200.

Samalikova, M. (1985). Characteristics of discontinuities and their influence on the mechanical behaviour of the rock mass. *Bulletin of the International Association of Engineering Geology*, **31**, 111–121.

Selby, M. J. (1987). Rock slopes. In Anderson, M. G. and Richard, K. S. (eds), *Slope Stability*, Wiley, Chichester, pp. 475–504.

Siever, R. and Woodford, N. (1979). Dissolution kinetics and weathering of mafic minerals. *Geochimica et Cosmochimica Acta*, **43**, 717–724.

Simmons, G. and Richter, D. (1976). Microcracks in rocks. In Strens, R. G. J. (ed.), *The Physics and Chemistry of Minerals and Rocks*, Wiley, Chichester, pp. 105–137.

Walker, G. P. L. (1951). The amygdale minerals of the Tertiary lavas of Ireland, (1) the distribution of chabazite in the Garron plateau area, Co. Antrim. *Mineralogical Magazine*, **29**, 773–791.

Walker, G. P. L. (1960). The amygdale minerals in the Tertiary lavas of Ireland, (3) regional distribution. *Mineralogical Magazine*, **32**, 503–527.

Whalley, W. B. (1984). Rock falls. In Brunsden, D. and Prior, D. B. (eds), *Slope Instability*, Wiley, Chichester, pp. 217–256.

Whalley, W. B., Douglas, G. R. and McGreevy, J. P. (1982). Crack propagation and associated weathering in igneous rocks. *Zeitschrift für Geomorphologie*, **26**, 33–654.

Yatsu, E. (1966). *Rock Control in Geomorphology*. Sozosha, Tokyo.

Yatsu, E. (1988). *The Nature of Weathering*. Sozosha, Tokyo.

6 Formation of Weathering Rinds on Andesite Blocks under the Influence of Volcanic Gases around the Active Crater of Aso Volcano, Japan

YUKINORI MATSUKURA and MITSUYOSHI KIMATA
University of Tsukuba, Japan

and

SHOZO YOKOYAMA
Kumamoto University, Japan

ABSTRACT

Reddish brown weathering rinds occur on the surface of andesite blocks distributed near Sensuikyo, Aso volcano. A cobble-sized rock sample with a size of 14 × 13 × 9 cm was collected for analysis. It has a gnamma-like depression on the top. Polarising microscope observations show that the innermost unweathered part (Zone I) of the sample consists of (1) plagioclases, clinopyroxenes and olivines which form phenocrysts and groundmass, and (2) magnetites forming microphenocrysts. The outer weathered parts, forming weathering rinds, are divided into two zones: (1) the inner zone with a thickness of 2 to 3 mm (Zone II) which has the same porphyritic texture as Zone I and (2) the outer thin coating layer with a thickness of 0.2 to 0.3 mm (Zone III). The results of several mineralogical analyses using an X-ray microdiffractometer, X-ray fluorescence spectrometer, electron microprobe and organic elementary analysers show that (1) Zone II is characterised by intensive leaching of cations such as alumina and silica from the phenocrysts and groundmass and crystallisation of jarosite $(KFe[SO_4]_2[OH]_6)$ and gypsum $(CaSO_4 \cdot 2H_2O)$, and (2) Zone III is characterised by the occurrence of amorphous hydrous silica. These findings suggest that sulphur contained in volcanic gases from Aso volcano is strongly connected with the formation of weathering rinds at this site.

INTRODUCTION

Andesite blocks with weathering rinds occur around the active crater of Aso volcano. The distribution of such blocks roughly corresponds to the area of (1) the spread of volcanic gases and (2) the resultant sparse vegetation. It is suggested that the volcanic gases cause the formation of weathering rinds.

The influence of volcanic gases on rock weathering has been recognised (e.g. Iwasaki, 1970, pp. 146–149). However, only a few geochemical investigations concerning the alteration of rocks surrounding hot springs have been performed (e.g.

Rock Weathering and Landform Evolution. Edited by D. A. Robinson and R. B. G. Williams
© 1994 John Wiley & Sons Ltd.

Anderson, 1935; Iwasaki *et al.*, 1964). Weathering rinds of andesitic and basaltic rocks located outside areas of active volcanoes have been studied on a mineralogical and chemical basis (e.g. Smedes and Lang, 1955; Coleman, 1982; Veldkamp *et al.*, 1990). The previous works indicate that the processes concerning weathering rind formation in areas of active volcanoes are still open to research. Taking mineralogical and chemical approaches, the present study attempts to elucidate the formative process of the weathering rinds of andesite blocks on the slopes around the active crater of Aso volcano.

GEOMORPHOLOGICAL SETTING AND THE SAMPLING SITE

Aso volcano, situated at the centre of Kyushu (Figure 6.1a), is one of the most active volcanoes in Japan, and is composed of a large caldera with a diameter of 25 km in

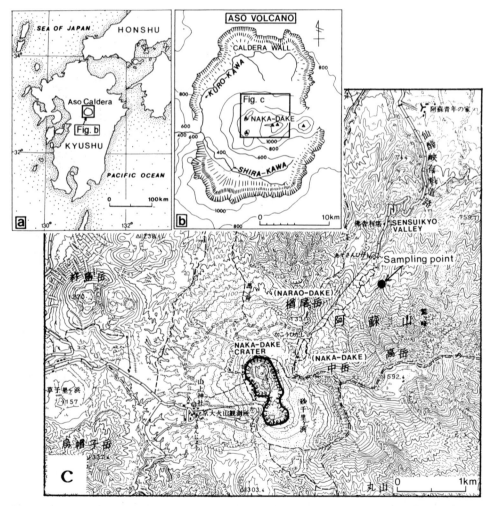

Figure 6.1 Location map showing sampling site, from the 1:50 000 topographic map ASO-SAN, published by Geographical Survey Institute. Reproduced by permission. The contour interval in map (c) is 20 m

a north to south direction and 18 km in an east to west direction, and many post-caldera volcanoes (Figure 6.1b). Among the post-caldera volcanoes, Naka-dake is the only active volcano and is located at the centre of the caldera. Eruptions have been repeated at the Naka-dake crater (Figure 6.1c) in historic times, resulting in volcanic ash falls and sometimes scoria falls from Strombolian eruptions (Japan Meteorological Agency, 1984, pp. 337–354).

Meteorological data from Aso Meteorological Station (32°53′N, 131°02′E, 1143 m a.s.l.), located about 3 km southwest of the study area, show that the mean air temperature is −2.1°C in the coldest month (January), 20.2°C in the warmest month (August), and 9°C through the year. The average annual precipitation is about 3300 mm.

Figure 6.2 Tor-like blocks of andesite in the investigated area

The sampling point is located on a ridge about 2 km northeast of Naka-dake crater (Figure 6.1c). The ridge consists of Washiga-mine volcano rocks which are a series of pyroclastic rocks, and lavas of basalt and of basaltic andesite erupted tens of thousands of years ago (Ono and Watanabe, 1985). The southerly wind makes volcanic gases from the Naka-dake crater pass through a col between the summits of Naka-dake and Narao-dake and flow toward Sensuikyo valley (Figure 6.1c).

The top of the ridge, where the sampling site is located, is composed of andesite blocks of various sizes (up to 3 m), and the larger blocks form tor-like features (Figure 6.2). All blocks are completely covered with reddish brown weathering rinds. Some blocks have hemispherical depressions facing upwards with diameters of 10–30 cm and depths of up to 15 cm (Figure 6.3). Such gnamma-like depressions have dish-like or bowl-shaped outlines. Rain-water pools form in some bowl-shaped deep depressions. Efflorescence of white powdery gypsum is observed on many of the surfaces of the dry depressions.

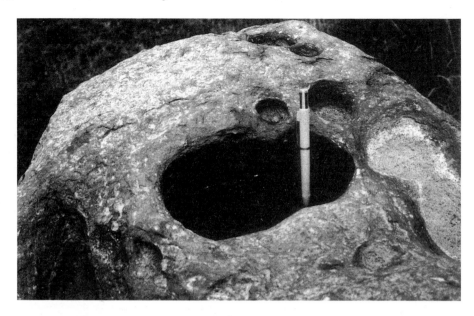

Figure 6.3 An andesite block with gnamma-like depression

ROCK SAMPLE FOR ANALYSIS

An andesite block (Figure 6.4a), 14 × 13 × 9 cm in size, with a gnamma-like depression on top was selected for mineralogical and chemical analyses. Figure 6.4b shows the cross-section of the middle part of the sample. The inner fresh-looking zone with a Mohs' hardness number between 6 and 7 appears black when wet and dark grey in dry conditions. Many micropores are observed on the cross-section. This sample is, therefore, recognised as vesiculate andesite.

The weathering rind of the sample has a thickness of 2 to 3 mm. Microscopic observation reveals many yellow phenocrysts scattered in a black groundmass. The rind is fragile in comparison with the inner part of the sample, and some parts of the rind can be easily shattered by fingernails. Waxy-brown (moderate brown, 5YR 3/4) coatings develop on the surface of the weathering rind, and also on the surface of the gnamma-like depression. The coating is too thin to recognise with the naked eye in the cross-section. The coating material is quite hard with a Mohs' hardness number between 9 and 10.

LABORATORY ANALYSES

Optical microscopic observations show that the texture of the weathering rinds does not change significantly in different parts of the sample. Thin sections (cut normal to the rock surface) and powdered samples for detailed laboratory analyses were taken from the bottom of the depression in the block (Figures 6.4b and 6.6a). The samples were examined using a polarizing microscope, X-ray microdiffractometer, X-ray

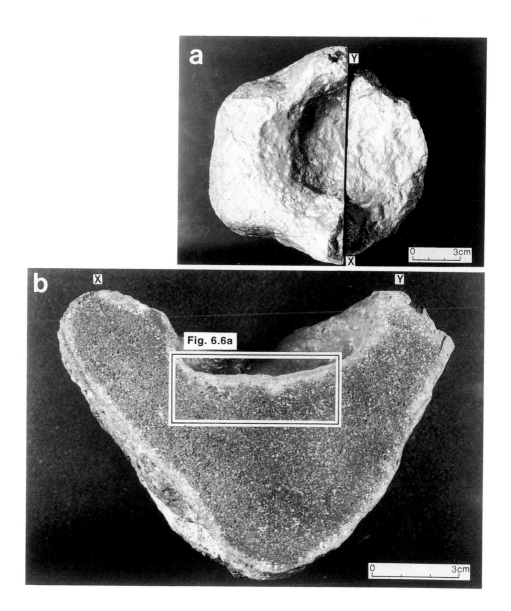

Figure 6.4 Rock sample for analysis. (a) overhead view (b) cross-section

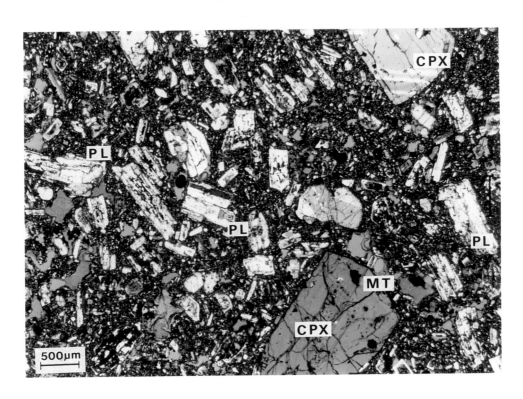

Figure 6.5 Photomicrograph of fresh host-rock zone in cross-polarised light. PL: Plagioclase, CPX: Clinopyroxene, MT: Magnetite

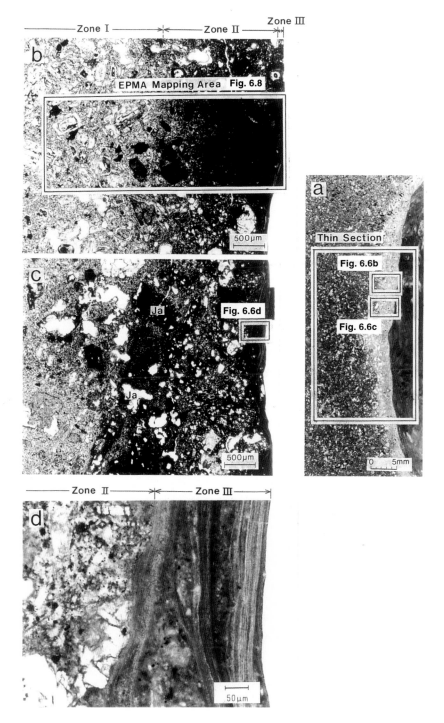

Figure 6.6 Photomicrograph of weathering rind and fresh zone. (a) the area of thin section; (b) and (c) weathering rinds and EPMA mapping area (see Figure 6.8); (d) coating layer in the outermost parts of weathering rinds

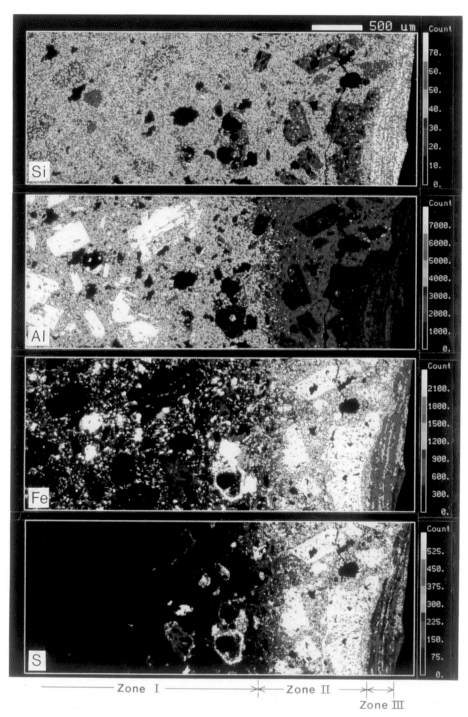

Figure 6.8 EPMA element maps of Si, Al, Fe and S. Mapping area is shown in Figure 6.6b. Scale bar in the upper part represents 500 μm (0.5 mm)

fluorescence spectrometer (XRF), electron-probe micro analysis (EPMA) and organic elementary analysis.

Optical Microscope Observation

The distinction between the fresh (unweathered) interior and the weathered crust is clear under the polarising microscope. The fresh interior has a porphyritic texture, the phenocrysts of which are composed primarily of lath-shaped zoned plagioclases with a maximum size of about 1 mm (Figure 6.5). The plagioclases contain dust inclusions and seem to have been altered along and/or within the zoning. Clinopyroxenes and olivines form subordinate phenocrysts. These minerals sometimes represent the glomerophyric texture. Magnetites are included as microphenocrysts. The reaction rims of olivines seem to be composed of plagioclases and clinopyroxenes with the same compositions as the phenocrysts. A small quantity of glass is observed in the groundmass. As a whole, the texture of the rock is hyalopilitic or pilotaxitic.

The weathered crust or rind looks black under the microscope (Figure 6.6b and c), and has a thickness of 2 to 3 mm. The inner parts of the rind display a porphyritic texture that is similar to that in the underlying unweathered rock. No definite boundary between the weathering rind and the unweathered rock is observable under the microscope. Apertures are more abundant in the weathering rind than in the unweathered rock. The apertures in the rind are surrounded by reddish brown materials (Figure 6.6c). Magnification of the outermost parts of the weathering rind reveals that (1) a hard coating is present with a thickness of 0.2 to 0.3 mm, (2) the coating is composed of pale-brown micrometer-scale laminations (Figure 6.6d), and (3) there is a sharp contact with the underlying zone. From the above observations the rock sample can be divided into three zones (Figure 6.6b and d): the fresh interior zone (Zone I), the weathered zone (Zone II) and the hard coating layer (Zone III).

X-Ray Diffraction Analysis

The minerals in each zone were examined by a newly improved X-ray microdiffractometer (Kimata et al., 1990). The examined specimen of less than 100 μm was exfoliated from a polished thin section. The paragenetic minerals in Zone I were identified as plagioclases, clinopyroxenes, orthopyroxenes, magnetites and olivines. Figure 6.7 shows the X-ray diffraction pattern of microcrystals in Zone II. Jarosite $(KFe_3(SO_4)_2(OH)_6)$ and gypsum $(CaSO_4 \cdot 2H_2O)$ were detected as weathering products, beside plagioclase as a primary mineral. The black/reddish brown microcrystals, shown in the centre of Figure 6.6c, are identified as jarosite. Anhedral crystals of jarosite occur throughout Zone II. Evidently, Zone III is mainly composed of amorphous materials, because many examined specimens in Zone III have no definite X-ray diffraction line and because its base line between 30° and 50° has been swollen. A small amount of jarosite is detectable in Zone III.

X-Ray Fluorescence (XRF) and Organic Element Analyses

Major-element analyses for bulk samples were conducted with a RIGAKU X-Ray Fluorescence Spectrometer 3270 by a tablet method. Results are shown in the left-

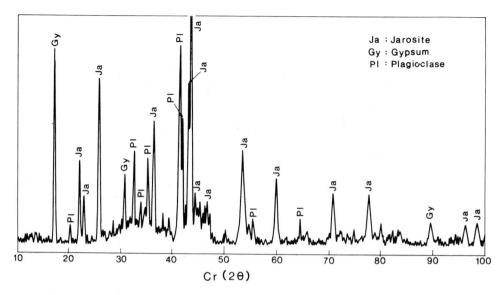

Figure 6.7 X-ray diffraction pattern of the microcrystals in the weathering rind

hand columns of Table 6.1. The chemical bulk composition of the fresh sample (Zone I: sample no. 1) indicates that the rock is tholeiitic andesite. The average bulk composition of the weathering rind (Zones II and III: no. 2) is much richer in SO_3 and poorer in Al_2O_3, MgO and Na_2O than Zone 1.

A Perkin-Elmer 2400 CHN element analyser was used to estimate the presence of nitrogen in the weathered bulk sample. This examination indicates the absence of ammonium in the jarosite present. Thus minor NH_4^+ substitution for K^+ is not possible.

Electron-probe Micro Analysis

Since the weathering rind is very thin, it is technically difficult to use XRF analysis to obtain the difference in chemical composition between Zones II and III. Moreover, XRF analysis gives no further detailed information about the distribution of chemical elements in each zone. For these reasons, electron-probe (EPMA) micro analysis was conducted in order to examine the chemical composition (quantitative point analysis) for micro spots, and its aerial variation (qualitative mapping) for each zone. Analyses were carried out on a JEOL JXA-8621, by a full ZAF-correction method; the electron probe was operated at 25kV and the electron beam was focused to about 10 μm in diameter.

Several representative results for point analysis in each zone are shown in the right-hand side columns of Table 6.1. Since the diameter of the beam spot is 10μm., data in Zone I show the chemical composition of each phenocryst mineral: nos. 3, 4 and 5 correspond to the phenocrysts of plagioclases, clinopyroxenes and olivines respectively. Zone II (nos. 6, 7 and 8) is characterised by a complex composition due to leaching and accumulation: (1) the amounts of MgO, Na_2O and CaO in this zone are

Table 6.1 Major-element compositions of the fresh zone and weathering rind (percentage by weight). The data for sample nos. 1 and 2 were obtained from the analyses with XRF for bulk specimens, and those for sample nos. 3 to 10 from the point analysis with EPMA

| | XRF (bulk) | | EPMA (spot) | | | | | | | |
| | Zone I | Zones II and III | Zone I | | | Zone II | | | Zone III | |
No.	1	2	3	4	5	6	7	8	9	10
SiO_2	53.0	54.0	54.15	52.41	37.75	62.63	48.12	37.21	66.77	72.87
Al_2O_3	21.0	6.1	27.62	0.63	0.01	13.40	2.10	5.47	3.80	4.07
TiO_2	0.87	0.55	0.07	0.32	0.02	0.70	0.70	0.39	0.57	0.48
FeO	9.4	8.1	0.82	20.32	25.20	6.08	6.08	20.36	8.30	4.32
MnO	0.15	0.08	0.01	0.76	0.45	0.05	0.69	0.00	0.02	0.02
MgO	2.3	0.82	0.07	20.34	36.92	0.82	17.02	0.27	0.57	0.69
CaO	7.8	11.0	10.34	3.99	0.14	0.74	3.63	0.25	1.21	1.38
Na_2O	3.0	0.80	5.43	0.08	–	3.01	0.21	0.37	0.46	0.54
K_2O	1.8	1.5	0.49	0.04	–	5.23	0.65	4.31	1.52	0.58
SO_3	0.13	17.0	0.03	–	0.04	2.73	2.59	13.52	5.02	1.74

smaller than those in Zone I, (2) SO_3 content is generally high exceeding 10% by weight in places (e.g. no. 8), and (3) K_2O of 3 to 5% by weight (e.g. nos. 6 and 8) is included in some places. On the other hand, Zone III has a relatively constant composition between 65 and 75% SiO_2 by weight and small amounts of Al_2O_3, FeO and SO_3. The total major element composition for the spots in Zone III are 85 to 88% by weight, and the remaining 12 to 15% is estimated to be the content of H_2O.

Figure 6.8 shows the coloured composition maps of Si, Al, Fe and S for the polished thin section, which extends from Zones I to III. The colour steps on the right of the figure indicate the relative content of each element: upper steps indicate larger quantities and the lowest one, black, corresponds to the pore space. This figure leads to the following summary. The content of Si decreases slightly from Zone I to Zone II, but then rises to a maximum in Zone III. The groundmass in Zones I and II is richer in Si than phenocrysts. The phenocrysts that are poor in Si in these zones are pyroxenes, olivines and magnetites. The content of Al decreases outward from Zone I to Zone II, and becomes extremely low in Zone III. Lath-shaped phenocrysts rich in Al occurring within Zone I are plagioclases. The content of Fe and S is highest in Zone II and is low in Zones I and III. It is noteworthy that S is absent in the inner part of Zone I, and Fe and S are concentrated in the same layers in Zone III. A high content of Fe is ascribed to magnetites, and a small amount of Fe to pyroxenes in Zone I. In summary, Zone I is rich in Si and Al, but devoid of S. Fe and S are accumulated but Al is reduced in Zone II, and Si is accumulated in Zone III.

DISCUSSION

Results obtained from the mineralogical and chemical analyses for Zones I and II are summarised as follows: (1) The porphyritic texture retained in Zone II indicates that the zone is an *in situ* weathered zone. (2) The smaller amounts of Si, Al, Mg, Ca and Na in Zone II than in Zone I indicate that these components in Zone II have been leached out from Zone II. (3) Intensive leaching of alumina has occurred both in the phenocrysts and the groundmass of Zone II. (4) In contrast to this, Fe and S have been accumulated in Zone II to form jarosite and a small amount of gypsum.

These results suggest the following weathering processes involving the formative processes of jarosite and gypsum in Zone II. Since no sulphur is contained in the host rock, it is clear that sulphur in Zone II has been derived from sources other than the host rock. The following field evidence suggests that the most probable source for sulphur will be the volcanic gases from the Naka-dake crater: (1) The rock sampling site corresponds to the area through which volcanic gases frequently pass. (2) Ossaka *et al.* (1984) observed that the volcanic gases from the crater of Naka-dake contain 8.7% SO_2 and 7.2% H_2S by volume in July 1958, and that the water of the crater lake of Naka-dake has a low pH (1.6–1.7) due to dissolution of these gases, and contains a high concentration of SO_4^{2-}, e.g. 12 000 mg l^{-1} in 1980 and 15 000 mg l^{-1} in 1982. (3) River water in the Sensuikyo valley, near the rock sampling site, has a pH of 4.8 and contains SO_4^{2-} of 120.9 mg l^{-1} (Y. Shimano, personal communication; data obtained on 12 December 1987).

It is considered that volcanic gases containing sulphur affect the leaching process in Zone II. Volcanic gases flowing down to the Sensuikyo valley are absorbed and/or dissolved into water within the surface layer of the rock. In addition, rain drops with dissolved volcanic gases attack the rock surface. In either case, the water becomes sulphuric acid with a low pH due to the dissolution of volcanic gases. The solubility of many substances is highly affected by pH: the leaching of cations such as Na^+, Mg^{2+}, and Ca^{2+} becomes more intensive with decreasing pH value. At very low pH (<4.0) alumina becomes more soluble than silica (e.g. Ollier, 1984, pp. 31–32). This is the most probable reason for the intense leaching of alumina in Zone II.

The iron compounds are immobile against weathering (e.g. Loughnan, 1969; Feijtel *et al.*, 1988), and the morphology of the jarosite in Zone II resembles that of the pyroxene, olivine and magnetite. This suggests that the Fe-rich minerals have been altered to jarosite; that is, the jarosite seems to be formed by combining immobile Fe (i.e. residue from the Fe-rich minerals), with sulphur supplied from volcanic gases. As for potassium, which is an essential element for the formation of jarosite, its particular source and the process of fixation to the Fe-rich minerals are unknown.

Gypsum found in Zone II is considered to be formed by the following sequential process: (1) Ca, leached out from plagioclase, combined in this zone in solution with sulphur supplied from volcanic gases. (2) H_2O evaporates from the solution.

The formative process of Zone III (the coating layer) is considered to be different from that of Zone II, because Zone III has a striped texture and is harder than the host rock. The X-ray diffraction analysis and EPMA analysis showed that Zone III contains a large amount of Si (about 65 to 75% by weight) and water (15%), and is composed of amorphous materials. This indicates that the main material of Zone III is amorphous hydrous silica. The sharp contact of Zone III with Zone II and the dissimilar chemical composition, suggest that Zone III (coating) is a product from exogenous material, and the amorphous silica layers indicate that solidification of gel materials occurred several times. It is considered that volcanic gases from the Naka-dake crater serve as a probable source of the Si forming the coating layer, although no available data have been obtained. Considering that relatively high concentrations of Si (maximum 52 ppm by weight) have been reported in the water vapour in high temperature gases from neighbouring Kuju volcano (Honda and Mizutani, 1968), a significant amount of Si is likely to be contained in the water vapour in the volcanic gases from the Naka-dake crater. The sequential processes of formation of the coating layer is considered to be as follows. (1) Water vapour in the volcanic gases is absorbed into the surface layer of the rock and on cooling becomes a liquid solution. (2) The water in this solution is evaporated to form a gel of amorphous silica. The repetition of the above process gradually increases the thickness of the lamination.

Gypsum is found in Zone II (Figure 6.7) and the white powder found on some gnamma-like surfaces is also identified as gypsum. It is, therefore, expected that salt weathering plays an important role in spalling the weathered zone and in forming the gnamma-like depression. However, in the case of the present study sample no enlargement process would occur because the surface of the depression is covered with a hard silica coating layer which protects the weathered zone from physical weathering and/or erosion.

CONCLUDING REMARKS

Chemical and mineralogical analyses of weathering rinds of an andesite block in Aso volcano revealed that intensive leaching of alumina and the formation of jarosite and gypsum occurred in the inner weathered zone. The outer layer of the weathering rinds is composed of amorphous hydrous silica, and a probable source of the silica is assumed to be the water vapour in the volcanic gases. These findings suggest that volcanic gases facilitate the formation of weathering rinds.

ACKNOWLEDGEMENTS

The authors are very grateful to Mr Norimasa Nishida of the chemical Analysis Center of the University of Tsukuba, for his help in the microprobe analysis. They are also much indebted to Dr Yasuo Shimano of Utsunomiya Bunsei College for providing valuable information concerning river-water quality in the Sensuikyo valley, and to Professor Tsuguo Sunamura of the Institute of Geoscience, University of Tsukuba, for critical reading of the manuscript. Part of this work is financially supported through the Science Research Fund of the Ministry of Education, Science and Culture.

REFERENCES

Anderson, C. A. (1935). Alteration of the lavas surrounding the hot springs in Lassen Volcanic National Park. *American Mineralogist*, **20**, 240–252.

Coleman, S. M. (1982). Chemical weathering of basalts and andesites: evidence from weathering rinds. *US Geological Survey Professional Paper*, **1246** , 1–51.

Feijtel, T. C., Jongmans, A. G., Van Breemen, N. and Miedema, R. (1988). Genesis of two planosol in the Massif Central, France. *Geoderma*, **43**, 249–269.

Honda, F. and Mizutani, Y. (1968). Silicon content of fumarolic gases and the formation of a siliceous sublimate. *Geochemical Journal*, **2**, 1–9.

Iwasaki, I. (1970). *Chemistry of Volcanoes* (Kazan Kagaku). Koudansha, Tokyo (in Japanese).

Iwasaki, I., Hirayama, M., Katsura, T., Ozawa, T. and Ossaka, J. (1964). Alteration of rock by volcanic gas in Japan. *Bulletin Volcanologique*, **27**, 65–78.

Japan Meteorological Agency (1984). *National Catalogue of the Active Volcanoes in Japan*. Japan Weather Association, Tokyo (in Japanese).

Kimata, M., Shimizu, M., Saito, S., Murakami, H., Ohkanda, T. and Shimoda, S. (1990). Rapid collection of the X-ray powder pattern from a single microcrystal by crystal movement of Gandolfi style. *Annual Report of the Institute of Geoscience, University of Tsukuba*, **16**, 63–68.

Loughnan, F. C. (1969). *Chemical Weathering of the Silicate Minerals*. American Elsevier Publishing Co., New York.

Ollier, C. (1984). *Weathering*. Longman, London.

Ono, K. and Watanabe, K. (1985). Geological map of Aso volcano. *Geological Survey of Japan, Geological Map of Volcanoes*, **4** (in Japanese).

Ossaka, J., Hirabayashi, J. and Ozawa, T. (1984). The geochemical observation at Aso volcano. *The Report of the 2nd (1981) Intensive Observations of Aso volcano*, 82–84 (in Japanese).

Smedes, H. W. and Lang, A. J., Jr (1955). Basalt column rinds caused by deuteric alteration. *American Journal of Science*, **253**, 173–181.

Veldkamp, E., Jongmans, A. G., Feijtel, T. C., Veldkamp, A. and van Breeman, N. (1990). Alkali basalt gravel weathering in Quaternary Allier River terraces, Limagne, France. *Soil Science Society of America Journal*, **54**, 1043–1048.

7 Problems in Assessing the Weathering Action of Lichens with an Example of Epiliths on Sandstone

HEATHER VILES
Oxford University, UK

and

ALLAN PENTECOST
King's College London, UK

ABSTRACT

Lichens play a variety of roles in rock weathering, but despite several recent studies on the mechanisms involved we are very far from a general understanding of the interaction between lichens, weathering and landform production. Specific geomorphological questions which still need answering include: the influence of lichen community development on weathering rates over time; the interaction of lichen weathering with other weathering processes; and the role played by lichen weathering in determining landforms at different scales. It is important, but very difficult, to monitor lichen weathering rates. Observations of epilithic lichen activity on sandstones from the Cedarberg Mountains in South Africa show that at the SEM scale lichen etching of the underlying rock is apparent. At the microscale there are no specific landforms created, but such activity, coupled with the protective lichen cover, seems to contribute to the development of macroscale ruiniform relief.

INTRODUCTION

The weathering action of lichens on a variety of substrates has been noted by many authors, with early reports by Sollas (1880) and Bachman (1890). More recently, Jones *et al.* (1981) and Wilson and Jones (1983) have investigated the role of lichens in mineral weathering and pedogenesis; Muxart and Blanc (1979), Ascaso *et al.* (1982), Danin and Garty (1983) and Viles (1987) have discussed lichen weathering of limestones and dolomites; and Hallbauer and Jahns (1977), Wessels and Schoemann (1988) and Cooks and Otto (1990) report on the role of lichens in weathering sandstones and other quartzitic rocks. Lichens have also been observed to be important agents of building stone deterioration, especially on marble surfaces (del Monte, 1991). Lichens are also involved in the deterioration of glass (Mellor, 1921; Harvey and King, 1971; Brightman and Seaward, 1977). Different types of

Rock Weathering and Landform Evolution. Edited by D. A. Robinson and R. B. G. Williams
© 1994 John Wiley & Sons Ltd.

lichens are involved in these processes, with some being epilithic (i.e. living on the rock surface), some endolithic (actively boring into the rock surface), and some chasmolithic or cryptoendolithic (living in hollows or fissures within the rock).

There has long been debate over whether lichens are also capable of protecting the underlying substrate from other weathering and erosional processes (see, for example, Giacobini *et al.*, 1986). Archibald Geikie claimed in 1893 that 'A crust of lichens doubtless on the whole protects the rock underneath it from atmospheric agents' (p. 47), and many later workers concur. Furthermore, lichens may play an important role in crust formation (i.e. the precipitation and entrapment of minerals on rock surfaces) as is shown by the development of lichen stromatolites on some terrestrial limestones (Klappa, 1979; Jones and Kahle, 1985). Lichen processes may also act synergistically with other denudational mechanisms; thus we find in the literature records of snail and lichen denudation in the Negev desert, Israel (Shachak *et al.*, 1987); and bagworm larvae and lichen denudation (Wessels and Wessels, 1991) in the Orange Free State, South Africa.

Table 7.1 is a compilation of some suggested mechanisms of lichen activity on rock and mineral surfaces, including both chemical and mechanical processes. Evidence for the operation of these processes comes mainly from detailed microscopic and microchemical analyses of the lichen : rock interface. As such, most studies have had a very small-scale focus. Furthermore, many of these studies have been carried out by geologists and/or botanists who may not be interested in the landforming potential of lichen mechanisms. Some potentially important lichen weathering mechanisms have received very little scientific attention. For example, because saxicolous lichens are of many different colours and their albedos diverge widely from that of the underlying rock, rock surfaces covered by a variety of lichens may experience a range of different temperatures, all different from that of the base rock. Deeply pigmented lichens with low albedos, e.g. *Acarospora fuscata* and the brown *Parmelias*, probably raise surface temperatures well above the air temperature (see Kershaw, 1983, for a review of this phenomenon in the Arctic). Lighter coloured lichens will produce lower surface temperatures. In lichen mosaics consisting of two or more species (or on a surface where one species is distributed patchily) 'hot' spots may be produced beneath dark lichens producing local stresses in the rock surface. In order to assess fully the

Table 7.1 Lichen weathering mechanisms and forms produced. Compiled from Syers and Iskandar (1973)

Chemical mechanisms	Results of chemical action
Chelation by extracellular, soluble compounds	Fruiting body pits
	Grooves at endolithic thalli interfaces
Attack by oxalic acid	Etching of minerals
Attack by water acidified by respired carbon dioxide	Precipitation of alteration products, e.g. calcium oxalate, which may or may not play a further role in weathering
Physical mechanisms	Results of physical action
Rhizine penetration	Exfoliation of rock surface layer
Thallus expansion and contraction on wetting and drying	Cracking of rock
	Increase in pore volume

geomorphological significance of lichen weathering we need to consider two major areas in more detail, i.e. the rates of lichen weathering processes, and the significance of lichen weathering at larger spatial and temporal scales than the simple action of a single lichen on a rock.

There have been a few important measurements, or estimates, of the rate of lichen weathering as presented in Table 7.2. We may ask why there have not been more such observations. Primarily, there is a problem of finding suitable methods with which to measure rates of lichen attack which, as shown in Table 7.1, may involve a wide range of processes. Danin and Garty (1983) in their study of lichen weathering of limestones in the Negev desert found that lichen fruiting body pits, grooves where two lichen thalli meet, and grooves caused by snails grazing on lichens were all present. Very different rates of formation may be involved with each of these micro-forms. Many of the methods used so far to measure lichen weathering have been somewhat esoteric, including the analysis of the mineral contents of the faeces of bagworm larvae and snails (Wessels and Wessels, 1991), and the measurement of the volume of endolithic (rock-boring) lichens (Wessels and Schoemann, 1988). Some workers have used the depth of the weathered layer below lichens on dated surfaces (in comparison with the weathered layer on bare surfaces) as a measure of their weathering activity (e.g. Jackson and Keller, 1970). In many situations this method may be difficult to apply because weathering depths are difficult to measure on some rocks, and because comparable bare and lichen-covered surfaces are not always available.

Many of the commonly used geomorphological techniques for measuring weathering rates are impossible to use for lichen weathering. Experimental studies in the laboratory are problematical because it is difficult to grow lichens under artificial conditions, and to speed up their weathering activity. The micro-erosion meter (MEM) has been used on lichen-covered surfaces (Trudgill, 1985), but this technique is probably highly influenced by lichen growth rates and water contents. To obtain a satisfactory estimate of the rate of lichen weathering over an area of bare rock one needs some estimate of the amount of weathering carried out by each individual lichen (or perhaps more importantly per unit area of lichen); an estimate of the percentage of the rock surface covered by lichens; as well as some idea of the growth rates of the lichens and the changing pattern of the community over time.

There have been several studies of lichen growth rates and community development which may be of use in the realistic assessment of lichen weathering action. Wool-house et al. (1985), for example, quote typical crustose lichen radial growth rates as 1–5 mm yr^{-1}. This, however, applies to horizontal increases in diameter, and vertical growth rates (down into the substrate) may be very different. In several endolithic species, growth downwards is concentrated in the vegetative stage; whereas during the reproductive phase lateral growth predominates (Wessels and Schoemann, 1988). Lichen communities change in species composition and growth rates over time and their weathering impact may change accordingly. Lawrey (1991) indicates that in lichen communities pioneers are not eliminated by competitors, but rather the number of species increases over time as addition, rather than replacement, of species occurs. Armesto and Contreras (1981) suggest that saxicolous lichen communities are non-equilibrium systems which are constantly changing. Pentecost (1980) shows how, despite growth and competition between lichen thalli, senescence of lichens

Table 7.2 Some lichen weathering rates and comparable rates in the absence of lichens

Place	Rock type	Rate (μm yr^{-1})	Reference
Lichens present			
Antarctica	Beacon Sandstone	5–30	Speculative rate; Hale 1987
South Africa	Clarens Sandstone	96	Endolithic lichen volume; Wessels and Schoemann, 1988
South Africa	Clarens Sandstone	0.02	Quartz content of bagworm larvae faeces; Wessels and Wessels, 1991
Israel	Jerusalem Limestone	5	Danin and Caneva 1990
Mendips, England	Limestone	112	MEM; Trudgill, 1985
Hawaii	Recent lava	0.51–8.4	Weathering crust thicknesses; Jackson and Keller, 1970
Lichens absent			
Australia	Sandstone	0.7	Debris collection, in rock shelters; Sullivan and Hughes, 1983
England	Sandstone	33–110	Photographic method; Pentecost, 1991
Mendips, England	Subsoil limestone	3–5	MEM; Trudgill, 1985
Hawaii	Recent lava	0–<0.16	Weathering crust thicknesses; Jackson and Keller, 1970

continually provides gaps for recolonization. A similar situation is recorded by Woolhouse *et al.* (1985) on a vertical gneiss rockface.

At a larger scale, it is of geomorphological importance to relate processes to form, and thus we need information on the variability of lichen weathering across landforms. Several studies report on the distribution of different lichen communities and the controls on their occurrence. McCarthy (1983), for example, describes the very different lichen communities found on different parts of a limestone pavement (such as clints, grikes and kamenitzas). On granite kopjes in Zimbabwe, Scott (1967) found that lichen distribution was primarily controlled by water availability, with secondary influences of nutrient status and rock texture. In a detailed study of the distribution of lichens across a rock pool in sandstone from the Transvaal, South Africa, Wessels and Büdel (1989) found light and water availability to be the major controlling factors. John and Dale (1991) studied lichen community pattern at three scales on a sandstone rock slide in the Canadian Rocky Mountains and found that the major factors controlling differences between individual rock faces were light, temperature and water availability as well as surface stability (controlled in this case by the degree of weathering). In the Negev desert, Danin and Garty (1983) tried to relate such controls to the variable occurrence of specific lichen weathering features (e.g. pits and grooves), but there have been few attempts to relate lichen activity to larger landforms.

In the rest of this paper we describe and assess the geomorphological significance of lichen weathering activity in a ruiniform sandstone landscape in the Cedarberg Mountains, Cape Province, South Africa, drawing on the work we have reviewed above.

STUDY AREA

The Cedarberg Mountains, which reach an elevation of just over 2000 m, are in the northern Cape Province (Figure 7.1). Composed largely of Cape System Sandstone of Palaeozoic age, plus some exposures of Upper Shales, the Cedarbergs form an anticlinal ridge. The sandstone (previously described by Rogers, 1905; du Toit, 1926; and Houghton, 1969) is a quartzarenite, containing medium to coarse quartz grains and very few feldspars. Dominantly white in colour and hard, ferrous compounds commonly become oxidised in the weathered surficial zone producing brown staining. The mountain slopes are strewn with large rocky outcrops (Figure 7.2) as well as large areas of gently dipping bare rock slopes. The sandstone is well-jointed and clearly bedded. Unequal weathering along joints and over surfaces produces a highly irregular topography. As du Toit (1926, p. 178) puts it:

> On weathered faces the rocks are very prone to become pitted and the most fantastic outlines may be developed, more particularly in the huge blocks which strew the slopes of the ranges; at the same time the surface takes on a grey tint due to the growth of lichen.

The climate here is Mediterranean with a winter rainfall regime. The vegetation is sclerophyllous Mediterranean scrub, often thorny, with *Euphorbia* and *Pachypodia* species. There are scattered small termite mounds, and bare rock surfaces are strewn with small animal droppings (possibly *Hyrax*).

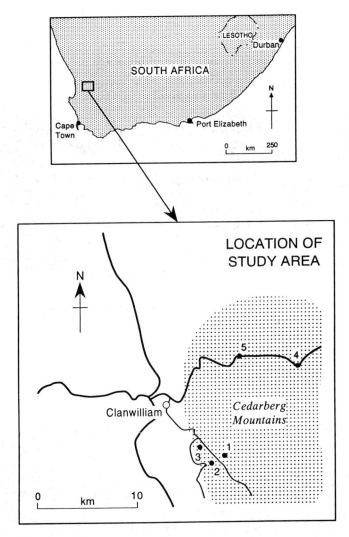

Figure 7.1 Location map of the Cedarbergs and sample sites

The dissected relief here, with many fantastically carved bare rock towers, may be called 'ruiniform' following the usage of Mainguet (1972). She described a range of sandstone landscapes from the Sahel with highly dissected surfaces which she called 'le modelé ruiniforme', in areas such as Hombori, Mali and Kiffa in Mauritania. Similar relief has also been identified in Australia in areas such as the Ruined City, Arnhem Land and the Bungle Bungles, East Kimberley, Western Australia. Jennings (1983) describes the former as a tower karst formed in silica-cemented quartz sandstone, and Young (1986, 1987 and 1988) has provided detailed descriptions of the latter, which is also developed in quartz-rich sandstones. The scale of relief in the Cedarberg Mountain ruiniform landscape is much smaller than in some of these

Figure 7.2   Cape System Sandstone forming ruiniform relief in the Cedarberg Mountains

Figure 7.3   A flat surface covered by lichens, including *Caloplaca cinnabarina*

Figure 7.4 A dissected surface covered by a range of different lichens

examples, with towers commonly less than 20 m high whereas Young (1987) describes towers of 100 m or more in the Bungle Bungles.

In the Cedarbergs individual towers themselves contain a highly variable micro-relief, with flat surfaces, etched bedding planes, pans and pitting all common (Figures 7.3 and 7.4). Many of these features look like karren, and may indeed be produced by dissolution of quartz (as found by Chalcraft and Pye, 1984, for similar forms in a humid tropical quartzite landscape at Roraima, southeast Venezuela). Pitting developed on sandstone tors in the Polish Carpathian Mountains has, however, been ascribed to the combined action of freeze–thaw and intense chemical activity (Alexandrowicz, 1989). Work from other parts of South Africa suggests that weathering basins in sandstone are initiated by mechanical processes, and enlarged by biochemical processes involving cyanobacteria and lichens (Cooks and Pretorius, 1987). Polygonal cracking also occurs on many flat, horizontal, usually lichen-covered, faces in the Cedarbergs. Such cracking has been observed in, amongst other areas, the Hawkesbury Sandstone, NSW, Australia, and Fontainebleau, France (Branagan, 1983; Robinson and Williams, 1989). The origin of polygonal cracking is under debate, but seems to occur in siliceous sandstones with case-hardened surfaces.

MATERIALS AND METHODS

Five towers, from 3 to 20 m high, were sampled at random at locations shown in Figure 7.1. On each tower, approximately five sub-sites characterised by distinctive micro-relief were studied in more detail. In all, 21 sub-sites, usually about 1 × 1 m in area, were studied. In each site qualitative observations were taken of relief (i.e. classified as smooth; rippled; dissected; smooth with polygonal cracks; pitted; flaked or stepped). On the flatter sites, lichen cover (in terms of percentage of surface covered by lichens of all species) was estimated visually, and at all sites the dominant colour(s) of the lichen mosaic recorded. These very simple field observations are shown in Table 7.3. At each sub-site at least one sample of rock and the covering lichens was taken for subsequent laboratory analysis. As far as possible, these samples were chosen to be representative of the sub-site. Back in the laboratory, these samples were observed with a hand-lens and dissecting microscope (up to × 30 magnification). First, the top surface was examined, and then the samples were fractured in order to inspect a cross-section through the lichen : rock interface. The percentage cover of all lichen species was estimated, and in cross-section the depth and colour of any 'weathered' or 'altered' zone recorded as well as the presence of any chasmolithic microorganisms (often green patches towards the base of the altered zone (see Figure 7.5). The lichens present were identified whenever possible.

Table 7.3 Field survey results of micro-relief, lichen colours and percentage cover

Sub-site no.[a]	Micro-relief	Lichen colours[b]	Lichen cover[c] (%)	Rock sample
1a	smooth	red/black	90	r1
1b	smooth	black/grey	70	r2
1c	smooth	black/grey	95	r3
1d	smooth	red/black		r4
1e	smooth	red/black		r5a,b
2b	smooth	red/grey		r7
3a	smooth	red/grey	80	r11a,b
3b	smooth	grey		r12
3e	smooth	red/grey		r15
4a	smooth	red/grey	90	r16
2c	smooth+pc[d]	red/grey	95	r14
2d	smooth+pc	red/grey	50	r9
3d	smooth+pc	red/grey	95	r8a,b
5c	smooth+pc	brown/grey	90	r20
1c	dissected	green/grey		r3
1e	dissected	grey/black		r5
2e	dissected	red/grey		r10
5a	dissected	varied		r18a,b
3c	dissected	varied		r13
3e	dissected	grey/green		
5b	dissected	black/grey/red		r19
5d	dissected	black/grey		r21
2a	rippled	grey		r6
4b	rippled	varied		r17
5e	stepped	red/grey		r22

[a]Thus 1a = site 1, sub-site a.
[b]Colours noted are the dominant colours of the rock surface lichens.
[c]Lichen covers are given as percentages of the rock surface covered by all lichen species present.
[d]pc = polygonal cracking.

a)

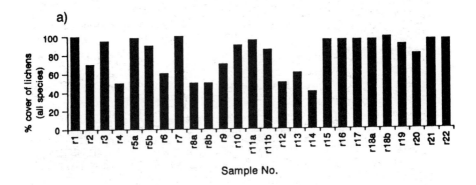

b)

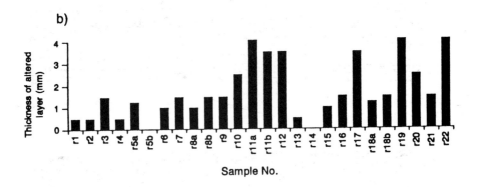

c)

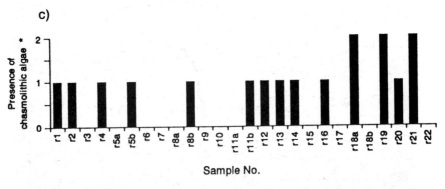

*0 = absent, 1 = present, 2 = abundant

Figure 7.5 Summary of rock sample observations. (a) % of rock sample surface covered by lichens of all species. (b) Thickness of altered zone in mm. (c) Presence of chasmolithic algae at base of altered zone

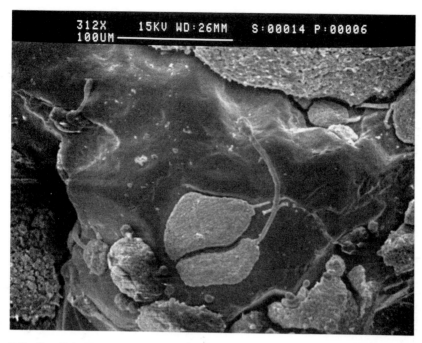

Figure 7.6 Small thalli of *Caloplaca cinnabarina* colonising a quartz grain. SEM photograph with scale bar of 100 μm

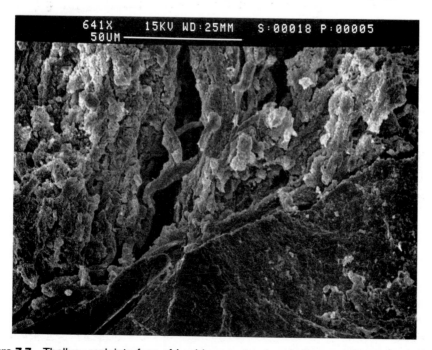

Figure 7.7 Thallus : rock interface of *Lecidea* sp. showing shallow etching on quartz surface by fungal hyphae (in centre of picture). SEM photograph with scale bar of 50 μm

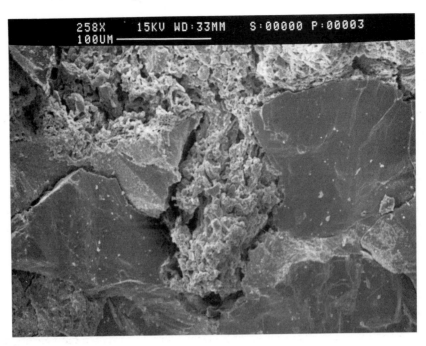

Figure 7.8   Rhizine of *Parmelia* sp. penetrating *c.* 200 μm into quartz grains. SEM photograph with scale bar of 100 μm

Finally, a smaller sample for SEM observations (*c.* 1 × 1 cm) was taken from each of the 26 rock samples (although six were damaged or unsuitable). Examples are shown in Figures 7.6–7.8. The lichens present on each of these were identified, and observations made where possible on thallus thickness. Sometimes, the thallus : rock interface was too rough to permit such measurements. Observations at a magnitude of *c.* × 1000 were made of the lichen : rock interface to identify any obvious lichen weathering features (such as etching or rhizine penetration). Four surface samples were also prepared with lichen cover removed by H_2O_2 to help clarify the nature of lichen imprints.

RESULTS

Results of the field survey in Table 7.3 show the diversity of relief at the tens of centimetres to metre scale (from flat to highly dissected) and lichen cover (often over 90% on flat, sunny surfaces) on the towers studied. A clear zonation of lichen communities with relief is evident, as flat surfaces and ridges are colonised by a mixture of red and grey coloured species (dominated by *Caloplaca cinnabarina*, *Buellia* cf. *stellulata*, *Acarospora fuscata* and *Lecidea* species) whilst more shaded pit sides and vertical faces host a diverse assemblage including a range of grey and greenish species (including *Buellia verruculosa*) and mosses. The periodically wetted floors of many pits are largely bare or sometimes covered with a sparse algal community.

Lichen community development does not seem to show any clear variation with micro-relief (at millimetre and centimetre scales). Thus, particular lichens are not associated with any specific surface features such as fruiting body pits or grooves. Observations from the rock samples collected show that a discoloured zone, usually between 1 and 5 mm thick, is present beneath many of the lichens (Figure 7.5b). This zone is usually white in colour, although sometimes there is a very dark band at the base. There does not seem to be any obvious correlation between the thickness of this zone and either the lichen species on top, the position of the sample, or the colour of the rock. At this scale, lichen cover on the samples collected ranges between 40 and 100% (Figure 7.5a). Chasmolithic microorganisms (probably cyanobacteria) are found on approximately one half of the samples, mostly in small patches, except on three samples from site 5 which have more extensive chasmolithic layers (Figure 7.5c). These three samples were all relatively dark-coloured, iron-rich rocks, whereas most of the other samples were much paler and dominated by quartz. Presence of chasmoliths does not correlate with a thick discoloured zone.

Dissecting microscope observations show that most lichen thalli were quite thin, mainly around 200 μm (range 50 to 750 μm), but very closely attached to the rock surface. Small thalli were seen to be colonising bare surface quartz grains (Figure 7.6) and in many cases seemed to be creating shallow depressions within the quartz grains. Chasmolithic microorganisms appeared to be located entirely in the spaces between quartz grains; there was no evidence that they were penetrating the grains in any way.

SEM investigations tell us more about the nature of the rock, and the effects that lichens are having on it. Most of the samples are formed of quartz grains with relatively little cement. The grains are held together by silica overgrowths which are interlocking and sometimes show evidence of etching and dissolution. Within the discoloured zone (white area) there is evidence that cement has been removed or replaced by silica. At the base of many lichen thalli there is clear evidence of etching of quartz grain and overgrowth surfaces (Figure 7.7). In some cases parts of the thallus extend down in 'rootlet' forms to about 150 μm into the rock between quartz grains (Figure 7.8). However, it is often difficult to interpret the lichen : rock interface. Samples treated with H_2O_2 show some clear, shallow depressions ranging from 5 to 100 μm in diameter which are assumed to be produced by lichen thalli, but there are no fruiting body pits or fungal boreholes. Figure 7.9 summarises the SEM observations from 20 samples with a range of different lichens. *Buellia* cf. *stellulata* and *Lecida* species were the most commonly observed species, and both showed much evidence of etching at the base of their thalli. 'Rootlet' penetration was observed occasionally under these two species, and also under *Parmelia* species.

Together, these observations at different scales suggest that on the Cedarberg Mountains sandstone ruiniform landscape the extensive, thin cover of epilithic crustose lichens is etching the quartz grains and removing cement from the rock on which it is growing. Any lichen weathering activity is concentrated within a very thin surface zone and does not result in any characteristic microscale landforms. More research is needed to show whether differential lichen weathering activity has contributed to the development of the varied relief at the tens of centimetres to metre scale.

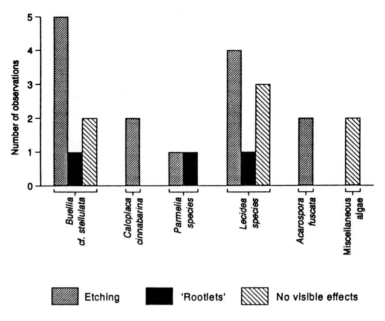

Figure 7.9   Summary of SEM observations

DISCUSSION

Our results can be compared both with other observations of lichen weathering activity, and also with explanations of the development of ruiniform sandstone landscapes. Most observations of lichen weathering on sandstone and other quartzitic rocks have focused on the action of endolithic species (e.g. Wessels and Schoemann, 1988; Cooks and Otto, 1990). These studies, especially those carried out on the Clarens Sandstone in South Africa, show relatively deep penetration of endoliths (up to 5.2 mm). The action of endolithic lichens is relatively clear, as quartz grains become detached from the rock and bound into the thallus, where they are etched and may subsequently be removed on death, or grazing, of the lichen. In Beacon Sandstone exposures in dry valleys in the Antarctic, Friedmann (1982) records extensive cryptoendolithic lichens growing entirely under the surface of the rock, often forming a layer some 10 mm thick. These lichens seem to aid exfoliation of the rock surface. In Antarctica these cryptoendoliths are found in areas too hostile for epiliths to grow, although epiliths are found elsewhere on Antarctic sandstone outcrops (Hale, 1987). In most places, however, epiliths and endoliths occur together (as recorded by Cooks and Otto, 1990) or epiliths alone are found (as found in this study). It is clear from our results that, in agreement with Danin and Garty (1983), epilithic lichens appear to leave no clear traces of their existence on the rock surface in terms of micro landforms. They do, however, play a role in rock weathering more generally, by etching grains and removing cementing material in the surface layer, and thus promoting surface lowering. Our work suggests that, unlike endolithic species, it is difficult to discriminate between the weathering action of the different epilithic species in our study; all seem to be involved to some degree and in similar ways.

Young (1986, 1987 and 1988) has written much on the development of similar ruiniform landscapes in Australia, and sees the presence of case hardening, coupled with long-term dissolution of quartz, as being the crucial factors producing these distinctive towers. In the Bungle Bungles he notes that most rock surfaces are covered with an algal layer, which he thinks may play a role binding and hardening the surface layer. Our results from the Cedarbergs indicate that similar processes may be operating here, with epilithic lichen cover etching quartz grains, but at the same time (until their death or removal by grazers) providing an overall protective cover a few hundreds of micrometres thick.

One important question for the interpretation of these landscapes is the nature and role of the altered zone. We have identified a largely white zone approximately 1–5 mm thick under many of the surface lichen growths. Other workers have referred to a 'bleached zone' a few millimetres thick. Wessels and Büdel (1989) found a bleached zone underneath the lichen *Peltula obscurans* on Clarens Sandstone in the Transvaal, South Africa. In a very different environment, Friedmann (1982) found a clear white zone 2–4 mm thick (with a 1 mm thick black zone on top, and a thin green layer below) just under the surface of Beacon Sandstone exposures in Antarctica. These zones were found to be all part of cryptoendolithic lichen growths. Similar altered zones have also been recognised as 'weathering rinds' on many exposures of sandstone. Colman (1981) reviews some studies of weathering rind thickness on a range of rock types, and they have been used as a relative dating technique, but there is no consideration of their genesis. Finally, other workers have recognised similar altered zones as 'case hardening'. Thus, Mustoe (1982) finds that Chuckanut Arkose exposures in Washington State commonly have a hardened layer (millimetres to centimetres thick), dark grey or reddish in colour, compared with light-coloured rock below. In comparison, Conca and Rossman (1982) find Aztec Sandstone from Nevada to have a case-hardened crust 0.5 to 5 mm thick (average 1.5 mm). This crust is lighter in colour than the parent rock, and is enriched in calcitic cement and kaolinite. In the Bungle Bungles the case-hardened layer is usually less than 1 cm thick, and characterised by the infilling of voids with kaolinite and amorphous clays (Young, 1987). Case hardening is generally seen to be a characteristic feature of arid regions, often occurring in association with tafoni although most authors do not comment on the presence of lichens. As noted earlier in this paper, case hardening may also be a characteristic part of polygonal weathering development.

There seems to be much scope for a reasoned comparison of these different altered zones, their genesis and importance for landform development. The role of lichens and other organisms in their development has been hinted at from many areas, including our observations here, but certainly requires more detailed study. Further investigations are also required on the weathering impact of combined epilithic and endolithic lichen communities, and the factors which control the distribution patterns of these two growth forms. Finally, it is of geomorphological interest to understand the role of lichens at different scales. As de Boer (1992) discusses, building on the seminal work of Schumm and Lichty (1965) and others, geomorphological systems are part of a nested hierarchy, and processes dominant at one scale may take on a very different role at other scales. Table 7.4 is a first attempt to collate lichen

Table 7.4 Lichen weathering and geomorphology at different scales

Scale	Landforms	Processes	Controlling factors
μm	Fungal boreholes and etching under lichen thalli	Chemical attack by lichen acids, carbon dioxide	Crystallography Lichen species Micro-environment
mm	Fruiting body pits Grooves Case hardening and altered zones	Chemical attack by lichens + grazing by snails etc. Chemical attack by lichens + inorganic chemical weathering	Microscale rock variability Aspect Micro-climate Lichen community dynamics
cm	Weathering basins	Biochemical attack + various inorganic weathering processes	Jointing and bedding planes Slope angle Lichen and algal zonation
m+	Ruiniform relief	Weathering and erosion	Case hardening Long-term etching of quartz Rock characteristics Climate and climatic change

influences on weathering at different scales using the literature reviewed above and the Cedarberg Mountains observations presented in this paper. More research is needed to confirm or refute some of the possible linkages suggested in Table 7.4.

ACKNOWLEDGEMENTS

We are very grateful to Amy and Andrew Goudie for their invaluable assistance in the field, and Anne Murray for her help in sample preparation, and an anonymous referee for helpful comments on an earlier draft.

REFERENCES

Alexandrowicz, Z. (1989). Evolution of weathering pits on sandstone tors in the Polish Carpathians. *Zeitschrift für Geomorphologie*, NF **33**, 275–289.

Armesto, J. J. and Contreras, L. C. (1981). Saxicolous lichen communities: non-equilibrium systems. *American Naturalist*, **118**, 597–603.

Ascaso, C., Galvan, J. and Rodriguez-Pascal, C. (1982). The weathering of calcareous rocks by lichens. *Pedobiologia*, **24**, 219–229.

Bachmann, E. (1890). The relation between calcicolous lichens and their substratum. *Berichte der deutsches Botanische Gesellschaft*, **22**, 101–104.

Branagan, D. F. (1983). Tesellated pavements. In Young, R. W. and Nanson, G. C. (eds), *Aspects of Australian Sandstone Landscapes*, Australian and New Zealand Geomorphology Group Special Publication No. 1, pp. 11–20.

Brightman, F. H. and Seaward, M. R. D. (1977). Lichens of man-made substrates. In Seaward, M. R. D. (ed.), *Lichen Ecology*, Academic Press, London.

Chalcraft, D. and Pye, K. (1984). Humid tropical weathering of quartzite in southeastern Venezuela. *Zeitschrift für Geomorphologie* NF **28**, 321–322.

Colman, S. M. (1981). Rock weathering rates as functions of time. *Quaternary Research*, **15**, 250–264.

Conca, J. L. and Rossman, G. R. (1982). Case hardening of sandstone. *Geology*, **10**, 520–523.

Cooks, J. and Otto, E. (1990). The weathering effects of the lichen *Lecidea* aff. *sarcogynoides* (Koerb.) on Magaliesburg quartzite. *Earth Surface Processes and Landforms*, **15**, 491–500.

Cooks, J. and Pretorius, J. R. (1987). Weathering basins in the Clarens formation sandstone, South Africa. *South African Journal of Geology*, **90**, 147–154.

Danin, A. and Caneva, G. (1990). Deterioration of limestone walls in Jerusalem and marble monuments in Rome caused by cyanobacteria and cyanophilous lichens. *International Biodeterioration*, **26**, 397–417.

Danin, A. and Garty, J. (1983). Distribution of cyanobacteria and lichens on hillsides of the Negev Highlands and their impact on biogenic weathering. *Zeitschrift für Geomorphologie* NF **27**, 423–444.

de Boer, D. H. (1992). Hierarchies and spatial scale in process geomorphology: a review. *Geomorphology*, **4**, 303–318.

del Monte, M. (1991). Trajan's column: Lichens don't live here any more. *Endeavour*, **15**, 96–93.

du Toit, A. L. (1926). *The Geology of South Africa*. Oliver and Boyd, Edinburgh.

Friedmann, E. I. (1982). Endolithic microorganisms in the Antarctic cold desert. *Science*, **215**, 1045–1053.

Geikie, A. (1893). *Text-book of Geology*, 3rd edition. Macmillan, London.

Giacobini, S., Nugari, M. P., Micheli, M. P., Mazzone, B. and Seaward, M. R. D. (1986). Lichenology and the conservation of ancient monuments: an interdisciplinary study. *Biodeterioration*, **6**, 386–292.

Hale, M. E. (1987). Epilithic lichens in the Beacon sandstone formation, Victoria Land, Antarctica. *Lichenologist*, **19**, 2269–2287.

Hallbauer, D. K. and Jahns, H. M. (1977). Attack of lichens on quartzite rock surfaces. *Lichenologist*, **9**, 119–122.

Harvey, J. H. and King, D. G. (1971). Winchester College stained glass. *Archeologia*, **103**, 149–177.

Houghton, S. H. (1969). *Geological History of South Africa*. Geological Society of South Africa, Cape Town.

Jackson, T. A. and Keller, W. D. (1970). A comparative study of the role of lichens and 'inorganic' processes in the chemical weathering of recent Hawaiian lava flows. *American Journal of Science*, **269**, 446–466.

Jennings, J. N. (1983). Sandstone pseudokarst or karst? In Young, R. W. and Nanson, G. C. (eds), *Aspects of Australian Sandstone Landscapes*, Australian and New Zealand Geomorphology Group Special Publication No. 1, pp. 21–30.

John, E. and Dale, M. R. T. (1991). Determinants of spatial pattern in saxicolous distribution and community development. *Lichenologist*, **23**, 227–236.

Jones, B. and Kahle, C. F. (1985). Lichen and algae: agents of biodiagenesis in karst breccia from Grand Cayman Island. *Bulletin of Canadian Petroleum Geology*, **33**, 446–461.

Jones, D., Wilson, M. J. and McHardy, W. J. (1981). Lichen weathering of rock-forming minerals; application of scanning electron microscopy and microprobe analysis. *Journal of Microscopy*, **124**, 95–105.

Kershaw, K. A. (1983). The thermal operating environment of a lichen. *Lichenologist*, **15**, 191–207.

Klappa, C. F. (1979). Lichen stromatolites: criterion for subaerial exposure and a mechanism for the formation of laminar calcretes (caliche). *Journal of Sedimentary Petrology*, **49**, 387–400.

Lawrey, J. D. (1991). 'Biotic interactions in lichen community development: a review. *Lichenologist*, **23**, 205–214.

Mainguet, M. (1972). *Le modelé de Gres*, 2 volumes. Institute Géographique National, Paris.

McCarthy, P. M. (1983). The composition of some calcicolous lichen communities in the Burren, western Ireland. *Lichenologist*, **15**, 231–248.

Mellor, E. (1921). L'action mécanique des lichens dans la dététioration des vitraux d'église. *Comptes rendus des séances de société de biologie*, **85**, 634–635.

Mustoe, G. E. (1982). The origin of honeycomb weathering. *Geological Society of America, Bulletin*, **93**, 108–115.

Muxart, T. and Blanc, P. (1979). Contribution a l'étude de l'alteration differentielle de la calcite et de la dolomite dans les dolomies sous l'action des lichens. Premières observations au microscope optique et au MEB. *Proceedings, International Symposium on Karst Erosion, Union Internationale de Spéleologie*, pp. 165–174.

Pentecost, A. (1980). Aspects of competition in saxicolous lichen communities. *Lichenologist*, **12**, 135–144.

Pentecost, A. (1991). The weathering rates of some sandstone cliffs, central Weald, England. *Earth Surface Processes and Landforms*, **16**, 83–92.

Robinson, D. A. and Williams, R. B. G. (1989). Polygonal cracking of sandstone at Fontaine-bleau, France. *Zeitschrift für Geomorphologie*, NF **33**, 59–72.

Rogers, A. W. (1905). *An Introduction to the Geology of Cape Colony*. Longmans, Green and Co., London.

Schumm, S. A. and Lichty, R. W. (1965). Time, space and causality in geomorphology. *American Journal of Science*, **263**, 110–119.

Scott, G. D. (1967). Studies of the lichen symbiosis. 3. The water relations of lichens on granite kopjes in central Africa. *Lichenologist*, **3**, 368–385.

Shachak, M., Jones, C. G. and Granot, Y. (1987). Herbivory in rocks and the weathering of a desert. *Science*, **236**, 1098–1099.

Sollas, W. J. (1880). On the action of a lichen on a limestone. *Report, British Association for the Advancement of Science*, p. 586.

Sullivan, M. E. and Hughes, P. J. (1983). The geoarcheology of the Sydney basin sandstones. In Young, R. W. and Nansen, G. C. (eds), *Aspects of Australian Sandstone Landscapes*, Australian and New Zealand Geomorphology Group Special Publication No. 1, pp. 120–126.

Syers, J. K. and Iskandar, I. K. (1973). Pedogenetic significance of lichens. In Ahmadjian, V. and Hale, M. E. (eds), *The Lichens*, Academic Press, New York, pp. 225–245.

Trudgill, S. T. (1985). *Limestone Geomorphology*. Longman, London.

Viles, H. A. (1987). A quantitative scanning electron microscope study of evidence for lichen weathering of limestone, Mendip Hills, Somerset. *Earth Surface Processes and Landforms*, **12**, 467–473.

Wessels, D. C. J. and Büdel, B. (1989). A rock pool lichen community in northern Transvaal, South Africa: composition and distribution patterns. *Lichenologist*, **21**, 259–278.

Wessels, D. C. J. and Schoemann, B. (1988). Mechanism and rate of weathering of Clarens Sandstone by an endolithic lichen. *South African Journal of Science*, **84**, 274–277.

Wessels, D. C. J. and Wessels, L. A. (1991). Erosion of biogenically weathered Clarens Sandstone by lichenophagous bagworm larvae (Lepidoptera; Psychidae). *Lichenologist*, **23**, 283–292.

Wilson, M. J. and Jones, D. (1983). Lichen weathering of minerals: implications for pedogenesis. In Wilson, R. C. L. (ed.), *Residual Deposits: Surface Related Weathering Processes and Materials*, Geological Society Special Publication No. 11, Blackwell Scientific, Oxford, pp. 5–12.

Woolhouse, M. E. J., Harmsen, R. and Fahrig, L. (1985). On succession in a saxicolous lichen community. *Lichenologist*, **17**, 167–172.

Young, R. W. (1986). Tower karst in sandstone: Bungle Bungle massif, north western Australia. *Zeitschrift für Geomorphologie*, NF **30**, 189–202.

Young, R. W. (1987). Sandstone landforms of the tropical East Kimberley region, north-western Australia. *Journal of Geology*, **95**, 205–218.

Young, R. W. (1988). Quartz etching and sandstone karst: Examples from the East Kimberleys, Northwestern Australia. *Zeitschrift für Geomorphologie*, NF **32**, 409–423.

Section 2

WEATHERING OF BUILDING STONES

8 Processes and Rates of Urban Limestone Weathering

R. J. INKPEN
University of Portsmouth, UK

R. U. COOKE
University of York, UK

and

H. A. VILES
Oxford University, UK

ABSTRACT

Recent concern over humanly-induced environmental change has highlighted the potential importance of geomorphology to identifying and understanding the weathering of historic and contemporary buildings in polluted environments. This paper outlines the basic physical and chemical processes of limestone weathering in urban areas and the temporal and spatial weathering trends produced based on research carried out in Oxford, London and other parts of southern Britain. Some problems of monitoring and interpreting these trends are considered and some suggestions for future geomorphological work put forward.

INTRODUCTION

The distinct nature of limestone weathering in urban environments was recognised as early as 1880 by Geikie, who identified faster weathering of gravestones in Edinburgh compared with surrounding rural areas. More recently, concern has grown across Europe over the economic and cultural costs of the decay of historic monuments and this has renewed interest in urban weathering. The Cathedrals Advisory Commission (now the Cathedrals Fabric Commission) estimates, for example, that £24.3 million are needed over the next three years to maintain and repair church buildings, a substantial proportion of which is needed to replace or protect stone. Mansfield (1988) estimated the cost of cleaning buildings in the UK to be about £78 million a year.

Traditional geomorphological skills have an important role to play within this context. Identification of weathering forms, measurement of weathering rates, past and present, and the investigation of the relations between forms and processes help to identify trends and causes of urban weathering. Similarly, the ability of the geomorphologist to interpret weathering as a complex system helps to develop

Rock Weathering and Landform Evolution. Edited by D. A. Robinson and R. B. G. Williams
© 1994 John Wiley & Sons Ltd.

critiques of current predictive damage functions based on simple physicochemical weathering models.

This paper outlines the urban weathering processes considered to be important, and then reviews some of the potential for geomorphological studies of urban weathering in calibrating temporal and spatial weathering rates and their potential use for criticising two recent simple, predictive physicochemical weathering models.

URBAN WEATHERING PROCESSES

'Normal' rainfall, a weak carbonic acid, has a pH of about 5.6. Pollutants, in particular sulphur dioxide, can reduce this pH to 4.5 or even as low as 3.5. Incorporation of sulphur dioxide as sulphate into precipitation provide the wet depositional route for pollutants to the stone surface. Pollutants may also be deposited by the dry depositional route, which includes both dry-fall deposition of particulate sulphate and the direct gas/solid reaction between sulphur trioxide and limestone (sulphation) (Figure 8.1). The chemical reactions associated with these processes are presented in Table 8.1. Most work on weathering in polluted urban environments has focused on the impact of sulphurous gases on accelerating limestone weathering.

WEATHERING RATES

Weathering processes have operated in urban areas at different rates through time, depending on the level of atmospheric pollution. Weathering rates would be expected to reflect changes in urban pollution levels. (Figure 8.2). There should also be a spatial difference in weathering rates between areas with high pollution levels, such as urban areas, and less polluted areas, such as rural areas. The magnitude of the urban/rural weathering differences should reflect the magnitude of their relative pollution levels (Figure 8.3).

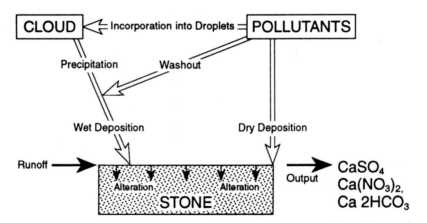

Figure 8.1 Depositional routes for atmospheric pollution. Modified from Inkpen (1989)

Table 8.1 Reactions associated with dissolution and sulphation

---

Atmosphere

Carbon dioxide
$$CO_2\ (g) \rightleftharpoons CO_2(aq)$$
$$CO_2(aq) + H_2O \rightleftharpoons H_2CO_3(aq)$$
$$H_wCO_3(aq) \rightleftharpoons H^+HCO_3^-(aq)$$
Reversible dissociation

Sulphur dioxide
$$SO_2(g) \rightleftharpoons SO_2(aq)$$
$$SO_2(aq) + H_2O(l){}^{1/2}O_2(aq) \rightarrow 2H^+SO_4^-(aq)$$
$$HO + SO_2 \rightleftharpoons HSO_3$$
$$HSO_3 + O_2 \rightleftharpoons H_2O + SO_3$$
$$SO_3(g) + H_2O(l) \rightleftharpoons 2H^+ + SO_3^-$$
$$SO_2(g) \rightleftharpoons SO_3(g)$$

Atmosphere/stone

Carbonic acid
$$Ca^{2+}CO_3^{2-}(aq) + H^+HCO_3^- \rightleftharpoons Ca^{2+}2(HCO_3^-)(aq) + H_2O(l) + C,2(aq)$$

Sulphuric acid
$$Ca^{2+}CO_3^{2-}(aq) + 2H^+ + SO_4^{2-}(aq) \rightarrow Ca^{2+}SO_4^{2-}(aq) + CO_2(g) + ,2O(l)$$

Sulphation
Basic reaction (Livingston, 1985)
$$SO_3(g) + CaCO_3(s) + H_2O(g) \rightarrow CaSO_4.\ 2H_2O(s) + CO_2(g)$$
First stage: $CaCO_3(s) + SO_2(g) \rightarrow CaSO_3(s)$
Second stage: $CaSO_3(s) + \tfrac{1}{2} O_2(g) \rightarrow CaSO_4(s)$
Overall reaction
$$CaCO_3(s) + \tfrac{1}{2} O_2(g) \rightarrow CaSO_4(s) + CO_2(g)$$

Sources: Skoulikidis and Charalambous (1981), Amoroso and Fassina (1983), Jaynes (1985), Livingston (1985).
g = aseous
aq = aqueous
l = liquid
s = solid

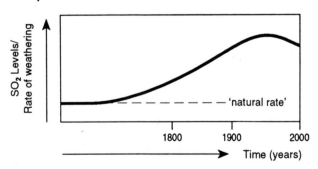

Figure 8.2 Temporal variations in weathering rates and sulphur dioxide levels. Modified from Cooke, 1989

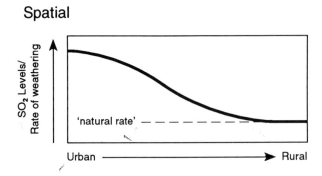

Figure 8.3 Spatial variation in weathering rates and sulphur dioxide levels. Modified from Cooke, 1989

Historical Patterns

Gravestones provide a datable rock surface which can be used to analyse the patterns of historical weathering experienced in urban areas. Despite some difficulties with methods employed (Livingston and Baer, 1988; Inkpen and Cooke, in preparation), they are still extensively used (Table 8.2). The lead lettering on the marble grave-stones is usually used to indicate the level of the original surface of the gravestone. In production the lead lettering and marble surface would have originally been polished flush with each other. Upon weathering the marble surface retreats at a significantly faster rate than the lead lettering, which hardly weathers at all.

Kupper and Pissart (1974) used a similar technique of measuring loss relative to a reference surface to calibrate the urban/rural contrast in weathering on marble

Table 8.2 Gravestone weathering rates (in μm yr^{-1})

Location	Rate	Reference
Durham, UK		Attewell and Taylor (1988)
Meadowfield (urban/industrial)	10	
Quebec Village	2	
(rural)		
Jarrow (urban/industrial)	9.5	
Philadelphia, USA		Feddema and Meierding (1987)
City centre	35	
20 km from city centre	5	
Eastern Australia		Dragovich (1986)
Industrial	2.5	
Urban	1.7	
Liege, Belgium		Kupper and Pissart (1974)
Industrial	3.65	
Urban	2.39	
Rural	1.20	

The measurement method was the lead lettering index on marble gravestones, except for Kupper and Pissart (1974) who used siliceous veins in the marble as the reference level.

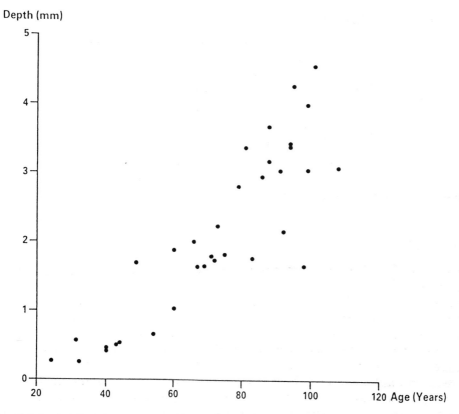

Figure 8.4 Relationship between depth of loss of marble and age of gravestone in a Swansea graveyard

gravestones in Liege, Belgium./Their data suggested that weathering rates in industrial areas were three times those in rural areas over the same time period. Similar studies in Australia (Dragovich, 1986) and the UK (Attewell and Taylor, 1988) also reveal urban/rural contrasts, although their magnitude varies (Table 8.2).

A study of gravestone weathering in Swansea (Cooke *et al.*, in preparation) reveals that weathering also varies over time. Both the amount of weathering and the rate of weathering increased significantly with the age of the gravestone (Figure 8.4, Table 8.3).

Such conclusions need to be qualified, as accurate atmospheric pollution records do not start in the UK until the early 20th century and reconstruction of pollution environments in the 19th century has to rely on other sources. These sources, such as coal records, are at best indirect and at worst misleading. They can often only illustrate trends in pollution levels and they are of little value in revealing extreme episodes that may be of crucial importance to weathering.

The initial increase in weathering rate over time, illustrated in the figure, may represent a prolonged adjustment period for the gravestones. The period of adjustment for gravestones is unknown and the higher levels of loss for the older gravestones may reflect higher pollution levels in the past. Alternatively, the amount of

Table 8.3   Regression of Swansea weathering data—amount versus age

	Coefficient	S.D.	*t*-ratio	*p*-value
Constant	−1.245	0.338	−3.69	0.001
Gradient	0.0466	0.00447	10.41	0.000
Amount lost $= 46~\mu\text{m}~\text{yr}^{-1}$				

material lost may not follow a simple linear relationship with pollution level; there may be critical levels of pollution which create inherent thresholds in the weathering system. For example, above a certain ambient pollution level, the marble can no longer absorb sulphur dioxide. Any additional pollution in the atmosphere will not be taken up by the marble and so will have little impact on the amount of material lost by the gravestone. Uncertainties as to how the gravestone weathering system evolves and responds to short-term pollution changes, and the inaccuracies in pollution reconstruction, limit the development of predictive models and the investigation of process from these data.

Current Spatial Patterns

Recent studies of urban weathering have used continuous, standardised pollution monitoring records of daily sulphur dioxide levels to quantify the relationship between weathering and pollution. Honeybourne and Price (1977, cited in Jaynes, 1985) suspended limestone tablets at two sites, Westminster (London) and Garston (a rural site near Watford), where pollution levels were monitored. Every six months the dry weight losses of the tablets were compared. Tablets at Westminster lost twice as much weight as those at Garston over the 10 years of exposure, although the sulphur dioxide levels at Westminster were five to seven times higher than those at Garston.

More recent studies have also used dry weight loss of stone tablets to assess weathering rates (Table 8.4). Jaynes (1985) in southeast England, and Inkpen (1989) in southern Britain, assessed the urban/rural differences and the relative contribution of wet and dry depositional processes to weathering loss. Some Portland Stone and Monks Park tablets were exposed to both wet and dry depositional processes (exposed tablets) and some only to dry depositional processes (sheltered tablets). Accumulated weathering products in the sheltered tablets were removed by leaching, after which the tablets were dried and weighed. The weight loss of the sheltered tablets between their initial dry weight and that upon leaching represented the weathering due to dry depositional processes. This amount could then be used to estimate the proportion of weight loss of exposed tablets accounted for by dry depositional processes. This proportion was roughly a third (Table 8.5), although the value seemed to vary with stone type.

The urban/rural differences in weathering rates varied with the urban/rural pair of sites considered. Recent falls in urban pollution levels resulted in a convergence of urban/rural pollution levels over time (Figure 8.5). That differences in weathering rates persist at all may point to other factors within urban environments, such as temperature differences or differences in frost conditions, as being important for the

Table 8.4 Urban and rural weathering rates (in $\mu m \ yr^{-1}$)

Location	Rate	Method	Stone	Reference
York[a]	11	Tablet carousel, 4 years' exposure	Portland Stone limestone	Butlin et al., 1992
London[a]	12			
Wells[b]	6			
Lincoln[b]	9			
Washington DC, USA	9	Runoff slabs	Limestone	Baedecker et al., 1992
	54	Test briquettes		
Westminster	8.5	Tablets, 10 years' exposure	Portland Stone limestone	Honeybourne and Price, 1977
Garston	3.3			
London (Victoria)	15.8	Tablet carousel, 2 years' exposure	Portland Stone limestone	Jaynes, 1985
Gravesend	7.3			
Wormdale (rural)	5.8			
Avonmouth (industrial)	19.3	Tablet carousel, 2 years' exposure	Portland Stone limestone	Inkpen, 1989
Rotherham (urban)	7.6			
Lincoln (urban)	8.2			
Hartgill (rural)	5.5			
Glentham (rural)	6.6			

All sites in UK unless otherwise stated
[a]Relatively polluted
[b]Relatively unpolluted

Table 8.5 Proportion of exposed tablet weight change due to sheltered processes

Site	Portland Stone	Monks Park
Avonmouth	0.41	0.28
Weston	0.39	0.23
Walsall	0.34	0.20
Lincoln	0.29	0.23
Glentham	0.32	0.14
Rotherham	0.34	0.20
Harthill	0.49	0.33
Mean	0.37	0.23
Standard deviation	0.07	0.06

Source: Inkpen (1989).

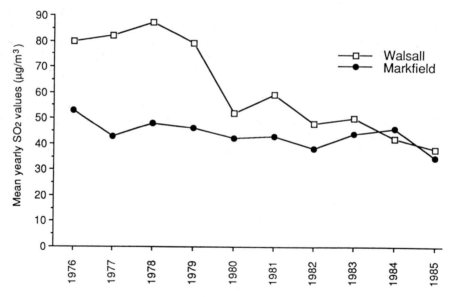

Figure 8.5 Example of convergence of urban/rural sulphur dioxide levels. From Inkpen (1989)

maintenance of the weathering differences. Moreover, rural pollution levels may not have been static; they may have increased as a result of an increase in long-distance transport of pollutants leading to a regional rise in pollution levels.

PHYSICOCHEMICAL MODELS OF WEATHERING

Some recent studies have developed empirical and theoretical models of weathering to predict weathering rates and distinguish between the contribution of carbonic and sulphuric acid dissolution and sulphation. Both Livingston (1992) and Webb *et al.* (1992) used the idea encapsulated in the expression:

$$\text{Stone loss} = \begin{array}{c} \text{'Natural'} \\ \text{Bicarbonate} \\ \text{Solubility} \end{array} + \begin{array}{c} \text{Rain Acid} \\ \text{Neutralization} \end{array} + \begin{array}{c} \text{Sulphate} \\ \text{from dry} \\ \text{deposition} \end{array}$$

Both used ideas of electro-neutrality (balance of ionic neutrality of rainfall and runoff) and carbonate equilibria to distinguish the relative contribution of each process. Livingston (1992) identified a chemical signature for each process. For dissolution by carbonic acid there is a relative increase in bicarbonate concentration of the runoff. For dissolution by 'acid rain' there is a reduction in H^+ concentration or an increase in pH of the runoff. For dry deposition, there is a net gain in the sulphate concentration in the runoff. Each process produces an increase in Ca^{2+} concentration which can be apportioned to each process using the ideas of electro-neutrality. Livingston expressed the relative contribution of each process on a triaxial graph (Figure 8.6).

Webb *et al.* (1992) used the same idea of ionic balance to derive a loss equation or damage function (Table 8.6). The terms in the equation represent the rate of dry deposition onto a surface area, A (deposition velocity), the carbonic dissolution of limestone, and the neutralisation of acid rainfall, all integrated over the period of sample exposure. Instead of dealing with chemical change event by event they were more concerned with using a chemically sound model to predict stone loss over a given period. To do this they operationalise their model terms to some measured environmental parameters.

They exposed sets of Portland Stone limestone cube (40 mm) at a number of sites in the UK where SO_2 and NO_x were continuously monitored. The cubes were dry weighed before and after exposure. Exposure periods ranged from 70 to 1065 days, with most sites having samples exposed for between 365 and 730 days. Their model

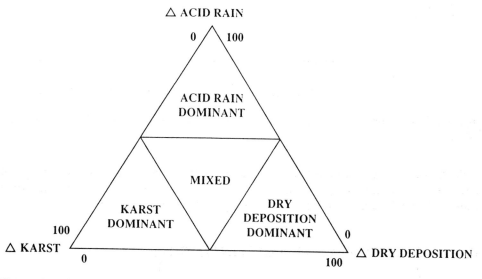

Figure 8.6 Triaxial plot of relative contribution of weathering processes. From Livingston (1992)

Table 8.6 National Power physicochemical model of rain-washed limestone dissolution

Stone cube loss (g m^{-2} day^{-1}) =

$$0.00037_{V_d} C_{SO2} + \frac{5.6 \times 10^{-11}}{[H^1]_r} \frac{\Sigma(FR - Evap)}{D} + 10[H^+]_i \frac{\Sigma FR}{D}$$

where,

V_d is the deposition velocity in mm s^{-1}
C_{so2} is the mean SO$_2$ concentration (ppb) during the exposure period
F is the rainfall interception factor
R is the rainfall in mm
$Evap$ is the evaporation from the surface during the exposure period in mm
D is the duration of the exposure period in days
$[H^+]_i$ is the volume-weighted hydrogen ion concentration of the rainfall in moles L^{-1}
$[H^+]_r$ is the volume-weighted hydrogen ion concentration of the runoff in moles L^{-1}

Source: Webb *et al.* (1992). Reproduced by permission.

suggests that at all sites dissolution by carbonic acid was the dominant process (50% of the loss in some cases) with dissolution by 'acid rain', i.e. rainfall of dilute sulphuric acid with a pH of below 5.6, being the least significant, accounting for between 5 and 10% of weathering loss. Sulphation and other dry deposition processes accounted for the remainder of the weathering.

Livingston (1992) used his model to describe runoff data from Washington DC (USA) and Mechelen (Belgium) cathedrals, concluding that for each, dry deposition was a major process of material loss, whereas acid rain again accounted for only 5–10% of the Ca^{2+} concentration in runoff, well within the error limits of the model.

MODEL LIMITATIONS

Potentially, physicochemical models have uses for conservation and in policy formulation and practices by identifying areas susceptible to damage by different pollution-related weathering processes. However, application of such models is subject to some important limitations. First, it is very difficult to compare the results of gravestones, tablets and runoff studies. Not only are the methods and weathering indices different, but each technique assesses weathering over different time periods. Rates derived from gravestone studies are not sensitive to the short-term changes in environmental variables that are likely to be of greater importance in tablet and runoff studies.

Secondly, the physicochemical models were derived from weathering data from very small runoff areas where contact time between rainfall and stone was short, and where the kinetics of chemical reactions may be important rather than equilibrium conditions. In addition, the samples used had relatively smooth surfaces, unlike the complex interconnected surfaces of buildings. On building surfaces irregularities may be areas where storage of weathering products can occur. The importance of storage of weathering products in the stone surface was recognised by both studies, but considered not to be significant over their period of study. This storage capacity may vary with the area considered, as large areas have the capacity to absorb runoff from some, brief rain events resulting in only a few, usually larger rain events producing measurable runoff. Such rain events may have a surplus of sulphate in runoff, but it may be the weathering product of acid dissolution in the past which has not yet been removed from the weathering system. Another large event relatively soon after the first will not remove as much because the store of weathered material is empty. This gives an erroneous impression of the weathering effectiveness of the second rain event. All these factors combined emphasise the importance of understanding the weathering history of a surface.

Thirdly, the physicochemical models assume that all weathering proceeds by chemical alteration of the limestone and its removal by runoff. Other forms of weathering are not considered. On exposed surfaces cracking, splitting and flaking, for example by frost, could lead to sudden increases in weathering rates over short periods. The relative importance of these forms of weathering and their relationships to changes in atmospheric pollution are relatively understudied.

Fourthly, areas sheltered from rainfall and runoff are not considered by the models. Such areas are likely to develop deposition or gypsum crusts not considered in the model, excluding a major weathering form from the calculation of weathering losses.

Lastly, the persistence of the urban/rural weathering difference in the face of declining urban sulphur dioxide levels and, according to the models, a low acid rain effect, may result from the continued presence of other distinctions between the two environments. Differences such as those in temperature profiles or rainfall characteristics between the two could contribute to the maintenance of the contrast.

CONCLUSION

Spatial and temporal variation of urban/rural weathering rates have been analysed using gravestones, stone tablets and runoff studies. Relations between pollution trends and weathering rates at a specific site are difficult to analyse because of problems in reconstructing pollution histories and understanding the short-term responses of the gravestone weathering system. Physicochemical models of weathering and recent stone tablet studies point to the importance of dry depositional processes for weathering and a relatively minor role for acid rain.

Results from runoff studies are often directly used to predict weathering behaviour of buildings. The complex weathering history of a surface or catchment is often not considered, which could significantly affect how results of such short-term studies are interpreted. Likewise, the contribution of other weathering forms to weathering rates and other variables to the persistence of the urban/rural contrast is not clear.

REFERENCES

Amoros, G. G. and Fassina, V. (1983). Stone decay and conservation. *Materials Science Monographs*, **11**, Elsevier, Amsterdam.
Attewell, P. B. and Taylor, D. (1988). Time-dependent atmospheric degradation of building

stone in a polluting environment. In Marinos, P. G. and Koukis, G. C. (eds), *Engineering Geology of Ancient Works, Monuments and Historical Sites*, Balkema, Rotterdam, pp. 739–753.

Baedecker, P. A., Reddy, M. M., Reimann, K. J. and Sciammarella, C. A. (1992). Effects of acidic deposition on the erosion of carbonate stone—experimental results from the US National Acid Precipitation Assessment program (NAPAP). *Atmospheric Environment*, **26B**, 147–158.

Butlin, R. N., Coote, A. T, Devenish, M., Hughes, I. F. C., Huhhens, C. M., Irwin, J. G., Lloyd, G. O., Massey, S. W., Webb, A. H. and Yates, T. J. S. (1992). A four year study of stone decay in different pollution climates in the UK. In Rodrigues, J. D., Henriques, F. and Jeremias, F. T. (eds), *Proceedings Seventh International Congress on Deterioration and Conservation of Stone, Lisbon, Portugal 15–18 June, 1992*, pp. 345–354.

Cooke, R. U., (1989). Geomorphological contributions to acid rain research: studies of stone weathering, *Geographical Journal*, **159**, 361–366.

Cooke, R. U., Inkpen, R. J. and Wiggs, G. F. S. (in preparation). Changing rates of weathering in polluted atmospheres of Britain.

Dragovich, D. (1986). Weathering rates of marble in urban environments, eastern Australia. *Zeitschrift für Geomorphologie*, **30**, 203–214.

Feddema, J. J. and Meierding, T. C. (1987). Marble weathering and air pollution in Philadelphia. *Atmospheric Environment*, **21**, 143–157.

Geikie, A. (1880). Rock weathering as illustrated in Edinburgh church yards. *Proceedings of the Royal Society of Edinburgh*, **10**, 518–532.

Honeybourne, D. B. and Price, C. A. (1977). *Air Pollution and the Decay of Limestones*. Building Research Establishment N117/77, Building Research Establishment, Garston.

Inkpen, R. J. (1989). Stone weathering and atmospheric pollution in a transect across Southern Britain. Unpublished Ph.D. Thesis, University of London.

Inkpen, R. J. and Cooke, R. U. (in preparation). Comparison of methods for measuring surface recession on gravestones.

Jaynes, S. M. (1985). Studies of building stone weathering in south-east England. Unpublished Ph.D. thesis, University of London.

Kupper, M. and Pissart, A. (1974). Vitesse d'érosion en Belgique des calcaires d'âge primarie exposés à l'air ou soumis à l'action de l'eau courante. In *Geomorphologische Prozesse und Prozesskombinationen in der Gegenwart unter verschiedenen Klimabedingen. Report of the Commission on Present-day Geomorphological Processes (International Geographical Union. Abhandlungen der Akademie der Wissenschaften in Gottingen*, **29**, 39–50.

Livingston, R. A. (1985). The role of nitrogen oxides in the deterioration of carbonate stone. In *Fifth International Congress on Deterioration and Conservation of Stone, Lausanne 25–27 September 1985*, pp. 509–516.

Livingston, R. A. (1992). Graphical methods for examining the effects of acid rain and sulfur dioxide on carbonate stones. In Rodrigues, J. D., Henriques, F. and Jeremias, F. T. (eds), *Proceedings Seventh International Congress on Deterioration and Conservation of Stone, Lisbon, Portugal, 15–18 June, 1992*, pp. 375–386.

Livingston, R. A. and Baer, N. S. (1988). The use of tombstones in the investigation of the deterioration of stone monuments. In Marinos, P. G. and Koukis, G. C. (eds), *Engineering Geology of Ancient Works, Monuments and Historical Sites*, Balkema, Rotterdam, pp. 859–867.

Mansfield, T. A. (1988). A cost–benefit analysis of building soiling in the UK. *The Soiling of Buildings and Deterioration in Urban Areas*. Middlesex Polytechnic, Centre for Urban Pollution Research.

Skoulikidis, Th. and Charalambous, D. (1981). Mechanism of sulphation by atmospheric SO_2 of the limestones and marbles of the ancient monuments and statues. *British Corrosion Journal*, **16**, 70–77.

Webb, A. H., Bawden, R. J., Busby, A. K. and Hopkins, J. N. (1992). Studies on the effect of air pollution on limestone degradation in Great Britain. *Atmospheric Environment*, **26B**, 165–182.

9 Breakdown Patterns of Quartz Sandstone in a Polluted Urban Environment, Belfast, Northern Ireland

B. J. SMITH, R. W. MAGEE and W. B. WHALLEY
The Queen's University of Belfast, UK

ABSTRACT

Research into building stone decay has increased markedly in recent years, but has been biased towards calcareous building stones and major, culturally important sites.

The city of Belfast contains a multitude of buildings constructed from local sandstone and presents an opportunity to study the weathering of non-calcareous stones in an urban environment that experiences local, high concentrations of anthropogenic pollution. Increases in pollution date to the construction of many of these buildings during the 19th century, thereby giving them a relatively uncomplicated and comparable pollution history.

The study uses a geomorphological approach to examine pollution uptake by sandstone at a number of buildings in Belfast's urban environment. Preliminary results indicate that decay is largely the result of salt weathering processes, and features such as granular disintegration, flaking and scaling are abundant. The mechanisms operate in response to inputs from a variety of sources—in addition to those from atmospheric pollution—including marine aerosols, groundwater and road de-icing. Anthropogenic pollution, however, is integral to the development of other common decay phenomena such as black crusts.

INTRODUCTION

Studies of stone decay, especially in the British Isles, have understandably concentrated upon limestones, and towns and cities that are now subject to effective atmospheric pollution control. However, by a concentration upon limestones, which are the favoured stones for major buildings in southern England, interest has centred upon gradual solution loss. Also, studies in urban environments that are much cleaner than 30 years ago make it difficult to relate present decay to current atmospheric chemistry and to identify previous conditions under which inherited decay and pollution uptake occurred. As one goes north and west within the British Isles, limestones become less important as a building stone and quartz sandstones are used extensively in northern England, Scotland and Northern Ireland. Of necessity, decay patterns experienced by these stones differ dramatically from those of limestones, and physical disruption is dominant. Some urban locations in these regions still continue

Rock Weathering and Landform Evolution. Edited by D. A. Robinson and R. B. G. Williams
© 1994 John Wiley & Sons Ltd.

(B)

(A)

Figure 9.1 (A) Thin black crust on 'Scottish' sandstone, Sinclair Seamen's Church, docks area of Belfast. Note extensive development of crust even over rainwashed areas and initial preservation of carved detail. (B) Badly decayed Scrabo Sandstone, Sinclair Seamen's Church. Decay has progressed to the point where the structural integrity of the tower is threatened and falling stone poses a safety hazard. Note earlier attempts (1950s) to stabilise the stone surface with a cement-based rendering

to experience elevated levels of atmospheric pollution/decay relationships that are frequently invoked as part of the so-called 'memory effect' when seeking to explain continued decay in cleaner environments.

In this study a variety of sandstone buildings in central Belfast are examined. The buildings experience a wide range of decay features from black crusts (Figure 9.1A) to severe physical disruption (Figure 9.1B) and are used to develop a typology of active sandstone decay features. These features are related to high levels of atmospheric pollution, especially of sulphur dioxide in winter, and are used to investigate decay sequences and their implications for assessment of long-term decay rates based upon short-term observations.

LOCATION

Pollution Background

Belfast lies to the northwest of mainland Europe and could be expected, therefore, to experience only low levels of background pollution. Nevertheless, the city is currently derogued from EC regulations on atmospheric pollution and experiences prolonged periods of poor air quality, particularly in winter. Explanations for high levels of local pollution include the late introduction of clean air legislation (1964), which has yet to be fully enacted across the city. Terrorist activity has also made enforcement of regulations difficult, and the location of the city in a deep valley makes it prone to temperature inversions, which trap pollutants under anticyclonic conditions during winter. Despite these factors, Belfast did see a progressive decline in annual smoke and sulphur dioxide levels during the 1960s and 1970s. Since the mid-1980s, however, sulphur dioxide levels have begun to rise again (Figure 9.2). There is no definitive explanation for this, but the rise did coincide with the closing of the local gas undertaking and increased use of oil and solid fuel for domestic heating. It may also correspond to an increased frequency of winter anticyclones (Smith *et al.*, 1991a). Clearly, therefore, atmospheric conditions in Belfast have improved, but air quality remains consistently worse than most urban centres in, for example, England (Smith *et al.*, 1991a).

Buildings and Building Stones

Eight sandstone buildings in and around the centre of Belfast were surveyed in detailed together with a visual survey of other buildings exhibiting specific decay features. The principal stone type is 'Scrabo Sandstone', a local Triassic, quartzitic non-calcareous sandstone of variable composition which can contain numerous clay lenses. Other stones include Glasgow Freestone used for the Customs House and a range of other imported Scottish sandstones of unknown origin. On most of the Scrabo buildings the smooth ashlar (if present) or the ornamental dressings such as string courses, window mullions or quoins are also constructed of a different stone, often a red quartz sandstone, which may be of Scottish origin (Larmour, 1987). On one of the buildings, All Soul's Church, Scrabo has been coupled with Doulting Limestone used as a dressing. Many of these buildings date from the rapid

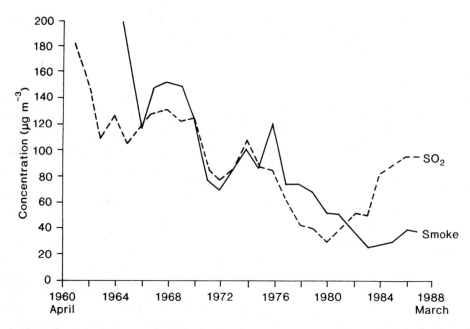

Figure 9.2 Atmospheric sulphur dioxide and smoke concentrations in south central Belfast. Data from Belfast City Council

expansion of Belfast in the mid-nineteenth century. Most, therefore, have an approximately coincident pollution history and, within the context of many studies of prominent historic buildings across Europe, their pollution history is relatively short and uncomplicated.

SANDSTONE DECAY FEATURES

There is now a comprehensive literature describing natural weathering of quartz sandstones—sufficient to suggest that there exists a range of features comparable in distinctiveness to the suites of weathering forms found on, for example, karstic limestones and granites (see Robinson and Williams, this volume, Chapter 22). Structure and mineralogy dictate that weathering is characterised by physical breakdown. On large outcrops this may be 'organised' to form phenomena such as polygonal cracking (e.g. Robinson and Williams, 1989; Williams and Robinson, 1989). Within individual joint blocks, breakdown is achieved primarily by granular disaggregation, flaking and scaling, although chemical alteration is not precluded. Silica dissolution is, for example, enhanced under saline conditions (Young, 1987), and the dissolution and reprecipitation of impurities from the stones or cements such as iron can produce case-hardened outer surfaces. Alternatively, the fixation of externally derived minerals, from wind-blown crusts for example, may lead to the formation of rock varnishes (Robinson and Williams, 1992). Equivalent features occur on quarried sandstones exposed to the combination of natural weathering environments and the anthropic additions and modifications experiences in polluted urban environments. What follows is a brief, illustrative review of some of the weathering features found on sandstone buildings in Belfast.

Black Crusts

Black gypsum crusts are extensive on the sandstone buildings of Belfast, although their occurrence on individual buildings is invariably patchy. Distribution, on a calcium-free substrate, is often related to a local calcium source. Crusts can occur, for example, below mortars (Figure 9.3A) or below limestone string courses (Figure 9.3B) from which either calcium carbonate or gypsum is washed in solution. An alternative gypsum source is directly from atmospheric deposition of particulate gypsum, blown on to stonework from adjacent surfaces, or from gypsum-containing particulate pollutants. In these circumstances crusts develop on surfaces protected from removal by rainwash, where they can develop into extensive encrustations (Figure 9.3c). Examination of crusts by scanning electron microscopy shows a wide range of constituents (Whalley et al., 1992). These include: inorganic airborne particulates (e.g. soil particles, street dusts, fly ash), organic airborne particles (e.g. plant remains, pollen), inorganic precipitates, perhaps produced in situ (e.g. gypsum) and organic growths (e.g. bacteria and fungae). In thin crusts (Figure 9.4A) crystalline gypsum may only occupy near-surface pores, but eventually (Figure 9.4B) gypsum extends over most protruding sand grains as a thin veneer. On both of these micrographs numerous carbonaceous fly ash particles derived from the combustion of both coal and oil can be seen adhering to stone and crust surfaces. As complete crusts develop, an open lattice-work of gypsum laths is formed, which enclose dust and fly ash particles (Figure 9.4C) as well as occasional halite crystals (Figure 9.4D) deposited possibly from marine aerosols.

The open structure of well-developed black crusts permits migration of gypsum into the underlying stone. Inward salt migration does not seem to be inhibited by, for example, reductions in porosity which might occur in calcareous stone either through calcite precipitation or conversion of grains to gypsum and the subsequent narrowing of pore entrances. The effects of this salt accumulation are seen when black crusts are breached (Figure 9.3D). Newly exposed stone can experience rapid, salt-induced breakdown, and it is possible that breaching of the crust may result from sub-crust pressures exerted by crystallised salt rather than through any external agency.

Case-hardening

Case-hardening, whereby material derived in solution from the stone interior is precipitated behind the exposed surface, is expected on limestones. It does, however, also occur on quartz sandstones in both natural outcrops (Robinson and Williams, this volume, Chapter 22) and on buildings. Invariably, this hardening includes the precipitation of iron oxides that are originally present either as coatings on quartz grains or as concretions within the sandstone. In Belfast, case-hardening of 1–2 mm depth only occurs on the more homogeneous Scottish sandstones. Although Scrabo Sandstone can be iron-stained in its quarried form, the clearly defined bedding and clay inclusions appear to preclude the formation of a uniform outer layer rich in reprecipitated iron oxide. Rupture of case-hardened layers can again lead to rapid mechanical breakdown of the sub-surface stone (Figure 9.5A) through granular

(A)

(B)

Figure 9.3   (A) Development of a gypsum black crust on Scrabo Sandstone below mortar joints, All Soul's Church.(From Webster 1992, reproduced by permission of The Robert Gordon University). (B) Development of a gypsum black crust on Scrabo Sandstone beneath a Doulting Limestone string course, All Soul's Church. (C) Growth of a gypsum/fly ash encrusta-

(C)

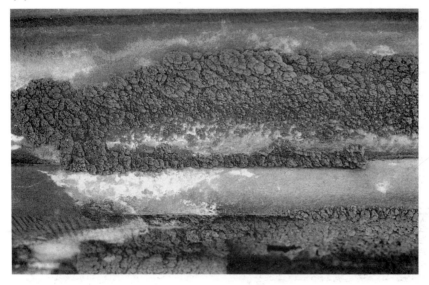

(D)

tion on a sheltered area of Scrabo Sandstone protected from rainwash, Sinclair Seamen's Church. (D) Rapid breakdown of Scrabo Sandstone by multiple flaking following breaching of a thin gypsum black crust, St Matthew's Church

(A)

(B)

Figure 9.4 Scanning electron micrographs of partial and complete gypsum crusts on Scrabo Sandstone from Sinclair Seamen's Church (white scale bars in μm). (A) Very limited crust development with gypsum crystals restricted to pores between quartz sand grains. Note the occurrence of perforated carbonaceous cenospheres produced from the combustion of oil-based fuels and smaller, complete spheres produced by coal combustion. Electron microprobe analyses show both types of fly ash to contain gypsum (Smith *et al.*, 1991a). (B) A more

(C)

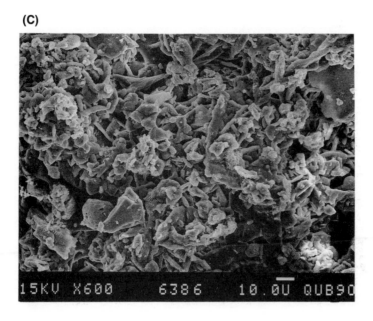

(D)

complete gypsum crust than (A), but still with irregular protrusions of quartz sand grains and numerous adhering fly ash particles. (C) and (D) Details of complete crusts comprising gypsum platelets incorporating fly ash and irregular mineral fragments (predominantly quartz) from street dusts or weathered debris. Note the open, porous structure of the crusts (C) and the cubic halite crystal (D).

(A)

(B)

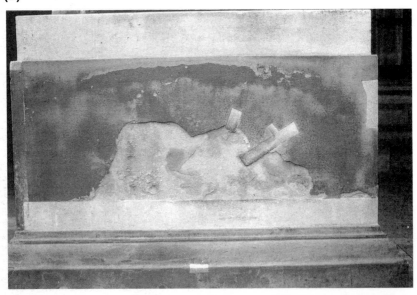

Figure 9.5 Iron-rich, case-hardened outer layer on a 'Scottish' sandstone, main building at Queen's University. Once the crust is breached, the sub-surface zone is rapidly exploited by granular disaggregation and flaking forming a hollow which further undermines the case-hardening. (B) Extensive contour scaling of a homogeneous sandstone; note earlier attempts

(C)

(D)

at piecemeal replacement of damaged zone. (C) Severe multiple flaking of Scrabo Sandstone, Sinclair Seamen's Church. (D) Widespread blistering of sandstone as a precursor to flaking, Lyttle's Warehouse

disaggregation and flaking (see later sections). Rapid breakdown could reflect structural weakening of the zone beneath the case-hardened layer due to loss of iron cement (Bluck, 1992). It may also derive from the long-term accumulation of salts by inward migration through a still-porous outer surface. Salt weathering beneath the case-hardened surface could feasibly generate widespread microfracturing through fatigue effects which would only be manifested once the structural support provided by the case-hardening is removed. Certainly, X-ray diffraction (XRD) analysis of the weathered debris (Smith *et al.*, 1991a) has identified major peaks for gypsum, but the structural condition of stonework beneath intact case-hardening has yet to be investigated.

Scaling

In the previous two sections, the catastrophic loss of surface scales has been described where, through black crust formation or case-hardening, a zone of structural weakness is created between the outer layer of a stone and the interior. Scaling of surface layers need not, however, be accompanied by obvious surface alteration (Figure 9.5B). Scales ($\geqslant 5$ mm) are evident on a range of buildings, although again their development appears inhibited on the heterogenous Scrabo Sandstone. These scales were described by McGreevy *et al.* (1983) who noted that, like the 'contour scales' first described by Schaffer (1932), their interior fracture surfaces frequently replicate surface detail. XRD and differential thermal analysis of efflorescences found on surfaces exposed by scaling show them to be predominantly gypsum (McGreevy *et al.*, 1983). Salt weathering simulation studies using Darney Stone (a Northumbrian Carboniferous sandstone used as a replacement in some Belfast buildings) have shown that scales can form when salts crystallise in a sub-surface zone related to a frequent wetting depth with dilute salt solutions (Smith and McGreevy, 1988). These experiments demonstrated that scales may break away from the stone in response to the cumulative stresses exerted by repeated expansion and contraction of microcrystalline salt that completely fills the pores. There is anecdotal evidence, however, that scaling may be prematurely triggered by, for example, a severe frost, although it should be noted that such frosts are relatively rare in 'maritime' Belfast.

Multiple Flaking

Once a contour scale, case-hardened layer or black crust has fallen away from a stone block, further breakdown can be very rapid. Frequently this is accomplished by multiple flaking. Each flake is characteristically less than 5 mm thick, less than approximately 10 cm^2 in area and frequently grouped with others as small 'booklets' (Figure 9.5G). As yet, the authors know of no definitive explanation for these flakes which are independent of and frequently normal to any bedding. Their formation, parallel to exposed surfaces, suggests that delimiting fractures are related to environmental cycles of heating/cooling and/or wetting/drying. It would seem to be special pleading to expect that each flake represents, for example, a particular wetting depth. Clearly, as with case-hardening, there is scope for further investigation. Although flaking is prevalent on newly exposed surfaces, it can also develop spontaneously without previous surface loss. In this instance its initial manifestation may be numerous fragile blisters (Figure 9.5D).

Granular Disaggregation

This is particularly evident in strongly bedded stones such as Scrabo Sandstone, where individual beds are exploited to produce a rock-meal of salt crystals and loose sand grains which accumulates at the base of walls (Figure 9.6A and B). Disaggregation of sandstones can be effectively achieved through salt crystallisation in pores near to exposed surfaces, which are then subject to heating/cooling, wetting/drying cycles (e.g. Smith *et al.*, 1987). Experiments and field observations have shown that clay-rich bands in Scrabo Sandstone are particularly susceptible to salt-induced breakdown (Smith and McGreevy, 1983). This may be a response to increased microporosity (<5 μm) in these bands which enhances salt weathering susceptibility (McGreevy and Smith, 1984). However, the presence of expanding-lattice clays (smectite) may also contribute to disruption when they expand on wetting (Bluck, 1992).

DISCUSSION: THE SPATIAL AND TEMPORAL DISTRIBUTION OF STONE BREAKDOWN

Overwhelmingly, the mechanical breakdown of quartz sandstones in Belfast is associated with salt concentration—primarily gypsum, but also limited amounts of halite. Simulation experiments have further established the effectiveness of salt weathering mechanisms in producing phenomena such as scaling and granular disaggregation. It is understandable, therefore, that breakdown principally occurs where salts are both available and retained, for example, near lime-based mortars or limestone decoration. Particulate pollutants (fly ash) are an alternative source of gypsum and their impact is concentrated in areas protected from rainwash (Figure 9.3C). There is some evidence, however, that in the current polluted environmental regime the supply and surface fixation of particulate pollution can exceed the rate of removal by rainwash. This is revealed in extensive, thin black crusts (Figure 9.1A and 9.4B) which mask large areas of many buildings. It is also common to see large, exposed expanses of buildings subject to active stone breakdown (Figure 9.7A). The rate of susceptibility to weathering on these walls is dictated by variations in stone characteristics (porosity, bedding, clay content, surface roughness) rather than variations in exposure.

Within each stone block, breakdown may also be spatially concentrated. Once protective outer surfaces are removed, rapid breakdown by granular disaggregation and/or flaking often produces a cavernous hollow or 'tafoni' (Figure 9.6C). Such features are common on natural exposures in salt-rich environments (coasts and deserts). Studies of these have shown that once formed, tafoni are protected from rainwash and become foci for salt accumulation and further salt weathering (Smith and McAlister, 1986). Such positive feedback is enhanced on buildings, where preferential decay of a particular block is guided back along mortared joints (Figure 9.6D).

Not all salts are, of course, derived directly or indirectly from atmospheric pollution or marine aerosols. Smith *et al.* (1991b) examined two particular sources: road de-

(A)

(B)

Figure 9.6 (A) and (B) Differential exploitation of bedding by granular disaggregation on Scrabo Sandstone, Sinclair Seamen's Church. Chalk stripes on (B) indicate selection for replacement during renovation. (C) Granular disintegration leading to the development of

(C)

(D)

cavernous hollows (tafoni) in a homogeneous sandstone. Gateposts at entrance to College Gardens. (D) Rapid retreat of Scrabo Sandstone through multiple flaking creating a 'hollow' protected from rainwash and salt removal, All Soul's Church

Figure 9.7 (A) Patchwork of black crusts and decay of Scrabo Sandstone, St Matthew's Church. (B) Blistering and efflorescence of salts derived from rising groundwater, Queen's University main building

icing salt and rising groundwater. In both instances subsequent weathering is concentrated near ground level and can be readily distinguished from more widespread damage related to atmospheric contaminants. The morphological expressions of decay are similar to those described in the previous sections, but road-salt-related breakdown shows a concentration of halite near ground level, and groundwater-related decay is frequently associated with extensive efflorescences during summer (Figure 9.7B).

The temporal pattern of breakdown is best described as 'sporadic'. Periods of relative inactivity, when surfaces are stabilised by the development of indurated crusts or layers, are succeeded by instantaneous loss as crusts are breached. Subsequently rapid breakdown may occur as weakened sub-surface layers are exploited. These sequences are illustrated in Figure 9.8, which shows a number of possible routes that breakdown can take. What must be stressed is that this diagram illustrates patterns of surface loss, and that, during periods of surface stability, sub-surface alteration and stress accumulation due to crystallised salt can still continue. It is this *cumulative* stress which may ultimately destroy surface indurations as internal strength thresholds are finally exceeded. Decay can also be triggered by external stresses including energetic stone cleaning or by extreme events such as a severe frost. From Figure 9.8 it can be seen that it is extremely difficult to predict long-term sandstone durability (denudation rate) from short-term observations, not least because future breakdown may take one of several possible routes. Future patterns of decay are also highly dependent upon future weathering environments, and can be independent of previous decay types. Furthermore, on any particular building individual blocks of stone, because of variations in exposure and/or structural and mineralogical differences, will occupy different positions within the weathering sequence at any one time (Figure 9.7A). Expressions of 'average rates of decay' based on short-term observations, therefore, become somewhat abstract concepts where breakdown is clearly episodic.

Salt accumulations found within the sandstones could conceivably be described as a 'memory effect', except that there is every evidence that they continue to accumulate at the present time. It is perhaps misleading, however, to isolate these memory effects as something peculiar that requires special explanation. The predominantly salt-related patterns of breakdown identified in the sandstones of Belfast are essentially a *threshold* phenomenon. There should be little expectation, therefore, that current weathering rates should correlate perfectly with present-day environmental conditions—outside of a coincidence between extreme events and short-lived episodes of rapid breakdown. In building stones that are prone to mechanical disruption and not subject to surface loss in solution, rates of decay will always reflect the cumulative effects of former weathering conditions. In the case of Belfast, widespread and severe sandstone decay demonstrates that, in polluted urban environments, accumulated stress can exceed strength thresholds over a relatively short timespan. The frequently rapid decay once these thresholds are exceeded highlights the difficulties of stabilising such stonework if, for example, outer black crusts are removed during cleaning. Because of this, complete stone replacements are often required during renovation work in the city. Finally, the current patterns of decay provide a useful insight into conditions which may have pertained prior to clean air legislation in other cities. If the supply of anthropic salts through atmospheric pollution were to be removed, rapid decay would still continue. A variety of natural

148

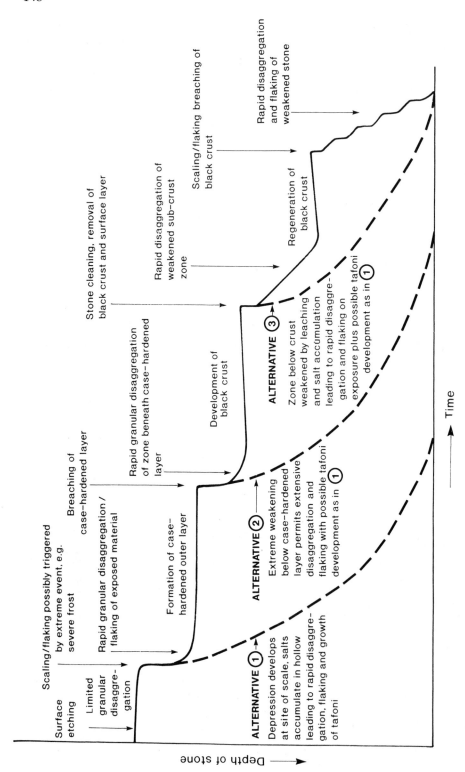

Figure 9.8 Schematic diagram illustrating possible pathways of sandstone decay. The sequences illustrated are not definitive in that, for example, not all stones develop case-hardened outer layers nor do all stones undergo energetic cleaning. It does, however, indicate the change of changes that can occur and emphasises the episodic nature of surface loss

salts would continue to be available (Smith *et al.*, 1991b) and there is now a considerable legacy of salts stored within stonework. Much of the sandstone has also been structurally weakened by earlier weathering. This weakened stone is now susceptible to a wide range of environmental stresses associated with heating/cooling and/or wetting/drying. Previous weathering has additionally created a range of morphological features, including cavernous hollows (tafoni), which create micro-environments amenable to salt concentration and continued decay.

ACKNOWLEDGEMENTS

The writers are grateful for the assistance provided by the Electron Microscope Unit of Queen's University and the support provided by the secretarial and technical staff of the School of Geosciences.

REFERENCES

Bluck, B. (1992). The composition and weathering of sandstone with relation to cleaning. In Webster, R. G. M. (ed.), *Stone Cleaning*, Donhead Publishing, London, pp. 125–127.

Larmour, P. (1987). *Belfast, An Illustrated Architectural Guide*. Friars Bush Press, Belfast.

McGreevy, J. P. and Smith, B. J. (1984). The possible role of clay minerals in salt weathering. *Catena*, **11**, 169–175.

McGreevy, J. P., Smith, B. J. and McAlister, J. J. (1983). Stone decay in an urban environment, examples from South Belfast. *Ulster Journal of Archaeology*, 1983, 167–171.

Robinson, D. A. and Williams, R. B. G. (1987). Surface crusting of sandstones in southern England and northern France. In Gardiner, V. (ed.), *International Geomorphology 1986*, Wiley, Chichester, pp. 623–635.

Robinson, D. A. and Williams, R. B. G. (1989). Polygonal cracking of sandstone at Fontaine-bleau, France. *Zeitschrift für Geomorphologie*, NF **33**, 59–72.

Robinson, D. A. and Williams, R. B. G. (1992). Sandstone weathering in the High Atlas, Morocco. *Zeitschrift für Geomorphologie*, NF **36**, 413–429.

Schaffer, R. J. (1972). *The Weathering of Natural Building Stones*. Building Research Establishment, Garston (reprint of 1932 edition).

Smith, B. J. and McAlister, J. J. (1986). Observations on the occurrence and origins of salt weathering phenomena near Lake Magadi, southern Kenya. *Zeitschrift für Geomorphologie*, NF **30**, 445–460.

Smith, B. J. and McGreevy, J. P. (1983). A simulation study of salt weathering in hot deserts. *Geografiska Annaler*, **65A**, 127–133.

Smith, B. J. and McGreevy, J. P. (1988). Contour scaling of a sandstone by salt weathering under simulated hot desert conditions. *Earth Surface Processes and Landforms*, **13**, 697–706.

Smith, B. J., McGreevy, J. P. and Whalley, W. B. (1987). Silt production by weathering of a sandstone under hot arid conditions: a simulation study. *Journal of Arid Environments*, **12**, 199–214.

Smith, B. J., Whalley, W. B. and Magee, R. W. (1991a). Background and local contributions to acidic deposition and their relative impact on building stone decay: a case study of Northern Ireland. In Longhurst, J. W. S. (ed.), *Acid Deposition: Origins, Impacts and Abatement Strategies*. Springer-Verlag, Berlin, pp. 241–266.

Smith, B. J., Whalley, W. B. and Magee, R. W. (1991b). Stone decay in a 'clean' environment: western Northern Ireland. In Baer, N. S., Sabbioni, C. and Sors, A. (eds), *Science, Technology and European Cultural Heritage*, Butterworth-Heinemann, Oxford, pp. 434–438.

Webster, R. G. M. (ed.), (1992). Stone Cleaning, Donhead Publishing, London, p. 229.

Whalley, W. B., Smith, B. J. and Magee, R. W. (1992). Effects of particulate air pollutants on materials: investigation of surface crust formation. In Webster, R. G. M. (ed.), *Stone Cleaning*, Donhead Publishing, London, pp. 27–34.

Williams, R. B. G. and Robinson, D. A. (1989). Origin and distribution of polygonal cracking of rock surfaces. *Geografiska Annaler*, **71A**, 145–159.

Young, A. R. M. (1987). Salt as an agent in the development of cavernous weathering. *Geology*, **15**, 962–966.

10 Spatial Variations in Intensity of Alveolar Weathering of a Dated Sandstone Structure in a Coastal Environment, Weston-super-Mare, UK

D. N. MOTTERSHEAD
Edge Hill College of Higher Education, Ormskirk, Lancs, UK

ABSTRACT

Marine structures of known age provide an opportunity to study rates of rock weathering under controlled conditions.

The sea walls of Weston-super-Mare, Avon, UK were completed in 1888 and capped by Forest of Dean Stone which now exhibits well developed alveolar weathering. Spatial variations in weathering intensity are revealed by mapping using a specifically designed scale.

Weathering intensity is shown to vary in relation to several variables—linear distance to HWM, height above MHWST, windward/leeward exposure, and azimuth. Marine salt penetration of the weathered rock is inferred from chloride determinations. The magnitude of salt deposition is assessed in relation to spatial variations in weathering intensity.

Cone indenter tests of rock strength permit estimates of the degree of weakening of the rock caused by weathering.

A maximum reduction of the initial surface of at least 110 mm is observed, indicating a minimum rate of weathering of 1 mm yr^{-1}. The confidence of Victorian engineers in the ability of Forest of Dean Stone to withstand marine weathering processes is shown to be misplaced.

INTRODUCTION

The sea walls at Weston-super-Mare, in the county of Avon, UK, were completed in 1888 (Anon, undated; Brown and Loosely, 1979; Beisley, 1988; Woodspring Museum, 1988) and are capped throughout their length with sandstone coping stones. These stones exhibit alveolar (honeycomb) weathering of varying intensity. The disposition of the sea walls, at varying distance from the shoreline, varying altitude above sea level, and with varying azimuth and exposure, permits variations in weathering intensity to be studied in relation to these geographical variables. The construction of the walls provides a baseline date for the development of the honeycomb features and therefore permits determination of the rate of weathering.

Rock Weathering and Landform Evolution. Edited by D. A. Robinson and R. B. G. Williams
© 1994 John Wiley & Sons Ltd.

Alveolar weathering consists of an array of small pits, commonly centimetres in depth and width, separated by an intricate network of narrow walls (Mustoe, 1982a). It is commonly regarded as a small-scale version of multiple tafoni. The terms honeycomb, stone lattice, stone lace and alveolar weathering have been used as synonyms (Selby, 1985; Trenhaile, 1987; Summerfield, 1991). The French equivalent appears to be 'maladie alveolaire' (Pauly, 1976).

Views on the formation of honeycomb have been summarised by Evans (1970), Mustoe (1982b), Smith (1982) and Trenhaile (1987). Recent studies have been carried out by Höllerman (1975), Kelletat (1980), McGreevy (1985), Smith and McAlister (1986) and Viles and Goudie (1992). Honeycomb is associated in particular with semi-arid and coastal environments, which are characterised both by a supply of salts and by wetting and drying conditions. In coastal environments it appears to have its strongest development above high water mark in the spray zone (Gill, 1973; Mustoe, 1982b; Trenhaile, 1987). The action of salts is the current preferred explanation of alveolisation, although debate exists as to whether these operate through salt crystallisation (McGreevy, 1985) or through hygroscopic changes associated with variations in humidity (Pauly, 1976).

Whatever the precise mechanism of weathering, the implication of the presence of salts, which in coastal environments are of marine origin, suggests that the rate of alveolisation will depend on availability of, and exposure to, a supply of salts. These in turn will be a function in part of atmospheric processes, such as wind strength and direction, and turbulence at the ocean surface, and in part on the geographical factors of location of the rocks in receipt of the salt supply. The nature of the weathering forms produced over a given period of time would also be related to these factors.

This paper investigates the effects of location factors on rock weathering by alveolisation in one particular coastal weathering environment. The aims of this paper are therefore to identify and document, by field mapping, spatial variation in intensity of alveolisation, and thereby to identify geographical factors responsible for that variation. The presence of marine salts indicative of salt action is determined, and related to geographical location and weathering intensity. Rock strength determinations permit estimates of the extent of rock weakening. Finally, the evidently rapid rate of alveolisation permits a minimum rate of weathering to be identified. It is not the intention of this paper to focus on the actual mechanism of weathering.

THE FIELD AREA

Weston-super-Mare lies some 100 km upstream of the mouth of the Bristol Channel. With a westerly aspect it is exposed to waves with a transatlantic fetch through a compass arc of 18°. Locally, however, there are considerable variations in both the plan configuration of the coast and the nature of the foreshore in front of the sea walls (Figure 10.1). These imply that there are substantial spatial variations in exposure to wage energy, marine spray and onshore winds, the agencies responsible for delivering marine salts onshore. At the southern end of the study area, in Weston Bay, the sea wall has a westerly aspect. The sea wall is fronted in this section, above high water mark (HWM, as defined by Mean High Water level 1975, indicated on the OS 1:1250 map) by a sandy foreshore which progressively narrows northwards.

At the north end of the bay, the wall swings round to a southerly aspect, and is protected in part from direct exposure to westwards by the local peninsula of Knightstone. High water mark lies at the sea wall whilst low tide exposes a sand beach. There appears to be a locus of concentration of marine energy delivered to the wall close to the junction of Knightstone Road and Royal Parade, opposite to the Technical College. This was illustrated in media coverage of storms in February 1990, and is evidenced by the damage to the sea wall and replacement of the original stone coping by artificial materials. The degree of exposure increases towards this

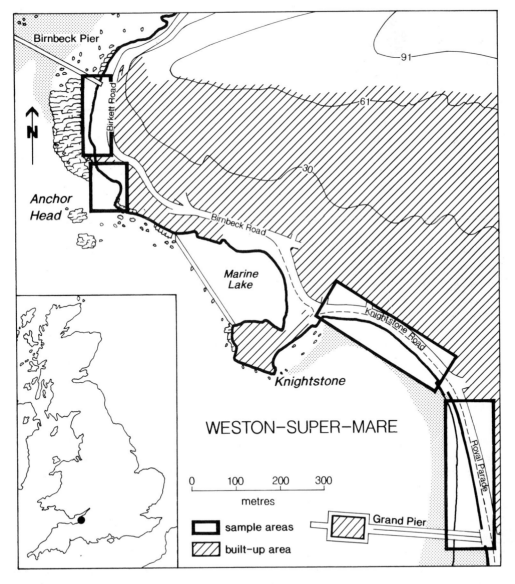

Figure 10.1   Weston-super-Mare, location map

zone from both directions. From the south the wall is protected by a progressively narrowing beach. From the west the wall emerges from the direct shelter of the Knightstone peninsula.

North of Knightstone the sea wall partially encircles the Marine Lake, where a causeway retains seawater at low tide, and the sea wall is fronted by a low level esplanade. North of Marine Lake, the sea wall trends around the headland of Anchor Head, formed by an outcrop of Carboniferous Limestone, to Anchor Cove. In this exposed section, the sea wall is fronted by a rocky foreshore; high water mark here lies against the sea wall, or not more than 5 m distant from it.

The tidal range in the field area exceeds 12 m and the level of the mean height of spring tides (MHWST) in 1991 was 6.2 m OD. The height of the top of the sea wall maintains a constant level of 8.7 m OD along the length of Royal Parade and Knightstone Road. Around Anchor Cove it ranges from 9.0 to 9.7 m OD; along the Birkett section it lies in the range 15.0 to 15.7 m OD. The geographical features of the sea walls are set out in Table 10.1.

Although the main body of the walls is composed of Carboniferous Limestone, the focus of interest in this study lies in the coping stones. The walls are capped by Forest of Dean Stone (Weston-super-Mare Gazette 13th Oct. 1883), a local facies of the Pennant Sandstone of Lower Carboniferous age from Gloucestershire (Warnes, 1926; Trotter, 1942; Welch and Trotter, 1961; Leary, 1986; Stone Industries, 1991). Forest of Dean Stone is a massive, fine-grained, slightly calcareous sandstone, buff to grey in colour. The matrix consists of quartz sand and orthoclase feldspar with small quantities of muscovite. The stone is bonded by silica with subsidiary cements of alkaline earth salts. Many grains are covered by a thin pellicle of iron oxide. SEM inspection reveals a significant interstitial mineral content.

The coping stones are 0.4 m wide and 0.2 m deep with length variable up to 2 m and commonly in the range 1.4–1.7 m. The original texture of the surface is hammer dressed, to create a rusticated finish. The surface exposed to weathering of newly laid stones would therefore be a surface formed by fracturing, and bearing hammer pick marks.

Throughout its century of existence the sea wall has suffered episodic damage and modification to its structure. It was breached by major storms in 1903, 1981 and 1990, and in several places the original sandstone coping stones have been replaced by artificial stone. Photographs available in Weston-super-Mare public library indicate that storms have dislodged some coping stones, which were subsequently replaced.

Table 10.1 Geographical variations in the sea walls sampled

	Elevation >MHWST	Exposure	Foreshore at HWM	Distance to HWM
Birkett	8.8–9.5 m	Exposed	Rocky	0–5 m
Knightstone W	2.5 m	Sheltered	Nil	0
Knightstone E	2.5 m	Exposed	Nil	0
Anchor	2.8–3.5 m	Exposed	Rocky	0–5 m
Royal Parade	2.5 m	Protected	Sand beach	10–50 m

The field analysis strongly suggests that they were replaced in their former positions. The section of wall skirting Marine Lake was partially protected by a canopied colonnade between 1930 and 1981.

Sites were therefore selected for study where (i) there is reasonable certainty that the wall bears coping stones as originally emplaced, and (ii) the conditions of exposure have not changed significantly since emplacement. These conditions appear to be met in the four sample sections indicated in Figure 10.1

THE FORMS OF ALVEOLAR WEATHERING

The forms of weathering present are variable, and appear to represent different degrees of development (Figures 10.2–10.5). Initially the rock may bear the indentations of the hammer marks, which tend to be linear in form with rectilinear corners, and are visible on some stones. Incipient weathering is indicated by the presence of isolated circular pits, conical or cylindrical in form, commonly <10 cm in diameter, and clearly distinct from the human working of the stone.

Forms which genuinely merit the term honeycomb are present when pits, commonly centimetres deep, are developed so close together as to be separated by a narrow wall only millimetres thick. Honeycomb thus defined may occupy an area only a few centimetres square, or may be so extensive as to cover the entire surface of a single stone.

Breakdown of honeycomb forms is indicated initially by a reduction in elevation of the walls separating the honeycomb cavities. Walls may become reduced to vestigial

Figure 10.2   Isolated circular pits (weathering grade 1)

Figure 10.3   Well-developed honeycomb (weathering grade 3). The alveoles have developed so as to be separated only by narrow well-defined walls

rounded humps which, at an advanced stage of honeycomb breakdown, may represent the only relief remaining. Some rock surfaces are totally smooth, usually freshly weathered in appearance, and lower than the initial surface.

These forms appear to represent the progress of honeycombing through a sequence of development and decay. The stripping of a honeycomb layer by denudation appears to have attracted only limited comment by previous authors. It is briefly described by Mottershead (1982) in respect of greenschist in south Devon, UK, and by Mustoe (1982b) in arkose in Washington State, USA, both in the spray zone of coastal environments. Mustoe provides a photographic illustration and the frontispiece of Levin (1990) displays a spectacular example. These examples demonstrate that the breakdown of the honeycomb layer appears to be a common phenomenon. This would appear to indicate that honeycomb is an inherently unstable form. The stripping of the honeycomb layer leads to the creation of a new rock surface substantially reduced in elevation in relation to the original surface. This implies a sequence of stages in the reduction of a rock surface by alveolisation, in which honeycomb represents just one stage in a transition.

The sequence outlined above, the stages of which are identifiable objectively on the basis of morphological typology, offers the opportunity to establish a scale of weathering intensity, which can then be employed as a mapping tool. Accordingly, a 10-point scale of weathering grades was developed (Table 10.2), combining the degree of weathering with its extent, which enables a weathering grade to be allocated to each individual coping stone. Weathering grade can then be mapped in the field on

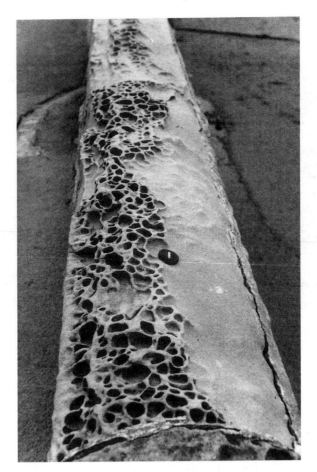

Figure 10.4 Shoreward (to the right)/landward contrast in weathering grade. Landward is honeycomb with some wall breakdown. Shoreward in the foreground is a completely stripped surface; midway along the coping stone is an area of vestigial reduced walls

Table 10.2 Classification of honeycomb weathering grades employed in this study

0	No visible weathering forms
1	Isolated circular pits
2	Pitting >50% of area
3	Honeycomb present
4	Honeycomb >50% of area
5	Honeycomb with some wall breakdown
6	Honeycomb partially stripped
7	Honeycomb stripping >50% of area
8	Only reduced walls remain
9	Surface completely stripped

Figure 10.5 Stripped honeycomb layer with vestigial wall remnants. The vertical slice of mortar at right, 11 cm deep, indicates the minimum initial elevation of the coping stone surface

a stone by stone basis. This quasi-qualitative scale can then be employed as a basis for quantitative analysis.

VARIATIONS IN WEATHERING INTENSITY

A total of 862 stones was individually mapped, in the four sample sections (Birkett, 205; Anchor, 109; Knightstone Road, 258; Royal Parade, 290). The mapped sections extend as far south as Grand Pier. The detailed mapping reveals substantial variability in weathering grade of adjacent stones. It is evident that separate stones may respond to a near-identical weathering environment with differing degrees of alteration. This is a feature reported orally by previous workers. In order to minimise this variability, the weathering grade values of consecutive groups of five stones were averaged. These average values were then mapped (Figures 10.6–10.9).

Comparison of the weathering grade values of different sections of the wall (Table 10.3) enable individual factors to be examined and their influence on weathering intensity to be analysed and identified, as follows.

Elevation

The effect of elevation above sea level is demonstrated by comparing wall sections of similar foreshore and exposure at Anchor Cove, 2.8–3.5 m above MHWST, and Birkett Road, 8.8–9.5 m above MHWST (Table 10.3). The non-parametric Mann-Whitney test is employed since the data are not of the interval type.

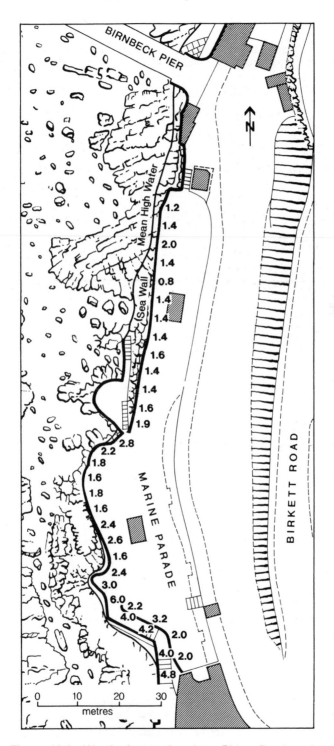

Figure 10.6 Weathering grade values, Birkett Road section

With $n_1, n_2, = 9{,}25$, $U = 2.0$; reject H_o at 0.001

Thus vertical proximity to sea level is a highly significant factor in determining weathering grade through the range of elevation represented in this test.

The exposed wall at Anchor Cove (2.8–3.5 m above MHWST) may also be compared with the exposed wall at the east end of Knightstone Road (Table 10.3), where HWM lies along the wall, and the elevation of the wall is similar (2.5 m above MHWST).

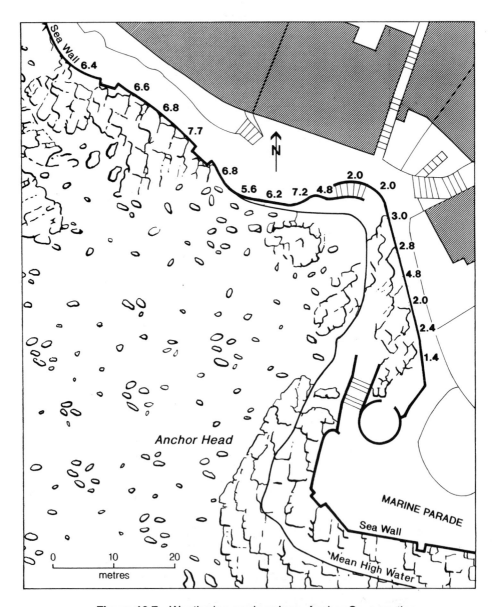

Figure 10.7   Weathering grade values, Anchor Cove section

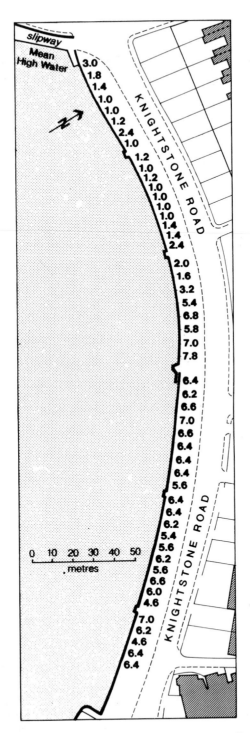

Figure 10.8   Weathering grade values, Knightstone Road section

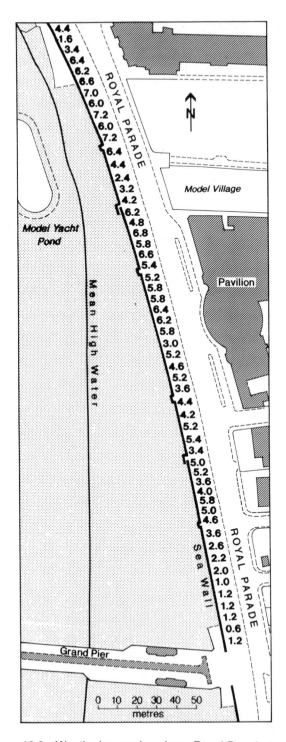

Figure 10.9 Weathering grade values, Royal Parade section

Table 10.3 Weathering grade values for sea wall coping stones in different sample locations. Each value is the mean of a run of five consecutive stones

Birkett	Anchor	Knightstone	
		East	West
exposed	exposed	exposed	sheltered
1.2	6.4	3.2	3.0
1.4	6.6	5.4	1.8
2.0	6.8	6.8	1.4
1.4	7.7	5.8	1.0
0.8	6.8	7.0	1.0
1.4	5.6	7.8	1.2
1.4	6.2	6.4	2.4
1.4	7.2	6.2	1.0
1.6	4.8	6.6	1.2
1.4		7.0	1.0
1.4	$\overline{6.45}$	6.6	1.2
1.6		6.4	1.0
1.8	sheltered	6.4	1.0
2.8		6.4	1.0
2.2		6.4	1.0
1.8	2.0	5.6	1.4
1.6	2.0	6.4	2.4
1.8	3.0	6.4	2.0
1.6	2.8	6.2	1.6
2.4	2.8	5.4	
2.6	2.0	5.6	$\overline{1.45}$
1.6	2.4	6.2	
2.4	1.4	5.6	
3.0		6.6	
6.0	$\overline{2.55}$	6.0	
		4.6	
$\overline{1.95}$		7.0	
		6.2	
		4.6	
		6.4	
		6.4	
		$\overline{6.12}$	

With $n_1, n_2, = 9,31$, $U = 101.5$; cannot reject H_o at 0.1

Thus over the small difference in elevation represented in this case, the difference in weathering grade is not statistically significant.

Exposure/Shelter

Both the Anchor and Knightstone sample sections embrace wall sections in somewhat sheltered locations (Table 10.3). Although these degrees of shelter are not susceptible to quantification, internal comparisons at each location permit the analysis of the effect of shelter at each in turn.

At Anchor Cove the south end of the wall rises in elevation from 2.8 to 3.5 m above MHWST, at the same time trending away from the line of HWM and extending behind the shelter of an elevated turret. Again the Mann-Whitney test was employed, as follows:

$$\text{With } n_1, n_2, = 9,8, \ U = 0.5; \text{ reject } H_o \text{ at } 0.001$$

The effects of increasing exposure on the Knightstone wall, as it emerges from the shelter from direct westerly winds offered by the Knightstone peninsula, can be examined by correlation of weathering grade with longshore distance representing increasing exposure. Spearman rank correlation yields:

$$r_s = +0.688 \ (n = 50); \text{ significant at } 0.001$$

In both cases weathering grade is thus shown to increase with increasing exposure to wind blowing directly onshore.

Linear Distance from High Water Mark

The influence of linear distance from HWM can be identified by examining the variation in weathering grade in the Royal Parade section (Figure 10.9). Here the wall maintains a constant elevation of 2.5 m above HWMST, and HWM diverges southwards away from the line of the wall. The average weathering grade value was correlated against rank distance from HWM representing decreasing exposure. The results are plotted in Figure 10.10. The relationship was tested using the Spearman rank correlation test, and yielded:

$$r_s = -0.585 \ (n = 55); \text{ significant at } 0.001$$

A strongly significant relationship is thus indicated, with weathering grade diminishing with increasing linear distance from HWM up to a distance of c. 55 m. to the south of the break in the sea wall at Grand Pier, where the sea wall commences again at a distance of 60 m from HWM, there is very limited evidence of alveolarisation. This suggests that 55–60 m represents the maximum distance from HWM that alveolisation has been able to develop in 105 years under the conditions of exposure at this section.

Azimuth

At Anchor Cove a rock wall of identical construction to the sea wall forms a circular turret whose coping is at an elevation of 5.9 m above MHWST. This circular wall permits observation of the influence of compass orientation on weathering grade (Figure 10.11). There is a marked variation in weathering grade exhibited by the top surface of the coping stones around the points of the compass.

The wall is at its closest point to HWM, c. 9 m, in the direction of 305°. The maximum weathering grade, however, defined by the running mean of five stones, exists where the wall faces southwest, centring on an azimuth of 215°. On this azimuth the line of HWM lies at a distance of 20 m from the wall.

The azimuth of 215° is, however, the direction from which the majority of strong winds blow. Weather records for 1991–92 from Priory School (UK NGR ST 369628) some 6 km to the east, reveal that of winds Force 3 and above, over 50% blow from

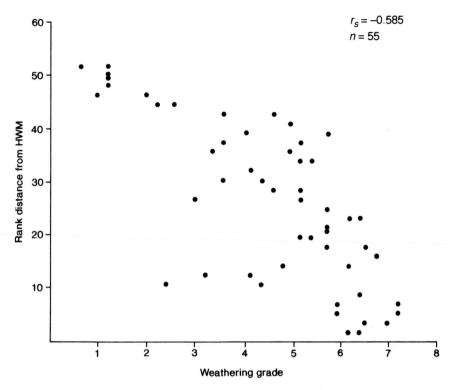

Figure 10.10 Plot of weathering grade against distance from HWM, Royal Parade section

the hemi-octal centring on 225°. In this case, then, the influence of distance to HWM on weathering grade is overridden by the direction of approach of storm winds. It is therefore concluded that exposure to spray and salt-laden winds is a more significant factor in determining weathering grade than linear proximity to HWM.

Aspect

The term aspect is used here in the sense of discriminating between seaward and landward exposure. It is employed in respect of the lateral faces of the coping stones. By comparing the two opposed faces of an individual stone the different effects of seaward and landward exposure can be examined. By using individual stones in this way, control is gained of the major variable of stone type on weathering grade. The weathering grade classification is equally applicable to the lateral faces. Their area is necessarily more limited than the top surface, but this does not impose any significant restriction on observation or interpretation.

Six samples were observed, each of 10 stones, at locations where both sides of the coping stones were readily accessible and observable. The results are set out in Table 10.4. In all cases the run of 10 stones exhibits a higher weathering grade on the seaward face. The mean difference exceeds one weathering grade in each case, rising to maximum differences in excess of 2.5 weathering grades. These comparisons confirm the role of aspect in respect of exposure in enhancing the honeycombing process.

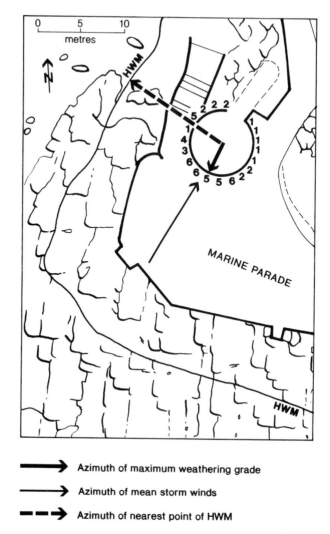

Figure 10.11   Weathering grade values and azimuth on the turret at Anchor Cove (values are for single stones)

MARINE SALT IN THE HONEYCOMBED STONES

The involvement of marine salts in honeycomb formation leads to a consideration of the salt content of the honeycombed rock in the present study. The effectiveness of wetting and drying in seawater, and specifically the effects of sodium chloride, the most abundant salt in seawater, has been demonstrated in coastal weathering elsewhere in southwest England (Mottershead, 1982 and 1988). Davison (1986) has employed chlorine, an element which is not an original component of most rock types, as an indicator of penetration by marine salts. This approach is adopted here.

Table 10.4 Comparison of weathering grade values of exposed (seaward) and sheltered (landward) faces of samples of 10 stones

A. By Knightstone slipway			B. Royal Parade by yacht pool			C. Royal Parade opp. York Ho		
E	S	D	E	S	D	E	S	D
6	2	−4	6	6	0	7	7	0
6	2	−4	7	6	−1	6	5	−1
5	1	−4	6	6	0	8	3	−5
5	1	−4	8	7	−1	2	1	−1
5	3	−2	7	7	0	6	5	−1
4	1	−3	6	3	−3	5	5	0
5	2	−3	7	6	−1	4	2	−2
2	1	−1	6	5	−1	5	5	0
1	1	0	6	5	−1	3	4	1
Mean		−2.7			−1.0			−1.1

D. Birkett steps			E. Steps by Royal Pier Hotel			F. Anchor Cove		
E	S	D	E	S	D	E	S	D
5	3	−2	0	3	3	6	2	−4
1	1	0	1	3	2	6	1	−5
6	6	0	7	3	−4	8	0	−8
2	1	−1	6	3	−3	7	7	0
2	0	−2	1	2	1	6	3	−3
4	2	−2	5	3	−2	9	3	−6
2	1	−1	7	2	−5	9	9	0
3	1	−2	6	3	−3	7	9	2
0	0	0	6	0	−6	8	5	−3
1	0	−1	6	5	−1	6	8	2
Mean		−1.1			−1.8			−2.5

E = exposed
S = sheltered
D = difference

Rock samples were taken from a range of locations throughout the study area, and analysed for chlorine. Standardisation was attempted by removing samples only from the surface 5 mm of the rock, although it should be remembered that the range of variation in relief of differently honeycombed rock may diminish the degree of standardisation achieved. For each sample, 1 g of powdered rock was shaken for 30 minutes in 100 ml of deionised water and then filtered. The filtrate was tested for chlorine by titration, using the method of MAFF (1986).

Chlorine contents of seaward face samples are shown to range from 9.5 to 42.4 ppm (Table 10.5). The distribution of chlorine content was examined for spatial

Table 10.5 Chlorine content in mg 1^{-1} of seaward face samples of weathered rock

Knightstone Parade (west → east)		Royal Parade (north → south)		Birkett		Anchor	
Sample	Cl	Sample	Cl	Sample	Cl	Sample	Cl
76	9.5	44	18.0	9	22.2	1	28.1
77	9.5	42	15.4	11	26.5	3	36.0
78	16.9	81	32.8	15	12.7	5	24.4
60	14.8	82	20.1	17	12.7	25	13.8
58	17.5	40	16.4	19	19.1	73	32.8
56	13.2	38	12.7	21	10.6	74	31.8
54	22.2	36	16.9	70	16.9	75	28.1
52	12.7	34	15.4	71	18.0	86	42.4
50	16.4	83	15.9	72	13.8		
79	21.2	84	19.2				
48	19.6	30	9.5				
46	29.1	85	10.1				

variations in a manner similar to those of the weathering grades, since these two phenomena might be expected, on *a priori* grounds, to be related.

Comparison can be made of the chlorine content of the walls of different exposure. The exposed low level Anchor Cove Wall can be compared with the exposed high level Birkett section. A Mann-Whitney test yielded:

With $n_1, n_2 = 9, 8$, $U = 6.5$; reject H_0 at 0.02

The Anchor wall can be compared with the partially sheltered Knightstone wall, again using Mann-Whitney:

With $n_1, n_2 = 8, 12$, $U = 11.0$; reject H_0 at 0.02

In both cases chlorine content is shown to be greater on the section with greater exposure or proximity to marine influences.

Both Knightstone Parade and Royal Parade walls possess a gradient of exposure along their length towards their confluence. Rank correlation of chlorine content with distance along the wall in the direction of increasing exposure yields the following:

Knightstone Parade:　$r_s = + 0.704$ ($n = 12$); reject H_0 at 0.002

Royal Parade:　　　　$r_s = + 0.526$ ($n = 12$); reject H_0 at 0.1

Overall these tests collectively demonstrate that the sections of wall with greater exposure to wind and marine spray are marked by higher concentrations of chloride, and therefore by inference marine salt content. Moreover, the pattern of chlorine concentration mirrors the distribution of weathering grades, with greater chlorine concentrations associated with sections exhibiting the higher weathering grades.

One unexpected phenomenon was the manner in which chlorine content varies between seaward and landward faces of the same stone. Of 28 pairs of samples thus collected, 20 pairs exhibited a higher chlorine content on the landward face, inversely related to the pattern of weathering grade distribution noted earlier. Commonly the chlorine content on the exposed face is c. 70% of that to landward. The cause of this difference must for the moment remain conjectural, but it could lie in either differential deposition or in differential retention. The former could come about by the eddying effect of the wind flowing onshore over the parapet and turning down over it. The latter could be caused by differential loss of the volatile chlorine on the windy exposed face after deposition.

Deposition of marine salts in coastal rocks can be caused by marine or atmospheric processes. The former include splash and spray by seawater directly on to coastal rocks. The latter embrace both washout and dry fallout processes. Estimates of rates of atmospheric chlorine deposition in this coastal environment can be gleaned from various sources. Eriksson (1960) discusses levels of chlorine deposition at the continental scale, and indicates deposition on exposed western coasts in Europe in the range 10–20 g m^{-2} yr^{-1} (equivalent to a daily rate of c. 25–50 mg m^{-2}). Locally, specific monitoring of rainfall and chlorine deposition were carried out by Vale (1991) at the island of Flatholm, c. 10 km offshore of Weston, for the period January 1989 to March 1990. In this study bulk deposition, by both wet and dry processes, was monitored for fortnightly sampling periods. In 15 of the 26 sampling periods during the calendar year 1989 chlorine concentration was blow the detection limit of 4 mg l^{-1}. By estimating a concentration of 2 mg l^{-1} for such periods a total load of chlorine deposition of 9.34 g m^{-2} yr^{-1} can be derived, delivered in a year in which the total precipitation amounted to 646 mm. (Sodium deposition over the same period amounted to 6.38 g m^{-2} yr^{-1}; the ratio of this mass to that of chlorine deposited is very close to that of the two elements in seawater, implying this as the common source of both elements.) Vale indicates that the sampling procedure, employing collectors at a height of 1.75 m above ground level may have resulted in an underestimate of 30%.

Dollard et al. (1983) in a study of upland Cumbria, UK, stress the role of occult precipitation, in the form of wind-driven mist and fog, in the deposition of atmospheric pollutants. Occult precipitation, that precipitation which escapes measurement in conventional rain collectors, is likely to have a higher concentration of pollutants than rain drops. By both contributing to a higher rainfall total and delivering a higher concentration of pollutants, occult precipitation is likely to make a significant additional contribution to pollutant deposition. It is estimated by Dollard et al. to increase observed rates of wet deposition by up to 20%. Occult precipitation is likely to be a significant factor on an exposed west-facing coast such as the study factor.

Vale's data also permit analysis of deposition patterns on shorter timescales. Within the 1989 annual total, 70% of deposition occurred within one two-week sample period. Indeed this annual load was exceeded within a single sample period in February 1990 (a period of damaging storms). The strongly episodic nature of atmospheric sea salt deposition is confirmed by a four-day experiment at Flatholm in May 1990, in which sodium deposition was monitored. In this experiment daily totals of 0.67 and 0.78 mg m^{-2} were recorded from two sites on a rainfall day, 15 May. The rain-free period 16–18 May yielded daily deposition rates of 0.013 and 0.010 mg m^{-2} from the same two sites. The prevalence of element deposition during rainfall strongly suggests that salt deposition by washout is the dominant process.

The relationship between chlorine deposition rate and rainfall can be tested by correlations of these two variables for the 11 samples for which Vale presents observed data for both. Spearman correlation yields:

$$r_s = +0.74 \ (n = 11); \text{ significant at } > 0.01$$

This confirms the conclusion of the short-term experiment described above that atmospheric chlorine deposition appears to be strongly associated with rainfall.

Vale states that high chlorine concentrations are associated with westerly winds, and that episodic high chlorine deposition rates probably result from the combination of westerly gales with rainfall. A larger body of data at the diurnal scale would be required fully to substantiate this. Marine deposition by splash and spray is likely to be most effective during storm conditions, when waves break over the walls, and is also an episodic process.

It is evident that the chlorine content of the coping stones displays a considerable variability. This study has shown that spatial variations in chlorine content conform with patterns of exposure to marine influences, and also with patterns of consequential weathering grades. This appears to be further confirmation of the role of marine salts in the development of honeycomb weathering at this site.

WEATHERING AND ROCK STRENGTH

Values of the mechanical strength of Forest of Dean Stone are indicated in the literature in a range of sources. With values of crushing resistance (uniaxial compressive strength) in the range 45–68 MPa, the sandstone is a rock of medium strength. Observations were made on the extent to which the fabric of the rock at the weathering surface had been weakened. However, in respect of honeycomb, a method of testing rock strength is required which is applicable to small samples of rock, rather than the decimetre-sized cube used in uniaxial testing. Such a method is provided by the cone indenter (NCB, 1977; Hall, 1987). This device was employed to obtain strength estimates of small chips of rock taken from the honeycomb. The instrument yields cone indenter (CI) values, which are linearly related to uniaxial compressive strength. CI values were determined for five samples of honeycomb rock and five samples of fresh rock with 10 replicates per sample (Table 10.6). The mean values per sample of honeycombed rock are shown to lie in the range 2.15–2.91, as compared to fresh rock in the range 2.79–4.47. These results indicate a mean loss of mechanical strength of 23% in the honeycombed rock tested, indicative of significant bond weakening by weathering processes.

RATES OF WEATHERING

Several authors have recorded the development of honeycomb weathering over relatively short periods (Table 10.7). The appearance of honeycombing on newly exposed rock in coastal environments appears to be measured in decades, from as few as 30–35 years (Gill, 1981). The evolution of the honeycomb described here, which has taken place over a period of 105 year, may be set alongside these.

Previous authors, however, make little reference to the degree of advancement of the honeycombing. Those described by Gill (1981) appear to be incipient forms. What

Table 10.6 Mechanical strength properties of fresh and weathered Forest of Dean Stone

(a) Values of crushing resistance in MPa of Forest of Dean Stone, from various sources.

Middleton, 1905	56.84
Warnes, 1926	67.57
	61.03
Stone Industries, 1991	56.70
	45.42
Forest of Dean Stone Firms Ltd	45.69

(b) Cone Indenter (CI) values for five samples each of fresh rock. Each value is a mean of 10 determinations.

	Weathered rock	Fresh rock
	2.91	3.01
	2.15	3.15
	2.70	3.19
	2.80	2.79
	2.23	4.47
Mean	2.56	3.32

Table 10.7 Rapid honeycomb formation in coastal environments, from various authors

Author	Location	Rock type	Time of formation (years)
Grisez, 1960	Loire-Atlantique, France	Crystalline schist	60
Winkler, 1975	Indiana, USA	Quartzitic sandstone	100
Gill, 1981	Victoria, Australia	Greywacke	30–35
Mustoe, 1982b	Washington, USA	Arkose Metavolcanic greenstone	<80 70
McGreevy, 1985	Co. Antrim, UK	Carboniferous sandstone	c. 140
Matsukura and Matsuoka, 1991	Honshu, Japan	Tuffaceous conglomerate	65
Viles and Goudie, 1992	Dorset, UK	Oolitic limestone	135
This paper	Avon, UK	Sandstone	105

this paper demonstrates, therefore, is that a honeycomb layer has developed and in several locations has advanced to the stage of decay and stripping over a period of 105 years.

An estimate of the rate of denudation by weathering can be gained by assessing the amount of vertical lowering undergone by the surface of the coping stones. At several points an indication of the level of the initial surface is provided by the top edge of the vertical slice of mortar between adjacent stones (Figure 10.5). The mortar commonly proves more resistant to weathering than the stones themselves, and stands proud of the surface of the stones. The maximum difference in level observed is a stone surface 110 mm lower than the adjacent mortar. This may actually underestimate the true rate of lowering of the stone, since at the places where negligible weathering is observed, the top surface of the stones tapers down toward the mortar-filled joint, and the top of the mortar is thus lower than the main area of the top of the stones. The weathering observed, 110 mm in 105 years, represents a mean lowering rate of 1.05 mm yr^{-1} and may therefore be regarded as a minimum estimate. It is, nevertheless, a very high rate of weathering, and replicates the rate of weathering observed by Grisez (1960) on micaschists on the Atlantic coast of France.

CONCLUSIONS

This study has presented a methodology for assessing variations in the intensity of alveolar weathering based on objectively defined grades of weathering. Weathering grade is then mapped of stones with varying disposition in relation to elevation above MHWST, plan distance from HWM, azimuth and aspect. These geographically varying dispositions represent gradients of severity of the weathering environment away from the line of high water.

It is concluded that the geographical variables investigated are significant controls on the development of alveolisation. Weathering grade is shown to decrease with increasing elevation above MHWST, and plan distance from HWM. Aspect also is shown to be an important determinant of weathering grade, both in relation to the specific direction of strong onshore winds, and to the shoreward/landward contrasts in individual stones. These conclusions confirm, in a systematic way, the general observations made previously in studies by Höllerman (1975), Kelletat (1980) and Mustoe (1982b).

The results of this study appear to confirm the dominance of the marine influence in determining the rate of alveolisation, as the source of the salts most likely to be responsible. Determinations of chlorine content of the stones, as indicative of the presence of marine salts, reveals significant concentrations. Chlorine contents tend to be higher where alveolisation is more intense.

The rusticated finish of the wall provides a surface apparently particularly receptive to weathering agents and susceptible to weathering processes. The original roughness of the rock surface would have provided loci for the initiation of alveolisation. Weathering appears to have progressed via the development of isolated pits, through a network of honeycombs, then to a stage of honeycomb decay and ultimately the stripping of the honeycomb layer. If this interpretation is correct, then honeycomb represents an intermediate point in a denudation sequence. It would appear to be a transient condition rather than an equilibrium form.

A rapid rate of weathering, >1 mm yr^{-1}, is indicated by reference to the baseline date represented by the construction of the sea wall. This weathering is associated with a marked reduction in the mechanical strength of the rock.

Engineers of the nineteenth century evidently had substantial confidence in the durability of Forest of Dean Stone. It is described as an excellent wearing and weathering stone (Middleton, 1905). The observations presented in this paper a century after the construction of the walls suggest that, in this particular weathering environment at least, this confidence appears to have been misplaced.

ACKNOWLEDGEMENTS

Edge Hill College provided a grant towards the cost of fieldwork. Kathy Coffey carried out the analytical work, and Ann Chapman drafted the diagrams. David Case and Jean Hilliard provided field assistance. The Meteorological Office, Bristol, and David Dennis of Priory School, Weston-super-Mare, provided climatic data. Nick Goff of Woodspring Museum assisted with the history of the sea walls. The Technical Services Department, Woodspring District Council, granted permission to work on the sea walls.

REFERENCES

Anon. (undated). A History of the Weston-super-Mare New Sea Front. Dare and Frampton, Weston-super-Mare.

Beisley, P. (1988). *Weston-super-Mare—A History and Guide*. Alan Sutton Publishing, Gloucester.

Brown, B. J. H. and Loosely, J. (1979). *The Book of Weston-super-Mare*. Barracuda, Buckingham.

Davison, A. P. (1986). An investigation into the relationship between salt weathering debris production and temperature. *Earth Surface Processes and Landforms*, **11**, 335–341.

Dollard, G. J., Unsworth, M. H. and Harve, M. J. (1983). Pollutant transfer in upland regions by occult precipitation. *Nature*, **302**, 341–343.

Eriksson, E. (1960). The yearly circulation of chloride and sulfur in nature: meteorological, geochemical and pedological implications, Part II. *Tellus*, **12**, 63–109.

Evans, I. S. (1970). Salt crystallisation and weathering: a review. *Révue de Géomorphologie Dynamique*, **19**, 153–177.

Gill, E. D. (1973). Rate and mode of retrogradation on rocky coasts in Victoria, Australia, and their relationship to sea level changes. *Boreas*, **2**, 143–171.

Gill, E. D. (1981). Rapid honeycomb weathering (tafoni formation) in greywacke, S.E. Australia. *Earth Surface Processes and Landforms*, **6**, 81.

Grisez, L. (1960). Alveolisation littorale de schistes metamorphiques. *Révue de Géomorphologie Dynamique*, **11**, 164–167.

Hall, K. J. (1987). The physical properties of quartz-micaschist and their application to freeze–thaw weathering in the maritime Antarctic. *Earth Surface Processes and Landforms*, **12**, 137–149.

Höllerman, P. (1975). Cavernous rock surfaces (tafoni) on Tenerife, Canary Islands. *Catena*, **2**, 385–410.

Kelletat, D. (1980). Studies on the age of honeycomb and tafoni features. *Catena*, **7**, 317–325.

Leary, E. (1986). *The Building Sandstones of the British Isles*. Building Research Establishment, Watford.

Levin, H. L. (1990). *Contemporary Physical Geology*, 3rd edition. Holt, Rinehart and Winston, Orlando.

MAFF (1986). *The Analysis of Agricultural Materials*, 3rd edition. Reference book 427, HMSO, London.

Matsukura, Y. and Matsuoka, N. (1991). Rates of tafoni weathering on uplifted shore platforms in Nojima-Zaki, Boso peninsula, Japan. *Earth Surface Processes and Landforms*, **16**, 51–56.

McGreevy, J. P. (1985). A preliminary scanning electron microscope study of honeycomb weathering of sandstone in a costal environment. *Earth Surface Processes and Landforms*, **10**, 509–518.

Middleton, G. A. T. (1905). *Building Materials: Their Nature, Properties and Manufacture*. Batsford, London.

Mottershead, D. N. (1982). Coastal spray weathering of bedrock in the supratidal zone at East Prawle, south Devon. *Field Studies*, **5**, 663–684.

Mottershead, D. N. (1983). Rapid weathering of greenschist by coastal salt spray, East Prawle, south Devon; a preliminary report. *Proceedings Ussher Society*, **5**, 346–353.

Mustoe, G. E. (1982a). Alveolar weathering. In Schwartz, M. L. (ed.), *The Encyclopedia of Beaches and Coastal Environments*, Hutchinson Ross, Stroudsburg, Pennsylvania.

Mustoe, G. E. (1982b). The origin of honeycomb weathering. *Bulletin Geological Society of America*, **93**, 108–115.

NCB (1977). *NCB Cone Indenter*. Mining Research and Development Handbook, 5.

Pauly, J.-P. (1976). Maladie alveolaire: conditions de formation et d'evolution. In Rossi-Manresi, R. (ed.), *The Conservation of Stone*, Centro per la conservazione delle sculture all aperto, Bologna, pp. 55–80.

Selby, M. J. (1985). *Earth's Changing Surface—An Introduction to Geomorphology*. Clarendon Press, Oxford.

Smith, B. J. and McAlister, J. J. (1986). Observations on the occurrence and origins of salt weathering phenomena near Lake Magadi, southern Kenya. *Zeitschrift für Geomorphologie*, NF **30**(4), 445–460.

Smith, P. J. (1982). Why honeycomb weathering? *Nature*, **298**, 121–122.

Stone Industries (1991). *Natural Stone Directory*, 8th edition. Ealing Publications, Maidenhead.

Summerfield, M. A. (1991). *Global Geomorphology—An Introduction to the Study of Landforms*. Longmans, Harlow.

Trenhaile, A. S. (1987). *The Geomorphology of Rock Coasts*. Clarendon Press, Oxford.

Trotter, F. M. (1942). *Geology of the Forest of Dean Coal and Iron-ore Field*. Memoirs Geological Survey GB, HMSO, London.

Vale, J. A. (1991). Atmospheric deposition of heavy metals to the Severn Estuary. Unpublished Ph.D. Thesis, University of Stirling.

Viles, H. and Goudie, A. S. (1992). Weathering of limestone columns from the Weymouth seafront, England. In Rodrigues, J. D., Henriques, F. and Jeremias, F. T. (eds), Proceedings 7th International Congress on *Deterioration and Conservation of Stone*, Vol. 1, Laboratorio Nacional de Engenharia Civil, Lisbon, pp. 297–304.

Warnes, A. R. (1926). *Building Stones: Their Properties, Decay and Preservation*. Ernest Benn, London.

Welch, F. B. A. and Trotter, F. M. (1961). *Geology of the Country around Monmouth and Chepstow*. Memoirs Geological Survey GB, HMSO, London.

Winkler, E. M. (1975). *Stone: Properties, Durability in Man's Environment*. Springer-Verlag, Wien.

Woodspring Museum (1988). *Weston-super-Mare: Historical Fact Sheet* (revised edition). Weston-super-Mare.

11 Erosion Rates of a Sandstone used for a Masonry Bridge Pier in the Coastal Spray Zone

KEN'ICHI TAKAHASHI and TAKASUKE SUZUKI
Chuo University, Japan

and

YUKINORI MATSUKURA
University of Tsukuba, Japan

ABSTRACT

Yayoi Bridge, connecting Aoshima island with Kyushu main island in Japan, is supported by four piers whose surface is composed of sandstone blocks. The four side walls of the piers, have a height of 3 m and a slope of 70°, and face approximately east, south, west and north. The bases of the piers are situated at Mean Tide Level (mean tidal range: 1.6 m). Each sandstone block has developed a dish- or bowl-like depression due to weathering and erosion. The greatest depth of each depression in all blocks was measured in 1971 and 1989 corresponding to 20 and 38 years, respectively, since the construction of the bridge.

The erosion features are summarized: (1) the erosion depth is the largest on the south-facing wall (maximum value is about 15 cm during 38 years), and gradually becomes smaller on the west- and east-facing wall, i.e. the erosion depth and the amount of insolation accepted are positively related; (2) the altitude of maximum erosion is situated at just High Tide Level on the south-facing wall, and at 3 m above MTL on the north-facing wall where the sea spray zone, i.e. wetting zone, is located higher due to prevailing waves. These findings suggest that the difference in erosion depth depending on aspect and erosion and altitude can be explained from the amount of insolation and sea spray. The erosion data indicate that the rate of erosion is not a linear function of time but an exponential function such as

$$D = A(1 - e^{-bt})$$

where D is the erosion depth, t is the time, and A and b are constants.

INTRODUCTION

It is important to estimate erosion rates of rock in the coastal spray zone in order to elucidate the formative processes of landforms such as shore platforms. Some erosion (solution) rates have been reported for soluble calcareous rocks situated in the

Rock Weathering and Landform Evolution. Edited by D. A. Robinson and R. B. G. Williams
© 1994 John Wiley & Sons Ltd.

intertidal or spray zones (e.g. Emery, 1941; Trudgill, 1976; Spencer, 1981), but only a few for insoluble rocks. Grisez (1960) reported that the surface of boulders of crystalline schists on a seawall had been honeycombed to a maximum depth of 66 mm in 62 years, i.e. at a maximum rate of about 1 mm yr^{-1}. Similar findings to Grisez's have shown that honeycomb structures form during a short period in the order of tens of years (e.g. Gill, 1981; Gill *et al.*, 1981; Mustoe, 1982). Takahashi (1975 and 1976) measured the depth of tafoni-like depressions formed in sandstone blocks used for a masonry pier of a bridge constructed in a shallow strait, and estimated that the maximum rate of deepening was 5 mm yr^{-1} for 20 years. Mottershead (1989) estimated the mean rate of surface lowering of greenschist in the supratidal zone to be 0.625 mm yr^{-1} based on repeated measurements conducted three times a year over seven years using a micro-erosion meter.

These previous papers treat the erosion rate as constant because of the limitations of the field measurements: in most cases, with exception of Mottershead's (1989) work, the amounts of erosion have usually been measured only once in the measuring period because they are extremely small. Recently, Matsukura and Matsuoka (1991) measured the depth of tafoni on the faces of marine cliffs with different ages of emergence, and concluded that the rate of tafoni deepening is not constant but decreases with time.

The problems to be solved, therefore, are (1) the reasons for the variability in the reported erosion rates, ranging from 0.6 to 5 mm yr^{-1}, as cited above, and (2) the time-dependence of the erosion rates. The present study tackles these problems, based on the re-measurement of the depth of tafoni-like depressions previously measured in 1971 by Takahashi (1975 and 1976).

THE STUDY SITE

Physical Setting

The pier studied by Takahashi (1975 and 1976) is one of the four piers of Yayoi Bridge connecting Aoshima island with Kyushu main island, Japan (Figure 11.1). The Aoshima island (maximum altitude: 5.7 m) is surrounded by shore platforms with an average width of about 200 m (Figure 11.2) and a mean height situated at mean tide level (MTL). The platforms are composed of rhythmic alternations of sandstone and mudstone of Pliocene age and are characterised by a corrugated surface with sandstone forming the ridges (Figure 11.2; Takahashi, 1975 and 1976). A cuspate spit is usually present on the platforms, extending from the main island toward Aoshima island.

Yayoi Bridge, 130 m long, was constructed in July 1951. It is aligned approximately east–west (exactly N79°E). The bridge is supported by four piers whose shape is the frustum of a pyramid. The piers stand on the shore platform. The top surface of the piers was initially wider than the deck of the bridge. When the bridge was repaired in March 1978 (27 years after construction), it was widened from 2.9 to 4.5 m.

Figure 11.3 summarises the meteorological data for the three decades (Japan Meteorological Agency, 1982, pp. 251–252), obtained at Miyazaki Meteorological Station located 15 km north of Aoshima island. According to the tidal record at

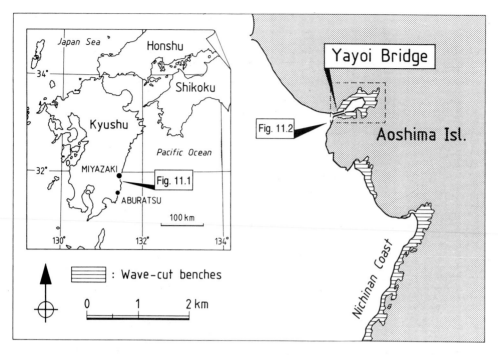

Figure 11.1   Location of Yayoi Bridge

Aburatsu Tidal Observatory, 30 km south of this island (see Figure 11.1), mean tidal range is about 1.6 m.

The Bridge Piers

The side walls of all four piers are formed of sandstone blocks, which are the same kind of sandstone as forms the ridges of the corrugated relief on the shore platforms. Each sandstone block has a size of about 35 × 25 × 35 cm. The joints between the blocks are filled with mortar. The physical and mechanical properties of sandstone similar to that used for the blocks are shown in Table 11.1.

Most of the sandstone blocks used for the piers have a depression in their surface, the maximum depth being generally located at the centre of the block (Figure 11.4). Some of the larger depressions possess a flat floor and resemble 'tafoni'. Sand grains produced by disaggregation generally occur on the surface of the depressions. These sand grains can be easily worn away by rubbing with a finger. Depressions are not observed on all blocks. The depth of the depressions varies according to the orientation of the wall and/or the elevation of the sandstone blocks. The surface of the mortar joints protrudes from the surface of the sandstone blocks except in the lowest part of each pier.

The depressions in the sandstone blocks are observed on all of the four piers. Takahashi (1975 and 1976) studied the second pier from the edge of Aoshima island for the following reasons: (1) since this pier is the highest, data for a wide range of heights covering the intertidal and spray zones can be obtained; and (2) since the pier

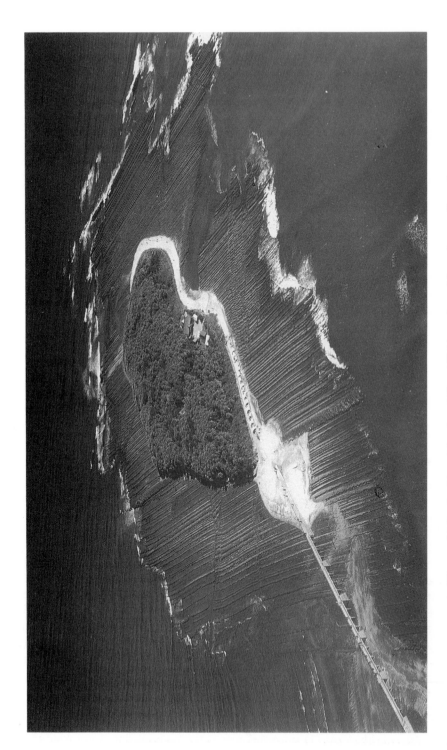

Figure 11.2 Aoshima island and Yayoi Bridge, viewed from the southwest

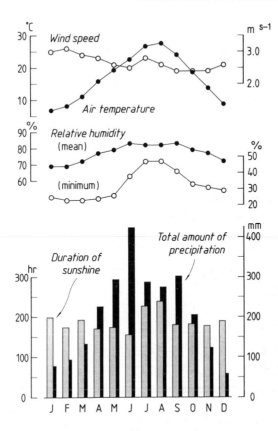

Figure 11.3 Meteorological data for the three decades 1951–1980 at Miyazaki Meteorological Station

Table 11.1 Some physical and mechanical properties of sandstone from Aoshima (after Takahashi, 1975 and 1976)

	Fine sandstone	Coarse sandstone
Apparent specific gravity	2.69	2.69
Bulk specific gravity	2.33	2.51
Porosity (%)	13.4	6.9
Maximum water content (%)	5.6	2.5
Maximum shrinking strain due to drying (%)	0.369	0.167
Thermal expansion coefficient ($\times 10^{-6}$)	6.80 (50–100°C)	9.62 (75–125°C)
Longitudinal wave velocity (km s^{-1})	2.55	3.16
Dynamic Young's modulus ($\times 10^4$ MPa)	1.52	2.51
Compressive strength (MPa): dry	64.3	99.0
wet	14.3	37.1
Tensile strength (MPa): dry	4.76	7.24
wet	0.68	2.51

is located in the middle of the bridge, it is possible to consider the effect of some external agents such as waves and wind on weathering and erosion. This pier is hereafter called the observed pier.

The sides of the observed pier slope at an angle of 70–75°. The four side walls facing east, south, west and north are called the E-wall, S-wall, W-wall and N-wall respectively. The top surface of the pier is called the horizontal surface (H-surface). There are 558 sandstone blocks forming the walls, which are set in 13 layers. The

Figure 11.4 Erosion features of the south-facing wall of the studied pier in 1971 (a) and in 1989 (b)

layers are numbered from the base to the top; the lowest being called the 1st and the uppermost the 13th layer.

The observed pier has a relative height of about 3.4 m, and its foundation, made of concrete resting on the shore platforms, has an altitude of 0.08 m below MTL. This means that the top of the foundation is situated at approximately MTL. At low tide, both the base of the pier and the shore platform are exposed, while at high tide the lower zone of the pier is submerged.

Mean high water of spring tide (MHWS) is 0.8 m above MTL. This height corresponds to the boundary between the 3rd and 4th layers. In this paper, the upper zone, above MHWS, is called the spray zone. The pier is, therefore, situated in the intertidal and spray zones.

Wave Conditions and Living Organisms

The incident direction of waves differs between normal and stormy weather conditions. Waves approach from the north in normal sea conditions: the N-wall is thus usually the most exposed to wave action. At high tide, waves attacking the N-wall often climb up to the crown of the pier (Figure 11.5a). The abundance of spray thrown up by such waves keeps the N-wall wet. Waves pass by the face of the E- and W-walls. The wave attack, therefore, is weak on these walls. The S-wall is immune from normal wave action due to its sheltered aspect (Figure 11.5b). On the other hand, storm waves usually approach from the southeast of Aoshima island and break in the area of submerged bedrock and/or the outer margin of the shore platforms. After breaking, bores advance across the shore platforms and reach the S-wall. Waves, in this case, strongly attack the lower part of the S-wall.

Littoral organisms are widely distributed on the N-wall: *Crassostrea gigas* (Ostreidae) are present in the lower half of the 1st layer, *Chthamalus challengeri* (Chthamalidae) are present in the zone between the middle of the 1st layer and the 4th layer and *Nodilittorina granularis* (Littorinidae) are scattered in the zone between the 5th and 9th layers. On the northern part of the 1st layer of the W- and E-walls, *Crassostrea gigas* and *Chthamalus challengeri* are observed. No organisms are observed on the S-wall. On the basis of their known zonation (e.g. Stephenson and Stephenson, 1949; Doty, 1957; Lewis, 1964), the distribution of these organisms on the observed pier indicates that the spray zone is higher on the N-wall due to its exposure to normal wave action. The layers between the 1st and 9th are exposed to an environment similar to intertidal zone and the 10th layer is situated in the supralittoral zone, i.e. lowest part of spray zone.

FIELD MEASUREMENTS AND RESULTS

Measurement of Depression Depth

The measurements were carried out twice, on 29 September 1971 and 2 August 1989, 20 years and 38 years after the construction of the bridge. For all sandstone blocks, the greatest depth of each depression, denoted as X, was measured from the surface of adjacent mortar joints, which is regarded as a datum plane for the measurements.

Figure 11.5 Waves attacking the N-wall (a) and the S-wall (b) at high tide under the normal weather conditions, viewed form Aoshima island

The reason for this is that (1) the surface of the mortar joints are flat, (2) on the surface of the mortar joints, the marks made by trowels at the time of the construction remain unchanged, and (3) the mortar joints have not eroded, irrespective of their aspect and height, except where they have crumbled due to the loss of support from neighbouring sandstone blocks as a result of the enlargement of depressions in these blocks.

According to the observations made in 1971, the surface of the sandstone blocks at the base of the pier (1st layer) protruded slightly from the surface of the mortar joints, and the surface of the blocks had no indications of erosion (Figure 11.4a). The original surfaces of these blocks at the time of bridge completion seemed to be completely preserved. The surfaces of all other sandstone blocks are, therefore, considered to have protruded slightly from the surface of the adjacent mortar joints at the time of completion of the bridge pier. It is, however, impossible to estimate the exact amount of protrusion of an individual sandstone block at that time. All that can be calculated is the mean protrusion length, P, based on measurements of 29 September 1971 on the protrusion of the 1st layer in each wall relative to the surface of the mortar joints. The results are: 1.9 cm (E-wall), 1.4 cm (W-wall), 2.3 cm (S-wall), and 1.5 cm (N-wall).

The depth of erosion and/or weathering d, in the period from the completion of the pier to the measurement time, can be obtained by:

$$d = X + P \tag{1}$$

where X is the measured depth of depressions.

Difference in *d*-value Due to Aspect

The d-value differs according to the orientation of the walls. The distribution of d-values in 1971 (Figure 11.6) generally shows two characteristics: (1) the d-value is generally largest on the S-wall and gradually becomes smaller on the W- and E-walls in this order, and is smallest on the N-wall, and (2) low d-values are found on the lowest two layers of all walls.

The 12th and 13th layers were removed when the bridge was repaired in 1978. Although most of the d-values obtained in 1989 are larger than those in 1971, the distribution of d-values in 1989 is similar to that in 1971 (Figure 11.6). A plot of the 1989 data in this figure indicates that: (1) the erosion depth is larger on the S-wall and gradually becomes smaller on the W- and E-walls in that order, and is smallest on the N-wall; (2) larger d-values (more than 25 cm) appear on the 4th, 5th, and 6th layers and the now uppermost 11th layer; (3) almost all the blocks on the N-wall have d-values of less than 5 cm, except the uppermost 11th layer; (4) the blocks on the W-and E-walls near the S-wall have a slightly larger value and those near the crown under the bridge have a small value; (5) the lowest two layers have a low d-value on all walls; and (6) the d-values of the blocks situated at the corners and upper edges of the pier are especially large, which is attributed to the 'edge effect' of weathering and/or erosion.

Altitudinal Change in *d*-value

The change in the depth of the depressions with height differs according to the orientation of the walls. The average d-value on each layer for both measurement times, denoted here as $\bar{d}_{20}$ and $\bar{d}_{38}$, was calculated (the subscripts 20 and 38 indicate the years elapsed from the construction of the pier in 1951). Data for both ends of each layer on a wall were omitted from the calculation. Figure 11.7 shows the results. The data for $\bar{d}_{20}$ on the 12th and 13th layers (only the 13th layer for the case of the N-wall) and $\bar{d}_{38}$ on the 11th layer clearly show that these suffered from the 'edge effect' of the crown. These layers are excluded from the following consideration.

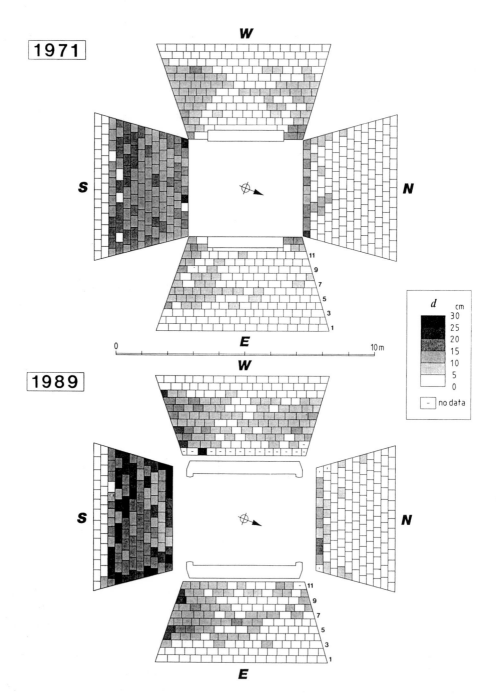

Figure 11.6   Erosion depth on the surface of each sandstone block, *d*, in 1971 and 1989

As shown in Figure 11.7 (1) the values of $\bar{d}_{20}$ and $\bar{d}_{38}$ of the 1st and 2nd layers are quite low on all walls: the surface of the blocks of these layers remains protruding slightly, and (2) the height of these layers corresponds to the upper half of the intertidal zone, up to 50 cm above MTL, and (3) this evidence shows that depressions are difficult to form in this zone.

The profiles of $\bar{d}_{20}$ above the 3rd layer vary with the orientation of walls. The maximum $\bar{d}_{20}$ on the S-wall is 10.4 cm on the 5th layer, and the neighbouring higher and lower layers, the 6th and 4th layers, have large values of 9.5 cm and 9.1 cm respectively. The value of $\bar{d}_{20}$ on the W-wall is largest in the zone between the 4th and 8th layers with the maximum value of 5.8 cm at the 7th layer; it decreases

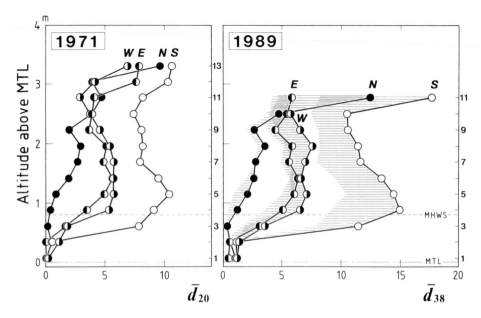

Figure 11.7 Average erosion depth for each layer of the stone masonry, $\bar{d}_t$, in 1971 and 1989. The subscript t of $\bar{d}_t$ means elapsed years from the construction of the pier, i.e. 1951. The figures on the vertical axis at the right side of both graphs indicate the number of the layers of the stone masonry above the foundation.

gradually above the 9th layer. The results on the E-wall show a similar pattern to that on the W-wall, although the values are slightly smaller. On the N-wall, the profile of values of $\bar{d}_{20}$ is different from that on the other walls; it gradually increases with increasing altitude. The 10th, 11th and 12th layers have large values of $\bar{d}_{20}$, and the 11th layer has the maximum value of 4.7 cm. This altitude, 2 m above MHWS, is about 1.5 m higher than the layer of the maximum value for $\bar{d}_{20}$ on the S-wall.

The profiles of $\bar{d}_{38}$ have a similar pattern to those of $\bar{d}_{20}$ in all respects. The larger d-values appear at different heights according to the wall orientation: at the 4th to 6th layers on the S-wall, at the 4th to 8th on the E- and W-walls, and at the 10th on the N-wall. These heights coincide with the supralittoral zone of each wall.

INFLUENCE OF INSOLATION AND SEA WATER SPRAY ON DIFFERENCES IN THE DEPTH OF THE DEPRESSIONS

Relationships Between the Depth of Depressions and Insolation

The results obtained from the field measurements suggest that the blocks with large depressions are situated on the faces receiving large amounts of insolation. Therefore, the relationship between the depth of the depressions and the amount of insolation will first be examined.

The maximum depths of depression, denoted as D_{20} and D_{38}, were calculated as the average value of the three layers with largest $\bar{d}_{20}$ and $\bar{d}_{38}$ value profiles for each orientation. The values of D_{20} and D_{38} thus calculated are respectively 9.7 cm and 14.2 cm on the S-wall (the 4th to 6th layers), 5.7 cm and 6.9 cm on the W-wall (the 5th to 7th layers), and 5.2 cm and 6.0 cm on the E-wall (the 5th to 7th layers). The value of D_{20} on the N-wall (the 10th to 12th layers) is 4.2 cm, and D_{38} is calculated to be 4.8 cm using data from only the 10th layer, excluding data from the 11th layer because of the 'edge effect'.

The amount of direct solar radiation on each wall at the start of each hour from sunrise to sunset on the 15th of every month was calculated using Uehara's (1961) equation; the summation of this quantity through one year was used as a substitute for the total amount of direct solar radiation per year, and is called the 'radiation index'. For the E- and W-walls, the shading effect of the bridge was estimated. The percentage ratio of the values of the radiation index to that on the horizontal surface was calculated., and is denoted as I_r (Figure 11.8). Radiation conditions were changed due to the repair of the bridge in 1978. The radiation index after 1978, denoted as I_r', was also calculated, which is shown in the lower part of Figure 11.8. This figure shows that the shade area increases due to the widening of the bridge.

To compare the total amount of direct solar radiation on each sandstone block for 20 years with that for 38 years, an integrating radiation index, denoted as I_{r20} and I_{r38} respectively, was obtained as follows:

$$I_{r20} = (I_r \cdot 20 \text{ yr})/20 \text{ yr} \tag{2}$$

$$I_{r38} = (I_r \cdot 27 \text{ yr} + I_r' \cdot 11 \text{ yr})/38 \text{ yr} \tag{3}$$

The average value of I_{r20} for the three layers which were used for the calculation of D_{20} is 85.7% for the S-wall, 30.7% for the W-wall, 20.8% for the E-wall and 12.3% for the N-wall. The average value of I_{r38} on the S- and N-walls was the same as that of I_{r20}, whereas it slightly decreases on the W- and E-walls, to 27.2% and 19.1% respectively.

Figure 11.9 shows the relationship between D_{20} and I_{r20} and that between D_{38} and I_{r38}. A significant relationship appears between the depth of depression and the amount of insolation: this shows that insolation plays an important role in the formation of the depressions.

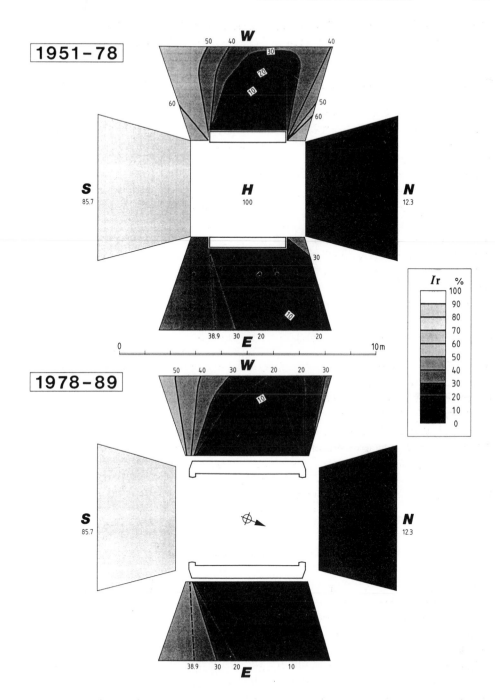

Figure 11.8 Distribution of the ratio of the relative annual amount of direct insolation on the four walls compared to that on the horizontal surface, I_r.

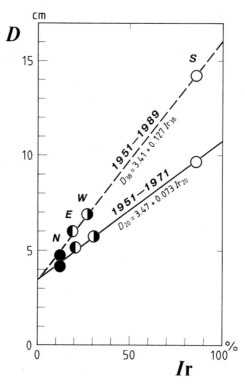

Figure 11.9 Relations between the maximum erosion depth on the four walls (D_{20} and D_{38}) and the average total amount of direct insolation (I_{r20} and I_{r38})

Effect of Sea Spray on Deepening of the Depression

The altitude of the maximum depth of depression differs with wall orientation; it is just at MHWS on the S-wall, and becomes gradually higher on the W- and E-walls, and is highest on the N-wall. The altitudes of the deepest depressions correspond with the variations in the altitude of the supralittoral zone under normal weather conditions; i.e. the spray zone on the N-wall is located at higher altitude due to its exposure to normal wave action. This fact strongly suggests that the difference in erosion depth with altitude depends on the supply of sea water spray.

Plentiful supplies of sea water and little insolation, conditions corresponding to the intertidal zone (the 1st and 2nd layers) of all walls and the zone below the 9th layer on the N-wall, result in a small depth of depressions. Deeper depressions have developed on the 3rd layer of the S-wall under conditions of abundant sea water and strong evaporation due to a high intensity of insolation. The deepest depressions have developed in the zone between the 4th and 6th layers of the S-wall under conditions of abundant sea water spray (not sea water) and strong evaporation. Above the 7th layer on the S-wall, the supply of sea water spray decreases with increasing height, resulting in shallower depressions compared with the 4th to 6th layers.

Thus, the depth of depressions is controlled by the frequency and intensity of both wetting by the supply of sea water spray and drying (evaporation) by insolation. In other words, the difference in the depth of depressions according to wall orientation and altitude is due to the combined effect of insolation and sea water spray.

EROSIONAL PROCESSES

Rates of erosion are generally high if the materials to be eroded offer less resistance or have only a low compressive strength (e.g. Sunamura, 1992, pp. 96–99, for wave abrasion; Suzuki and Takahashi, 1981, for wind abrasion). In the present case, however, the resistant sandstone (64–100 MPa in compressive strength; Table 11.1) is eroded to form the depressions, whereas the less-resistant mortar is nowhere eroded, irrespective of aspect and altitude (the strength of mortar used for joints cannot be measured, but the maximum compressive strength of common mortar used in Japan is about 50 MPa in the dry state (Nakamura, 1968, p. 202). This indicates that (1) these depressions never form on fresh (unweathered) sandstone, and (2) they form only after the surface layers of the sandstone blocks have suffered some loss of strength due to physical weathering. The deepening of the depressions is, therefore, weathering-controlled erosion.

Among the weathering processes, insolation (thermal) weathering and frost shattering are not effective, because (1) mortar, having a similar value of thermal expansion coefficient (7–14×10^{-6}) to that of sandstone (6–10×10^{-6}; Table 11.1), is not weathered/eroded and (2) the mean air temperature scarcely ever falls below 0°C (Figure 11.3). The possibility of wetting–drying weathering is examined next. Wetting –drying weathering occurs on rocks having both a high value of the wetting–drying slaking index (strength reduction index; Takahashi, 1976), i.e. more than 0.8, and a high value of maximum shrinking (expansion) strain, i.e. more than 0.5%, as shown in the chart of the relationships between both values (cited by Sunamura, 1992, Figure 4.19). In the study area, the wetting–drying index is found to be 0.78 for fine sandstone and 0.63 for coarse sandstone. Maximum shrinking strain is 0.369% for fine sandstone and 0.167% for coarse sandstone (Table 11.1). Plotting these values on the chart indicates that these sandstones are largely resistant to wetting–drying weathering.

As already pointed out, field evidence shows that (1) the depth of the depressions is controlled by the frequency and intensity of both wetting by the supply of sea water spray and drying (evaporation) by insolation and (2) the shape of some depressions is similar to 'tafoni' and the sand grains produced by disaggregation are found on the surface of depressions. Tafoni have recently been suggested to be formed by salt weathering (e.g. Johannessen et al., 1982; Mottershead, 1982; Mustoe, 1982; McGreevy, 1985; Matsukura and Matsuoka, 1991). These findings indicate that the salt weathering in the superficial part of the sandstone blocks plays the most important role in the strength reduction and resulting formation of the depressions.

The formation and deepening of these depressions is caused by the sequential process (weathering-controlled process of erosion) of granular disintegration and deterioration due to salt weathering of the surface of sandstone blocks, followed by removal of the detached sand grains and weakened sub-surface layer of sandstone through the action of such external agents as wind and waves.

EROSION RATES

The area shaded with thin horizontal lines in the right-hand graph of Figure 11.7 shows the increment in $\bar{d}_t$ for the 18 years from 1971 to 1989. This figure indicates that the erosion rate is not constant: the erosion rate for the first 20 years is larger than that for the subsequent 18 years (1971–1989). The 5th layer on the S-wall, for instance, has an average erosion rate of 0.52 cm yr^{-1} for the initial 20 years, whereas the erosion rate for the next 18 years is 0.22 cm yr^{-1}.

Matsukura and Matsuoka (1991) proposed the following equation for tafoni development in a coastal environment:

$$D = D_c(1 - \exp(-\beta t)) \tag{4}$$

where D is the tafoni depth (in centimetres), t is the time (in years), D_c and β are constants. This equation will be applied to the present case because the process of depression formation is similar to that of tafoni.

Since, in the present case, D in Equation (4) represents the erosion depth of a depression, the erosion rate will be discussed using the D_{20} and D_{38} data. The equations obtained for a best-fit curve for each wall are shown in Figure 11.10. Since the coefficient D_c indicates the critical depth of erosion, the possible maximum value of the future erosion depth can be estimated for each wall. For example, the average critical depth of erosion on the S-wall is expected to attain a large value of 19.6 cm. The critical depth is 7.3 cm for the W-wall, 6.2 cm for the E-wall, and 4.9 cm for the N-wall.

The decrease in the rate of erosion with time may be due to increasing depth of the depressions. If salt weathering is essential for depression development, the increasing depth may lead to the inner walls of the depressions becoming more difficult to desiccate due to decreasing exposure to sun and wind, as suggested by Matsukura and Matsuoka (1991). The value of β, which indicates the degree of decrease in the rate of erosion, is estimated to be 0.077, 0.088 and 0.099, for the W-, E- and N-walls respectively; these are larger than that for the S-wall (0.034). The largest β-value for the N-wall stems from a problem in selecting data: D_{20} was obtained from three layers (10th, 11th and 12th), whereas D_{38} was obtained from only one layer (10th). Possible reasons why the value of β on the W- and E-walls is

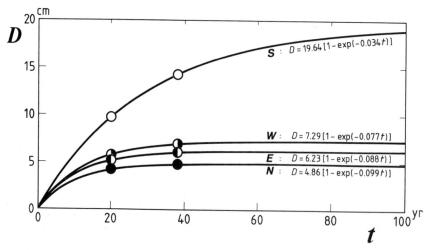

Figure 11.10   Increase in the erosion depth on each wall, D, with time, t

larger than that on the S-wall is the increase in the shade area, or the decrease in I_r, due to the repair of the bridge in 1978. This is indicated by the fact that I_{r38} on these two walls is slightly smaller than I_{r20} (Figure 11.9).

CONCLUDING REMARKS

The depth of depression of the surface of sandstone blocks used for a masonry bridge pier was measured in 1971 and 1989. The results show that the difference in the depth of depression according to wall orientation and altitude is due to the combined effect of insolation and sea water spray. This fact suggests that salt weathering plays an important role in forming and deepening the depressions. The deepening rate decreases with time as expressed by an exponential function.

Further studies are necessary to confirm the functional relationships of the rate of deepening and to elucidate the mechanisms of deepening. These studies include (1) measurements of the temporal change in water content of the surface layer of the sandstone blocks, (2) examination of the relationships between the subtle difference in rock properties and the erosion rate, and (3) evaluation of the abrasive force of waves acting on the rock surfaces.

ACKNOWLEDGEMENTS

The authors would like to express their sincere appreciation to Professor Tsuguo Sunamura of the University of Tsukuba for reading the manuscript and providing constructive criticism. This study was financially supported through the Grant-in-Aid for Personal Research of the Chuo University (1990 and 1991) and the fund of University of Tsukuba Project Research.

REFERENCES

Doty, M. S. (1957). Rocky intertidal surfaces. *Geological Society of America Memoir*, **67**(1), 535–585.

Emery, K. O. (1941). Rate of retreat of sea cliffs based on date inscriptions. *Science*, **93**, 617–618.

Gill, E. D. (1981). Rapid honeycomb weathering (tafoni formation) in greywacke, SE Australia. *Earth Surface Processes and Landforms*, **6**, 81–83.

Gill, E. D., Segnit, E. R. and McNeill, N. H. (1981). Rate of formation of honeycomb weathering features (small scale tafoni) on the Otway coast, SE Australia. *Proceedings of the Royal Society of Victoria*, **92**, 149–154.

Grisez, L. (1960). Alvéolisation littorale de schistes métamorphique. *Revue de Géomorphologie Dynamique*, **11**, 164–167.

Japan Meteorological Agency (1982). *Climatic Table of Japan, Part 2, Monthly Normals by Stations (1951–1980)*. Japan Meteorological Agency, Tokyo.

Johannessen, C. L., Feiereisen, J. J. and Welles, A. N. (1982). Weathering of ocean cliffs by salt expansion in a mid-latitude coastal environment. *Shore and Beach*, **50**, 26–34.

Lewis, J. R. (1964). *The Ecology of Rocky Shores*. English Universities Press, London.

Matsukura, Y. and Matsuoka, N. (1991). Rates of tafoni weathering on uplifted shore platforms in Nojima-zaki, Boso Peninsula, Japan. *Earth Surfaces Processes and Landforms*, **16**, 51–56.

McGreevy, J. P. (1985). A preliminary scanning electron microscope study of honeycomb weathering of sandstone in a coastal environment. *Earth Surface Processes and Landforms*, **10**, 509–518.

Mottershead, D. N. (1982). Coastal spray weathering of bedrock in the supratidal zone at east Prawle, south Devon. *Field Studies*, **5**, 663–684.

Mottershead, D. N. (1989). Rates and patterns of bedrock denudation by coastal salt spray weathering: A seven-year record. *Earth Surface Processes and Landforms*, **14**, 383–398.

Mustoe, G. E. (1982). The origin of honeycomb weathering. *Geological Society of America Bulletin*, **93**, 108–115.

Nakamura, A. (1968). Cement. In Aoki, K. *et al.* (eds), *Handbook of Materials for Civil Engineering Use*, Sankaido, Tokyo (in Japanese).

Spencer, T. (1981). Micro-topographic change on calcarenites, Grand Cayman Island, West Indies. *Earth Surface Processes and Landforms*, **6**, 85–94.

Stephenson, T. A. and Stephenson, A. (1949). The universal features of zonation between tide-marks on rocky coast. *Journal of Ecology*, **37**, 289–305.

Sunamura, T. (1992). *Geomorphology of Rocky Coasts*. Wiley, Chichester.

Suzuki, T. and Takahashi, T. (1981). An experimental study of wind abrasion. *Journal of Geology*, **89**, 509–522.

Takahashi, K. (1975). Differential erosion originating washboard-like relief on wave-cut bench at Aoshima island, Kyushu, Japan. *Geographical Review of Japan*, **48**, 43–62 (in Japanese with English abstract).

Takahashi, K. (1976). Differential erosion on wave-cut bench. *Bulletin of the Faculty of Science and Engineering, Chuo University*, **19**, 253–316 (in Japanese with English abstract).

Trudgill, S. T. (1976). The marine erosion of limestone on Aldabra Atoll, Indian Ocean. *Zeitschrift für Geomorphologie*, NF Supplement, **26**, 164–200.

Uehara, M. (1961). Physical and meteorological studies on the cultivation and utilization of slope land. *Memoirs of Faculty of Agriculture, Kagawa University*, No. 7, 1–107 (in Japanese with English abstract).

12 Salt Weathering and the Urban Water Table in Deserts

R. U. COOKE
University of York, UK

ABSTRACT

The towns of Khiva, Bukhara and Samarkand lie in the heart of Uzbekistan's irrigated cotton belt and contain numerous outstanding examples of Islamic architecture. These are of major ethnic, religious, cultural and research importance, and are being destroyed by salt weathering. This paper presents a summary of reconnaissance field observations and consequent laboratory analyses, and explores some possible solutions to the problem. Weathering appears to be related to capillary rise, a rising water table due to over-irrigation and increasing groundwater salinity.

GENERAL MODELS OF SALT WEATHERING

Several recent studies have explored the nature of salt weathering of building materials at or near the ground surface in urban, coastal desert areas. In Egypt (Bush *et al.*, 1980), Bahrain (Doornkamp *et al.*, 1980), Dubai (Halcrow Middle East, 1977), and most recently in Ras Al Khaimah (Cooke and Goudie, 1992), it has been suggested that damage caused by salts depends mainly on the position of the water table in relation to the ground surface and the foundations of buildings and other structures, and on the height of the capillary fringe above the water table (Figure 12.1, 12.2 and 12.3). The second major factor is the chemical composition of the groundwater. Observations of the relationships between these factors have led to the general proposition that, for a given material, damage is directly proportional to the shallowness of the water table and the salinity of groundwater (e.g. Cooke *et al.*, 1982). Within this context, it has been possible to recognise several zones of weathering intensity that can serve successfully as the basis for planning decisions and regulations, and construction decisions (see also, for example, Fookes and French, 1977).

This simple model is dynamic. It varies both spatially and temporally in two principal ways: through the movement of the water table and the capillary fringe associated with it; and through changes to the chemical composition of the groundwater. Figure 12.4, which is based on a diagram by George (1992), attempts to summarise the dynamics of the model. The position of the water table reflects, of course, a balance between the numerous and varied inputs and outputs. In the middle

Rock Weathering and Landform Evolution. Edited by D. A. Robinson and R. B. G. Williams
© 1994 John Wiley & Sons Ltd.

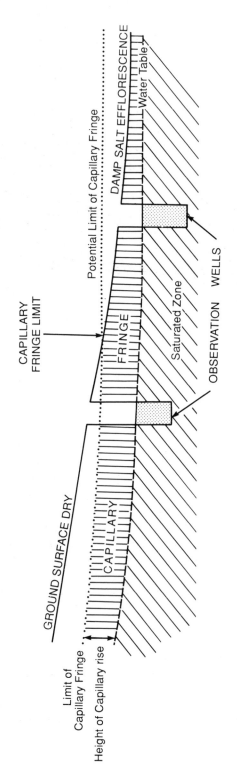

Figure 12.1 The capillary fringe and related groundsurface and groundwater phenomena

Figure 12.2 The effects of salt attack on the capillary fringe: destruction of a wall in Ras Al Khaimah (Arabian Gulf)

east, one of the major causes of increased salt weathering has been the recent rises of water tables in urban areas. These have often been the consequence of (a) an increase of runoff from new, relatively impermeable urban surfaces; (b) the return of excess, often imported irrigation, industrial and domestic waters; (c) the leakage of pipes in the system, often because of inadequate maintenance; (d) the reduction of groundwater extraction in favour of imported or desalinised water; and (e) poor subsurface drainage (e.g. George, 1992). Such urban water table rises contrast with the falls of water tables in many rural areas as a result of local over-extraction for irrigation and domestic water purposes.

The salinity of the groundwater, the zone of capillary rise, and near-surface sediments also vary spatially and over time, as shown in Figure 12.4. Each of the subsurface inflows can influence, either positively or negatively, groundwater salinity. For instance, leakages and returns from the urban water transport system to groundwater may either increase or decrease salinity. The same is true of surface runoff and throughflow. There are three major additional ways in which the salinity of groundwater and the zone of capillary rise may be altered: (a) by aerosol deposition on the surface, an especially important factor along desert coasts; (b) by the incorporation of extraneous salts into mortars (e.g. by the use of saline sands and aggregates), and the use of salt-rich stone in construction (e.g. beach rock); and (c) by the incursion of sea water as a result of freshwater abstraction, according to the Ghyben–Herzberg relationship, which states, *inter alia*, that the level separating fresh and salt water is raised by approximately 40 times the amount of water table lowering (e.g. Ward,

1967). Equally important in the context of salt weathering, these diverse changes can lead to differences in the chemical composition of the increasingly saline water. For instance, leaked irrigation water might introduce nitrates, intrusive sea water increases chlorides, saline mortars may add sulphates, and runoff may augment carbonates, depending on local circumstances. All experimental studies of salt weathering now suggest that the processes are strongly influenced by the types of salts involved (e.g. Cooke *et al.*, 1992).

Figure 12.3 The effects of salt attack in the capillary fringe: destruction of a telegraph pole in California (Photograph: A. S. Goudie)

Evidence is now accumulating that these models, developed essentially in the Middle East, are of wider applicability. For instance, they are relevant to understanding the weathering of foundations and tombstones in temperate areas. And they can also be applied, with modifications, to urban areas in deserts that are not coastal, such as the cities of Uzbekistan.

Key

 Changes to salinity

▓ Extraneous sources of salt

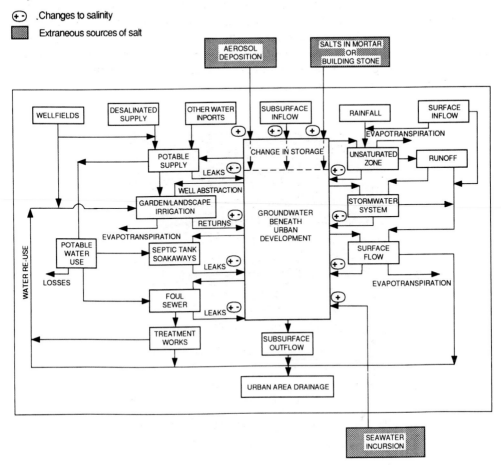

Figure 12.4 A model of a typical groundwater budget, and associated salinity dynamics, beneath an urban coastal development in the Middle East (after George, 1992, with modifications and additions)

SALT WEATHERING IN UZBEKISTAN

For at least a millennium, Islamic buildings—including mausolea, minarets, mosques and madrassas, some of them architectural masterpieces of unique distinction—have survived the rigours of the desert environment in central Asia. Ancient irrigation-based communities along the Silk road in Uzbekistan, such as Bukhara and Samarkand, contain some of the finest examples of this rich Islamic heritage. Today they are shrines visited by over half-a-million tourists each year, and they are a vital source of foreign exchange for the Republic. Khiva has already been declared a World Heritage City by UNESCO; Bukhara may follow.

Figure 12.5 Surface distribution and salt efflorescence in Char Minor mosque, Bukhara. Photographic evidence suggests that the upper limit of dampness has risen in recent years

Not all the major buildings have stood the test of time. Some were destroyed by the Mongol invasions of the early thirteenth century and during other wars; others suffered irreparably from earthquakes or floods; and some certainly deteriorated as result of weathering damage. But of those that remain, there is clear evidence of serious damage due to salt attack, and attempts at solving the problem appear largely to have failed (Figure 12.5). Urgent action is required to understand the nature of the problem and to develop a strategy for successful conservation.

The damage to buildings at present, excluding that caused by ground settlement or earthquakes, and that associated with the removal of tiles from roofs, is over-whelmingly to be found below a line around the base of buildings that is up to 3.5 m above ground level. Within this zone, there is extensive evidence of the flaking and disintegration to powder of the loess-based bricks, brick fracturing, extrusion of bricks and mortar from walls, break-up of brick patinas, the disruption of alabaster and other surface coatings, and salt efflorescence. The damage is clearly occurring at present because it affects new and recently restored buildings (Figure 12.6).

Figure 12.6 Recent damage and salt efflorescence in Char Minor mosque, Bukhara, following repairs in 1980

The ground-level weathering zone is characterised by both dampness and salt efflorescence. Figure 12.7 shows an example of the relative dampness profile in one building, based on a survey using the Protometer Survey-master Slimline. (Because the meter is based on a conductivity principle, its reading in this arid region probably reflect both dampness and saltiness.) The upper limit of dampness is variable, but it only rarely exceeds 3 m, and it is sometimes higher on east-facing, sheltered walls than on walls receiving higher insolation. Within the zone, the more permeable mortar tends to give higher readings than the finer-grained bricks. Salt efflorescences in the zone include a rich variety of salts (Table 12.1): a sample from Khiva includes a high proportion of halite; samples from Bukhara include magnesium and sodium sulphates, two salts that are known to be especially effective agents of salt weathering (e.g. Cooke, 1979); two samples include, perhaps surprisingly, potassium nitrate, that may indicate a fertiliser source; and gypsum is also found.

The weathering zone appears, on first impressions, to be associated with the capillary fringe above the water table, and its upper limit appears to coincide with

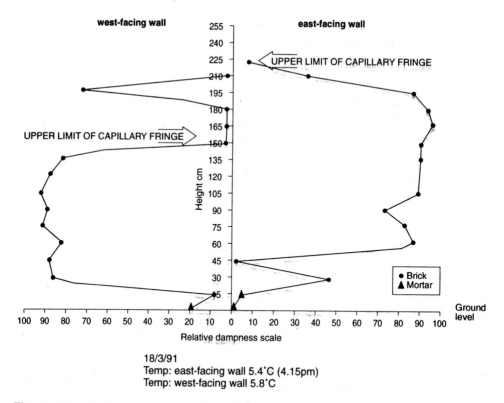

Figure 12.7   Profiles of 'dampness' in walls of the 12th century Kalyan minaret in Bukhara

the upper limit of the capillary fringe (Figure 12.1). Here, as in deserts elsewhere, it seems groundwater and its salts are drawn upward through surface material and buildings above the water table by 'evaporative pumping' that is fundamentally related to high surface temperatures, the nature of sub-surface temperature gradients, and the grain size of the materials through which the moisture passes. In general the height of rise is proportional to the surface temperature, the steepness of sub-surface temperature gradients, and the fineness of the material. The key to understanding salt efflorescence on the outside of walls lies in the differences between external air temperature and external wall temperature, that allows surface evaporation, and the fact that there is normally a daytime temperature gradient from the cool inside of buildings to the outside, which encourages moisture movement to the outside of the wall. As the moisture is evaporated from the external exposed surfaces within the weathering zone, so salts crystallise and cause some disruption; and, once emplaced, certain salts (such as magnesium and sodium sulphate), which can be hydrated in ambient temperature conditions, may change in volume to cause further disruption.

There is, however, an unexpected problem with this interpretation. In a few locations, the thickness of the capillary zone is abnormally high, in places exceeding 6 m, whereas a thickness of 2–3 m might be expected. This prompts the suggestion, developed by R. A. Legg (personal communication), that there may be two zones of

Table 12.1 Analysis of salt samples from Uzbek Islamic monuments. Major salts (%)

KHIVA: madrassa Mohammed Amin Khan, efflorescence (120 cm above ground level, inside)	$CaCO_3$, calcium carbonate $NaCl$, halite KNO_3, saltpetre	10% 70% 5%
BUKHARA: Samanid mausoleum, efflorescence (34 cm a.g.l., west side)	$MgSO_4 \cdot 6H_2O$, hexahydrate—major $MgSO_4 \cdot 7H_2O$, epsomite—minor Na_2SO_4, thenardite—minor	c. 60% 60% c. 20%
BUKHARA: Magoki-Attari mosque, efflorescence (7 cm a.g.l., south side)	$MgSo_4 \cdot 6H_2O$, hexahydrate KNO_3, saltpetre $NaNO_3$, soda nitre possible Na_2SO_4	
BUKHARA: Chor Minor mosque, efflorescence (1 m a.g.l., south side)	$CaSO_4 \cdot 2H_2O$, gypsum—major $NaCl$, halite—present Na, K, and Mg and sulphates but not mirabilitite or thenardite	65% 5%

The analyses were undertaken by Dr J. McArthur of the Department of Geological Sciences, University College London.
a.g.l. = above ground level

capillary rise, one immediately above the water table (as normal), the other at or above ground level, the two separated by a relatively dry zone (Figure 12.8). The suggested mechanism for pumping water across the 'dry zone' is as follows (Figure 12.9): in winter, when the surface is relatively cool and there is a marked temperature gradient from the water table to the surface, water will evaporate from the ground-water at depth, and the warm vapour will pass upwards through the dry zone to condense in the relatively cool surface layers. Such condensed water could be supplemented by rainfall, snowmelt, local runoff, irrigation water and floodwater. In the condensation zone, capillary rise could proceed into finer-grained material of the buildings from the surrounding soil.

This local variant of the general model may help to explain weathering in some buildings. But in most it is not necessary because, although the water table varies in depth, it is today commonly sufficiently close to the surface for the capillary fringe immediately above it to penetrate walls and foundations without there being any need for a second capillary zone.

These explanations do not account for the fact that the weathering seems to be a relatively recent phenomenon. It seems possible that the urban water tables, and their associated capillary fringes, have been rising in recent years, although the evidence is limited: historical photographs of Chor Minor in Bukhara indicate a possible rise over the last 25 years; local hydrologists certainly believe that rises have occurred; and surface flooding is reported to be increasing in places like Nukus. An explanation is readily to hand. Irrigation has increased rapidly in this region since the Russian conquest in the late nineteenth century, and especially since 1945, to sustain cotton production. Groundwater in the region may have been increasingly nourished by the leakage from unlined canals, by percolation of surplus irrigation water, and by water used to flush salts from fields (Akiner *et al.*, 1992). Thus, the gain of water in this region is probably the Aral Sea's loss: it could be that the increased weathering of Islamic monuments in Uzbekistan is yet another manifestation of the environmental crisis in the Aral Sea Basin.

It is also possible that urban groundwater, which is certainly quite saline (Akiner *et al.*, 1992), is becoming more saline, or that salts are increasingly at or near the groundsurface. There are several possible reasons. First, although it cannot be proved, the application of fertilisers might increase groundwater salinity and perhaps provide an explanation of the nitrates (Table 12.1). Secondly, salts might be blown in from the newly exposed salty margins of the Aral Sea. This is possible at Khiva,

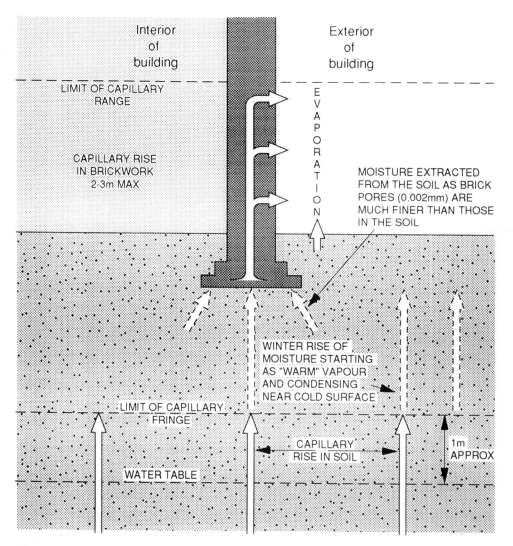

Figure 12.8 A possible explanation of the capillary rise in buildings several metres above the water table (R. A. Legg, personal communication)

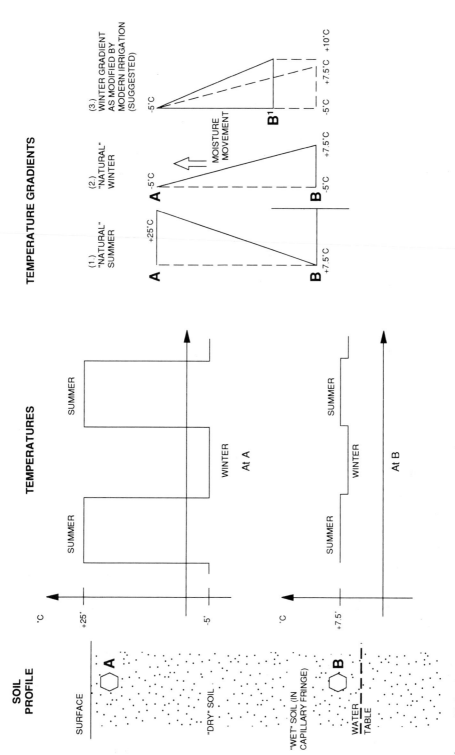

Figure 12.9 An explanation of the double zone of capillary rise in terms of temperature gradients (R. A. Legg, personal communication)

but it is rather unlikely that such salt could be transported in significant quantities over 600 km to Bukhara. Thirdly, and more likely, salts could be deflated locally from extensive salinised fields surrounding the towns. Fourthly, some salts could be introduced through the aggregates used in mortars. Local conservators believe that traditional mortars used carefully desalinated water, whereas water used today is not treated and is quite saline (e.g. water being used in repairs at the emir's citadel in Bukhara in 1991 had an electrical conductivity of 2700 μS cm^{-1}): if this is true, the change reflects a decline in quality control of construction. Finally, flooding occurs in some of the lower-lying areas, such as northern Bukhara, and may introduce salts into the near-surface water system.

The solution to the problem of salt weathering in Islamic buildings in Uzbekistan is quite clear: the capillary fringe and the salts associated with it must be removed from the buildings. This could be achieved in several ways. First, through more effective water management. This approach could include the installation of effective pumping wells locally, the more effective removal of irrigation water, and the use of groundwater for irrigation. Secondly, by preventing access of capillary water to specific buildings by emplacing effective barriers. Damp proof courses are an obvious solution. There is clear evidence of their selective use in some buildings since the eighteenth century (and this evidence itself may indicate that the weathering problem is not new). Some old courses are ineffective today, because efflorescence occurs above them; and some new damp proof courses have clearly failed. The potential for and viability of damp proof courses has been investigated by Myers (1991). These approaches to a solution should only be implemented after a thorough regional survey of the buildings at risk, the water table and the capillary fringe, and of contemporary changes to the last two features.

ACKNOWLEDGEMENTS

The reconnaissance study of Uzbekistan was undertaken with Dr S. Akiner and Dr R. A. French, and was generously sponsored by the Uzbek Ministry of Culture. Analysis of salts shown in Table 12.1 was kindly undertaken by Dr J. McArthur, Department of Geological Sciences, UCL. Figures 12.8 and 12.9 were kindly provided by R. A. Legg. Robert Myers contributed to discussions on the salt hazard in Uzbekistan.

REFERENCES

Akiners, S., Cooke, R. U. and French, R. A. (1992). Salt damage to Islamic monuments in Uzbekistan. *Geographical Journal*, **158**, 257–272.

Bush, P., Cooke, R. U., Brunsden, D. Doornkamp, J. C. and Jones, D. K. C. (1980). Geology and geomorphology of the Suez city region, Egypt. *Journal of Arid Environments*, **3**, 265–281.

Cooke, R. U. (1979). Laboratory simulation of salt weathering processes in arid environments. *Earth Surface Processes and Landforms*, **4**, 347–359.

Cooke, R. U. and Goudie, A. S. (1992). Salt weathering hazard in the Julfar area, Ras Al Khaimah (unpublished report).

Cooke, R. U., Brunsden, D., Doornkamp, J. C. and Jones, D. K. C. (1982). *Urban Geomorphology in Drylands*. Oxford University Press, Oxford.

Cooke, R. U., Warren, A. and Goudie, A. S. (1992). *Geomorphology in Deserts*. UCL Press, London.

Doornkamp, J. C., Brunsden, D. and Jones, D. K. C. (eds) (1980). *Geology, Geomorphology and Pedology of Bahrain*. Geobooks, Norwich.

Fookes, P. G. and French, W. J. (1977). Soluble salt damage to surfaced roads in the Middle East. *Journal of the Institute of Highway Engineers*, **24**(12), 10–20.

George, D. J. (1992). Rising groundwater: a problem of development in some urban areas of the Middle East. In McCall, G. J. H. Laming, D. J. C. and Scott, S. C. (eds), *Geohazards, Natural and Man-made*. Chapman and Hall, London, pp. 171–182.

Halcrow Middle East (1977). *Report on Geomorphological Investigations in Dubai, April 1977* (prepared by J. C. Doornkamp, D. Brunsden, D. K. C. Jones, P. Bush, M. J. Gibbons and R. U. Cooke), 3 vols.

Myers, R. (1991). *Salt Damage to Important Islamic Monuments in Uzbekistan: Report on a Visit to Tashkent, Bukhara and Khiva to Assess the Viability of Introducing Damp Proof Courses into Historical Buildings Suffering from Salt Attack*. Price and Myers, London.

Ward, R. C. (1967). *Principles of Hydrology*. McGraw-Hill, London.

Section 3

WEATHERING SURFACES AND DATING

13 Micro-mapping as a Tool for the Study of Weathered Rock Surfaces

JAN O. H. SWANTESSON
University of Karlstad, Sweden

ABSTRACT

A device for accurate micro-mapping of surfaces up to 40×40 cm is currently under development. It is based on a commercially available laser gauge probe that can measure height values with a resolution of 0.025 mm. A frame for movements of the probe in the x- and y-directions has been constructed. The time needed to measure about 100 000 values is less than 2 hours. The data is collected directly in the field on a computer.

Until now the device has mainly been used to characterise rock surfaces with varying degrees of weathering along the Swedish west coast. Statistical processing of the data collected allows roughness values for different kinds of surfaces to be calculated. Differences not visible to the naked eye can thereby be demonstrated. The progress of weathering as a function of the altitude can be followed. Due to the present isostatic land rise the height above sea level is directly related to the time of exposure to weathering agents. The data is also presented in 3-D models and contour maps.

Measurements of rock carvings threatened by weathering have also been performed. Even in homogeneous rocks there are great differences in how well they are preserved. By repeated measurements on the same site the device can be used as an accurate micro-weathering or micro-erosion meter. The method can be applied to most problems where accurate mappings of small areas are needed.

When studying minor weathering phenomena there is often a need for a detailed and objective micro-topographical description of rock surfaces. It is also of paramount interest to be able to follow the progress of weathering and micro-erosion. For this purpose an accurate micro-mapping device for surfaces up to a size of 40×40 cm is currently under development.

TECHNICAL DESCRIPTION

The equipment is based on a commercially available laser gauge probe that can measure height values with a resolution of 0.025 mm within a measurement range of 100 mm. The principle of the probe is illustrated in Figure 13.1. It has a weight of 0.6 kg and the light source consists of a GaAs laser diode with a typical wavelength of 850 nm. The light spot is circular with a diameter of 0.4 mm on the surface. It is

Rock Weathering and Landform Evolution. Edited by D. A. Robinson and R. B. G. Williams
© 1994 John Wiley & Sons Ltd.

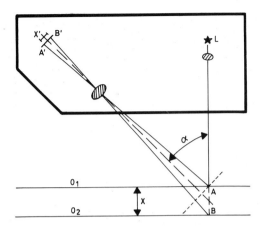

Figure 13.1 The measurement principle of the laser gauge probe. L = Light source. When the light hits the surface at A the reflection is focused through a lens at point A' on a detector (x'). If the beam hits the surface at B the reflection is instead focused at point B'. The distance X, between the surface heights O_1 and O_2 is the maximum measurement range. α is the angle between the light beam and the optical axis of the detector. If parts of a measurement site are inclined at more than this angle, no height measurements can be obtained from these parts

indicated by a visible red light spot. Non-linearity and inaccuracy can together give an error of 0.1% of the measurement range. With a good calibration routine the precision of height values is better than 0.1 mm. Laser gauge probes with higher resolutions are also commercially available. However, they have a smaller measurement range and cannot, therefore, be efficiently used on most natural rock surfaces.

For movements of the laser gauge probe in the x- and y-directions a special frame has been constructed. The material is mainly aluminium and all four legs are adjustable. As seen in Figure 13.2, a carriage for movements in the y-direction is mounted onto the frame. It is controlled by a stepping motor. On this carriage a smaller one for movements in the x-direction is placed. Apart from the laser gauge probe, a motor and an encoder for control of the position on the x-axis is mounted on the x-direction carriage. Movable parts are connected to supporting steel spindles by concave rolls with ball bearings. The stepping motor has a rack and pinion gearing against the frame. The other motor and the encoder have the same type of gearing against the y-direction carriage.

During field measurements a car accumulator is used as power supply. The two motors, the encoder and the laser gauge probe are connected to control boards placed in a box. A portable computer is used for measurement data collection, data processing and as the main unit of control. An IBM-compatible computer with at least a 286 processor and place for a full-length expansion board will fulfil the requirements. With the present device it is possible to map areas with a size of up to 40 × 40 cm in a time of less than 2 hours. The minimum interval between height measurements is approximately 0.25 mm in both the x- and the y-direction. This interval can be chosen in the software that has been written purposely for the micro-

Figure 13.2 The complete micro-mapping device during measurements of a well-preserved rock carving of a horse. The carving is situated at Aspeberget in the parish of Tanum in the northern part of the province of Bohuslän, southwest Sweden

mapping device. Since all parts of the equipment are portable, measurements can be performed also at some distance from roads.

The current (1992) price of the laser gauge probe is approximately £10 000. An equal amount of money is then needed for the computer, the construction of the frame and other peripheral devices. A more detailed technical description of a prototype of the device, as well as results from test measurements, are provided by Swantesson (1989, pp. 131–147 and pp. 184–193, and 1992a). If desired, it is possible to mount another laser gauge probe with, for example, a higher resolution on the frame. Future developments of the frame will improve the precision in both the x- and y-directions. It will also allow measurements on surfaces with all inclinations.

For repeated measurements on the same site the method for relocation of the micro-mapping device is of crucial importance. One possible approach is proposed by High and Hanna (1970). Three studs were fixed to the rock surface. One of the three legs of their micro-erosion meter had a conical depression, the second had a V-notch depression, and the third leg had a flat bottom. This system made relocations simple and fast.

However, in many places it is not ethically possible to drill studs into the rock surface. This is the case, for example, where Bronze Age rock carvings are present. In such places another system of ground control points has to be used. On each measurement site at least three points have to be marked. On, for example, a granite this can be discreetly accomplished by painting certain quartz grains. The quartz is less likely to weather away than other minerals. The exact position (x, y, z) of the ground points is measured with the device before each micro-mapping. Calculations can then be performed to ensure correct registration of maps derived from measurements taken on different occasions. With this method, it is also easy to compute the loss of material from the rock surface.

Other Precision Measurements of Small Surfaces

Different kinds of mechanical instruments for precision measurements of small surfaces are described in the literature. Dahl (1967), for example, used a simple apparatus for measuring deterioration near well-preserved quartz veins. It consisted of two bars with four adjustable legs. A plate, with a pressure pin for height measurements, could be moved between the bars. The micro-erosion meter described by High and Hanna (1970) was equipped with an engineer's dial gauge, calibrated to 0.002 mm. McCarroll (1992) also presents an instrument for field measurement of rock surface roughness in which the height values are obtained using a dial gauge.

The most obvious advantages of relatively simple mechanical instruments are that they are cheap and that their weights are low. They can, therefore, be carried into remote areas. The precision of single point measurements can also be very high. However, when large numbers of height values are to be measured in a short time, these types of instruments cannot compete with the micro-mapping device described in this article. Another disadvantage of mechanical instruments is that the measurements are not contact-free. Neither is the data directly available in digital form on a computer. This makes all calculations more time-consuming.

In industry a device that automatically moves a stylus, in a profile, over a surface is used. It can be applied when the surface finish of different products has to be controlled. The precision is equal to or better than with a dial gauge, and many hundreds of height values can be measured on a profile of a few centimetres length. The price of this equipment is approximately the same as for the micro-mapping device. For field use adjustments are needed. One disadvantage is that the measurements are not contact-free. It is, therefore, impossible to use this instrument on heavily weathered rocks or on soft soil material. Difficulties may also arise on very rough surfaces, such as many natural rock outcrops.

A device, in many aspects similar to the apparatus in this article, is described by Freij (1990). However, it is mainly intended for very small areas (up to a few square centimetres). When larger areas are to be measured, the time needed is considerable. Photogrammetry is often used for mapping. In contrast to the micro-mapping device, with a laser gauge probe. several steps are needed before data are available in digital form. The price of the equipment is also much higher. Furthermore, it is doubtful whether the precision is as good as with the method described here.

PROCESSING OF THE DATA

Calculations and other processing of the data can be performed immediately after the measurements are finished. Profile lines, for example, can be drawn on the computer monitor while measuring.

More than 100 000 height values are often available from one measurement site. Statistical methods are, therefore, often useful to characterise different types of surfaces. It is, for example, possible to calculate the specific area, correlations between similar surface elements and the height difference between the highest and the lowest point in unit areas. Inclinations and direction of slopes in unit areas can also be calculated.

The method used here has proved valuable in distinguishing different kinds of rock surfaces from each other. Unit areas with a size of 10 × 10 height values have been used. Their true size is 11.8 × 11.9 mm. The best fit of a sloping plane was calculated for each area, as well as the regression coefficient. The standard deviation of the height values from this plane was used as a measure of surface roughness. It is suggested that this roughness index should be standardised to, for example, 100 values in a unit area of exactly 1 square centimetre. This has, however, not yet been possible

Table 13.1 Statistics based on measurements of rock surfaces with and without rock carvings. The measurement sites and the calculations are described in the text. A = Number of unit areas used for the calculations. x = Average surface roughness in mm. σ = Standard deviation of the surface roughness in mm. L = Lowest surface roughness on the measurement site in mm. H = Highest surface roughness on the measurement site in mm

Measurement site	A	x	σ	L	H
Sillvik I 1.8 m a.s.l.	460	0.213	0.090	0.093	0.849
Sillvik II 8.0 m a.s.l.	460	0.468	0.220	0.145	1.872
Sillvik III 18.5 m a.s.l.	460	0.476	0.273	0.112	2.246
Huds Moar	228	0.373	0.087	0.261	0.985
Tegneby-Litsleby	138	0.217	0.048	0.115	0.331
Aspeberget—Bull	520	0.341	0.114	0.177	0.961
Aspeberget—Horse	672	0.177	0.065	0.090	0.708
Aspeberget—Ship	546	0.311	0.074	0.172	0.713
Kalleby	570	0.331	0.089	0.145	0.741
Massleberg	270	0.209	0.163	0.095	1.316

with the prototype frame used. Roughness indices for different types of rock surfaces are found in Table 13.1. Since each measurement site comprises several unit areas, further statistical analysis can easily be made. For example, the standard deviation, skewness and kurtosis of the roughness distribution can be calculated (Swantesson, 1992b).

The measured surfaces can be visualised by presenting them in the form of contour maps or three-dimensional (3-D) models. For these methods a plotter or a printer of good quality is necessary. In the case of maps, the scale and the contour interval can be varied. For flat surfaces a contour interval of 0.2 mm, or less, can often be used. 3-D models can be rotated and tilted in order to obtain the best picture of the mapped area. It is also possible to vary the vertical exaggeration. For most surfaces an exaggeration of × 2 is enough. However, to be able to see details on very smooth surfaces it sometimes has to be up to × 10.

EXAMPLES OF MEASUREMENTS

To date, the micro-mapping device has been used to characterise rock surfaces exhibiting different types of weathering phenomena and to follow the progress of weathering with increasing altitudes along the Swedish west coast. Measurements have also been performed on rock carvings in the county of Göteborg och Bohus, southwest Sweden. The carvings are an important cultural heritage and their survival is often threatened by weathering. The location of the sites described here are shown on the map in Figure 13.3.

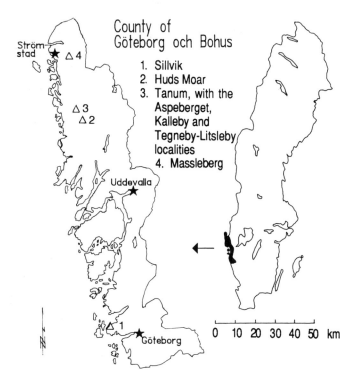

County of Göteborg och Bohus

1. Sillvik
2. Huds Moar
3. Tanum, with the Aspeberget, Kalleby and Tegneby-Litsleby localities
4. Massleberg

Figure 13.3 Map showing the location of the sites described in the text

Characterization of Weathered Rock Surfaces

A classification of mainly micro-weathering forms occurring in Southern and Central Sweden has been made by Swantesson (1992c). Such forms are excellently represented by contour maps or 3-D models. An example of pitting in an amphibolitic metabasite from central Göteborg is shown in Figure 13.4. The weathering form consists of rounded holes up to 1 cm in diameter and 1 cm deep. It is most common in basic, fine-grained rocks. The dissolution of opaque minerals is often the primary reason for pitting.

A 3-D model from a zone with intense granular disintegration is reproduced in Figure 13.5. The zone stretches N–S over a distance of 7 km in the city of Göteborg.

Pitting, Johanneberg, central Göteborg

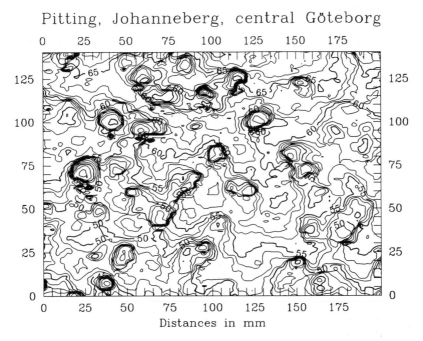

Figure 13.4   Contour map of a surface with pitting in a metabasite. The contour interval is 1.25 mm

The rock is a medium-to coarse-grained, gneissic, alkalic granite. A clay mineralisation of montmorillonite along the grain boundaries is present where rapid recent granular disintegration occurs (Samuelsson, 1973). In Figure 13.5 it can be seen that the upper surfaces are relatively flat. They are only partially affected by visible weathering. However, the rock on the steep sides can be detached by hand. In front of the slopes there is a surface with loose, poorly cemented mineral fragments. The roughness is directly correlated to the size of the fragments.

Measurements at Different Altitudes

The areas along the Swedish west coast offer good opportunities to study the progress of micro-weathering in crystalline rocks during the Holocene. Bare rocks are common and due to isostatic land rise the length of time that the rocks have been exposed to atmospheric conditions is directly related to their altitude above sea level. Although the best preserved rock surfaces are found beneath loose deposits, such as clay (Samuelsson, 1964), the weathering rate beneath the sea surface is very low. The present land rise near Göteborg is 1.5 mm yr^{-1} (Miller and Robertsson, 1988) and the marine limit lies at 110 m a.s.l. (Norin and Swantesson, 1989). The relationship between weathering and the rate of land rise is worthy of more research, since we can thereby learn more about the natural rate of weathering and how it works.

Micro-mappings have been performed at three altitudes at Sillvik (no. 1 on the map in Figure 13.3). The rock is a fairly inhomogeneous, grey, plagioclase-rich gneiss,

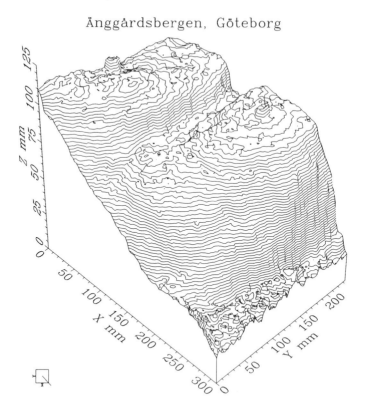

Figure 13.5 3-D model with stacked contour lines in a perspective projection, of a surface with granular disintegration. The vertical exaggeration is × 2. Two nuts placed on the site for relocation purposes are seen on the flat upper surfaces

traversed by quartz veins (Sandegren and Johansson, 1931). As seen in Table 13.1 and Figure 13.6, the average roughness value is more than twice as large at 8.0 m a.s.l. as at 1.8 m a.s.l. The reason is that micas at the surface are weathered away at the higher altitude as a result of the longer time of exposure to weathering agents. An increased surface roughness due to limited mineral weathering is often one of the first signs of rock deterioration, but it cannot always be observed with the naked eye. The roughness values are almost the same at 18.5 m a.s.l. as at 8.0 m a.s.l. (Table 13.1), which suggests that, when a certain degree of irregularity is reached, slow surface lowering continues without a further increase of the micro-relief. Only larger quartz nodules and veins proceed to increase their height in relation to the rest of the rock. At 18.5 m a.s.l. they project about one centimetre. Maps and 3-D models from the measurements at Sillvik are found in Swantesson (1992a).

At three locations where glacial striations are present in granite, comparative measurements have been made at two different altitudes. In all locations the surface roughness increases with the height above sea level and length of the time of exposure to postglacial weathering (Swantesson, 1989, pp. 114–145, and 1992b). The micro-relief gradually increases by the detachment of single mineral grains, or fragments

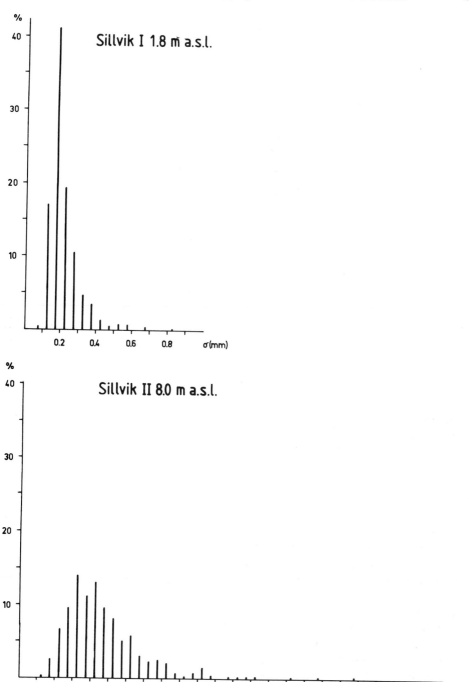

Figure 13.6 Diagrams of the distribution of roughness values on rock outcrops at two different altitudes at Sillvik, in the vicinity of Göteborg. The elevation is indicated in each diagram. The standard deviations of height values are plotted on the x-axis and the per cent of unit area, on which the calculations are based, on the y-axis

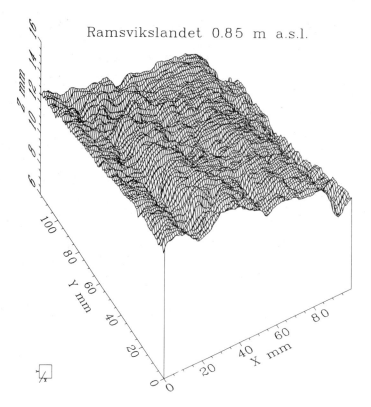

Figure 13.7   3-D model in orthographic projection of a granite surface with glacial striations on Ramsvikslandet, 90 km north of Göteborg. The vertical exaggeration is × 7.5

and at the same time the glacial striations become shallower and wider. At last they disappear entirely. The reason is that the mineral weathering seems to be faster at their borders than elsewhere. A 3-D model of a rock surface with striations is reproduced in Figure 13.7.

Huds Moar is situated along the main road from Göteborg to Oslo at an altitude of 100 m a.s.l. (No. 2 on the map in Figure 13.3). The area suffered a severe forest fire less than two decades ago. As an effect of the fire, the present rock breakdown is fast and takes the form, according to the terminology used by Swantesson (1992c), of a shallow granular disintegration. Large amounts of grus are trapped at the borders between rock outcrops and vegetation (Figure 13.8). The bedrock is Bohus granite, which covers huge areas in the county of Göteborg och Bohus. Its composition is limited to the field of monzogranite, but it varies widely in texture and colour (Eliasson, 1987). At the measurement site the rock is equigranular and medium-grained. The roughness value (Table 13.1) is probably the highest that can be attained in this type of rock, if major quartz veins or fissures are not present. The value is lower than at the two highest altitudes at Sillvik. This is because the maximum micro-relief is largely dependent on the grain size of the rock experiencing weathering.

Figure 13.8 Grus accumulations at the border between a Bohus granite outcrop and vegetation at Huds Moar, 105 km north of Göteborg. Rapid weathering has been induced by an intense fire in the area

Measurements on Sites with Rock Carvings

The micro-mapping method has been used on several sites with rock carvings in the northern part of the county of Göteborg och Bohus. All the mapped sites are on Bohus granite. Many of the carvings are deteriorating. The method is useful to assess where protection measures are needed and to monitor the decay exactly. More than 1000 sites with over 40 000 rock carvings are reported from this part of the county by Bertilsson (1987). He describes the region as the main rock art area of all Scandinavia, and perhaps even of Europe. The carvings are of very varied form, showing, for example, ships, humans, foot-prints, animals, circles, ring-crosses and cup-marks. The majority of the carvings are from the Bronze Age (Coles, 1990), which began in 1800 BC, when the sea level was about 20 m higher than today, and ended 500 years BC. The depth of the rock carvings is usually greater than one millimetre. In some cases they can reach 10 mm deep. This is, for example, common for cup-marks.

Three of the measured sites are from the Aspeberget locality in the parish of Tanum (No. 3 on the map in Figure 13.3). It is situated on quite steep slopes facing north and east. As seen in Table 13.1, the sites display varying roughness values. The horse carving is almost unaffected by weathering and cut-marks can still be observed. The rock is smooth, and micro-striations are present in the granite. Mineral weathering is limited to only a few places.

The site with the ship carving displays several signs of recent damage. The rock surface is rusty coloured and has a granular appearance. The iron-rich impregnations

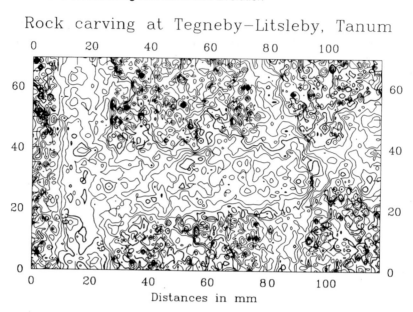

Figure 13.9 Contour map of part of a rock carving with mounted warriors at Tegneby-Litsleby, Tanum. The contour interval is 0.2 mm. Where the rock carving is present the contour lines are less dense than elsewhere

and alterations in the upper few millimetres of the rock cause flaking. The flakes are approximately 1 mm in thickness, and loose edges are common. The rock beneath the flakes is the original colour. As seen in Table 13.1, the roughness is approaching the value on the heavily weathered granite at Huds Moar. The bull carving is situated under a roof, built in the hope of protecting the figures at this site. The roughness values are high here also (Table 13.1). This indicates that just a little further weathering will cause the carvings to disappear. In many places rock fragments are loose and the red paint of the bull is flaking off. A further reason for the fast weathering is that, despite the roof, water regularly trickles over the granite.

The rock carvings at Tegneby-Litsleby are situated 1 km southwest of Aspeberget. They consist of a group of mounted warriors on horses, armed with spears and rectangular shields (Figure 13.9). The flat outcrop is well preserved, although the weathering has gone somewhat further than at the site with the horse at Aspeberget (Table 13.1). The outcrop at Massleberg, No. 4 on the map in Figure 13.3, is also well preserved, and micro-striations are still present. Flaking has occurred near the rock carvings, but it is not active at present. The higher standard deviation of roughness values at this site is due to a fissure crossing the mapped area (Table 13.1). The rock surface at Kalleby, 4.5 km south of Aspeberget, has roughness values similar to the earlier described ship and bull carvings (Table 13.1). The figures are difficult to recognise, and the outcrop is extensively covered by lichens. Mineral weathering is present, as well as some damage due to flaking.

The low roughness values on three of the investigated rock carving sites, correspond to an elevation of less than 2 m a.s.l., when they are compared with measurements from other places. This implies an exposure to postglacial weathering for less than

1000 years. One explanation for these smooth surfaces could be that the rock is less prone to weathering than elsewhere. However, this is not plausible since carvings at other sites are also on homogeneous rocks and in similar topographical settings. A better explanation is that the well-preserved carvings have been covered by sufficiently thick loose deposits after they were made.

The other three carvings seem to have roughness values that are normal for the altitude and their time of exposure to atmospheric conditions. With a decay rate of 1–1.5 mm per thousand years in crystalline rocks, they will finally disappear (Swantesson, 1992b). Unfortunately, the decay rate looks set to rise as a result of increased pollution and 'acid' rain. This means that there is an even greater urgency to protect the carvings. Indeed, it is possible that many rock carvings have already been destroyed by weathering, particularly on unfavourable sites and in rocks 'softer' than Bohus granite.

CONCLUSIONS

Measurements with the micro-mapping device show promising results. It is suitable to trace decay on all types of stone material. To give a complete picture of the weathering status the method should be accompanied by photographs and descriptions of visible damage and lichen cover, and analyses of the mineralogy and local climate. Measurements on more sites will ultimately make it possible to construct a model of how Holocene micro-weathering works in different kinds of rocks. Thereby it will be possible to predict future changes under varying sets of environmental circumstances. The device can be used in all cases where a detailed mapping of small surfaces is desired and not just in a rock weathering context. Minor forms caused by glacial erosion, structures and wind ripples are some examples.

ACKNOWLEDGEMENTS

I thank the Swedish Conservation Institute of National Antiquities for financial support. Hans Alter has helped me to construct the prototype frame. Björn Oldberg is helping me with programming and electronics. Rolf Nyberg critically read a draft of the paper.

REFERENCES

Bertilsson, U. (1987). The rock carvings of northern Bohuslän—spatial structures and social symbols. *Stockholm Studies in Archaeology*, **7**, 1–203.
Coles, J. (1990). Images of the past. *Bohusläns museum och Bohusläns hembygdsförbund—Uddevalla*, pp. 1–92.
Dahl, R. (1967). Post-glacial micro-weathering of bedrock surfaces in the Narvik district of Norway. *Geografiska Annaler*, **49A**, 155–166.
Eliasson, T. (1987). Trace element fractionation in the late Proterozoic Bohus granite, SW Sweden. *International Symposium on Granites and Associated Mineralizations*, Abstracts, Salvador, Brazil, pp. 211–214.
High, C. and Hanna, F. K. (1970). *A Method for the Direct Measurement of Erosion on Rock Surfaces*. British Geomorphological Research Group, Technical Bulletin 5, pp. 1–24.

Freij, H. (1990). Dokumentation och analys av arkeologiska ytstrukturer. In Arrhenius, B. (ed.), *Laborativ Arkeologi 4*, Arkeologiska forskningslaboratoriet—Stockholm, pp. 49–56.

McCarroll, D. (1992). A new instrument and techniques for the field measurement of rock surface roughness. *Zeitschrift für Geomorphologie*, NF **36**, 69–79.

Miller, U. and Robertsson, A. M. (1988). Late Weichselian and Holocene environmental changes in Bohuslän, southwestern Sweden. *Geographia Polonica*, **55**, 103–111.

Norin, J. and Swantesson, J. (1989). Is the marine limit in the Göteborg area too low? *Geologiska Föreningens i Stockholm Förhandlingar*, **111**, 7–10.

Samuelsson, L. (1964). Nya fynd av subglacialt bildade kalkstenar. *Geologiska Föreningens i Stockholm Förhandlingar*, **85**, 414–427.

Samuelsson, L. (1973). Selective weathering of igneous rocks. *Sveriges Geologiska Undersökning*, **C 690**, 3–16.

Sandegren, R. and Johansson, H. E. (1931). Beskrivning till kartbladet Göteborg. *Sveriges Geologiska Undersökning*, **Aa 173**, 1–141.

Swantesson, J. (1989). *Weathering Phenomena in a Cool Temperate Climate*. Göteborgs Universitet, Naturgeografiska Institutionen—GUNI Rapport 28, pp. 1–193.

Swantesson, J. (1992a). *Mikrokartering av naturstensytor*. Högskolan i Karlstad, Naturvetenskapligt forskningsforum—Arbetsrapport, 92: 6, 1–36.

Swantesson, J. (1992b). A method for the study of the first steps in weathering. In: Kuhnt, G. and Zölitz-Möller, R. (eds). *Beiträge zur Geoökologie*, Kieler Geographischen Schriften, 85, pp. 74–85.

Swantesson, J. (1992c). Recent microweathering phenomena in Southern and Central Sweden. *Permafrost and Periglacial Processes*, **3**, 275–292.

14 Diagnosis of Weathering on Rock Carving Surfaces in Northern Bohuslän, Southwest Sweden

RABBE SJÖBERG
Umeå University, Sweden

ABSTRACT

As a part of a project concerning the recent weathering of Bronze Age rock-carvings in Bohuslän, which is presumed to be caused by increased acidity in the area, a number of granite rock surfaces with rock art have been tested using the Schmidt hammer. The aim of this study was to determine the weathering status of the rock surfaces surrounding the carvings and to investigate possible controlling factors. The results showed that the sites could be divided, according to the weathering status, into three groups where Group I was only slightly weathered, and Group III was highly weathered. The results showed that rock surfaces on greyish-red, medium-grained and mainly even-grained granite have a tendency to weather somewhat faster than those on other types of granite in the area. Proximity to larger roads had no significant impact on the degree of weathering, but the existence of (mostly recently planted) spruce forests, which develop a very acidic litter, seems to increase the rate of weathering on the rocks, as does water from this biotope trickling over the rock surface. There is also a tendency for granite rock surfaces that are stepped on by visitors to become secondarily polished, and this polishing seems to delay chemical weathering.

INTRODUCTION

Weathering, together with the processes of erosion and transport, is responsible for the denudation of land-surfaces. This normal denudation may cause severe problems, when, as in the province of Bohuslän, rock surfaces are decorated with cultural monuments in the form of rock-carvings which are thousands of years old. The denudation, in the long term, implacably leads to the destruction of these rock-carvings. Has the increased acidity, which especially has affected the granite- and gneiss-regions in western Sweden, increased the speed and intensity of the weathering? What effect does this acidity have on the carved rock surfaces? Are the unique cultural monuments disappearing because of this increased acidity? These questions have worried the authorities taking care of the cultural monuments.

At the Center for Arctic Research at Umeå University, a method of quantifying weathering on bedrock and raised boulder-beaches has been investigated and developed during the latter part of the 1980s (Sjöberg, 1990a, 1991a and 1991b; Sjöberg and

Rock Weathering and Landform Evolution. Edited by D. A. Robinson and R. B. G. Williams
© 1994 John Wiley & Sons Ltd.

Broadbent, 1991). This method has also been shown to have applications within other fields of research. This development led Dr Gustav Trotzig at the Central Board of National Antiquities (CNBA) to ask the Center if the method could be used to diagnose the state of weathering of rock-carving surfaces of Bohuslän.

The fieldwork took place in September 1988 and May 1989, when a total of 30 rock-carving surfaces chosen by Dr Ulf Bertilsson (CBNA) were investigated. The results of these investigations are presented and discussed in this paper.

RESEARCH AREA, TYPE OF BEDROCK AND WEATHERING OF BOHUS GRANITE

The research area is situated within Strömstad, Tanum and Lysekil communities, in the province of Bohuslän, in the southwestern part of Sweden (Figure 14.1). All the area lies within the outcrop of what is specifically identified as Bohus granite, which, for Sweden, is a relatively young bedrock. Within the research area Bohus granite can be separated into three main types: (1) red, medium-grained, mainly porphyritic

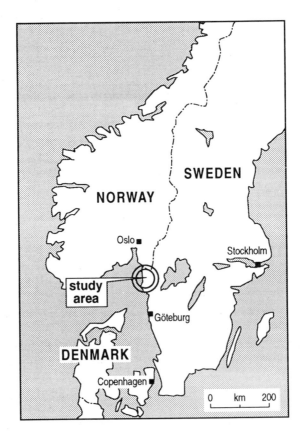

Figure 14.1   The research area in the northern part of the province of Bohuslän, southwest Sweden

granite, or 'wild' granite; (2) red, medium-grained, mainly even-grained granite; and (3) greyish-red, medium-grained, mainly even-grained granite (Asklund, 1947).

Bohus granite mainly has the following mineral content: quartz (20–30% by volume), feldspars such as microcline and plagioclase (60–70% by volume) and dark minerals such as mica, magnetite etc. (Asklund, 1947; Hillefors, 1983).

Bohus granite is comparatively rich in quartz (Hillefors, 1983), and thus it ought to have been affected by chemical weathering and frost-shattering to a higher degree than other types of Precambrian rocks, which generally have a lower content of quartz. The chemical weathering of Bohus granite is mostly shown in the form of weathered gravel, but also in the form of a discoloured weathered upper zone, 5–10 cm deep. This so-called 'barkskikt' has been discoloured by weathering of the dark minerals, and is normally interpreted as the depth of weathering that has taken place since the last glaciation (Asklund, 1947, p. 78), but it may also be the roots of a much older zone of deep weathering.

Weathering of the Bohus granite is characterised by flaking of the corners and faces of joint blocks. According to Hillefors (1983), the reasons for this foliation include lingering tension, and unloading in the bedrock, released at the surface. Along the joints, vertical weathering zones can be found with a breadth of a couple of centimetres (Asklund, 1947). Out of these vertical fissures, iron-rich minerals have been dissolved and redeposited, evidenced by brownish-black trickle paths of iron oxide on the surfaces (see Figure 14.6 for examples). All these phenomena have recently been investigated by Swantesson (1989).

During the Quaternary glaciations the granite-landscape was morphologically transformed. This transformation was probably not as great as previously interpreted (Lidmar-Bergström, 1987 and 1988). According to these theories the roches moutonnées are the base of a former deep-weathering zone (Lindström, 1988), which explains, in part, the discoloured weathering zone. According to this theory, these forms might be remnants of the rock surface preserved below the regolith (Rudberg, 1973; So, 1987). During glaciation, these rounded rocks were ground into bright polished surfaces by the plasticity of the ice (Figure 14.2), and were cut by glacial striae. These striae together with flaking, and later the carvings (Figure 14.3), initiate the weathering of the rock-carving surfaces. The estimated eustatic land-uplift in the area is 2.6 mm y^{-1} (Mörner, 1979).

The progress of weathering is clearly visible on the rock surface. First, the lower parts of the polished surface start to foliate. These foliation surfaces become larger and larger, and in time form a granular surface, out of which harder minerals, such as feldspar crystals, protrude (Figure 14.4). The planar upper surface of these crystals mostly show glacial micro-striae, which indicate the level of the rock surface during the melting of the glaciers. If these micro-striae are found level with the rock surface, this is an indication that the rock has weathered very slightly in post-glacial time. The above-mentioned granular surface in time develops into an undulating coarse granular surface as the dark minerals in the bedrock become more weathered and are eroded away. This leads to an increase in the area of the bedrock surface, which itself increases the speed of the weathering.

Thus, it is possible to determine whether a bedrock surface is weathered from simple visual analysis. However, it is not easy to judge the state of weathering and thus not possible to compare one surface with another.

Figure 14.2 Bright polished bedrock surface with glacial striae and rock carvings at Litsleby (Tanum 75)

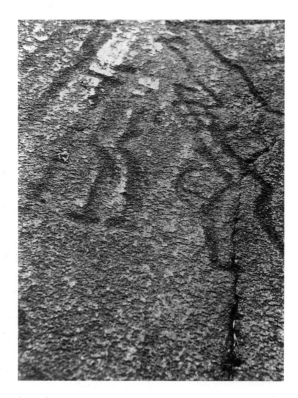

Figure 14.3 Striated, severely weathered bedrock surface at the rock carving site of Aspeberget (Tanum 12)

Figure 14.4 Macrophoto of weathered bedrock surface at Aspeberget. The site is situated above the rock-carving Tanum 23. *In situ* feldspar crystals with glacial micro-striae protrude a couple of millimetres above the weathered bedrock surface and show the level of the original late-glacial bedrock surface

METHOD

To measure the weathering, the Schmidt Test-hammer Type L was used, which has an impact energy of 0.735 Nm. With this relatively small impact energy and the well-rounded top of the impact plunger, this type is a better choice when testing easily damaged surfaces, such as the rock-carvings, than the commonly used Type N instrument with an impact energy of 2.227 Nm. The instrument's error and the sources of error when using the Type N instrument have been studied by McCarroll (1987 and 1989), who also made suggestions to minimise their effect.

Since the method depends on a statistical treatment of data, a minimum of five to seven impacts must be done on every tested surface (Proceq, 1977). For each of the 30 rock surfaces investigated in this study, the mean value of five subsamples has been used for a diagnosis of the state of weathering. Ten impacts were done on each subsample, and then the mean value, median value and standard deviation were calculated. The overall state of weathering is taken to be the mean value (*Rm*) of 50 impacts.

During the first field season, in 1988, twelve of the best known sites with rock-carvings in the region were tested (Sjöberg, 1988). The main goal was to determine the weathering status of the rock surrounding the carvings in terms of *Rm* values, but the altitude of the carvings, as read by a calibrated aneroid barometer, and the orientation of the carvings, as shown by a Suunto compass, were also studied. In 1989 another 18 sites were tested using the same method, complemented by a more thorough study of other factors which could influence the weathering, including the type of granite bedrock in the area, the inclination of the rock surfaces, cracks and joints in the rock, the distance from larger roads, and the present vegetation in the area.

RESULTS

In 1988 the following sites with rock-carvings were studied: Hede (Kville 124 and 125), Karlsund (Kville 181 and 190), Krokbräcke (Kville 5), Litsleby (Tanum 75), Tegneby (Tanum 72), Aspeberget (Tanum 12 and 23), Vittlycke (Tanum 1) and Fossum (Tanum 260). Detailed data of the weathering status etc. are found in Sjöberg (1988) and are summarised in Table 14.3.

In 1989 complementary studies were done on some of the above-mentioned sites and on new sites, at Fossum (Tanum 260, 253 and 254), Tegneby (Tanum 33), Litsleby (Tanum 51 and 66), Bro Utmark (Tanum 192), Kalleby (Tanum 405), Torp (Kville 216), Brastad-Bracka (Brastad 1, 14 and 18), Vese (Bro 26a) and Massleberg (Skee 614). Some of the results from these sites are described below and are summarised in Tables 14.1 and 14.3.

1. Fossum (Tanum 260)

This is one of the most famous, and, thus, most visited rock-carvings in the area. The rock surface is situated at 50 m a.s.l., faces southeast (145°) and has a dip of 14–17° toward a gravel road which carries heavy traffic in summer. The surrounding vegetation consists of mixed forest, dominated by juniper, birch and young stands of spruce. The rock surface is crossed by joints in a northwest–southwest direction. Seepage water trickling across part of the surface has created paths of black coatings at the joints. To the naked eye, the rock surface looks well preserved with an even, polished to granulated morphology (Figure 14.6).

The R values obtained are shown in Table 14.1. The Rm value indicates a very slight degree of weathering. However, the low value of subsurface 3, as well as the high s-value on this subsurface, partially indicates an initial severe weathering.

2. Tegneby (Tanum 33)

This rock surface, which is situated at an altitude of 25 m, has a dip of 10°, and faces east–northeast. The surface is covered by a thin layer of seepage water that trickles from a dung-hill. The morphology of the carved rock surface is classified as polished to granular. The part of the surface most affected by the dung-water is coarsely granular. The contrast between the well-preserved, polished to granular surface of the carved part of the surface, and the coarsely granular surface of areas covered by trickling seepage water is a significant feature of the site. The carved surface is crossed by a dyke of feldspar. The surrounding vegetation is meadow.

3. Litsleby (Tanum 66)

This bedrock surface is situated on the outer part of a small peninsula, at an elevation of 20 m a.s.l., surrounded by meadows and tilled fields. The bedrock surface dips 11°

toward the southeast (138°). The surface is crossed by several joints striking north–south, and has a larger, central exfoliation surface caused by fire. The morphology of the surface is even to granular. A couple of brownish-black trickle paths are found on the surface.

The bedrock surface has relatively high and even R values. The values from the foliated area do not differ markedly from those of the other subareas, but the trickle paths yield lower values.

4. Bro Utmark (Tanum 192)

This is a relatively steep (18°) bedrock surface, exposed towards the northeast (48°). It is situated at an altitude of 25 m, in a spruce forest with occasional pines, just above ploughed land. The surface is kept constantly wet by trickling water, and most of the surface is covered with needles. The structure of the surface is polished with visible micro-striae.

The surface has a relatively low Rm value, which indicates a certain softening of the surface. The reasons may be the constant thin flow of water, and acid products in the form of rotting spruce needles.

5. Kalleby (Tanum 405)

This bedrock surface, which faces southeast (128°), and is almost horizontal (4°), is kept constantly wet by trickling water. It is situated at an altitude of 30 m, in spruce forest below a sparsely vegetated granite hill. This surface is methodologically interesting in that, in the middle of the 1980s, a cast was taken of the rock carving using silicon material. This cast surface is clearly visible on the bedrock surface in the form of a light square. The original structure of the surface varies from polished to granular, with a foliated area in the northwest corner. At this location, first the original surface and then the cast surface were tested. The subsurfaces of the cast part of the bedrock surface are situated immediately adjacent to corresponding subsurfaces of the original surface.

The investigation clearly shows that casting with silicon can damage rock carvings, because the cast surface has been loosened and become more easily weathered than the original bedrock surface.

6. Brastad-Backa (Brastad 1)

These are the most northerly and most visited rock-carvings of the three investigated sites at Brastad-Backa. The bedrock surface is situated at 45 m a.s.l., near a gravel road that carries heavy tourist traffic in the summer. The surface dips 7° towards the southeast (122°). It is surrounded by mixed juniper forest, and has an even to coarse, granular structure. The most even parts, subsurfaces 3–5, are situated on the part of the bedrock surface which seems to be most visited by tourists, because of the famous rock-carving known as 'Skomakaren' (the Shoemaker).

The values from the more even subsurfaces are clearly separable from those from coarse granular subsurfaces.

Table 14.1 Investigated localities, subsamples, morphology, mean value (m), range (R), standard deviation (s), and weathering coefficient (Rm)

Locality	Subsample	Morphology	m	R	Rm	s
1. Fossum (Tanum 260)	1	even	54.4	48–62		3.9
	2	even with path	55.6	43–61		7.8
	3	granulated	39.0	0–58		16.8
	4	granulated with path	44.6	28–52		7.7
	5	even	55.6	35–61	49.8	8.3
2. Tegneby (Tanum 33	1	polished	57.1	45–62		5.2
	2	polished-gran.	54.8	40–60		6.7
	3	polished-gran.	59.8	56–62		2.3
	4	polished-gran.	54.5	44–60		4.9
	5	polished-gran.	50.9	41–59	55.3	6.5
	6	coarse granular	33.3	15–48		9.4
3. Litsleby (Tanum 66)	2	even	45.6	24–57		12.6
	3	even-granular	44.7	26–54		11.8
	4	polished-even	45.7	28–58	47.4	12.9
	5	exfoliation surf.	42.8	22–55		11.4
	6	trickle path	35.4	0–49		16.2

4. Bro Utmark (Tanum 192)	1	polished	44.4	30–57		12.5
	2	polished	45.5	26–60		11.8
	3	polished	46.1	28–56		10.3
	4	polished	30.9	15–52	40.2	13.0
	5	polished	34.1	24–56		11.6
5. Kalleby (Tanum 405), original surface	1	polished-gran.	42.9	20–54		13.7
	2	foliated	17.2	0–44		17.9
	3	even	44.5	17–55		11.5
	4	granular	26.0	17–52	32.6	9.5
cast surface	1	polished-gran.	25.7	17–50		9.9
	2	foliated	20.8	0–47		16.3
	3	even	22.8	14–42		7.9
	4	granular	17.5	0–39	21.7	14.1
6. Brastad-Bracka (Brastad 1)	1	even-granular	44.6	34–52		6.7
	2	coarse granular	32.6	24–49		7.3
	3	even-granular	48.3	36–55		7.3
	4	even-granular	51.2	44–55		3.6
	5	even-granular	44.7	38–54	44.3	6.0

Table 14.2 Rank and *Rm* values of investigated sites

Rank	Site	*Rm*
I	Tanum 33	55.3
	Bro 26A	52.9
	Tanum 72	51.4
	Tanum 75	50.8
	Tanum 260	49.8
	Tanum 51	49.8
	Tanum 1	49.7
II	Tanum 254	47.7
	Tanum 66	47.4
	Tanum 23	45.8
	Kville 124	45.4
	Kville 5	45.2
	Brastad 1	44.3
	Tanum 253	43.7
	Kville 156a	42.3
	Kville 125	32.1
III	Tanum 12 S	40.4
	Tanum 192	40.4
	Tanum 12 N	38.7
	Tanum 66a	37.3
	Kville 190	36.7
	Kville 181	36.1
	Kville 216	35.9
	Brastad 14	33.8
	Kville 156b	33.0
	Tanum 405	32.6
	Brastad 18	27.7

DISCUSSION

From the *RM* values, the state of weathering on the different bedrock surfaces can be evaluated and the surfaces ranked (Table 14.2). The surfaces are here ranked in three classes, where I indicates a very slight degree of weathering, and III a high degree of weathering.

What factors affect the weathering on the carved rock surfaces? Recent acidity has been suggested as an important reason why several of the rock-carvings are disappearing, but how it is possible to find objective, measurable factors showing that the acidity is solely the reason for this weathering? In the investigation of 1988, factors such as the altitude and the exposure of the surfaces were correlated against the weathering value (*Rm*). No correlation was found. In 1989 a number of new factors were investigated. These were type of bedrock, the dip of the bedrock surface, the distance from the nearest road carrying vehicular traffic, and the nature of the surrounding vegetation.

The different values of these factors are listed in Table 14.3, and are discussed below.

Table 14.3 Investigated possible factors determining the weathering of bedrock surfaces

Site	Orientation (°)	Elevation (m a.s.l.)	Rm	Dip (°)	Distance from nearest road[a]	Vegetation	Bedrock[b]
Skee 614	78	31	49.4	13	3	meadow	2
Tanum 1	115	25	49.7	17.5	2	mixed	1
Tanum 12N	0	25	38.7	21	3	pine	1
Tanum 12S	45	25	40.4	21	2	pine	1
Tanum 23	135	25	45.8	14	3	pine	1
Tanum 33	62	25	55.3	10	3	meadow	1
Tanum 51	55	45	49.8	8	3	pine	1
Tanum 66	138	20	47.4	11	3	meadow	1
Tanum 66a	140	20	37.3	9	3	meadow	1
Tanum 72	55	35	51.4	2	2	pine	1
Tanum 75	135	45	50.9	7	1	pine	1
Tanum 192	48	25	40.2	18	3	spruce	1
Tanum 405	128	30	32.6	4	3	spruce	1
Tanum 253	170	45	43.7	0	1	mixed	2
Tanum 254	180	45	47.7	12	1	mixed	2
Tanum 260	145	50	49.8	16	1	mixed	2
Kville 5	135	35	45.2	6	2	deciduous	1
Kville 124	225	50	45.4	9	2	spruce	3
Kville 125	90	60	42.1	12	1	mixed	3
Kville 156a	157	30	42.3	11	3	deciduous	3
Kville 156b	157	30	33.0	11	3	deciduous	3
Kville 181	45	30	36.1	25	3	deciduous	3
Kville 190	112	25	36.7	13	3	deciduous	3
Kville 216	145	25	35.9	23	1	pine	3
Bro 26a	43	25	52.9	11	2	meadow	3
Brastad 1	122	45	44.3	27	1	mixed	3
Brastad 14	71	40	33.8	11	1	mixed	3
Brastad 18	67	45	27.7	17	1	mixed	3

[a]See test for three classes of distance.
[b]See text for description of three types of bedrock.

The Altitude of Investigated Bedrock Surfaces

As can be seen from Figure 14.5a, the altitude of the investigated surfaces varies from 20 m up to 65 m a.s.l. A large number of the surfaces are situated between 20 and 35 m a.s.l. Another group is distributed between 45 m and 50 m a.s.l., with a single surface at an altitude of 60 m. The spread of the *Rm* values is great in both these groups. The correlation between altitude and *Rm* value is very low ($R_{xy} = 0.12$, $p = 0.5$, i.e. not statistically significant). Thus, over the range of altitudes investigated, the degree of weathering of the Bohus granite does not increase with altitude. Within the study area the fact that post-glacial transgressions reach the 35 m level (Hessland, 1943) must also be taken into consideration.

The Exposure of Investigated Bedrock Surfaces

From Table 14.3 it can be seen that most of the bedrock surfaces are exposed within a sector from north 45° to south, with single surfaces facing north, west and southwest. The exposure of the surfaces may be of importance regarding the original function of the rock-carvings. In the present context, it is more important to note that most of the surface are directly exposed to sunshine only during the morning hours, and that, already at noon, they are shadowed, and less affected by direct heating from sunshine, which otherwise could initiate sun-cracking of the bedrock surface. This may explain why statistical correlation between exposure and *Rm* values is very low (R_{xy}-0.13, with $p = 0.5$, i.e. not statistically significant).

The Inclination of Investigated Bedrock Surfaces

A Suunto clinometer was used to measure the inclination or dip of the bedrock surfaces, and the instrument was read to the nearest full degree. As can be seen from the graph (Figure 14.5b), most of the surfaces have a dip of 8–15°. The correlation coefficient between dip and *Rm* is low for the full data set ($R_{xy} = -0.34$, which is marginally significant at $p = 0.007$). However, Figure 14.5b shows that all surfaces with a dip of 20° or greater have *Rm* values below 40. This indicates that bedrock surfaces with a relatively steep inclination are more affected by weathering than surfaces with a lower inclination. The explanation may be that the transport processes are faster on steeper slopes, and that this faster transport of weathered products leads to a faster increase of the surface area, which increases the speed of weathering.

Figure 14.5 Diagrams showing correlation of *Rm* values with other features. (a) *Rm* is not correlated with the altitude of the investigated bedrock surfaces with rock carvings. (b) A weak negative correlation is found between the inclination of the investigated bedrock surfaces and *Rm*. Steeply inclined bedrock surfaces (>20°) seem to have relatively low *Rm* values. (c) The distribution of *Rm* statistical means and standard deviations of the bedrock surfaces of the three different types of Bohus granite that have been investigated. (d) The distribution of *Rm* values on bedrock surfaces surrounded by different types of vegetation. Mean values and standard deviations are shown

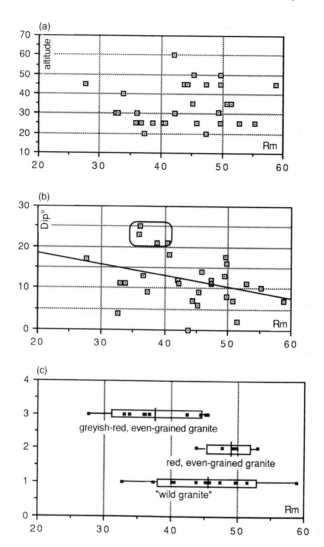

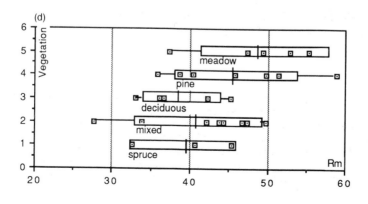

Different Types of Granite Forming the Investigated Bedrock Surfaces

Since weathering can be defined as the disintegration of rocks into products that are in greater equilibrium with the present environment (Swantesson, 1989), it is of interest to find out if the different types of granite in the region react differently to weathering.

As already mentioned, the investigated region is totally within the group of granites called the Bohus granite. Asklund (1947) divided the Bohus granite in the investigated region into the following units:

1. Red, medium-grained, mainly porphyritic granite, 'wild granite'.
2. Red, medium-grained, mainly even-grained granite.
3. Greyish-red, medium-grained, mainly even-grained granite.

Granite of Type 1 is mainly found within the parish of Tanum, and Type 2 in restricted bodies scattered throughout the granite area. Type 3 is the most common of the Bohus granites. The investigated sites are evenly distributed among the different types of Bohus granite. Are there any differences in weathering among these three types of granite? From Figure 14.5c, it can be seen that Bohus granite of Types 1 and 3 seems to be more easily weathered than Type 2. A very weak tendency for Type 3 to be more easily weathered than Type 1 can also be seen on the graph.

Distance of the Bedrock Surface from the Nearest Road

This factor is taken into consideration since the acidity caused by cars has been emphasised in several contexts, and was considered as of possible importance by the CBNA. The data has been divided into three classes:

1. The surface is exposed to, and within a close distance from, a road.
2. The surface is situated less than 100 m from a road.
3. The surface is situated more than 100 m from a road.

No correlation was found between the weathering of the bedrock surfaces and the distance to roads. There are three possible explanations for this result:

1. With the method used, it is not possible to differentiate weathering caused by acidity on the bedrock surface.
2. Up to 1989, acidity caused by car traffic had not increased the weathering of the investigated surfaces.
3. The fear that acidity caused by cars may damage the rock carvings is over estimated.

Vegetation

The basic- and acid-buffering capacity of the litter is of importance for chemical weathering. The concentration of acids formed in the ground depends on its water content. In the weathering process, these acids are active, as long as they are not neutralised. This means that the amount of weathering is dependent on the composition of the litter. Acid litters contribute to the acidification of water, and to the weathering potential of soil moisture. Decisive factors for the effect are partly the pH caused by the litter, and partly the content of basic matter in the litter, which can have a buffering effect on the degree of the acidity (Stålfeld, 1960, p. 181).

Stålfeld also refers to an earlier investigation by Hesselman (1926), who studied the pH of different kinds of litter from broadleaved deciduous and evergreen needle-bearing plants. As examples of plants forming acidic litter, Hesselman mentions spruce, pine, juniper, and other species common in spruce and pine communities. They have a pH of 3.7–4.7, and produce soil moisture with a low pH value. Examples of trees with a moderate degree of acidity and a relatively high degree of basic buffering material are birch, alder, aspen and beech, and other species that are associated with these communities or occur in meadows. These have a pH of 4.6–6.3. Examples of trees and bushes with a high degree of basic matter are elm and hazel, with pH values in the litter of 6.6–7.3. More complicated is the group of plants and trees that produce both acid and basic buffering materials. Among these, Hesselman mentions maple, oak and larch. In the context of this study, it must be emphasized that, in Bohuslän, spruce forest is mostly found in more humid areas and is more dependent on a high ground-water table than pine forest, which is found more often in areas with thin, drier layers of soil.

Thus, the different types of vegetation around the investigated bedrock surfaces have been divided into: spruce forest, pine forest, mixed forest, deciduous forest and meadows. Figure 14.5d shows the mean value of Rm for the investigated bedrock surfaces for each type of vegetation. Two interesting relations can be observed. First, bedrock surfaces close to meadows, which have a high pH value, also have very high Rm values. Secondly, surfaces surrounded by spruce forest, with a low pH value, have significantly lower Rm values. This relationship is so clear that the mean value for the surfaces surrounded by meadows lies above the upper limit of surfaces classified as Group I, indicating a very slight degree of weathering, while the mean value of surfaces in spruce forest lies below the upper limit of Group III, indicating a high degree of weathering.

The relatively high values for surfaces in pine forest may be explained by the low humidity within this biotope. Similarly, the low values on surfaces in deciduous forest may be explained by the high humidity within this biotope. Mixed forest grows mostly on relatively dry, morainic soils, in association with pine, deciduous trees, and invading spruces, which may explain the intermediate position of this category.

It is clear from this investigation that bedrock surfaces surrounded by spruce forest have a high rate of weathering. New plantations of spruce forest must therefore be avoided in the vicinity of sites with rock-carvings.

Black Coatings

During fieldwork in 1988 (Sjöberg, 1988), black coatings (Figure 14.6) were noted on several of the investigated bedrock surfaces. These coatings were mostly in the form of trickle paths starting at major joints. A question was raised whether or not these coatings increased the speed of weathering. Measurements gave very ambiguous results.

During the present investigation these coatings have again been observed. They appear mostly on smooth surfaces, less often on granular, and rarely on coarse granular surfaces. This observation raises the question whether the coating, in fact,

Figure 14.6   A part of the famous rock carving at Fossum (Tanum 260). Note the paths of black coating, and the polished to granular surface of the bedrock

Table 14.4   *Rm* values on black coatings

Surface near	*Rm*	Even, coated surface	Granular coated surface
Tanum 260	49.8	58.5	44.6
Bro 26A	52.9	54.6	
Brastad 1	44.3	49.5	

is protecting the surface against weathering. Therefore, three surfaces with coatings (Tanum 260, Bro 26a and Brastad 1) were chosen for a minor comparative study of structure, coating and *Rm* value.

In all cases smooth surfaces with coatings have values above the mean value of the surface as a whole (Table 14.4). The samples are too small to be the basis of a general statement, but the problem requires further investigations.

Anthropogenic Polishing

During fieldwork in 1988, it was also noted that the common factor for bedrock surfaces classified as belonging to Group I was that the glacially polished surface was preserved to a very high degree. It was noted that this might depend on several circumstances. A second common factor regarding these surfaces was that they were easily accessible, and intensively visited by tourists. They have a dip which makes it possible for visitors to walk on the surfaces to have a closer look at the rock-carvings. Generally it has been suggested that this anthropogenic wearing has a negative effect on the carvings. The results of this investigation show that most of the surfaces commonly visited by tourists belong to Group I. Therefore a hypothesis was formulated, suggesting that wearing conserves the polished character of the bedrock surface, and delays the weathering processes. This assumption was built on the fact that the granite is harder than sandstone and limestone, and is, therefore, not mechanically as easily worn.

To test this hypothesis, a smooth granite surface of local bedrock, of known age, with anthropogenic wearing had to be investigated. Suitable objects for investigation are outdoor stairs. People generally step in the middle of stairs, this part thus theoretically suffering more mechanical erosion than the outer parts, which ought to show signs of normal weathering. Therefore, this hypothesis was first tested on the church in Fjällbacka, which was built of cut, local bedrock in 1892. Here, first the upper stair of the main entrance was investigated. One area in the middle part of the stair, and one area in the corner, close to the wall, were tested with 15 impacts on each area. Secondly, the stairs leading up to the entrance of the sacristy were tested in the same way. The stairs to the main entrance ought to be more worn than those of the sacristy. Therefore, the middle parts of the latter ought to be less weathered and show higher values than those of the main entrance.

The results obtained are shown in Table 14.5. They show that the middle parts of the stairs clearly have higher R values, and are thus less weathered, than the corners. The low values of the standard deviation (s) show the homogeneity of the bedrock. The middle part of the stair to the sacristy shows values in between these extremes and a higher s value. The R value from the corner is comparable with that of the main entrance.

The same type of test was repeated on the old church of Svenneby, built in the 12th century. First the narrow stairs to the old, original entrance to the sacristy was tested. A second test was done on the stone leading to the present main entrance to the church, most probably opened in the 16th century. As this step is very short, it is most probably often stepped over by visitors, and thus it is little worn. The stairs leading up to the gate-house were also tested. This stair is not dated, but gives an impression of not being very old.

The results (Table 14.6) show the same tendencies as those of Fjällbacka church. The most used and worn parts of the stairs show the highest R, and the lowest s, values. That these values from Svenneby are also lower than those of Fjällbacka is explained by the fact that Svenneby church is at least 700 years older.

These tests seem to verify the hypothesis that anthropogenic wearing of polished-even granite surfaces delays weathering. Thus, the following controversial conclusion

Table 14.5 Fjällbacka church. *Rm* and *s* values on stairs

Site	Place	*Rm*	*s*
Main entrance	corner	40.6	6.2
	middle	60.5	2.78
Sacristy	corner	38.3	9.3
	middle	51.7	8.8

Table 14.6 Svenneby church. *Rm* and *s* values of investigated parts

Site	Place	*Rm*	*s*
Main entrance	middle	42.5	4.9
Sacristy	corner	32.3	6.7
	middle	45.7	4.6
Gate-house	corner	28.1	12.4
	middle	46.1	7.3

can be drawn: polished to smooth bedrock surfaces and rock-carvings in those surfaces do not seem to be damaged by the wear caused by visitors' shoes. On the contrary, it seems as this wear (polishing) conserves the polishing of the surface, and delays the weathering.

This is a problem, however, since several rock-carvings on granular to coarse granular bedrock surfaces definitely are severely weathered. It is very difficult to teach visitors to separate the structure of the rock-carving surfaces they visit. Therefore the Central Board of National Antiquities has recommended that rock-carving surfaces must not be trodden on.

CONCLUSIONS

- Through the investigation of 29 bedrock surfaces with rock-carvings, using the Schmidt Test-hammer, it has been noted that the variability of weathering on these surfaces is high. Some of these surfaces are only slightly affected by weathering, while others are severely weathered.
- A weathering succession of the bedrock surfaces, starting from a polished initial position, through a granular medium position by flaking, to a present coarse granular stage has been established during the fieldwork.
- Surfaces of greyish-red, medium-grained, mainly even-grained granite tend to weather somewhat faster than other types of granite in the investigated area.
- Given the method used, thorough examination of anthropogenic factors which have increased the rate of weathering, as well as the effects of increased air- and water-acidity, has not been possible. There is no evidence that the distance from a road is of any importance, but there is a clear indication that surrounding planted spruce forest, combined with the above-mentioned factors, eventually increases the rate of weathering of the bedrock surfaces.
- Factors such as the altitude, within the observed limits, and the exposure of the bedrock surfaces do not seem to affect the rate of weathering.
- The weathering processes seem to be delayed by glacial polishing of the surfaces.
- Polishing can also be caused by anthropogenic wear, e.g. by visitors walking on the bedrock surfaces.
- Silicon casting of bedrock surfaces dramatically increases the rate of weathering, by the roughening of the bedrock surface as a result of the detachment of mineral grains which become included in the silicon cast.

- Most affected by weathering are surfaces of greyish-red, medium-grained, mainly even-grained granite, surrounded by spruce forest, which are wetted by trickles of water originating from the above-mentioned biotope.

REFERENCES

Asklund, B. (1947). Svenska stenindustriområden I–II. Gatsten och kantsten. *Sveriges Geologiska Undersökning, Ser. C.*, **479**, 37–109.

Day, M. J. and Goudie, A. S. (1977). Field assessment of rock hardness using the Schmidt hammer. *British Geomorphological Research Group, Technical Bulletin*, **18**, 19–29.

Hessland, I. (1943). Marine Schalenablagerungen Nord-Bohusläns. *Bulletin Geological Institute of Upsala* **XXXI**.

Hillefors, Å. (1983). Fjord och sprickdalslandskapet i mellersta Bohuslän. *Naturinventeringar i Göteborgs och Bohus län.* Länssyrelsen NE, 1983, 5, *Göteborg*.

Lidmar-Bergström, K. (1987). Berggrundsformer i Skåne—resultatet av en lång utveckling. *Svensk Geografisk Årsbok*, **63**, 42–59.

Lidmar-Berström, K. (1988). Exhumed Cretaceous landforms in south Sweden. *Zeitschrift für Geomorphologie.* NF Supplementband **73**, 21–40.

Lindström, E. (1988). Are roches moutonnées mainly preglacial forms? *Geografiska Annaler*, **70A**(4), 323–331.

McCarroll, D. (1987). The Schmidt Hammer in geomorphology: five sources of instrument error. *British Geomorphological Research Group, Technical Bulletin*, **36**, 6–27.

McCarroll, D. (1989). Potential and limitation of the Schmidt hammer for relative-age datings: field tests on neoglacial moraines, Jotunheimen, southern Norway. *Arctic and Alpine Research*, **21**(3), 268–275.

Mörner, N.-A. (1979). Earth movements in Sweden, 20 000 BP to 20 000 AP. *Geologiska Föreningens i Stockholm Förhandlingar*, **100**(3), 279–286.

Proceq, S. A. (1977). *Operation Instructions Concrete Test Hammer Types L and LR.* Zurich.

Rudberg, S. (1973). Glacial erosion forms of medium size—a discussion based on four Swedish case studies. *Zeitschrift für Geomorphologie*, NF Supplementband **17**, 33–48.

Sjöberg, R. (1988). Diagnos av vittring på ristande hällytor i Tanums kommun, Bohuslän. *Center for Arctic Cultural Research, Ulmeå University, Research Reports* No. 11.

Sjöberg, R. (1990a). Measurement and calibration of weathering processes and lichenometric investigations on a wave washed moraine, Bådamalen, on the upper Norrland coast, Sweden. *Geografiska Annaler*, **72A**(3–4), 319–327.

Sjöberg, R. (1990b). Diagnos av vittring på ristade hällytor i norra Bohuslän. Del 2, *Center for Arctic Cultural Research, Umeå University, Research Reports* No. 19.

Sjöberg, R. (1991a). Weathering studies on pseudokarst-caves along the northern Swedish coast. *Zeitschrift fü Geomorphologie*, **35**(3), NF 305–320.

Sjöberg, R. (1991b). Relative datings with the Schmidt test-hammer of terraced house-foundations in Forsa parish, Hälsingland, Sweden. *Laborativ Arkeologi*, **5**, 93–99. Arkeologiska forskningslaboratoriet Stockholms Universitet.

Sjöberg, R. and Broadbent, N. (1991). Measurement and calibration of weathering, using the Schmidt hammer, on wave washed moraines on the upper Norrland coast, Sweden. *Earth Surface Processes and Landforms*, **16**(1), 57–64.

So, C. L. (1987). Coastal forms in granite, Hong Kong. In Gardiner, V. (ed.), *International Geomorphology 1986*, Part 1, pp. 1213–1229.

Stålfelt, M. (1960). *Växtekologi*, 2nd edition, Stockholm.

Swantesson, J. O. H. (1989). *Weathering Phenomena in a Cool Temperate Climate.* GUNI Rapport 28, Göteborg.

Swedish Geological Survey (SGU) (1945). Bergrundskarta över Bohusläns granitområde. *Sveriges Geologiska Undersökning, Ser C.* **479**.

15 The Influence of Feldspar Weathering on Luminescence Signals and the Implications for Luminescence Dating of Sediments

ROMOLA PARISH
University of Sussex, Brighton, UK

ABSTRACT

The effect of chemical weathering on the luminescence signal of feldspars has been recognised from the analysis of waterlain sediments of Holocene age. The sediments fell into three groups based on their stratigraphic position, sedimentological and luminescence characteristics.

The observed relationship between low signal intensities and weathered grains was tested by a simulated weathering experiment on one sample which previously demonstrated luminescence signals which were stable and of high intensity. The results show a reduction in signal intensity the appearance of instability in the luminescence signal. This supports the proposal that chemical weathering selectively destroys the defects at which trapped charge is held. This has severe implications for dating such sediments.

INTRODUCTION

The work presented in this paper formed part of a study in which sedimentological analyses and luminescence dating were used to assess the effects of transport and depositional processes and post-depositional changes in sediments on the luminescence characteristics of the samples, and their potential for dating. The sediments were fine-grained (mainly 1–60 μm), waterlain deposits of Holocene age from Britain and the Netherlands.

The intercalation of the inorganic sediment samples with archaeological material (such as occupational debris and structures) and sediments resulting from sea-level changes which had been independently dated by radiocarbon provided a chronological control for the samples.

The application of luminescence dating techniques to sedimentary deposits is an important development in the establishment of absolute chronologies for inorganic sedimentary sequences, which may lack suitable material for radiocarbon dating, or contain material which exceeds the radiocarbon age range. Luminescence techniques are also important for sediments which fall within periods where the radio-

Rock Weathering and Landform Evolution. Edited by D. A. Robinson and R. B. G. Williams
© 1994 John Wiley & Sons Ltd.

carbon calibration curve is flat and different radiocarbon ages can produce the same calibrated age range.

Luminescence dating is an absolute dating technique with present limits of up to 200 000 years, although recent work has claimed limits of up to 800 000 years (Berger *et al.*, 1992). The crystals used for measurement are usually quartz and feldspar, which are prevalent in most sedimentary environments. This enables luminescence techniques to be applied to a wide variety of sedimentary deposits.

Feldspars are of primary concern in this project, and with quartz are the important minerals for dating. This is due to the bleaching of the luminescence signals of these minerals by sunlight, the high intensity of their luminescence emissions, and their ability to hold absorbed charge in traps or defects in the crystal structure for long periods of time, i.e. longer than the limits of the dating technique (Wintle and Huntley, 1982; Godfrey-Smith *et al.*, 1988). The limits of the dating techniques are constrained by the saturation of the mineral crystals with dose which is in part determined by the level of activity of the sedimentary environment.

This paper presents the results of a study of the effects of 'active' weathering of feldspars on the luminescence characteristics of a sample, and its significance for the dating of feldspar-rich deposits. Samples undergoing 'active' weathering in this case refers to sediments containing feldspars which are currently being broken down, as opposed to sediments in a state of 'interrupted' weathering, where feldspars may have been weathered in the past, but in their present environment are in a stable state. The results in this paper indicate that chemical weathering is a cause of loss of signal and reduced potential for dating.

LUMINESCENCE DATING OF SEDIMENTS

Luminescence dating is based on the measurement of trapped charge held in defects in crystals. The charge is derived from the effects of ionizing radiation from naturally occurring elements (uranium, thorium and potassium). The defects in crystals act as traps in which this charge is held until evicted by heating or exposure to light. Sediment dating is based on the fact that exposure to sunlight empties the stored charge from some of the traps. The crystals, usually quartz and feldspar, accumulate trapped charge according to the annual dose-rate and time. The event dated is the last exposure to sunlight.

The recombination of the trapped electrons with opposite-charged 'holes' may be either radiative (i.e. accompanied by the emission of luminescence) or non-radiative. The eviction of the charge prior to recombination is achieved either by heating the sample or by exposure to light. In the laboratory, measurements are made of the luminescence by heating in the case of thermoluminescence (TL), or by exposure to light of specific wavelengths, known as optically-stimulated luminescence (OSL). The light sources used in OSL include a green laser for stimulation of quartz (Huntley *et al.*, 1988; Rhodes, 1990), or infrared diodes for feldspars (Hütt and Jaek, 1989; Poolton and Bailiff, 1989). The technique in which infrared diodes are used is called infrared stimulated luminescence (IRSL) and this was applied along with TL to samples in this project.

The age of sediments is calculated by the equation: age = equivalent dose (ED)/

annual dose-rate, where the ED is measured in the laboratory by giving an artificial radiation dose which induces a luminescence signal of equivalent intensity to that emitted from the natural sample. The dose-rate is evaluated by measuring the alpha, beta and gamma components of the bulk sediment. Tests are made for the stability of the luminescence signal, particularly with respect to fading (loss of signal intensity during storage). Fading after the application of an artificial radiation dose to samples would result in incorrect evaluation of the ED and therefore the age of the sample.

THE LUMINESCENCE DATING AND SEDIMENT ANALYSIS PROJECT

The project consisted of the detailed analysis of six core samples from selected sites. The sites were selected on the basis of the information available at each site relating to its environmental history and chronology, and included Flag Fen, Peterborough, Britain (Pryor et al., 1986) and Hazendonk and Slingeland in the Netherlands (Louwe-Kooijmans, 1974). All samples were fine-grained polymineral waterlain (mainly fluvial and lagoonal) sediments.

For each core a detailed examination of the sedimentology was made, including particle size analysis, mineralogy and scanning electron microscopic (SEM) examination of the grain surfaces. In addition, fine-grained (6–11 μm) sample fractions were extracted and their TL and IRSL characteristics examined. This included fading studies, signal intensity and reproducibility, and bleaching under sunlight and filtered white light (light from a Tungsten Halogen lamp passed through a BG-39 filter). Of a total of 27 samples, 17 were successfully dated by one or both luminescence methods (Parish, 1992). However, the remaining samples could not be dated because of problems of instability and poor signal intensity. These were defined by failure of the TL plateau test, fading of the TL and IRSL signals, and IRSL signal-to-noise ratios of less than 4.0. The signal-to-noise ratio is determined by the detected signal divided by the noise level (signal emitted from a glowed out disc).

A relationship was observed between samples which could not be dated and samples which demonstrated significant chemical weathering of the grain surfaces under SEM examination. This is discussed below.

IDENTIFICATION OF THE PROBLEM

Three types of sample were recognised, based on their combined sedimentological and luminescence characteristics.

> *Group 1* contained samples with clean grains of quartz and feldspars (mixed potassium and plagioclase), a low abundance of mixed-layer clay minerals (Figure 15.1a), and other clay minerals dominated by kaolinite, which is the stable end-product of the weathering of potassium feldspar. The natural IRSL signal was high with initial intensities over 160 counts in the first second. The signal exhibited a linear growth with added dose, and a low (<5%) scatter in the normalised data. These samples were dated (Parish, 1992) and showed good agreement between TL and IRSL ages (correlation coefficient, $r = 0.99$). The agreement between luminescence and radiocarbon ages was less close ($r = 0.56$ for TL and $r = 0.67$ for IRSL). Error values of between 8.6% (IRSL) and 20.9% (TL) were recorded.

(a)

(b)

Figure 15.1   Grains from a Group 1 sample; clean faces, clearly defined grains (scale bar = 10 μm). (b) Grains from a Group 2 sample; grains are visibly degraded with abundant amorphous clay minerals. (scale bar = 10 μm)

Exact agreement between ages determined by different dating techniques is not expected as different materials are used for radio carbon than for luminescence, and the techniques relate to different events. However chronological consistency (demonstrated by increasing age with depth) was obtained for both radiocarbon and luminescence techniques for the sequences sampled (Parish, 1992).

Group 2 samples contained a proportion of weathered feldspar grains (Figure 15.1b), but these were not as visibly degraded as those of Group 3. The quartz grains tended to be unweathered. The clay content was dominated by kaolinite, with small amounts of illite and montmorillonite in some samples, but mixed-layer clays were rare. These sedimentological characteristics indicate that weathering of the grains had occurred, possibly prior to deposition in their present context, but that their present environment was relatively stable with respect to active degradation. This is illustrated by the preservation of fine laminations formed during deposition.

The luminescence characteristics were similar to Group 1, and dates produced were in good agreement with expected ages, although there was a slightly increased scatter in the data for some samples, increasing the error to a maximum of 38.9%. This may be related to the weathering regimes that the sediments had undergone in the past. However, these samples showed no indications of instability in their luminescence signals.

Group 3 samples were characterised by severely degraded grains, showing the etch pits and delineation of cleavage associated with weathering of feldspars (Figures 15.2a and 15.2b). The clays were dominated by transitional clay products of weathering, illites and montmorillonites (from plagioclases), with only relatively small amounts of kaolinite (from potassium feldspar). Mixed-layer clays were abundant. These characteristics suggest that the environment is one of active weathering. The profiles of these sediment strata tended to be more mixed vertically (as indicated by the organic and water contents, and the particle size distribution of the strata measured at 5–10 cm intervals). No sedimentary structures were preserved. The luminescence characteristics of these samples included extremely weak signals (signal-to-noise ratios of less than 4.0), and a weak response to added dose. The TL and IRSL signals of several samples faded, even after storage at 50°C for 6 weeks. This instability was not removed by preheating at 160°C for 2 hours and 100°C for 3 days.

Signal intensity varied widely as a result of the age of the sample, the number of grains on the disc (i.e. weight of fine-grain sample discs), and preparation methods. For the purposes of this study, where all the samples which were dated were of Holocene age and were prepared using a standard method (see below), the signal intensity is used as a qualitative characteristic of the samples examined.

There are a number of causes of weak luminescence signals including the sensitivity of the sample related to mineralogy (Akber and Prescott, 1985), and proportion of non-luminescent material on the sample discs (Questiaux, 1991). Pure, colourless quartz is not stimulated by infrared, but all the samples in this study contained at least 40% feldspar (as indicated by semi-quantitative analysis of X-ray diffraction (XRD) which would emit substantial IRSL signals.

The equipment used to detect the IRSL signals contained a GG-420 filter which transmits the blue emissions from feldspars. The proportion of blue emissions is much greater from alkaline feldspars than from plagioclases, although this depends on the elemental composition of individual feldspar samples (Huntley *et al.*, 1988). Samples containing predominantly plagioclases, such as albite, have significant emission in the orange–red wavebands rather than the blue–green, and the detected IRSL signal using the equipment given above is, therefore, relatively weak.

(a)

(b)

Figure 15.2 Potassium feldspar grain, severely degraded, showing the preferential exploitation of cleavage planes. ($\times$ 10 000; scale bar = 0.5 μm). (b) A sub-rounded feldspar grain showing distinct etch pits on the surface, formed by dissolution at 'sites of excess energy'. Notice the clay flakes adhering to the eroded surfaces. ($\times$ 10 000; scale bar = 0.5 μm)

The mineralogy of the samples was evaluated semi-quantitatively by XRD analysis. The samples investigated consisted primarily of alkaline feldspars (amounting to >40% of the total minerals identified by XRD) indicating that the signals were likely to be dominated by blue–green emissions detected by the equipment used (Parish, 1992). The correlation between weathered grains and indicators of active weathering processes *in situ*, together with the luminescence characteristics typical of Group 3 suggests that chemical weathering of feldspars selectively exploits the defects (see section below) from which the luminescence signal arises, resulting in a loss of signal.

The preceding section indicates the importance of employing sedimentary analysis in support of luminescence dating studies. It also demonstrates the need for detailed investigation of the sedimentary environment, and processes operating during deposition and burial, which may have significant effects on the measured luminescence signals.

CHEMICAL WEATHERING OF FELDSPARS

Chemical degradation of feldspars occurs by hydrolysis and by the substitution of elements within the lattice, as a result of chemical disequilibrium between the surface of the solid and the surrounding soil solution. The soluble elements (such as Ca^{2+}) are replaced by lower valency cations (such as Na^+), and K^+ is replaced by H^+. This occurs primarily at 'centres of excess energy' (Berner and Holdren, 1977) such as cleavage and other planes of weakness, and at defects in the crystal lattice, resulting in the characteristic pitting of grains (Figure 15.2a and 15.2b). Similar surface features are produced by the acid etching of quartz (Bell and Zimmerman, 1978).

The cations mentioned above are soluble at normal soil pHs (4.0–9.0), but Al^{3+} is not, and is a major structural element of feldspars. The presence of organic acids (humic and fulvic) derived from organic matter in a sediment or soil significantly increases the rate of feldspar dissolution because the solubility and mobility of Al^{3+} is facilitated by the process of chelation, in which the formation of complexes between metal and organic ions renders the metals (including Al^{3+} and Fe^{2+}) soluble at normal soil pHs. The proximity of the samples to strata rich in organic matter, such as peat, needs to be considered when selecting samples for dating. The high mobility of Ca^{2+} contributes to the fact that plagioclases tend to break down more rapidly than potassium feldspars, independent of climatic conditions (James *et al.*, 1981).

Feldspars break down into clay minerals by hydrolysis and carbonation. An example of the transformation of albite and formation of kaolinite is given below:

$$2NaAlSi_3O_8 + 2CO_2 + 11H_2O {=\!=} Al_2Si_2O_5(OH)_4 + 2Na^+ + 2HCO_3^- + 4H_4SiO_4$$

The formation of particular clay minerals depends on the availability of ions in solution. If only part of the potassium is removed, K-feldspar may form illite $(KAl_2(AlSi_3)O_{1)}(OH)_2)$. The type of clay mineral present in the sample may be indicative of the stability of the sedimentary environment with respect to chemical weathering. Kaolinite tends to form as a stable end-product of feldspar weathering either directly from the feldspar or via intermediate clay minerals such as illite and montmorillonite. The clay composition of the sample can reflect the weathering status of the sample.

The rate of weathering of feldspars depends on the abundance of defects and the structure of the grain, with respect to vulnerable cleavage planes, and thus the rate is not directly affected by grain size or surface area (Holdren and Berner, 1979); Holdren and Speyer, 1985 and 1987). Weathering continues until an equilibrium state is established between the exposed grain surface and the surrounding solution, after which the grains remain in a stable state (as hypothesised for the Group 2 samples) until further disequilibrium occurs. Ions contained in percolating groundwater are an important weathering agent as the removal of free cations by leaching maintains a state of disequilibrium.

A WEATHERING EXPERIMENT

A fine-grained, feldspar dominated sample from Group 1 was subjected to artificial weathering following the procedure used by Wollast (1967). The sample was gently dispersed into individual grains. Buffer solutions of pH 6.0 and pH 8.0 (as described in Wollast, 1967) were made up to 1 l in a polythene bottle and 50 g of the sample was added. In addition, a control sample was made with 50 g of sample in deionised water (pH 7.0). These were kept in a water bath at 20°C for 6 weeks, covered to prevent bleaching under the red light conditions of the laboratory, and agitated frequently.

Subsamples of *c.* 50 ml were abstracted at 10-day intervals from each bottle. The 6–11 μm fraction was separated by sedimentation in acetone for 2 minutes and for 20 seconds from each subsample. This fraction was settled onto abraded aluminium discs. Each disk had a mean weight of ±0.5 mg. The remainder of the subsample was dried and used for XRD analysis (using a CoKα tube and orientated samples). An SEM examination was made of the surface of the prepared discs. These were gold coated and magnified up to × 10 000 at a working distance of 6–7 mm.

TL and IRSL measurements were made of the natural signal of the 6–11 μm fraction (three discs for each) using a Risø automated system (Bøtter-Jensen, 1988; Bøtter-Jensen *et al.*, 1991); for TL a Corning 7-51 filter and a heat rate of 10°C s^{-1} were used, and for IRSL, 880 nm diodes and an exposure time of 150 s. The TL glow curves represent the signal intensity plotted against temperature up to 500°C (e.g. Figure 15.7). The IRSL curves represent the decay of the detected signal from an initial maximum with time of exposure to the diodes (e.g. Figure 15.5).

The three discs from each subsample were normalised using a 1 second IR exposure prior to measurement. No preheating treatment was given, except in the case of the fading samples, discussed below. The natural signals, and mineralogy were compared to the characteristics of the original material, and the control sample. A fading test was carried out by giving each of the glowed-out TL and IRSL discs 30 GY from a Sr^{90} source, and storing at 50°C for up to 4 weeks. Measurements of the signal intensity were taken in pairs (one TL disc and one IRSL disk) at weekly intervals for up to 4 weeks, using the same measurement conditions as the measurements of the natural signal.

EXPERIMENTAL RESULTS

The XRD analysis showed a significant change in the proportions of the identified minerals as a result of exposure to the buffer solutions. There was a fall in the amount of feldspar and quartz and an increase in the amount of clays, which were dominated by illite. An example is shown in Figure 15.3. The changes observed were similar for both acid and alkaline buffers, whereas no significant change was observed in the control sample.

Under SEM analysis, the grains of the control sample remained clearly defined, without significant abundance of clay flakes. After 40 days in the pH buffers, the grains exhibited some of the characteristic etch pits associated with weathering, and the development of flaky clay coatings associated with mineral breakdown. The change between the original material and those exposed to the buffers for 40 days is shown in Figures 15.4a and 15.4b.

The initial intensity of the natural IRSL signal (i.e. that which was detected in the first 3 seconds of IRSL exposure for each subsample measured) fell dramatically (from 470 to 170 photon counts) with the duration for which the sample was held in the buffered samples. No decrease was observed for the control sample (Figure 15.5), thus indicating that this reduction is not due to external effects, such as bleaching by laboratory light. This reduction reached levels as low as 27% of the original signal with signal-to-noise ratios falling to <45.0.

Besides showing a reduction in initial intensity, the IRSL signal was observed to exhibit a change in the form and rate of the decay curve. This was detected by plotting the ratio of the decay curve of the $I(N+\beta)$ divided by the $I(N)$ IRSL decay curves, where $I(N+\beta)$ was the decay curve of the natural signal. The ratio curves are shown in Figure 15.6. Curve A is the ratio curve for the subsample in the control buffer, and curve B is the ratio curve for the subsample exposed to the pH 6.0 buffer for 40 days.

In addition, the natural TL signals were measured. The signal intensity was reduced and a significant change in the 280°C peak shape was observed for samples from both acid and alkaline buffers (Figure 15.7). The flattening out of the peak indicates some charge transfer may be taking place, which may result from changes in the surface distribution of luminescence centres associated with weathering (Bailiff, personal communication) or from destruction of the defects during weathering, which may also be a cause of the signal loss. The selection of the 6–11 μm fraction should have minimised the effect of masking by clay minerals. Their presence on the surfaces of the feldspar grains may have contributed to the effects observed in the TL and IRSL signals.

Another significant observation was that fading could be induced by exposure of the sample to acid and alkaline buffer solutions when the sample had previously shown no sign of instability. As shown in Figure 15.8, the control sample continued to remain stable, but the IRSL signal of the buffered samples faded significantly, losing up to 45% of the signal after storage. The TL and IRSL signals were also observed to face by similar proportions.

The sample, therefore, not only showed a reduction in the natural signal, but also loss due to fading was observed after the addition of an artificial radiation dose to the measured discs, and storage for up to 4 weeks. The percentage loss of signal due to

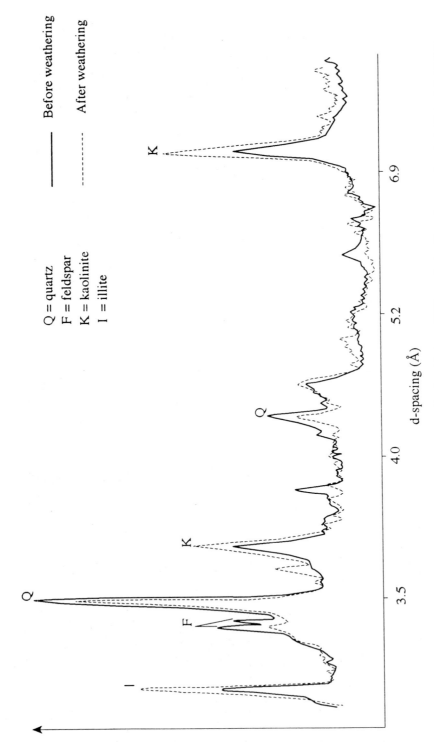

Figure 15.3 Effect on mineralogy of exposure to acid buffer for 40 days; XRD analysis. The changes in mineralogy are similar for both acid and alkaline buffers; a reduction in the proportion of feldspar, and increase in clays, particularly illite

(a)

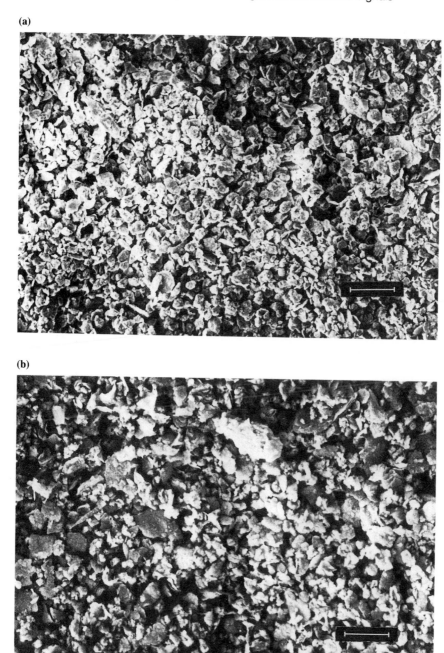

(b)

Figure 15.4 SEM photographs for the surfaces of discs prepared from the original material before weathering treatment (a) and after 40 days in the acid buffer (b). Note the change in surface texture and appearance of the grains, which are less clearly defined after weathering. The increase in clay minerals is also significant. The effect of the alkaline buffer was similar to that of the acid and the control sample was unchanged. (scale bar = 10 μm)

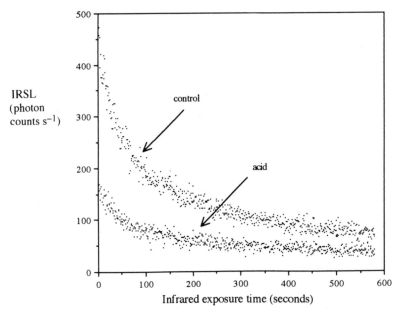

Figure 15.5 Reduction in IRSL intensity after weathering. Decay curves are shown for a control and an acid buffered sample after 40 days. The initial intensity of the latter is reduced from 470 to 170 photon counts

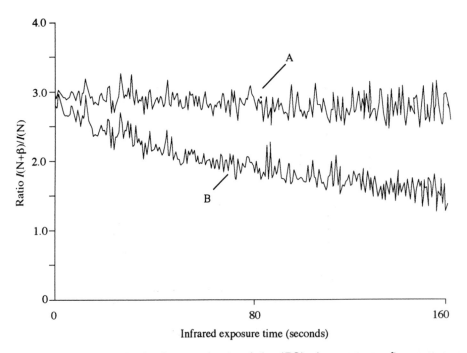

Figure 15.6 Changes in the form and rate of the IRSL decay curve after exposure of subsamples to buffer solutions for 40 days. The ratio plot is described in the text. Curve A represents the subsample in the control solution (pH 7.0) and curve B the subsample in the acid buffer (pH 6.0)

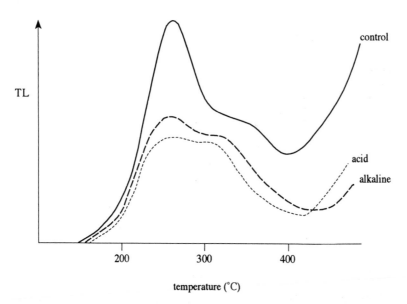

Figure 15.7 Changes in the shape of the 280°C TL glow curve peak after 40 days in the buffer solutions. The glow curve for the control sample was unchanged from that of the original sample (not shown)

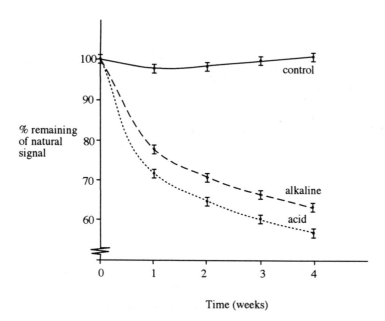

Figure 15.8 Loss of IRSL initial signal intensity (i.e. photon counts in the first 2 seconds) with storage under different conditions)

fading increased with the time for which the samples were exposed to the buffer solutions.

CONCLUSIONS

Chemical weathering selectively exploits the defects in feldspar crystals, and causes a loss of signal, the onset of instability (manifest as fading of the TL and IRSL signals) and changes in the structure of the TL peak.

A sample, which was characterised by strong luminescence emissions and stability of the signal, was exposed to buffer solutions of pH 6.0 and 8.0 for 40 days. As a result of this exposure, the luminescence characteristics of the sample were significantly changed. The control was stored in de-ionised water and was unchanged, retaining the characteristics of the original material.

Artificial weathering, by inducing disequilibrium between a buffer solution and the sample, has been observed to cause a number of changes. Indicators of active weathering develop, i.e. the accumulation of mixed-layer clays and etching of feldspar grains. With respect to the luminescence signal, a significant loss in intensity is seen, together with progressive signal loss.

Earlier correlation between weathering indicators and poor luminescence signals demonstrated by waterlain sediments has been confirmed in this study. A state of active weathering induced by exposing a sample to acid and alkaline buffer solutions has caused a weakening in signal intensity and stability and the development of weathering indicators as observed in the undated samples studied previously.

IMPLICATIONS FOR DATING

The importance of the identification of actively weathering environments needs to be stressed. The use of quartz for dating such layers may also not be reliable as the XRD analysis indicates weathering of the quartz fraction, though to a lesser extent than for feldspars. Detailed analyses such as the examination of preserved microstructures can indicate whether weathering is active. When using luminescence dating to support stratigraphic analysis it is important to examine individual grains microscopically for evidence of degradation and formation of weathering-related clay coatings (as opposed to those resulting from the translocation of fines within a profile).

In the case of sediments classified as Group 2 in this study, the grains may have been affected by weathering, but the sedimentary structures such as fine laminates have been preserved in the sedimentary section, indicating that leaching and mixing associated with active weathering has not occurred.

Other indicators of active weathering are in the clay mineralogy and the presence of mixed-layer clays. Clays dominated by mixed-layer or expanding lattice types such as illite and montmorillonite are less stable than clays such as kaolinite, and more easily affected by further alteration.

It is the combination of these characteristics that is most useful, rather than properties taken individually. It is the active weathering environment which appears to affect the dating potential of sediments; a stable environment containing either weathered or unweathered grains is not, in these samples, a cause for concern.

ACKNOWLEDGEMENTS

This project was funded by SERC at the University of Durham. The work was accepted as part of a PhD thesis and I acknowledge the assistance of my three supervisors at Durham: Ian Bailiff, Anthony Harding and Michael Tooley. I also thank Helen Rendell and Ann Wintle for their comments on an earlier draft of this paper.

REFERENCES

Akber, R. A. and Prescott, J. R. (1985). Thermoluminescence spectra in some feldspars: early results from studies of spectra. *Nuclear Tracks and Radiation Measurements*, **10**, 575–580.

Bell, W. T. and Zimmerman, D. W. (1978). The effect of HF etching on the morphology of quartz inclusions for thermoluminescence dating. *Archeometry*, **20**, 63–65.

Berger, G. W., Pillans, B. and Palmer, A. S. (1992). Dating loess up to 800 ka by thermo-luminescence. *Geology*, **20**, 403–406.

Berner, R. A., and Holdren, G. R. (1977). Mechanism of feldspar weathering: some observational evidence. *Geology*, **5**, 369–372.

Bøtter-Jensen, L. (1988). The automated Risø TL dating reader system. *Nuclear Tracks and Radiation Measurements*, **14**, 177–180.

Bøtter-Jensen, L., Ditlefsen, C. and Mejdahl, V. (1991). Combined OSL (infra-red) and TL studies of feldspars. *Nuclear Tracks and Radiation Measurements*, **18**, 257–263.

Godfrey-Smith, D. I., Huntley, D. J. and Chen, W. H. (1988). Optical dating studies of quartz and feldspar sediment extracts. *Quaternary Science Reviews*, **7**, 373–380.

Holdren, G. R. and Berner, R. A. (1979). Mechanism of feldspar weathering—II. Observations of feldspars from soils. *Geochimica et Cosmochimica Acta*, **43**, 1173–1186.

Holdren, G. R. and Speyer, P. M. (1987). Reaction rate–surface area relationships during the early stages of weathering.—II. Data on eight additional feldspars. *Geochimica et Cosmochimica Acta*, **51**, 2311–2318.

Huntley, D. J., Godfrey-Smith, D. I., Thewalt, M. L. W. and Berger, G. W. (1988). Thermoluminescence spectra of some mineral samples relevant to dating. *Journal of Luminescence*, **39**, 123–136.

Hütt, G. and Jaek, I. (1989). Infrared stimulated photoluminescence dating of sediments. *Ancient TL*, **7**, 48–51.

James, W. C., Mack, G. H. and Suttner, L. J. (1981). Relative alteration of microcline and sodic plagioclase in semi-arid and humid climates. *Journal of Sedimentary Petrology*, **51**(1), 151–164.

Louwe-Kouijmans, L. P. (1974). The Rhine–Meuse delta area. *Analecta Praehistorica Leidensia*, **7**.

Parish, R. (1992). The application of sedimentological analysis and luminescence dating to waterlain deposits from archaeological sites. Unpublished Ph.d. Thesis, University of Durham.

Poolton, N. R. J. and Bailiff, I. K. (1989). The use of LEDs as an excitation source for photoluminescence dating of sediments. *Ancient TL*, **7**, 18–20.

Pryor, F., French, C. and Taylor, M. (1986). Flag Fen, Fengate, Peterborough. 1: discovery, reconnaissance and initial excavation 1982–85. *Proceedings of the Prehistoric Society*, **52**, 1–24.

Questiaux, D. G. (1991). Optical dating of loess: comparisons between different grain size fractions for infrared and green excitation wavelengths. *Nuclear Tracks and Radiation Measurements*, **18**, 133–139.

Rhodes, E. J. (1990). Optical dating of quartz from sediments. Unpublished D.Phil. Thesis, University of Oxford.

Wintle, A. G. and Catt, J. A. (1985). Thermoluminescence dating of soils developed in Devensian loess at Pegwell Bay, Kent. *Journal of Soil Science*, **36**, 293–298.

Wintle, A. G. and Huntley, D. J. (1982). Thermoluminescence dating of sediments. *Quaternary Science Reviews*, **1**, 31–53.

Wollast, R. (1967). Kinetics of the alteration of K-feldspar in buffered solutions at low temperature. *Geochimica et Cosmochimica Acta*, **31**, 635–648.

16 Pedogenic Weathering and Relative-age Dating of Quaternary Alluvial Sediments in the Pindus Mountains of Northwest Greece

JAMIE C. WOODWARD
Manchester Metropolitan University, UK

MARK G. MACKLIN
University of Leeds, UK

and

JOHN LEWIN
UCW Aberystwyth, UK

ABSTRACT

A chronosequence of soil profiles is described from the terraced alluvial sediments of the Voidomatis River in northwest Greece. Pedogenic weathering and profile development include increases in soil depth, organic matter translocation, leaching of calcium carbonate, clay illuviation, oxidation of ferrous iron minerals and changes in the magnetic susceptibility of different soil horizons. The pedogenic weathering data form several relative-age indicators which are used to supplement the interpretation of the primary depositional record for the basin. It is concluded that the northern Pindus mountains were probably glaciated before the last major (Late Würm) ice advance and that at least two major phases of glacial activity can now be recognised in the Greek Pleistocene.

INTRODUCTION

The study of pedogenic weathering and soil formation is of fundamental importance to geomorphology. A detailed understanding of the processes and products of mineral weathering is an essential requirement for both contemporary studies of material transfer and for longer-term investigations of sediment diagenesis and landsurface lowering. In many environments, prior to entering a mass movement or fluid transport system, sediment will reside in a soil system and it is the processes of weathering and pedogenesis operating there which largely control the quality of solutes and sediments eventually transported from an area (Statham, 1977). Even though many of these processes act only very slowly, pedogenic weathering forces the alteration and mobilisation of surficial materials and is thus a major element in the relief-forming

Rock Weathering and Landform Evolution. Edited by D. A. Robinson and R. B. G. Williams
© 1994 John Wiley & Sons Ltd.

mechanisms of any climatic zone (cf. Büdel, 1982). It is also apparent that the study of pedogenic weathering profiles is a most useful tool for investigating landscape history and environmental change, especially during the Quaternary Period (Birkeland, 1984; Catt, 1986; Bull, 1991).

During pedogenesis, as mineral breakdown proceeds and solute and particle transfers take place, weathering profiles can stockpile a variety of environmental clues. Consequently soil profiles, and especially soil chronosequences, offer a rich source of information on rates and processes of mineral weathering and may aid in establishing regional chronologies for Quaternary landform development. Soils have provided a major input to many palaeoenvironmental investigations in three main ways: (1) As soils represent episodes of relative geomorphological stability, they mark the position of buried or relict landsurfaces (e.g. Kukla, 1987), and may assume considerable stratigraphical importance over wide areas (e.g. Rose et al., 1985). (2) Their weathering products are often preserved in sufficient detail to form a record of the environmental conditions responsible for their formation—thus providing useful information that cannot be obtained from the primary depositional record (Kemp, 1987). (3) Since many pedogenic features can be related to profile maturity, soils can be used as a means of relative-age dating (e.g. Birkeland, 1982; McFadden and Hendricks, 1985; Colman and Pierce, 1986). Furthermore, in the absence of datable materials and diagnostic fossils, evidence obtained from pedogenic weathering may provide the only available means of differentiating between otherwise similar deposits. The last two decades have witnessed both a growing awareness of the value of such data sources and an ever increasing number of researchers seeking to integrate information from soil profiles into more traditional, primary-sediment-based studies of the Quaternary rock record (cf. Birkeland, 1984; Boardman, 1985; Weide, 1985; Bull, 1991).

The authors have recently conducted detailed field investigations in the Voidomatis River basin of northwest Greece in order to establish the nature of the Quaternary sedimentary record. This work formed part of a wider geoarchaeological survey of Late Quaternary environmental change and Palaeolithic settlement, allied to recent excavations at the Klithi rockshelter (Bailey et al., 1984 and 1986). While the alluvial sequence provided the main focus for our initial investigations (Lewin et al., 1991; Woodward et al., 1992), the project also included an investigation of soil profile development in the terraced alluvial sediments of the Voidomatis River. These terrace surfaces are generally well preserved, forming prominent landscape features throughout most of the catchment. This chronosequence presented a rare opportunity to observe patterns of pedogenic weathering and soil profile genesis during the late Quaternary in a hitherto little studied region of southern Europe.

In terms of Quaternary landform development, the region is of particular interest as it represents one of the most southerly locations in Europe with evidence of extensive Pleistocene glacial activity (Woodward, 1990). Furthermore, the annual rainfall currently exceeds 2000 mm, and climatic conditions in this humid alpine Mediterranean zone promote comparatively rapid rates of pedogenic weathering by most European standards. This paper summarises the results of our soil profile work by describing the nature of the pedogenic weathering regime and illustrating the use of several relative-age indicators to supplement and refine our interpretation of the primary depositional record.

GEOLOGICAL AND GEOMORPHOLOGICAL SETTING

The Voidomatis Basin

The study area is situated in the Epirus region of northwest Greece approximately 40 km northeast of the provincial capital of Ioannina. The Voidomatis River (384 km^2) drains part of the high-relief karst terrain of the Pindus Mountain Range (Figures 16.1 and 16.2). Elevations within the basin range from <450 m on the Konitsa Plain (Figure 16.3) to over 2400 m along the watershed of the Voidomatis and Aoos Rivers. The catchment is developed in resistant Jurassic and Eocene limestones which are capped in places by thick flysch deposits of Late Eocene to Miocene age. The hard limestone rocks support the formation of deep gorges and steep-sided tributary ravines (Figure 16.2). In contrast, the erodible flysch deposits support sub-catchments of lower relief with comparatively higher drainage densities and present-day sus-pended sediment yields. The physiography of the catchment is presented in greater detail in Woodward *et al.* (1992). The five soil profiles under discussion are located along a 6 km reach of the Voidomatis River which incorporates all of the Lower Vikos Gorge (Vikos to Old Klithonia Bridge) and the most southerly part of the Konitsa Plain (Figures 16.1 and 16.2).

Climatic Regime

The Pindus Mountains trend roughly NNW–SSE and lie parallel to the modern Ionian coastline of northwest Greece. The Epirus region contains many of the highest mountains in Greece with several peaks exceeding 2400 m. These mountains present a major relief barrier to the convectively unstable, moisture-laden air masses associ-ated with the Mediterranean winter, and the study area falls within the 'Mountain climate' zone of Walter and Lieth (1960), and is located on the boundary between the 'Rain all year' and 'Winter and Autumn' Mediterranean rainfall regimes described by Huttary (1950). Mean annual rainfall frequently exceeds 2000 mm (Furlan, 1977). The summer months are generally hot and dry although heavy thunderstorms are not uncommon during July and August. Mean July temperatures of 15°C are typical in the highest central belt of the northern Pindus Mountains and this may rise to 20°C at intermediate altitude.

The Quaternary Alluvial Sequence in the Voidomatis Basin

The alluvial deposits and landforms of the Voidomatis River have recently been described in detail by Lewin *et al.* (1991). This work documents five major Quaternary alluvial units (Figure 16.1) and their major features are summarised in Table 16.1. The weathering profiles developed on the terraced alluvial units which postdate the Kipi unit form the basis of this study. From the information presented in Table 16.1 and Figure 16.4 it is clear that these alluvial units are not of uniform composition. Lithological contrasts are apparent in both the coarse (8–256 mm) and fine (<63 μm) elements of the parent alluvial sediments, and these result from a series of marked shifts in catchment sediment sources which took place during the Late Quaternary (Woodward *et al.*, 1992) and are briefly described below.

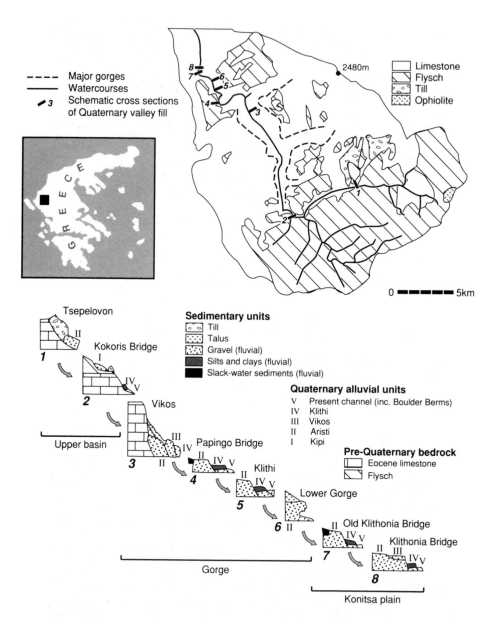

Figure 16.1   A simplified geological and drainage map of the Voidomatis River basin showing the location of eight key sites where the stratigraphic relationships in the Quaternary alluvial sequence are particularly well exposed. The lower diagram shows schematic sections of the alluvial succession and associated terrace surfaces at each of these sites. Site 1 is located approximately 1.5 km south of the village of Tsepelovon in the glaciated portion of the catchment. The main Vikos Gorge lies between Kokoris Bridge (2) and Vikos (3), and the Lower Vikos Gorge lies between Vikos (3) and the Old Klithonia Bridge (7). The Voidomatis River joins the Aoos River approximately 5 km north of site 7

Figure 16.2   SPOT satellite image of the Lower Vikos Gorge and the Konitsa Plain to illustrate the high-relief, block-faulted terrain of the study region. The western end of the Vikos Gorge is also shown and six of the sites shown in Figure 16.1 are also indicated. This scene covers an area of approximately 20 km × 14 km. The Voidomatis and Aoos Rivers both exit fault-bounded limestone gorges onto the Konitsa Plain and their confluence is approximately 10 km from the Albanian border

Figure 16.3 The southern part of the Konitsa Plain looking upstream (eastwards) into the Lower Vikos Gorge with the western end of the Vikos Gorge in the background. The course of the modern stream is marked by the line of trees to the right. Soil profiles B and D are located 40 m and 100 m respectively downstream of the end of the Lower Vikos Gorge. An 8 m high right-bank section in Aristi unit gravels is shown in the middle of this photograph

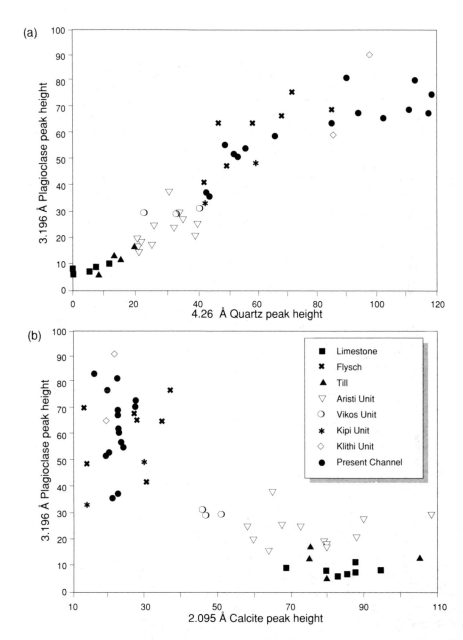

Figure 16.4 Peak height data from X-ray diffraction traces showing the broad mineralogical composition of the <63 μm component of the parent alluvial sediments. (a) Plot showing the strong positive correlation between quartz and plagioclase in the basin parent materials. (b) Plot showing calcite and plagioclase relationships. The Aristi unit sediments (soil profiles A and B) are poor in plagioclase and quartz and rich in calcite reflecting their origin from limestone-rich glacial (till) sediments. In contrast the Klithi unit sediments (soil profiles D and E) are rich in non-carbonate (flysch-derived) materials. The Vikos unit sediments (soil profile C) are intermediate between these two groups

Table 16.1 Altitudinal relationships, lithological properties, depositional environments (including major suspended sediment sources) and ages of the Quaternary alluvial units in the Voidomatis basin. The dates were obtained by the following methods: a) AMS ^{14}C; b) ESR; c) Thermoluminescence

Alluvial unit	Height of terrace surface above river bed level (m)	Maximum observed thickness of unit (m)	Clasts N	Samples N	Clast lithological (bedload) composition (8–256 mm)			
					% Limestone	% Flysch	% Flint	% Ophiolite
Present channel			8388	7	72.7	26.6	0.5	0.2
Klithi unit	Mean = 3.2 S.D. = 0.7 Range = 1.8–4.5	4.5	1139	2	69.3	29.6	1.0	0.1
Vikos unit	Mean = 6.8 S.D. = 1.7 Range = 3.9–9.7	8.3	695	2	82.3	12.8	0.6	4.3
Aristi unit	Mean = 12.4 S.D. = 3.9 Range = 6.7–25.9	25.9	5680	9	94.6	3.1	2.2	0.1
Kipi unit	56	22.9	361	1	18.7	36.7	0.9	44.0

Alluvial Sediment Sources

During the last glaciation the limestone headwaters of the Voidomatis basin (Tsepelovon district, Figure 16.1) supported valley glaciers, and huge amounts of limestone-rich sediment were supplied to the river system. The Late Würm Aristi unit is thus dominated by limestone gravels (94.6%) whereas the coarse sediment fraction of the late Holocene Klithi unit contains almost 30% flysch (Table 16.1). X-ray diffraction analyses have also highlighted the marked lithological contrast between the fine sediment loads of the cold stage, full-glacial Voidomatis (Aristi unit), which is rich in $CaCO_3$, and the fine sediment loads of the Klithi unit and modern fluvial systems which are composed largely of flysch-derived silts (Figure 16.4). The Vikos unit sediments are intermediate in composition between those of the Aristi unit and the modern floodplain sediments. The lithological composition of the post-Aristi alluvial units reflects the waning glacial input and the increase in erosion and sediment delivery from those parts of the catchment underlain by flysch rocks and soils. It can be shown that the relative proportions of flysch- and limestone-derived sediment in each of the alluvial units is a significant soil-forming factor and has implications for parent material composition and pedogenic weathering rates.

DATA COLLECTION

Field Sampling

Following a basin-wide geomorphological survey, five alluvial soil profiles (A to E) were selected for detailed study. These were located along a 6 km reach of the

Table 16.1 (Continued)

Coarse (C)/ fine (F) sediment member ratio	Munsell colour of <63 μm fraction	Fluvial sedimentation style and (in parentheses) the dominant source of the suspended sediment load	Age of unit (years BP)
C > F	yellowish brown 10YR 5/8	Incising, confined meandering gravel bed river. Low suspended sediment load. (Flysch sediments)	<30
C ⩽ F	yellowish brown 10YR 5/8	Aggrading, high sinuosity gravel bed river. High suspended sediment load. (Flysch sediments)	1000 (±50)[a]–30
C ⩾ F	brownish yellow 10YR 6/8	Incising wandering gravel bed river. Low suspended sediment load. (Flysch and glacial sediments)	24 300 (±2600)[b]– 19 600 (±200)[c]
C ⩾ F	very pale brown 10 YR 8/7-4	Aggrading, low sinuosity, coarse sediment river system. High suspended sediment load. (Glacial sediments)	28 200 (±7000)[c]– 24 300 (±2600)[b]
C > F	yellowish brown 10YR 5/8	Aggrading (?) low sinuosity, coarse sediment river system (Flysch sediments)	>150 000[c]

Voidomatis River in the Lower Vikos Gorge and the southernmost margin of the Konitsa Plain (Figure 16.1 and Table 16.2). Profiles B and D are located approximately 40 m and 100 m respectively immediately downstream of the Old Klithonia Bridge at which point the Voidomatis leaves the confines of the Lower Vikos Gorge. To minimise any site-dependent variations in soil profile character, the sections under investigation are all situated in right-bank, free-draining portions of terraced alluvial sediments away from the main gorge walls, major talus formations and tributary junctions. Soil samples (c. 150–200 g) were collected, mostly at 10 cm intervals, from the terrace surface down to unweathered or least altered parent material (C horizon). Sixty-two sediment samples were collected from five soil profiles which ranged in depth from 0.6 to 2 m (Table 16.2).

Table 16.2 Location of the soil profiles discussed in this chapter. The Quaternary alluvial units which form the parent material at each of these sites (A to E) are shown in Figure 16.1 and their lithological properties are summarised in Table 16.1

Soil profile	Parent material	Location of profile (see Figure 16.1)	Depth of sampled profile (cm)	Number of soil samples
A	Aristi unit	Papingo Bridge (4)	200	20
B	Aristi unit	Old Klithonia Bridge (7)	140	14
C	Vikos unit	2 km downstream of Vikos (3)	100	10
D	Klithi unit	Old Klithonia Bridge (7)	60	3
E	Klithi unit	Klithi (5)	150	15

Laboratory Methods

All the sediment samples were air dried in the laboratory at room temperature for 48 hours prior to screening through a 1 mm mesh sieve. Calcium carbonate ($CaCO_3$) and organic carbon (loss on ignition) determinations were made on bulk (<1 mm) samples following methods described by Gross (1971). The particle size characteristics of the fine sediment fraction (<63 μm) were measured using the computer-interfaced SediGraph 5000ET system reported by Jones *et al.* (1988). Low field magnetic susceptibility measurements were carried out on a mass specific basis using a standard Bartington system (Thompson and Oldfield, 1986) and total ferric iron present in oxyhydroxide phases such as haematite and goethite (Fe_2O_3d) was determined using the dithionite–citrate–bicarbonate procedure described by Mehra and Jackson (1960).

PEDOGENIC WEATHERING AND PROFILE DEVELOPMENT

Weathering takes place in soil profiles because minerals are only stable in the environment in which they form—following deposition in a new environment, minerals break down and liberate products which are in equilibrium with ambient soil chemistry (Pye, 1983).

Climatic Regime and Profile Development

Climate is often the dominant influence upon the pedogenic weathering regime—controlling both the rate and degree of mineral alteration and thus ultimately regulating the properties of many soils. This climatic control can be viewed in terms of moisture delivery to the soil profile and near-surface temperature conditions. The volume of moisture to actually percolate into the soil profile (moisture throughput) controls the degree of edaphic activity, and regulates the intensity of weathering and leaching conditions, whilst temperature can influence the rate of chemical and biological processes in the soil. Increased leaching may result in the removal of bases and build-up of iron oxides, while a rise in temperature will accelerate the rate of formation of a particular mineral and therefore its abundance in profiles of a particular age (Catt, 1988). In contrast, in exceptionally arid environments, even the most mobile profile constituents can be retained in the upper soil and the development of illuvial horizons is severely retarded or prevented. Climate has a direct effect upon the type of soil profile that is formed.

Parent Material Composition and Profile Development

Since different mineral species vary markedly in their resistance to surface weathering (cf. Curtis, 1976), the physical and chemical properties of parent sediments are also crucial factors in chronosequence development. Parent material lithology exerts an important influence upon the rate of operation of all weathering processes and the overall pace of profile development. For example, sediments containing easily weathered iron-bearing minerals may redden very quickly, while those containing more resistant iron minerals may take longer to attain an equivalent degree of redness (Pye, 1983). Similarly, under free-draining, leaching conditions, parent materials with a comparatively low $CaCO_3$ component may become decalcified fairly rapidly, allowing the pedogenic weathering front to penetrate downwards relatively quickly into the primary sediment, and will also contain a greater proportion of non-carbonate minerals available for subsequent modification. In other words, sediments, with many weatherable components will show more rapid weathering rates than sediments with less weatherable components.

By comparing profile depth functions between soils of different ages in the same locality, it is possible to observe the genesis and development of pedogenic features. A number of closely related processes operate to produce the pedogenic weathering regime of the Voidomatis Basin and these are described below.

Organic Matter Translocation

Soil organic matter is derived from numerous sources including the terrace surface litter supply, the dissolved or suspended organic load from stemflow and the *in situ* decomposition of root material and soil organisms. The processes involved during the incorporation of organic material into the weathering profile are numerous and complex. In the Voidomatis alluvial soils, as is generally the case in most soil types, organic content tends to reach a fairly stable level first in the upper profile. The parent alluvial sediments contain little organic material. For example, unweathered C horizon sediments at the base of the Pleistocene sections have organic components of <2.5%. All the profiles are richer in organic matter at the soil surface, although these values are not related to soil age. Under certain conditions, however, the depth of organic-enriched horizons can provide a good indication of profile maturity. The accumulation of organic matter deep in the weathering profile appears to be encouraged by a climatic regime in which a period of moisture deficiency promotes aggregate shrinkage and vertical fracturing allowing finely comminuted humus and particulates to be translocated as colloidal suspensions from the upper soil. Following decalcification of the upper soil, this mechanism appears to be the dominant agency for the downward diffusion of organic matter in this environment. Only profiles A and B contain >7% organic material below 50 cm (Figure 16.5), while the younger Vikos and Klithi unit soils exhibit comparatively low values (<4%). This is because the pedogenic weathering front has not yet penetrated so deeply into their respective substrates. The build-up of organic matter in the illuvial *B* horizon appears to be a useful measure of relative age (Table 16.3).

Leaching of Calcium Carbonate

The presence of $CaCO_3$ in the soil matrix retards chemical weathering processes and restricts the vertical movement of fine particles. $CaCO_3$ is progressively leached from the upper soil and this is the first major pedogenic alteration of the alluvial parent material. The effect of the leaching process over time is well illustrated by the increasing depth to carbonate in all five sections (Table 16.3). Profiles A to E are all free-draining sites which do not favour the reprecipitation of $CaCO_3$ in the lower profile. The extent and depth of $CaCO_3$ removal exerts an important control upon

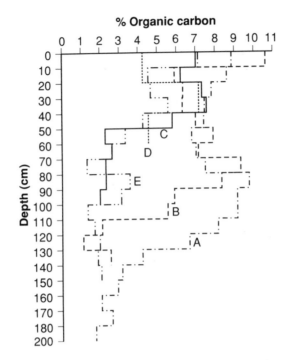

Figure 16.5   Changes with depth in organic carbon content for all five profiles

soil pH and, therefore, upon the rate of breakdown of residual, non-carbonate mineral species. Secondary deposition of $CaCO_3$ has not been observed in any of the studied profiles. Under the modern humid climatic regime, the high annual rainfall totals (>2000 mm) and the unconsolidated nature of the host sediment gravel matrix promote a vigorous and effective leaching zone in the upper soil. The amount of carbonate removal is greatest in the Pleistocene soils and, taking into consideration the original $CaCO_3$ content of the parent sediment, there appears to be a straight-forward relationship between depth to carbonate and soil age. The carbonate-free horizon at profile A is much deeper than the equivalent feature in the other soils suggesting that this soil is considerably older than the rest of the sequence.

Clay Illuviation

According to Birkeland (1984), argillic horizon development is largely time-dependent since mineral breakdown and clay translocation are relatively slow proces-ses. In the present study only the soils developed on the Vikos and Aristi alluvial units display B horizons containing clay-grade (<2 μm) material (Figure 16.6). Clays are absent from the parent materials for all the soils and the B horizon clays represent secondary accumulation. It is to be expected that soils developing in different parent materials will not produce B horizons at the same rate (Borchard and Hill, 1985) even if the starting materials in each case are clay-free. The crucial factors here are the mineralogical composition of the parent material and the clay-production potential

Table 16.3 Selected soil parameters to illustrate the extent of pedogenic weathering in each of the profiles and the significance of parent material composition and time in profile development

Soil profile	Depth to $CaCO_3$ (cm)	Thickness and depth of argillic B horizon (cm)	Finest median size (ϕ) of the fraction <63 μm (and depth in cm)	Maximum organic content of B horizon (%)	Profile dithionite extractable iron maxima (mg kg^{-1}) (and depth in cm)	Mean magnetic susceptibility of top 60 cm (m^3 kg^{-1})
A	110	140 (0–140)	8.98 (80–90)	9.76	24 090 (100–110)	238.5
B	60	50 (30–80)	6.23 (70–80)	9.33	18 170 (70–80)	95.2
C	40	30 (10–40)	6.98 (20–30)	7.56	18 610 (30–40)	128.7
D	20	0	5.52 (20–40)	7.17	12 590 (20–40)	65.2
E	0	0	5.32 (20–30)	No B horizon	7250 (30–40)	27.3

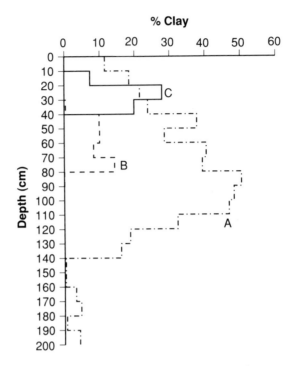

Figure 16.6 Downprofile variations in clay content illustrating the extent of argillic *B* horizon development in the Pleistocene alluvial soils

of this mineral suite. Parent materials rich in $CaCO_3$ (such as those of the Aristi unit) may contain only a comparatively low stock of clay-yielding minerals. Since in the present study, the parent materials are all free of clay-grade ($<2\ \mu m$) material, the extent of argillic *B* horizon development would seem to offer a useful guide to relative age. Only the pre-Holocene soils contain argillic horizons and these are shown in Figure 16.6. Several processes are involved in the development of such horizons, and the amount of illuvial clay in argillic horizons and their thickness is dependent on several factors. In general, however, a plot of clay content against depth usually shows an increase in the amount and in the thickness of *B* horizon clay with time (Birkeland, 1984). It is likely that the clay-rich *B* horizons evident in profiles A, B and C are the result of a number of processes including the *in situ* production of clay through mineral alteration as well as the incorporation of aeolian dust. The relative import-ance of these processes is difficult to determine; however, the extent of clay illuviation at site A is clearly much greater than at sites B and C, and provides further evidence to suggest that this soil is considerably older than the soils at the other two sites.

It is instructive to compare all five profiles using the median grain size of the fraction $<63\ \mu m$. These depth functions are presented in Figure 16.7 and serve to highlight the time-dependent nature of fine sediment illuviation. The five median size profiles illustrate the evolution of the argillic *B* horizon as both the proportion of fines increases and the illuvial horizon thickens and moves deeper into the substrate over time. In very mature soils this development may also be accompanied by a thinning

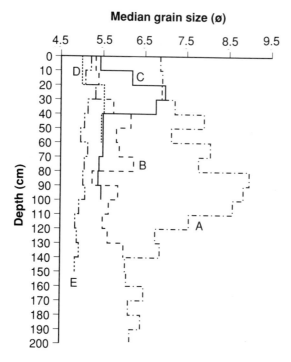

Median grain size (ø)

Figure 16.7 Downprofile changes in median grain size of the fraction <63 μm

of the *A* horizon (cf. McFadden and Weldon, 1987). This process could be indicative of a significant contribution from aeolian dust especially when the theoretical clay-production potential of elluvial horizons has been exceeded.

Soil Iron Transformations

Numerous workers have documented the time-dependent nature of pedogenic iron accumulation in weathering profiles (e.g. McFadden and Hendricks, 1985). The terraced alluvial sediments of the Voidomatis basin promote free-draining conditions, and chemical weathering by oxidation takes place as ferrous iron-bearing minerals convert to insoluble ferric oxides or hydroxides. The results of this process are clearly evident in the Pleistocene soils of sites A, B and C which all show very considerable contrasts in total ferric iron (Fe$_2$O$_3$d) content between the *C* and *B* horizons (Figure 16.8). Although profile E shows no evidence of such alteration, profile D evidences some iron oxide production in the *B* horizon (20–40 cm). The amount of Fe$_2$O$_3$d in profiles A, B, C and D is greatest in the *B* horizon and usually towards the base of this feature (Figure 16.8). Over time the ferrous iron component in the parent mineral suite is progressively converted to ferric iron and in the Voidomatis soils total ferric iron and the depth and thickness of the iron-enriched illuvial horizon seems to provide a useful measure of relative age (Table 16.3).

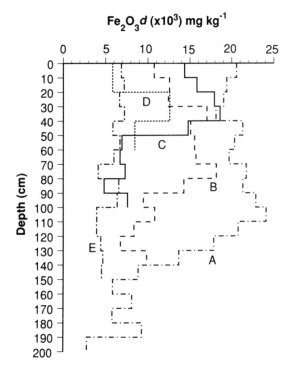

Figure 16.8 Downprofile changes in total iron oxide (Fe_2O_3d) content in each of the studied profiles

Magnetic Susceptibility Horizonation

There is a strong positive correlation between the amount of ferric iron present and the magnetic susceptibility of the soil samples. Indeed, the observed pattern of magnetic enhancement is closely related to the pedogenic transformation of iron-bearing compounds (see Dearing *et al.*, 1985). The production of magnetic minerals by pedogenic weathering is similarly time-dependent and a positive relationship exists between magnetic susceptibility and profile age. This general trend is again modified by the influence of parent material composition. Those substrates richer in flysch-derived materials (profiles C, D and E) contain a greater proportion of non-carbonate minerals available for the *in situ* production of magnetic minerals by pedogenesis. Profile E on the Klithi unit has a relatively constant downprofile magnetic signature (range 18 to 30.3 m^3 kg^{-1}) with a mean value of 23.8 m^3 kg^{-1}. Profile B (Late Pleistocene) has a 50 cm thick clay-rich *B* horizon, yet profile D (late Holocene) displays a very similar magnetic susceptibility maxima to that of B at depths between 20 and 40 cm (Figure 16.9).

The Voidomatis soils also display the characteristic feature of topsoil magnetic enhancement which has been documented by several authors, although the actual processes responsible are still not fully understood (Thompson and Oldfield, 1986). Providing the magnetic properties of the parent sediments are known, the magnetic

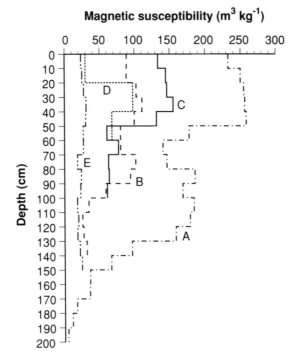

Figure 16.9 Downprofile changes in low frequency magnetic susceptibility

susceptibility of illuvial horizons provides a useful indication of soil age (Singer *et al.*, 1992).

THE VOIDOMATIS CHRONOSEQUENCE

The Klithi Alluvial Unit

The sediments at profile E were deposited within the present century (Lewin *et al.*, 1991) and show no evidence of pedogenic weathering. Apart from this most recent alluvial unit, all the soils have decalcified upper horizons, and the respective depths to carbonate are shown in Table 16.3. The magnetic susceptibility and median size data for profile E show no marked downprofile variation. In contrast, profile D, located on the southernmost part of the Konitsa Plain, has a decalcified upper horizon (20 cm), and the magnetic susceptibility and total ferric iron depth functions indicate some pedogenic alteration, although mineral weathering has not been sufficient to liberate clay-grade (<2 μm) material. Organic deposits in an equivalent section approximately 100 m upstream of site E have been radiocarbon dated as modern (102.7 ± 1.2%, OxA-1747). However, Klithi unit sediments on the Konitsa Plain have yielded ^{14}C dates of 800 ± 100 (OxA-192) and 1000 ± 50 (OxA-191) years BP

(Woodward *et al.*, 1992). The pedogenic weathering profiles thus corroborate the ^{14}C dating evidence by indicating the continued development of the Klithi unit until quite recent times.

The Vikos and Aristi Alluvial Units

The soil profile on the Vikos alluvial unit at site C includes an argillic *B* horizon 30 cm in thickness and pedogenic weathering is evident only in the upper 50 cm of this profile. Due to the significant flysch component in the parent material (Figure 16.4 and Table 16.1), the magnetic susceptibility and clay values in the *B* horizon actually exceed those of the profile at site B. However, on both stratigraphical and morphological grounds, as well as TL and ESR datings (see Figure 16.1 and Table 16.1) the alluvial sediments at site B clearly predate those at profile C. This scheme is supported by their respective weathering profiles. The argillic *B* horizon at site B is thicker and located deeper in the profile than at site C, and organic material has been translocated down to a depth of 110 cm. The observed differences in degree of pedogenic weathering are thus attributable primarily to age and secondly to contrasts in parent material composition. Some further examples of this pattern are shown in Table 16.3 and Figure 16.10. Profiles C and D have developed in parent sediments with a significant flysch component, and therefore contain a greater proportion of non-carbonate minerals available for pedogenic modification. The availability of a greater proportion of weatherable minerals allows illuvial horizons to develop relatively rapidly.

It is clear from Figure 16.6 that the argillic *B* horizon of profile A is much thicker and richer in clay than the other soils in the basin. This supports the observed pattern of $CaCO_3$ removal by also suggesting that soil profile A is considerably older than profile B. Indeed, it is unlikely that the large difference in clay accumulation could be attributed solely to site-dependent factors since both profiles are developed in free-draining, glacio-fluvial gravels in very similar topographic settings. The exposures at both sites A and B are capped by a distinctive, strongly weathered fine member approximately 70 cm in thickness, which immediately overlies coarse, cobble- to boulder-sized limestone gravels (Figure 16.11). The pedogenic weathering front at profile A is, however, considerably deeper and has penetrated the base of this fine member.

The extent of parent material alteration in each of the studied profiles may be further illustrated by means of the plots shown in Figure 16.12. These portray the maximum and minimum values of the parameters shown for each of the five profiles. The gradient of the line for each profile provides a useful illustration of the difference between the parent material and the most strongly altered horizons. In each case profile E displays a relatively flat line indicating little or no alteration, whereas profile A always evidences the greatest difference. The data for profile E provide a useful guide for the possible degree of at-a-site primary variability in the data set. It is interesting to note that in each of the examples shown, the maximum values lie in the same sequence: A, C, B, D and E, whereas the actual age-scheme from oldest to youngest is A, B, C, D and E. This further highlights the influence of parent material composition on these three parameters for this particular alluvial chronosequence.

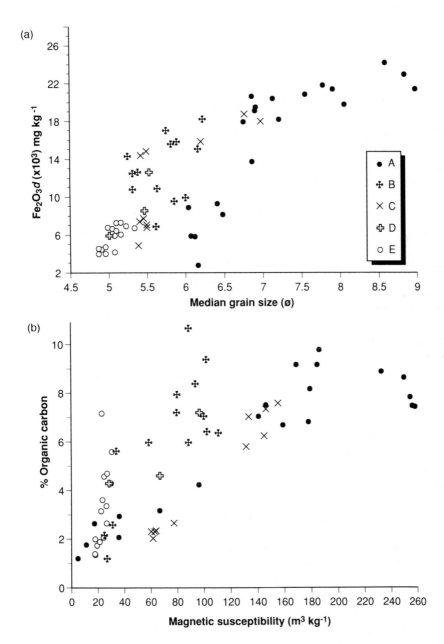

Figure 16.10 Bivariate plots showing the relationship between (a) Fe$_2$O$_3$d content and median grain size, and (b) % organic carbon and magnetic susceptibility in all five weathering profiles. These plots further highlight the dual influence of time and parent material composition on the pattern of pedogenic alteration

Figure 16.11   Soil profile A in the Lower Vikos Gorge. The fine-grained member capping this section is approximately 70 cm in thickness. This photograph shows a typical section in the coarse, limestone-rich glacio-fluvial gravels of the Aristi unit

In a recent paper Veldkamp and Feijtel (1992) have attempted to evaluate the influence of parent material composition on subsoil weathering rates in the Allier terrace chronosequence of central France. They simulated long-term (several hundred thousands of years) weathering patterns using a simple process model and concluded that the role of parent material controlled weathering is only prominent in 'young' sediments with many weatherable fragments and that, after prolonged weathering, this influence diminishes. Such a process may partly explain the comparatively rapid development of the Vikos unit (C) and Klithi unit (D) soils, each of which contains a much greater proportion of flysch-derived fine sediment than the Aristi unit profiles (Figure 16.4).

A Pedogenic Weathering Index

Using the iron oxide data presented in Figure 16.8 a simple weathering index has been developed incorporating information on both the degree of pedogenic transformation within each profile (relative to parent material composition) and the thickness of the weathered zone. This index is the product of the difference between the maximum Fe_2O_3d values and the least-altered or unweathered sample values at the base of each profile (mg kg^{-1}) and the total thickness (m) of the strongly weathered horizon. These values have been plotted against the ages of the alluvial units (Table 16.1) and are

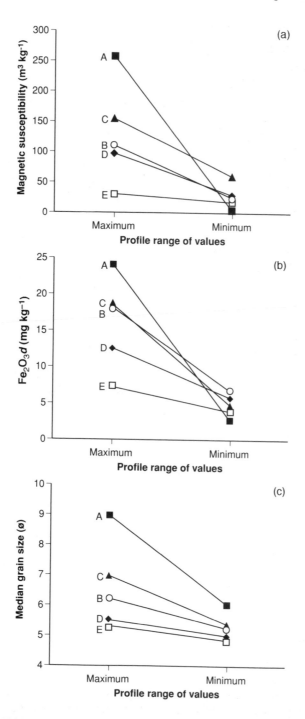

Figure 16.12   Maximum and minimum values of (a) low frequency magnetic susceptibility, (b) iron oxide (Fe$_2$O$_3$$d$ × 10^3) and (c) median grain size of the fraction <63 μm for each of the soil profiles

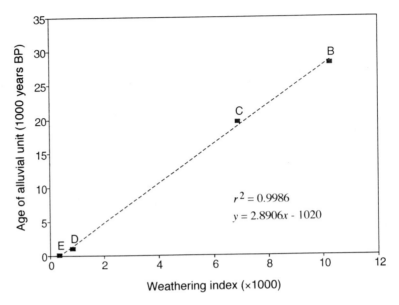

Figure 16.13   The relationship between soil profile age (see Table 16.1) and the weathering index derived from the iron oxide (Fe$_2$O$_3$d) data (see text for explanation)

shown in Figure 16.13. Using this relationship the index score for profile A gives an estimated age of c. 85 000 years BP which places this aggradational phase and the glacial activity to which it relates early in the last cold stage. However, this age estimate must be regarded as a minimum as we do not know if the relationship between this weathering index and profile age is truly linear in this environment for profiles older than profile B. Indeed, many studies have highlighted how weathering rates tend to slow down with the passage of time (e.g. Birkeland, 1984; Bull, 1991; Swanson *et al.*, 1993). We would like to stress that this age-estimate is preliminary and should be regarded as a minimum. It is quite possible that profile A actually predates the Last Interglacial, and at present it is only possible to say with certainty that profile A represents an episode of glacial activity which predates the last major (Late Würm) ice advance in the Pindus Mountains.

IMPLICATIONS FOR THE GLACIAL HISTORY OF NORTHWEST GREECE

The alluvial sediments at profile B (Old Klithonia Bridge) have been dated to between 28 000 and 24 300 years BP (Table 16.1), indicating that deposition took place towards the end of the last glacial period (Bailey *et al.*, 1990; Lewin *et al.*, 1991). In contrast, the alluvial sediments at profile A (Papingo Bridge) have not been dated by TL or ESR and the soil developed at this site is clearly the oldest of the five studied profiles. Moreover, the extent of pedogenic weathering at this site (relative to site B) implies that these river sediments are considerably older than the last glacial maximum. All the measured parameters indicate that profile A is considerably older than the rest of the sequence. To date, despite extensive field survey, profile A is the only location in the catchment where we have found evidence of such strongly weathered Aristi unit sediments.

This is perhaps not surprising, since these deposits are located on the inside of a large meander some distance (250 m) from the contemporary floodplain system in the widest section of the Lower Vikos Gorge (Figure 16.2). This body of sediment has thus been protected from fluvial erosion. Elsewhere in the basin the main gorge is much narrower and the modern river is vigorously reworking Pleistocene and more recent alluvium, as well as actively undercutting limestone bedrock walls at many locations. It seems likely that most of the earlier alluvium has been eroded away. Residence times for alluvium in such high-energy gorge environments can be relatively brief—rapid reworking is common, especially in narrow bedrock reaches where increases in discharge during major flood events are accompanied by large increases in river stage and stream power.

The most significant outcome of this pedogenic weathering investigation is the subdivision of the Aristi unit into two distinct alluvial units. Each of these is composed of glacially-derived sediments with similar lithological properties. It has been suggested previously that the large amount of clay present in profile A may partly reflect the incorporation of aeolian dust as well as *in situ* breakdown of clay-yielding minerals. This seems to be a plausible mechanism since this terrace surface could have been exposed to sub-aerial processes throughout the last glacial period.

CONCLUSIONS

During the Quaternary Period, phases of valley alluviation in the Voidomatis basin were followed by episodes of downcutting, and pedogenic weathering took place on stable terrace surfaces. The alluvial soils described in this paper have formed under long periods of seasonal leaching in a free-draining, oxidising environment. These soil profiles display significant differences in decalcification depth, amount of clay illuviation, total iron oxide content, and *B* horizon organic matter accumulation. All these features can be attributed primarily to differences in soil profile age—although interprofile contrasts in parent material composition have been sufficient to slightly modify this trend.

The relative-age sequence derived from soil profile development corroborates the model of alluvial history reported by Lewin *et al.* (1991). However, there is now evidence to suggest that the alluvial parent material at soil profile A may derive from an episode of Pleistocene glacial activity which predates the Late Würm Aristi unit at soil profile B (*c.* 28 200 to 24 300 years BP). It now seems likely, therefore, that the northern Pindus Mountains were also glaciated before the last major (Late Würm) ice advance and that at least two major phases of glacial activity can now be recognised in the Greek Pleistocene. Further work is in progress to test this implication.

In a wider context, this study has attempted to demonstrate how, in certain environments, sediments and landforms of markedly different ages can often be indistinguishable when only primary lithological properties and altitudinal criteria are employed. Despite recent major advances in the application and refinement of physical dating techniques such as ESR, luminescence and AMS ^{14}C, in the absence of datable materials, weathering profiles may provide the only available means of differentiating between otherwise similar deposits. Increasing awareness of the inherently homotaxial nature of the Quaternary rock record (Bowen, 1991) should ensure that relative-age estimates based upon measurable patterns of pedogenic weathering will continue to play an important role in studies of Quaternary landscape evolution.

ACKNOWLEDGEMENTS

This work was carried out while JCW held a SERC Ph.D. studentship at Darwin College and the Subdepartment of Quaternary Research (Godwin Laboratory) at the University of Cambridge. We would especially like to thank Geoff Bailey and all the members of the Klithi Archaeological Project for their support and also IGME (Athens) for permission to undertake fieldwork in Epirus. We also thank Watts Stelling for iron oxide analyses, Simon Robinson for assistance with magnetic susceptibility measurements and the sedimentology laboratory in the Department of Earth Sciences in Cambridge for generously allowing access to the SediGraph and XRD facilities. The support of the British Geomorphological Research Group through the provision of a research grant to JCW for work on SPOT satellite imagery of NW Greece is also gratefully acknowledged. Terry Bacon and Andrew Teed of the Department of Geography at the University of Exeter kindly prepared the diagrams and photographs.

REFERENCES

Bailey, G. N., Lewin, J., Macklin, M. G. and Woodward, J. C. (1990). The "Older Fill" of the Voidomatis Valley Northwest Greece and its relationship to the Palaeolithic archaeology and glacial history of the region. *Journal of Archaeological Science*, **17**, 145–150.

Bailey, G. N., Carter, P. L., Gamble, C. S., Higgs, H. P. and Roubet, C. (1984). Palaeolithic investigations in Epirus: the results of the first season's excavations at Klithi, 1983. *Annual of the British School of Archaeology at Athens*, **79**, 7–22.

Bailey, G. N., Gamble, C. S., Higgs, H. P., Roubet, C., Sturdy, D. A. and Webley, D. P. (1986). Palaeolithic investigations at Klithi: preliminary results of the 1984–1985 field seasons. *Annual of the British School of Archaeology at Athens*, **81**, 7–35.

Birkeland, P. W., (1982). Subdivision of Holocene glacial deposits, Ben Ohau Range, New Zealand, using relative-dating methods. *Geological Society of America Bulletin*, **93**, 433–449.

Birkeland, P. W. (1984). *Soils and Geomorphology*. Oxford University Press, New York.

Boardman, J. (1985). *Soils and Quaternary Landscape Evolution*. Wiley, Chichester.

Borchard, G. and Hill, R. L. (1985). Smectitic pedogenesis and late Holocene tectonism along the Raymond Fault, San Marino, California. In Weide, D. L. (ed.), *Soils and Quaternary Geology of the Southwestern United States*, Geological Survey of America Special Paper No. 203.

Bowen, D. Q. (1991). Glacial sediment systems. In Ehlers, J., Gibbard, P. L. and Rose, J. (eds), *Glacial Deposits in Great Britain and Ireland*, Balkema, Rotterdam, pp. 3–11.

Büdel, J. (1982). *Climatic Geomorphology*. Princeton University Press, Princeton.

Bull, W. B. (1991). *Geomorphic Responses to Climatic Change*. Oxford University Press, Oxford.

Catt, J. A. (1986). *Soils and Quaternary Geology: A Handbook for Field Scientists*. Clarendon Press, Oxford.

Catt, J. A. (1988). Soils of the Plio-Pleistocene: do they distinguish types of interglacial? In Shackleton, N. J., West, R. G. and Bowen, D. Q. (eds), *The Past Three Million Years: Evolution of Climatic Variability in the North Atlantic Region*, Royal Society of London, pp. 539–557.

Colman, S. M. and Pierce, K. L. (1986). Glacial sequence near McCall, Idahoe: Weathering rinds, soil development, morphology and other relative age criteria. *Quaternary Research*, 25, 25–42.

Curtis, C. D. (1976). Stability of minerals in surface weathering reactions: a general thermo-chemical approach. *Earth Surface Processes*, 1, 63–70.

Dearing, J.A., Maher, B.A. and Oldfield, F. (1985). Geomorphological linkages between soils and sediments: the role of magnetic measurements. In Richards, K.S., Arnett, R.R. and Ellis, S. (eds), *Geomorphology and Soils*, Allen and Unwin, London, pp. 245–266.

Furlan, D. (1977). The climate of southeast Europe. In Wallen, C. C. (ed), *Climates of Central and Southern Europe*, Elsevier, Amsterdam, pp. 185–223.

Gross, M. G. (1971). Carbon determination. In Carver, R.E. (ed.), *Procedures in Sedimentary Petrology*. Wiley, New York, pp. 541–569.

Huttary, J. (1950). Die Verteilung der Niederschlage auf die Jahreszeiten im Mittelmeergebiet. *Meteorologische Rundschau*, 3, 111–119.

Jones, K. P. N., McCave, I. N. and Patel, P. D. (1988). A computer-interfaced sedigraph for modal size analysis of fine-grained sediment. *Sedimentology*, 35, 163–172.

Kemp, R. A. (1987). The interpretation and environmental significance of a buried Middle Pleistocene soil near Ipswich Airport, Suffolk, England. *Philosophical Transactions of the Royal Society of London*, B317, 365–391.

Kukla, G. (1987). Loess Stratigraphy in Central China. *Quaternary Science Reviews*, 6, 191–219.

Lewin, J., Macklin, M. G. and Woodward, J. C. (1991). Late Quaternary fluvial sedimentation in the Voidomatis Basin, Epirus, northwest Greece. *Quaternary Research*, 35, 103–115.

McFadden, L. D. and Hendricks, D. M. (1985). Changes in the content and composition of pedogenic iron oxyhydroxides in a chronosequence of soils in southern California. *Quaternary Research*, 23, 189–204.

McFadden, L. D. and Weldon, R. J. (1987). Rates and processes of soil development on Quaternary terraces in Cajon Pass, California. *Geological Society of America Bulletin*, 98, 280–293.

Mehra, O. P. and Jackson, M. L. (1960). Iron oxide removal from soils and clays by a dithionite-citrate system buffered with sodium bicarbonate. *Proceedings of the 7th National Conference on Clays and Clay Minerals*, Pergamon, New York, pp. 317–327.

Pye, K. (1983). Red beds. In Goudie, A. S. and Pye, K. (eds), *Chemical Sediments and Geomorphology*, Academic Press, London, pp. 227–263.

Rose, J., Boardman, J., Kemp, R. A. and Whiteman, C. (1985). Palaeosols and the interpretation of the British Quaternary stratigraphy. In Richards, K. S., Arnett, R. R. and Ellis, S. (eds), *Geomorphology and Soils*, Allen and Unwin, London, pp. 348–375.

Singer, M. J., Fine, P., Verosub, K. L. and Chadick, O. A. (1992). Time dependence of magnetic susceptibility of soil chronosequences on the California coast. *Quaternary Research*, 37, 323–332.

Statham, I. (1977). *Earth Surface Sediment Transport*. Oxford University Press, Oxford.

Swanson, T. W., Elliott-Fisk, D. L. and Southard, R. J. (1993). Soil development parameters in the absence of a chronosequence in a glaciated basin of the White Mountains, California –Nevada. *Quaternary Research*, 39, 186–200.

Thompson, R. and Oldfield F. (1986). *Environmental Magnetism*. Allen and Unwin, London.

Veldkamp, A. and Feijtel, T.C. (1992). Parent material controlled subsoil weathering in a chronosequence, the Allier terraces, Limagne rift valley, France. *Catena*, 19, 475–489.

Walter, H. and Leith, H. (1960). *Klimadiagramm-Weltatlas*. Jena.

Weide, D. L. (1985). *Soils and Quaternary Geology of the Southwestern United States*. Geological Survey of America, Special Paper No. 203.

Woodward, J.C. (1990). *Late Quaternary Sedimentary Environments in the Voidomatis Basin, Northwest Greece*, Unpublished Ph.D Thesis. University of Cambridge.

Woodward, J. C., Lewin, J. and Macklin, M. G. (1992). Alluvial sediment sources in a glaciated catchment: the Voidomatis Basin, northwest Greece. *Earth Surface Processes and Landforms*, 16, 205–216.

Section 4

WEATHERING AND LANDFORM DEVELOPMENT IN TROPICAL AND ARID ENVIRONMENTS

17 Ages and Geomorphic Relationships of Saprolite Mantles

MICHAEL F. THOMAS
University of Stirling, UK

ABSTRACT

Uncertainty concerning rates of saprolite formation persists and confusion between the age of a weathering profile and the rate of saprolite formation has obscured many arguments about weathering and relief development. In the tropics and sub-tropics, orogenic zones are characterised by high rates of weathering and denudation whereas planate surfaces often have low rates of weathering, and are drained by rivers with low solute loads.

Published rates of saprolite formation from humid warm temperate areas, such as the Appalachians, give similar values ranging from 2 to 50 m Ma^{-1}, with a mean close to 20 m Ma^{-1}. This places most saprolite depths within the range of Quaternary weathering, though climatic oscillations may have reduced the speed of alteration by 50% during colder/drier periods.

Alteration in humid tropical regions, where rainfall inputs are 2–5 × those of temperate areas and soil/groundwater temperatures 2–3 × higher, may have been at least 3 × as rapid as in warm temperate areas. This implies rates of saprolite deepening of 6–150 m Ma^{-1}. The higher figure is debatable, but more than 50 m of saprolite could have formed in the last 1 Ma of the Quaternary. Neogene uplift has facilitated this rapid decay.

While deep profiles in continental interiors may have evolved over 10–100 Ma to produce ferritic or bauxitic crusts, the impetus for renewed decay comes wherever deep water flow is maintained by uplift and fracturing of rocks or climatic crises trigger saprolite removal. The dynamics of these weathering systems are central components of an understanding of Quaternary as well as longer term landscape evolution, and this principle also applies to higher latitude landscapes.

INTRODUCTION

Much has been written about the supposed 'ages' of saprolite mantles and their attendant laterites and bauxites, and many authors retain essentially 'cyclic' models for the explanation of differences in mineralogy between weathered mantles of different elevation and age (Grandin, 1976; Aleva, 1984). Although these ideas have been criticised, notably by McFarlane (1983b), there remains a tendency, based in part on some evidence from Australia (Bird and Chivas, 1988), to regard most profiles as the products of palaeoweathering, dating to the late Mesozoic or early Cenozoic.

Rock Weathering and Landform Evolution. Edited by D. A. Robinson and R. B. G. Williams
© 1994 John Wiley & Sons Ltd

On the other hand, interest in rates of weathering and saprolite formation has been stimulated by innovative work from scientists working in the Appalachian and Piedmont regions of the USA (Pavich, 1985, 1986 and 1989; Velbel, 1985, 1986; Dethier, 1986; Cleaves, 1989). The application of these ideas to the dynamics of weathering systems and the formation of weathering profiles in passive continental margin locations within humid areas of the tropics and sub-tropics raises interesting questions in geomorphology, concerning the conditions under which weathering penetration and chemical denudation are maximised, and how these biogeochemical systems relate to the physical systems effecting surface erosion.

   Uplift and rejuvenation of terrain may accelerate rates of chemical alteration and weathering penetration beneath interfluves, well away from linear zones of rapid surface denudation. Such rapid rates of weathering, combined with high rates of chemical as well as physical denudation are characteristic of orogenic zones, where carbonate and other sediments and fissile rocks occur. By contrast, low rates of weathering appear typical of many planate and crystalline terranes, where solute loads of rivers are equally low. For example, lowland rivers in Amazonia, such as the Negro, Xingu and Tapajos, have average solute concentrations in the range 5–8 mg l^{-1}. Such figures contrast, on a world-wide basis, with a weighted mean concentration of 120 mg l^{-1} (Walling and Webb, 1986). Such 'electrolyte poor' waters form a separate class, distinct from the normally carbonated waters which contain significant Ca^{2+} and HCO_3^-. In these waters 80% of the solute load is comprised of HCO_3^-, SO_4^-, Ca^{2+}, and SiO_2. But in the blackwaters of the Rio Negro and other forest streams, the high acidity (pH 4.5–5.1) leads to a cation–anion imbalance with H^+ ions making up a significant proportion of the specific conductance, along with Fe and Al probably in non-ionic form as organo-metallic complexes (Furch, 1984).

CAUSES OF LOW SOLUTE LOADS IN HUMID TROPICAL RIVERS

Low solute loads in tropical rivers can be a consequence of several factors, including:

1. Drainage from already leached terrains, typically underlain by deep kaolinised mantles or by quartz sands. Such old landsurfaces predominate on the Gondwana continents and strongly influence the figures for large drainage basins such as the Amazon, Orinoco, Zaire and Zambezi. The blackwater rivers which usually drain white sand regions come under this heading. Elsewhere the landsurface may be 'stabilised by a layer of indurated cation-poor material' according to Stallard and Edmond (1983).
2. Long-term storage of both particulate and solute loads in floodplain sediments, especially in the vast basinal tracts of alluviated land that occur on the Gondwana cratons which have been tectonically warped into large basins and swells (Niger, Zaire and Amazon).
3. Continuous mineral recycling in vegetation which has significant capacity to absorb silica and other cations (Lovering, 1959) and which recycles a high proportion of the nutrients from fallen and decayed trees. Stark and Jordan (1978) estimated that >99% of Ca and P is assimilated into the biomass, while amounts of K, Ca

and Mg in the total biomass are typically three to five times those found in soil A and B horizons from which leaching into rivers takes place.

4. Drainage from crystalline rocks which are widely exposed on the upswells of the cratonic regions will be poorer in solutes than drainage from orogenic regions, where carbonate and evaporite rocks, together with fissile greywackes and schists, plus volcanic suites, will all weather rapidly to supply ions into the river systems.

RATES OF WEATHERING PENETRATION

Both particulate and solute loads of rivers have been shown to increase with basin relief and annual runoff, and these factors clearly also influence the rate of weathering. Studies of weathering rates from the Appalachian region of the USA, pioneered by Cleaves et al. (1970 and 1974), have led to important results from more recent research (Pavich, 1985 and 1986; Velbel, 1985 and 1986; Cleaves, 1989). These are humid temperate forested areas of the Appalachian and Piedmont provinces (latitude 36–39°N), receiving rainfall in excess of 1000 mm yr^{-1}, and experiencing summer temperatures between 30 and 40°C, but cold winters and average annual temperatures of c. 10°C. No similar studies are known to the writer from humid tropical regions. Table 17.1a summarises some of these results, where 'rate' refers to the calculated rate of saprolite formation by isovolumetric weathering. Some other estimates of rates of weathering from tropical areas are given in Table 17.1b.

One salient observation emerges from these figures (if the exceptional value for tephra weathering is omitted), and that is that, whatever method is used, the figures obtained have a similar range from a minimum of 2 m Ma^{-1} to a maximum of around 50 m Ma^{-1}, while maximum figures for the long-term weathering of crystalline rocks appear to lie in the range 20–50 m Ma^{-1}. This is sufficient for most observed profiles (where other age criteria do not apply) to have been formed during the Quaternary.

On the other hand, where lateritic or gibbsitic residues are involved, estimates often range between 2 and 9 m Ma^{-1}, and since most laterite profiles attain thicknesses of >20 m, their periods of formation may need to be >10 Ma. If the lower estimate is used for profiles of c. 50 m, then the period of formation would be 25 Ma (Kronberg et al., 1982).

POSSIBLE RATES OF ALTERATION IN HUMID TROPICAL AREAS

What these figures do not indicate are possible rates of weathering for per-humid, tropical rainforest regions, where soil temperatures are two to three times those of the temperate areas, and rainfall amounts two to five times greater. If we make a crude assumption that increased base-flow will accelerate the weathering rate by a factor of times two, and that this might be increased to times three by the temperature increase, then 60–150 m Ma^{-1} of saprolite might be formed. The higher figure of 150 m appears high when compared with published accounts and evidence, particularly since figures so far obtained from tropical regions do not exceed 60 m Ma^{-1}. However, the enhanced biological productivity (leading to partial pressure of CO_2 of perhaps 100 times that of the atmosphere) and solute fluxes of the forest climates

Table 17.1 Results of studies of weathering rates

(a) Rates of saprolite production

Experimental area	(Author(s))	Method used	Rate
Smoky Mountains	Velbel (1985)	Mass balance	38 m Ma^{-1}
Southern Blue Ridge	Velbel (1986)	Mass balance	37 m Ma^{-1}
Pacific NW, USA	Dethier (1986)	Mass balance	33 m Ma^{-1}
Masanutten Mtn, Virginia	Afifi and Bricker (1983)	Mass balance	2–10 m Ma^{-1}
Southern Piedmont USA	Pavich (1986)	Solution loss	4 m Ma^{-1}
	Pavich (1989)	Residence time	20 m Ma^{-1}
Baltimore Piedmont	Cleaves et al. (1970)	Mass balance	5–8 m Ma^{-1}
	Cleaves (1989)	Equil. model	25–48 m Ma^{-1}

(b) Estimated rates of alteration

Area	Authors	Comment on calculation	Rate
Ivory Coast	Leneuf (1959)	Ferralitic leaching of Si, Ca, Mg, Na, K	5–50 m Ma^{-1}
	Leneuf and Aubert (1960)		13–45.5 m Ma^{-1}
	Boulangé (1983)		14 m Ma^{-1}
Chad	Gac (1979)		13.5 m Ma^{-1}
New Caledonia	Trescases (1973)	Ultramafic rocks	29–47 m Ma^{-1}
Uganda	Trendall (1962)	Ferricrete accumulation	9 m Ma^{-1}
Senegal	Nahon (1977)	Nodular ferricrete	c. 2–3 m Ma^{-1}
	Fritz and Tardy (1974)	gibbsitic alteration	3 m Ma^{-1}
	Ruxton (1968)	Profile depth in 650 Ka^{-1} volcanic ash	
Papua New Guinea	Hanntjens and Bleeker (1970)	Alteration of volcanic rocks	58 m Ma^{-1}
		—skeletal	c. 5 ka^{-1}
		—smectite	5–20 ka^{-1}
		—kandite	>20 ka^{-1}

Table 17.2 Some possible rates of weathering

$$
\begin{array}{llll}
25\,\text{mm ka}^{-1} & \equiv 25\,\text{m Ma}^{-1} & \equiv 80\,\text{m } 3.2\,\text{Ma}^{-1} & (\times 2 = 6.4\,\text{Ma}^{-1})^a \\
50\,\text{mm ka}^{-1} & \equiv 50\,\text{m Ma}^{-1} & \equiv 80\,\text{m } 1.6\,\text{Ma}^{-1} & (\times 2 = 3.2\,\text{Ma}^{-1}) \\
\mathbf{100\,mm\ ka^{-1}} & \mathbf{\equiv 50\,m\ 0.5\,Ma^{-1}} & \mathbf{\equiv 80\,m\ 0.8\,Ma^{-1}} & \mathbf{(\times 2 = 1.6\,Ma^{-1})^b} \\
250\,\text{mm ka}^{-1} & \equiv 50\,\text{m } 0.2\,\text{Ma}^{-1} & \equiv 80\,\text{m } 0.32\,\text{Ma}^{-1} & (\times 2 = 0.64\,\text{Ma}^{-1}) \\
1\,\text{m ka}^{-1} & \equiv 50\,\text{m } 50\,\text{Ka}^{-1} & \equiv 80\,\text{m } 80\,\text{ka}^{-1} & (\times 2 = 0.16\,\text{Ma}^{-1}) \\
\end{array}
$$

[a]Figures in the right hand column have been doubled as an arbitrary way of acknowledging periods of aridification, though these will have been perhaps 10–20% for many humid tropical areas.
[b]Rate considered a minimum for uplifted, humid tropical areas.

could lead to rates of alteration in excess of those quoted, and there is an urgent need to find measures of these processes under optimal conditions.

Table 17.2 indicates some hypothetical rates which might be considered. Such rates would be unlikely to remain constant over long time periods or during the formation of highly evolved, deep profiles, but could be maintained by repeated uplift and dissection of terrain.

It is interesting to reflect, that a rate of 0.1 mm yr^{-1} (100 mm ka^{-1}) would imply the formation of 100 m of saprolite during 1 Ma, but would only form 1 m during the Holocene (10 ka). Some studies of Quaternary erosion and sedimentation in eastern Brazil (Lichte, 1989) have implied very much higher rates of weathering to account for saprolite development beneath erosional stone-lines, but the evidence is not conclusive. The possibility of very high rates of weathering, approaching 1 mm yr^{-1}, appears limited by two sets of field evidence: (1) widespread mantle thicknesses of more than 100 m which might result are rare, and (2) long-term erosion and sedimentation rates derived from off-shore sedimentary volumes do not support correspondingly high rates of removal of saprolite which could account for the surviving profiles. Nevertheless, our perceptions of likely rates of weathering are probably influenced too much by figures derived either from temperate or highly seasonal climates or from beneath planate landsurfaces, and a rate of 100 mm ka^{-1} should not be thought unlikely within the humid tropics, where solute fluxes and hydraulic gradients are favourable.

WEATHERING RATES AND PRODUCTS

Stallard and Edmond (1987, p. 8300), have concluded for the Amazon Basin that,

> where weathering is sufficiently rapid, cation and silica concentrations are high, kaolinite forms, and quartz is stable. Where weathering rates are lower, perhaps because of thicker soils . . . silica concentrations decrease, and quartz dissolution can start contributing additional silica . . . eventually, with sufficiently low concentrations of silica, kaolinite is no longer stable, and any rock that is exposed to weathering can presumably have all its silica dissolved, leaving either gibbsite, or nothing.

In the lowland rivers of Amazonia, the proportions of Fe and Al to other elements are close to those in the crystalline rocks of the subjacent shield, and this suggested to Stallard and Edmond (1987), 'that the bedrock is actually dissolving completely'.

Such observations from lowland Amazonia can be compared with calculations of volume loss in the conversion of saprolite to soil. Pavich (1989) observed that the classical convex divides, arising from soil creep processes under forest, as described by Gilbert (1909), appear flattened in the southern Appalachian Piedmont landscape of low relief and gentle slopes (<8°). This flattening was attributed to three main processes of groundsurface lowering: volume reduction during the compaction of saprolite into soil; loss of mass in dissolved solids draining the B soil horizon; and loss of mass from the soil surface by erosion of clay. These processes combine to achieve the continual lowering of the so-called 'peneplain' surface, and can also result in relief inversion, leading to river gravels perched on residual hills.

The study of base-flow stream geochemistry alone led Pavich (1986 and 1989) to calculate a minimum rate of saprolite production of 4 m Ma^{-1}, but profile mineralogy indicates a total loss of mass during chemical alteration of saprolite to clay of as much as 75%, as indicated by ZrO and clay fraction increases. Pavich (1985) also took cores from the saprolite and measured the ^{10}Be content. ^{10}Be is formed in the atmosphere by cosmic ray spallation reactions and is then washed into the soil by rain, accumulating in the soil and saprolite over periods of 10^4–10^6 years. This provided a new rate of production and removal of saprolite of 20 m Ma^{-1}. Calculation of the loss of soil by surface erosion produced a figure of only 5 m Ma^{-1}, and this led Pavich (1986) to conclude that the other process of soil reduction must accomplish a volume reduction of 75%. His final figure of 20 m Ma^{-1} for saprolite production and removal, over a period of 70 Ma, would bring about a steady-state lowering of the landscape of around 1400 m, implying 'a balanced system of saprolite development, erosion and associated uplift' (due largely to isostasy) (Pavich, 1985).

The importance of this work, in the present context, is its agreement with much field evidence for surface lowering by similar processes in the tropics. It implies a downward movement of both the weathering front and the soil/saprolite boundary. The continuity of these processes is attested by the relatively constant thickness of the soil profiles on undulating upland surfaces, where they are less developed than on level terraces with an age of less than 1 Ma (Markewich *et al.*, 1987).

EVIDENCES OF GROUNDSURFACE LOWERING IN TROPICAL LANDSCAPES

Some specific mechanisms of groundsurface lowering by combined chemical and physical processes can be identified in tropical landscapes, including:

1. The accumulation of Fe, Ni and other heavy metals in weathering profiles (Esson and Surcan dos Santos, 1978; McFarlane, 1983; Aleva, 1991).
2. The accumulation of gravel layers on interfluves and upper slopes (Teeuw, 1991).
3. Non-abrasive lowering of valley floors by etching of bedrock and loss of fines and solutes (Thomas and Thorp, 1985).
4. The ferrolysis (breakdown) of clays in hydromorphic environments with formation of white sands (Brinkman, 1970).
5. The dissolution of rocks and the formation of karst hollows in crystalline terrain (McFarlane and Twidale, 1987).

In addition the bedrock control over topographic detail in erosional terrain can be construed as evidence for downwearing as against lateral planation (Hack, 1979).

Mineral exploration has commonly found that dispersion of geochemical anomalies away from dyke zones or other specific targets has been very limited, while at the same time concentration of minerals in the weathering profile by downward movement of both the weathering front and the groundsurface is common. This process accompanies ferricrete formation, and in some laterites leads to the accumulation of nickel. According to Esson and Surcan dos Santos (1978), 10.7 m of nickeliferous saprolite over a serpentinite at Liberade, Brazil, represents the residue of as much as 240 m of rock weathering. This figure may be questionable, but the principle is important. Aleva (1991) has recorded the accumulation of heavy minerals such as cassiterite (sp. gr. 7.0), monazite (sp. gr. 5.1) and zircon (sp. gr. 4.6) at the base of sandy weathering products derived from Permo-Triassic sandstones on Belitung Island, Indonesia. He claimed that this took place under gravity following elutriation of clay from the system. There is of course some movement away from interfluves towards adjacent valleys by mass movement which is recorded by stone-lines (Thomas and Thorp, 1985) and by hydromorphic dispersion, recorded by geochemical anomalies (Govett, 1987). The existence of gossans (weathered rock profiles containing sulphides) also requires a hypothesis for the concentration of Fe_2O_3 in the upper profile. This could occur by the McFarlane (1983) model, but Ollier and Galloway (1990) favour lateral movement of Fe^{2+} into valley floors where it becomes indurated with time and subsequently forms ridge tops by relief inversion. However, if the iron enrichment directly overlies the sulphide ore body, this latter explanation is unlikely to apply.

All this evidence suggests that rates of weathering penetration and saprolite production can match or exceed rates of surface denudation in many areas of relief formation, and that surface lowering involves a volume loss (>70%) due to (1) volume reduction during compaction of saprolite to soil; (2) loss of mass in dissolved solids draining the soil B horizon; and (3) loss of mass from the soil surface by erosion of clay (Pavich, 1989).

THE AGES OF WEATHERED LANDSURFACES

Recognition that many weathered landsurfaces started forming during the Mesozoic has had a pervasive influence on geomorphological thinking. By plotting $\delta^{18}O$ values for Australian regolith samples against both present latitude and ages assigned on the basis of geological evidence (overlying sediments or lavas), Bird and Chivas (1988) have argued that many regoliths are probably much older than formerly thought, some being pre Late Mesozoic. Schmidt and Ollier (1988) derived palaeomagnetic evidence of Late Cretaceous to Early Tertiary weathering on the upland plateau surfaces which form part of the Eastern Highlands of Australia, and similar conclusions have been reached for southern New South Wales by Pillans (1977). Ollier (1985) and Pain (1985) have summarised much of the evidence from Australia concerning the antiquity of the continental landsurfaces. According to Fairbridge and Finkl (1980) the Yilgarn landsurface has not been effectively lowered by denudation over 1.5×10^9 yr, and has been little affected by dissection during emergence of 250–300 m since the Eocene.

Recent work from Australia (Gale, 1992) has tended to reinforce this view of the continental interior of Australia, but it is equally clear that these results do not apply to the marginal areas of seaward drainage in Australia, and they cannot be applied in a general way to landscapes of other continents. A major issue to be resolved is therefore the contrast between the evolution of weathering profiles on the older landsurfaces in continental interiors, and the formation of possibly 'new' regoliths in the marginal zones of continents.

In the former, gradual vertical lowering, largely by chemical denudation, is reflected in the occurrence of ferricrete and alucrete caps, together with heavy metal concentrations in saprolites on the one hand, and the formation of white sands on the other. In the latter, a neglected suite of features can be identified that indicate the operation of rapid weathering sufficient to maintain thick saprolites, despite the formation of high relief and steep slopes. Within these warped or faulted marginal zones, it is possible to argue that the 'old' regolith of the interior is partially preserved and responsible for the appearance of weathered rock in such highly dissected zones. But, although there will be some degree of inheritance, there are good reasons to suppose that the bulk of the saprolite is a result of weathering that is broadly contemporaneous with the history of uplift. A similar argument can be applied to the deepening of profiles beneath deeply dissected inland plateaus (Thomas, 1989).

CHARACTERISTICS OF WEATHERED MANTLES IN MARGINAL ZONES

There have been few detailed studies of weathering in the marginal zones of continents, but the data which we have are pointers towards some preliminary interpretations:

1. Deep profiles exceeding 50 m are found, associated with dissected multi-convex topography. The humid tropical to sub-tropical Atlantic margin of southeast Brazil illustrates these features, being the type area for multi-convex, 'meias laranjas' topography. At Itaipova (Santa Caterina) a measured section of 80 m has been exposed by quarrying. Similar features can be seen in Cameroon and west Kalimantan (Borneo).
2. Deep weathering penetration occurs between high, rocky hills in areas of high relief, as along the Serra do Mar of southeast Brazil, in Hong Kong, the Freetown Peninsula, Sierra Leone, and in many other locations.
3. Saprolites are rarely associated with *in situ* duricrusts, but may be protected by stone-lines and transported gravels (Kadomura and Hori, 1990).
4. The alteration is often less advanced than on 'old' plateau surfaces, with fersiallitic weathering over considerable depths, and sometimes abundant corestones. Profiles in southeast Brazil, southeast Australia, Hong Kong, peninsular Malaysia and many other locations demonstrate this feature.
5. Where weathering has advanced to a ferrallitic stage, profile features, such as bauxitised horizons, often appear to conform with the present-day relief, as at Tayan in west Kalimantan (Rodenburg, 1984).
6. Massive landslides are found resting against weathered rock in a configuration that suggests a continuity of active weathering and morphogenesis. These are particularly evident in southeast Brazil.

The Serra do Mar in southeast Brazil forms the hinterland between Rio de Janeiro and Florianopolis over a distance of several hundreds of kilometres. The basement is a sheared and faulted continental margin, each fault zone exhibiting strong scarp relief with prominent 'sugar loaf' domes in granite and gneiss. The treads of the fault blocks are characterised by a multi-convex relief of weathered compartments, sometimes giving way to sedimentary basin fills. Weathering penetrates between the granite domes in areas of deep dissection. Corestone profiles are spectacular and common and these denote rapid weathering, rather than prolonged alteration below a stable landsurface. The well-known weathering profiles of Hong Kong also conform to this type. Comparable features are found developed within the basic igneous complex that gives rise to mountainous relief in coastal Sierra Leone. Deep weathering follows lines of faulting and more susceptible rock. In all these areas landsliding is a repeated phenomenon and a serious hazard. If landsliding is seen as a major and integral landforming process in these and similar areas, then renewal of the saprolite cover within the denudation system must be achieved.

Although it would be possible to argue that the weathering in these locations represents the roots of ancient profiles developed below the so-called South American Surface (or African Surface) of Late Cretaceous to Miocene age, such a hypothesis on its own does not satisfy the field observations. What may be true is that the profiles were deepened within or below the older saprolites in the Early Cenozoic, but what is seen today is predominantly a function of the period of relief development along the continental margins, and relics of these ancient profiles may be few.

Further evidence comes from Kalimantan (Indonesian Borneo), where it is supposed that an ancient Sundaland Continent has subsided to form the Sunda Sea between Borneo and Sumatra. The regolith of this continent is apparently buried by sediments in submarine locations (Batchelor, 1979), and it can be suggested that profiles on land are developed from this mantle. Again, inheritance from the older saprolite is probable, but the present-day profiles have been renewed by landward uplift in an equatorial climate. For example, bauxite profiles have been mapped in the low convex hills around Tayan on the Kapuas River and, according to Rodenburg (1984), these 'were developed by the *in situ* weathering of a dissected, uplifted peneplain'.

Comparisons with the southeast coastal zone of southern Africa are instructive. Both King (1972) and, more recently, Partridge and Maud (1987) have argued for strong Pliocene uplift in this area though their evidence is equivocal. Deep weathering in this region is less pervasive than in Brazil, but it is not absent. The strong uplift, deep dissection, and lower rainfall and temperatures of this region have all influenced landform outcomes. Deep weathering occurs in the granites beneath the Table Mountain Sandstones and the present author argued (Thomas, 1978) for deep weathering to be initiated along the retreating Table Mountain Sandstone escarpment in this region.

The situation in southeast Australia is no less intriguing. On the Monaro Plateau of New South Wales there are extensive, stripped landsurfaces of tors and shallow granite basins, yet at lower elevations, below the eastern escarpment, a multi-convex landscape of weathered compartments occurs in the Bega basin. Controversy has developed regarding the origins of these mantles. While Dixon and Young (1981) argued that hydrothermal alteration preceded subaerial weathering to produce the sandy regoliths of this area, Ollier (1983) has argued against this thesis and for pre-basalt (53–36 Ma) weathering (Ollier and Taylor, 1988; Ollier, 1991; personal communication). Diagnosis of hydrothermal alteration may be difficult here, but the dominantly sandy weathering makes it likely that the profiles have undergone continuous development as a result of uplift and warping, accompanied by down-wearing of the landscape within the Bega batholith which appears susceptible to weathering. According to Gale (1992) the long-term Cenozoic denudation rate for this part of Australia has been close to 10 m Ma^{-1}, which is well within the possible rates for saprolite production during this period (see Table 17.1).

There are, of course, important variations between areas, but the lessons that can be learned from studies of Appalachian USA (Table 17.1) suggest that we should reappraise our predominant view of weathering penetration during landscape dissection along the marginal zones of continents.

GEOMORPHIC FACTORS IN THE EVOLUTION OF WEATHERING PROFILES

These observations cannot be made without reference to the major evolutionary stages of the landscape and, with respect to the former Gondwana fragments, it is possible to point to the following areas of broad agreement:

1. Evidence for the widespread survival of a Gondwana (pre-continental drift) landscape is lacking.
2. Claims for a Cretaceous–early Miocene polycyclic planation are substantiated from a number of areas and include the African, Sul-Americana and Australian surfaces.
3. This planation took place at different elevations as the Great Escarpment came into being during the rifting of Gondwanaland.
4. Much of the deep weathering and associated lateritic alteration has been attributed to this period.
5. Miocene uplift (c. 18 Ma), followed by aridification of climate (in Australia and Chile at c. 15 Ma) terminated this phase.
6. Partial planation (illustrated by the Post-African 1 cycle) may have occurred during the mid-Miocene to late Pliocene.
7. Late Pliocene uplift is claimed for southern Africa (possibly 900 m in the southeast) but is disputed. There is little evidence for this event in eastern Australia.
8. Deep incision has characterised the high, faulted marginal zones, with only localised planation during the Pleistocene.

This pattern of events was not followed in an identical manner in all the continents, and claims for strict correlations must be greeted with scepticism (Thomas and Summerfield, 1987; Gilchrist and Summerfield, 1990 and 1991). The ability of weathering systems to maintain saprolite covers during marginal zone dissection has clearly been influenced by local factors such as the rate and amount of uplift, faulting and tilting of the basement, shearing of rock fabrics, and by the humidity and warmth of climate.

THE BEHAVIOUR OF WEATHERING PROFILES THROUGH TIME

Such events, upon which have been superimposed important changes of climate and environment, lead to varying behaviour in the denudation system. One aspect of this variation that has been neglected relates to the behaviour of the weathering profiles. Some ways in which weathering profiles may evolve in response to landforming events are listed in Table 17.3 (modified from Thomas, 1989).

Profile lowering will occur if weathering penetration and surface denudation are delicately balanced (when averaged over time periods of 10^2–10^4 yr). This can permit the descent of the column without substantial modification, but if Fe^{3+} is precipitated as goethite segregations which evolve towards crystalline hematite nodules, then these will accumulate in the soil B horizon, while resistant clasts of vein quartz (plus quartzite, tourmaline, corundum, etc.) may also accumulate as a stone-line. This twin process is probably self-limiting as the profile becomes armoured by a developing duricrust.

Profile deepening will depend on a minimal rate of surface lowering during continued weathering penetration. This can occur beneath plateaus lacking significant slope or drainage systems, but more commonly it will result from the development of a duricrust capping which protects the saprolite from further removal. These are conditions for pallid zone or bauxite zone formation, providing that free drainage is maintained by uplift and dissection.

Profile thinning is a process that results from groundsurface lowering, when weathering-resistant rock compartments are encountered at depth. The rate of erosion may be slow, but the column, in effect, becomes compressed from below. In this way a highly evolved upper horizon (probably a duricrust) may be seen to rest on almost fresh rock.

Table 17.3 Behaviour of weathering profiles over long time periods

Profile behaviour	Denudation balance[a]	Geomorphic outcome
Lowering	Balanced WP:SD (vertical lowering)	Mantled etchsurfaces; FE_2O_3 accumulation
Deepening	WP > SD (stable landsurface)	Weathered landsurfaces
Thinning	SD > WP (resistant rock at depth)	Rock exposure, etchsurfaces
Truncation	SD ≫ WP (lateral erosion)	Slope pediments, rock exposure
Collapse	SD and WP minimal (hydromorphic pedogenesis)	White sands, depressions
Burial	SD and WP zero (sedimentation, lava flows)	Sedimentary cover

[a]WP indicates rate of weathering penetration; SD rate of surface denudation.

Profile truncation is distinct from thinning, and arises from an acceleration of surface erosion that removes the more highly evolved topsoil and upper saprolite. This is usually caused by climatic crises leading to opening of the vegetation canopy and increased runoff, but may also be due to the progress of dissection. This process can result in fersiallitic materials being exposed where the present climatic parameters suggest development of a ferrallitic mantle.

Profile collapse appears to occur where ferrallitic residues become subject to hydro-morphic conditions, and widespread leaching of Al^{3+}, Fe^{2+}, and SiO_2 takes place, leading to the breakdown of kaolinite clay and the formation of podzolised white sands.

Profile burial can occur if sedimentation is not accompanied by erosion of the pre-weathered saprolite. Circumstances favouring this situation include tectonic subsidence, though some trimming or truncation of the profile is likely to occur. This has taken place in Sundaland, Indonesia.

Profile renewal probably continues alongside most other processes, except burial. However, soil systems energised by the dissection of old landsurfaces, under the protective cover of rainforest, may display well-organised profiles in sympathy with the prevailing relief, but developed in an ancient saprolite. This is the 'two cycle' theory of tropical pedology described by Ollier (1959).

When rates of erosion or accretion converge with rates of energy and mass fluxes in the soil, horizonation as an aspect of pedogenesis becomes opposed by the rates of denudation (morphogenesis) and rates of pedogenesis will be delicately balanced as envisaged with profile lowering. But, as has been shown, profiles also evolve through time, with accumulation of resistant clasts and oxides (mainly of Fe^{3+}). These may eventually prevent significant surface lowering, but encourage further deepening and horizonation of profiles.

The accumulation of stone-lines and ferricretes is an example of a feedback mechanism at work, preventing surface erosion from continuing uninterrupted across the landscape, unless the energy of surface transport can remove the larger and denser (Fe_2O_3) clasts. When surface erosion is accelerated, as when the vegetation canopy is opened up by desiccation and fire, truncation may occur for a period, to be followed by a renewal of the profile, when the canopy is restored. This took place over periods of 10^3–10^4 yr during the Quaternary glacial–interglacial (or interstadial) cycles.

CONCLUSIONS

It should be clear from this discussion that all weathering profiles cannot be regarded primarily as relict features of old continental landsurfaces. All remain in a dynamic state, although the ancient regoliths in arid continental interiors, such as Australia, may have altered only very slowly since their main periods of formation in the Mesozoic. In these instances, surface denudation has periodically overtaken weathering penetration to create stripped etchsurfaces (Mabbutt, 1961 and 1988; Fairbridge and Finkl, 1980). Elsewhere, the flux of water and solutes through the soil has been encouraged by uplift and dissection, and further research needs to be undertaken to establish more clearly the evolution of weathering profiles in such areas which include the passive continental margins. Many of these landscapes, especially along the eastern coastal areas, have evolved under predominantly humid forest climates, whether in temperate, sub-tropical or tropical latitudes.

REFERENCES

Afifi, A. A. and Bricker, O. P. (1983). Weathering reactions, water chemistry and denudation rates in drainage basins of different bedrock types: 1—sandstone and shale. In *Dissolved Loads of Rivers and Surface Water Quantity/Quality Relationships*, International Association of Hydrological Sciences Publication, **141**, 193–203.

Aleva, G. J. J. (1984). Lateritisation, bauxitisation and cyclic landscape development in the Guiana Shield. In Jacob, L. (ed.), *Bauxite*, Proceedings 1984 Symposium, Los Angeles, Society of Mining Engineers, American Institute of Mining, Metallurgical and Petroleum Engineers Inc., New York, pp. 111–151.

Aleva, G. J. J. (1991). Tropical weathering, denudation and mineral accumulation. *Geologie en Mijnbouw*, **70**, 35–38.

Batchelor, B. C. (1979). Geological characteristics of certain coastal and offshore placers as essential guides for tin exploration in Sundaland, Southeast Asia. *Geological Society of Malaysia Bulletin*, **11**, 283–313.

Bird, M. I. and Chivas, A. R. (1988). Oxygen isotope dating of the Australian regolith. *Nature*, **331**, 513–516.

Boulangé, B. (1983). Les formations bauxitique latéritiques de Côte d'Ivoire. Thesis, University of Paris.

Brinkman, R. (1970). Ferrolysis, a hydromorphic soil forming process. *Geoderma*, **3**, 199–206.

Cleaves, E. T. (1989). Appalachian piedmont landscapes from the Permian to the Holocene. *Geomorphology*, **2**, 159–179.

Cleaves, E. T., Godfrey, A. E. and Bricker, O. P. (1970). Geochemical balance of a small watershed and its geomorphic implications. *Geological Society of America Bulletin*, **81**, 3015–3032.

Cleaves, E. T., Fisher, D. W. and Bricker, O. P. (1974). Chemical weathering of serpentinite in the eastern Piedmont of Maryland. *Geological Society of America Bulletin*, **85**, 437–444.

Dethier, D. P. (1986). Weathering rates and the chemical flux from catchments in the Pacific Northwest, USA. In Colman, S. M. and Dethier, D. P. (eds), *Rates of Chemical Weathering of Rocks and Minerals*, Academic Press, New York, pp. 503–530.

Dixon, J. C. and Young, R. W. (1981). Character and origin of deep arenaceous weathering mantles on the Bega batholith, southeastern Australia. *Catena*, **8**, 87–109.

Esson, J. and Surcan dos Santos, L. C. (1978). Chemistry and mineralogy of section through lateritic nickel deposit at Liberade, Brazil. *Transactions Institute of Mining and Metallurgy*, **87 Sec. B**, B53–B60.

Fairbridge, R. W. and Finkl, C. W. Jr (1980). Cratonic erosional unconformities and peneplains. *Journal of Geology*, **88**, 69–86.

Fritz, B. and Tardy, Y. (1974). Étude thermodynamique du système gibbsite, quartz, kaolinite, gaz carbonique. Application à la genèse des podzols et des bauxites. *Sciences Géologiques Bulletin*, **26**, 339–367.

Furch, K. (1984). Water chemistry of the Amazon basin: the distribution of chemical elements amongst freshwaters. In Sioli, H. (ed.), *The Amazon*, Junk, Dordrecht, pp. 167–199.

Gale, S. J. (1992). Long-term landscape evolution in Australia. *Earth Surface Processes and Landforms*, **17**, 323–343.

Gilbert, G. K. (1909). The convexity of hilltops. *Journal of Geology*, **17**, 344–350.

Gilchrist, A. R. and Summerfield, M. A. (1990). Differential denudation and flexural isostasy in formation of rifted-margin upwarps. *Nature*, **346**, 739–742.

Gilchrist, A. R. and Summerfield, M. A. (1991). Denudation, isostasy and landscape evolution. *Earth Surface Processes and Landforms*, **16**, 555–562.

Govett, G. J. S. (1987). Exploration geochemistry in some low-latitude areas—problems and techniques. *Transactions Institute Mining and Metallurgy (Series B Applied Earth Sciences)*, **96**, B97–B116.

Grandin, G. (1976). *Aplanissements Cuirassés et Enrichessement des Gisements de Manganèse dans Quelques Régions d'Afrique de l'Ouest*. Mémoir ORSTOM, **82**, Paris.

Haantjens, H. A. and Bleeker, P. (1970). Tropical weathering in the territory of Papua New Guinea. *Australian Journal of Soil Research*, **8**, 157–177.

Hack, J. T. (1979). Rock control and tectonism—their importance in shaping the Appalachian highlands. *US Geological Survey Professional Paper*, **1126–B**.

Kadomura, H. and Hori, N. (1990). Environmental implications of slope deposits in humid tropical Africa: evidence from southern Cameroon and western Kenya. *Geographical Reports, Tokyo Metropolitan University*, **25**, 213–236.

King, L. C. (1972). *The Natal Monocline: Explaining the Origins and Scenery of Natal, South Africa*. University of Natal Press, Durban.

Kronberg, B. I., Fyfe, W. S., McKinnon, B. J., Couston, J. F., Filho, B. S. and Nash, R. A. (1982). Model for bauxite formation: Paragominas (Brazil). *Chemical Geology*, **35**, 311–320.

Leneuf, N. (1959). L'altération des granites calco-alcalins et des granodiorites en Côte D'Ivoire forestière et les sols qui en sont dérivés. Thesis, University of Paris.

Leneuf, N. and Aubert, G. (1960). Essai d'évaluation de la vitesse de ferrallitisation. *Proceedings 7th International Congress of Soil Science*, pp. 225–228.

Lichte, M. (1989). Arid processes in the SE-Brazilian relief evolution during the last ice-age. *International Symposium on Global Changes in South America During the Quaternary, São Paulo, 1989, Special Publication*, **1**, 60–64.

Lovering, T. S. (1959). Significance of accumulator plants in rock weathering. *Bulletin of the Geological Society of America*, **70**, 781–800.

Mabbutt, J. A. (1961). A stripped landsurface in Western Australia. *Transactions of the Institute of British Geographers*, **29**, 101–114.

Mabbutt, J. A. (1988). Land-surface evolution at the continental time-scale: an example from interior Western Australia. *Earth Science Reviews*, **25**, 457–466.

Markewich, H. W., Pavich, M. J., Mausback, M. J., Hall, R. L., Johnson, R. G. and Hearn, P. P. (1987). Age relations between soils and geology in the coastal plain of Maryland and Virginia. *US Geological Survey, Bulletin*, **1589a**.

McFarlane, M. J. (1983a). Laterites. In Goudie, A. S. and Pye, K. (eds), *Chemical Sediments and Geomorphology*, Academic Press, London, pp. 7–58.

McFarlane, M. J. (1983b). The temporal distribution of bauxitisation and its genetic implications. In Melfi, A. J. and Carvalho, A. (eds), *Lateritisation Processes*, Proceedings II International Seminar on Lateritisation Processes, University of São Paulo, Brazil, pp. 197–207.

McFarlane, M.J. and Twidale, C.R. (1987). Karstic features associated with tropical weathering profiles. *Zeitschrift für Geomorphologie, Supplementband*, **NF64**, 73–95.

Nahon, D. (1977). Time factor in iron crusts genesis. *Catena*, **4**, 249–254.

Ollier, C. D. (1959). A two cycle theory of tropical pedology. *Journal of Soil Science*, **10**, 137–148.

Ollier, C. D. (1983). Weathering or hydrothermal alteration. *Catena*, **10**, 57–59.

Ollier, C. D. (1985). Morphotectonics of passive continental margins: Introduction. *Zeitschrift für Geomorphologie, Supplementband* **NF54**, 1–9.

Ollier, C. D. and Taylor, D. (1988). Major geomorphic features of the Kosciusko-Bega region. *BMR Journal of Geology and Geophysics*, **10**, 357–362.

Ollier, C. D. and Galloway, R. W. (1990). The laterite profile, ferricrete and unconformity. *Catena*, **17**, 97–109.

Pain, C. F. (1985). Morphotectonics of the continental margins of Australia. *Zeitschrift für Geomorphologie, Supplementband* **NF54**, 23–35.

Partridge, T. C. and Maud, R. R. (1987). Geomorphic evolution of southern Africa since the Mesozoic. *South African Journal of Geology*, **90**, 179–208.

Pavich, M.J. (1985). Appalachian piedmont morphogenesis: weathering, erosion, and Cenozoic uplift. In Morisawa, M. and Hack, J. T. (eds), *Tectonic Geomorphology*, George Allen and Unwin, London, pp. 27–51.

Pavich, M. J. (1986). Processes and rates of saprolite production and erosion on a foliated granitic rock of the Virginia Piedmont. In Colman, S. M. and Dethier, D. P. (eds), *Rates of Chemical Weathering of Rocks and Minerals*, Academic Press, New York, pp. 552–590.

Pavich, M. J. (1989). Regolith residence time and the concept of surface age of the Piedmont 'peneplain'. *Geomorphology*, 2, 181–196.

Pillans, B. J. (1977). An early Tertiary age for deep weathering at Bredbo, southern NSW. *Search*, 8, 81–83.

Rodenburg, J. K. (1984). Geology, genesis and bauxite reserves of West Kalimantan, Indonesia. In Jacob, L. (ed.), *Bauxite*. Proceedings of the 1984 Symposium, Los Angeles, Society of Mining Engineers, American Institute of Mining, Metallurgical and Petroleum Engineers Inc., New York, pp. 603–618.

Ruxton, B. P. (1968). Measures of the degree of chemical weathering of rocks. *Journal of Geology*, 76, 518–527.

Schmidt, P. W. and Ollier, C. D. (1988). Paleomagnetic dating of Late Cretaceous to Early Tertiary weathering in New England, NSW, Australia. *Earth-Science Reviews*, 25, 363–371.

Stallard, R. F. and Edmond, J. M. (1983). Geochemistry of the Amazon. 2. The influence of geology and weathering environment on the dissolved load. *Journal of Geophysical Research*, 88 (C14), 9671–9688.

Stallard, R. F. and Edmond, J. M. (1987). Geochemistry of the Amazon. 3. Weathering chemistry and limits to dissolved inputs. *Journal of Geophysical Research*, 92 (C8), 8293–8302.

Stark, N. M. and Jordan, C. F. (1978). Nutrient retention by the root mat of an Amazonian rainforest. *Ecology*, 59, 434–437.

Teeuw, R. M. (1991). A catenary approach to the study of gravel layers and tropical landscape morphodynamics. *Catena*, 18, 71–89.

Thomas, M. F. (1978). The study of inselbergs. *Zeitschrift für Geomorphologie*, Supplementband NF31, 1–41.

Thomas, M. F. (1989). The role of etch processes in landform development II. Etching and the formation of relief. *Zeitschrift für Geomorphologie*, Supplementband NF33, 257–274.

Thomas, M. F. and Summerfield, M. J. (1987). Long-term landform evolution: key themes and research problems. In Gardiner, V. (ed.), *International Geomorphology 1986 Part II*, pp. 935–956.

Thomas, M. F. and Thorp, M. B. (1985). Environmental change and episodic etchplanation in the humid tropics of Sierra Leone: the Koidu etchplain. In Douglas, I. and Spencer, T. (eds), *Environmental Change and Tropical Geomorphology*, Allen and Unwin, London, pp. 239–267.

Trendall, A. F. (1962). The formation of 'apparent peneplains' by a process of combined lateritisation and surface wash. *Zeitschrift für Geomorphologie*, NF6, 183–197.

Trescases, J. J. (1973). Weathering and geochemical behaviour of the elements of ultramafic rocks in New Caledonia. *Bureau of Mineral Resources, Geology and Geophysics, Canberra*, 141, 149–161.

Velbel, M. A. (1985). Geochemical mass balances and weathering rates in forested watersheds of the southern Blue Ridge. *American Journal of Science*, 285, 904–930.

Velbel, M. A. (1986). The mathematical basis for determining rates of geochemical and geomorphic processes in small forested watersheds by mass balance: examples and implications. In Colman, S. M. and Dethier, D. P. (eds), *Rates of Chemical Weathering of Rocks and Minerals*, Academic Press, New York, pp. 431–451.

Walling, D. E. and Webb, B. W. (1986). Solutes in river systems. In Trudgill, S. T. (ed.), *Solute Processes*, Wiley, Chichester, pp. 251–327.

18 Regolith and Landscape Development in the Koidu Basin of Sierra Leone

R.M. TEEUW
University of Hertfordshire, Hatfield, UK

M.F. THOMAS
University of Stirling, UK

and

M.B. THORP
University College Dublin, Eire

ABSTRACT

Textural, micromorphological and mineralogical properties of regolith and sediments sampled from landform units developed on granite/gneiss in the forest/savanna zone of Sierra Leone are described for profiles containing one or more layers of topsoil, residual top gravel, colluvial/alluvial sediments, basal gravel and saprolite. Results indicate at least seven major paths for iron sesquioxide and clay movement, and that lateral eluviation in solution and suspension is a major component of the humid tropical denudation system. Given the continuity of these processes when compared with the episodic nature of surface erosion, their importance in landsurface lowering deserves further evaluation. Available evidence points to dynamic etchplanation as the prevailing mode of landscape development in the study region.

INTRODUCTION

This paper examines the composition and distribution of regolith in the Koidu region (8°38′N, 11°03′W) of Sierra Leone, a granite–gneiss plateau at 370–410 m above sea level, fringed by steep-sided ridges and hills of mainly meta-sedimentary rocks that rise to 810 m (Figure 18.1). The climate is hot, humid and monsoonal, with 80% of the annual rainfall (2355 mm) falling within six months. Until recent clearance, the region supported semi-evergreen rainforest, although the forest–savanna boundary lies only 50–100 km to the northeast.

The Koidu landscape is characterised by convex, nearly flat, interfluves and by a drainage network of channelless, swampy headwaters: river valleys with channels account for only 20% of the drainage system. Similar landscapes have been described from other regions of the seasonally humid tropics with granite–gneiss bedrock (Thomas and Goudie, 1985), including the 'demi-orange' interfluves and 'bas fond'

Rock Weathering and Landform Evolution. Edited by D. A. Robinson and R. B. G. Williams
© 1994 John Wiley & Sons Ltd

304

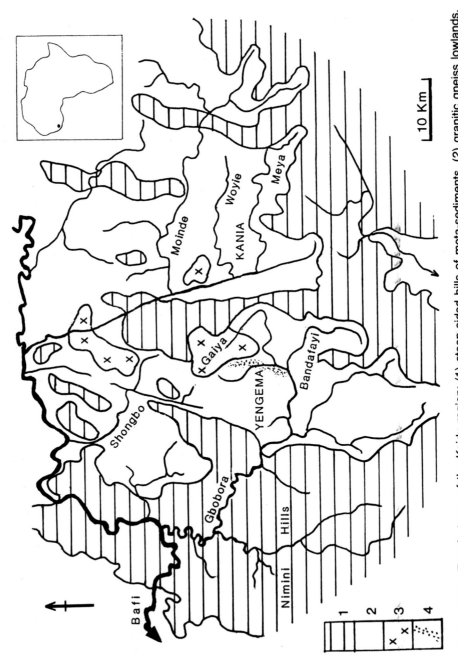

Figure 18.1 Terrain types of the Koidu region: (1) steep-sided hills of meta-sediments, (2) granitic gneiss lowlands, (3) granite inselbergs, (4) granodioritic gneiss with rare ferricrete mesas

valley swamps of the Ivory Coast (Avenard, 1973) and Togo (Leveque, 1979), and the 'inselberg/sohlenkerbtal' landscape of Tanzania (Louis, 1964) and Sri Lanka (Bremer, 1981).

Previous geomorphological research in the Koidu region has been directed partly towards an understanding of the fluvial deposits which contain important diamond placers (Thomas and Thorp, 1980; Thomas et al., 1985), and this work has raised a number of broader issues concerning the development of terrain in this environment (Thomas and Thorp, 1985).

The style of landscape development appears similar to the cratonic regime proposed by Fairbridge and Finkl (1980), whereby prolonged downwasting dominated by chemical eluviation and low energy transfers of fine sediment under a forest cover (biostasie), has alternated with shorter periods of more rapid surface erosion (rhexistasie). According to the model of a kimberlite pipe proposed by Hawthorne (1973), the kimberlite dyke zone (Cretaceous, 92 Ma) exposed in the Koidu area is indicative of deep erosion of the granite–gneiss basement during the Cenozoic, while the heavily duricrusted schist belt ranges that partially enclose the Koidu area stand 350–450 m higher and have been lowered far less. During this period, relief differentiation has become accentuated (Thomas, 1989b), and lateral planation has been confined to the formation of localised glacis or pediments within individual drainage basins.

The glacis truncate older weathering profiles and may have led to some basal sapping of rock hillslopes. The age of these features is uncertain, but pre-glacial (Pliocene) aridity may have been important, before the climate fluctuations of the Quaternary brought about more recent changes in the landscape. Local information on these Quaternary oscillations of climate is limited to the last 40 ka radiocarbon years. Analysis of river sediments indicates very wide fluctuations of discharge in the rivers and also major variations in the dynamics of hillslope processes (Thomas and Thorp, 1992).

During the progress of prolonged downwasting, etching of the bedrock occurs beneath both interfluves and stream channels, with sediment transfers consisting mainly of fine-grained materials. This type of regime allows the accumulation of heavy clasts in the surface and near-surface soil and regolith layers. These clasts include:

1. vein quartz, tourmaline and other resistant minerals from the weathering of subjacent rock, including diamond and corundum;
2. ferric iron sesquioxide pisoliths and nodules segregated by pedogenic processes as the weathered profile is lowered (McFarlane, 1983);
3. iron duricrust (ferricrete) pieces fragmented from formerly more continuous plinthite horizons in the soil;
4. fluvially rolled pebbles, mainly of quartz or quartzite, presumably incorporated from ancient river channels, long since destroyed.

In addition, important transfers of ions in solutions take place and some elements (especially Fe and Al) become complexed by chelating substances. Reducing conditions within the swampy valley floors lead to the leaching of Fe^{2+} from nodules, pisoliths and ferricrete fragments that enter the valley floors by processes of erosion and mass movement along the valley sides.

The processes by which lateral transfers of particulate material take place in a planate landscape are not always clear. However, valleyside sections display stone-lines within the top 0.25–1.0 m of soil (Thomas and Thorp, 1985) and these can be traced into the adjacent swamps, where the iron segregations become leached. Strong creep has been indicated by Aleva (1989) and Moeyersons (1989) as a major process affecting stonelines on steeper slopes and it is supposed that the process will continue to operate, if more slowly, on slopes of lesser inclination.

On the other hand, the stonelines appear to denote both the accumulation of clasts during downwasting (etchplanation) and surface lag deposits resulting from strong surface sheet erosion. Evidence for the latter was adduced by Thomas and Thorp (1985) and by Teeuw (1991a) in the form of valley-head deposits that revealed two superposed colluvial layers, each associated with stony material. Evidence also exists for the action of surface water on the interfluve areas between drainage systems, where surfaces patinated with iron sesquioxides and concentrations of quartz pebbles and heavy minerals into palaeo-rills have been described (Teeuw, 1989).

A more detailed investigation of the regolith properties is clearly called for, if the processes outlined previously are to become better understood. This study presents the results of one such programme of research.

METHODOLOGY

In order to characterise the regolith properties associated with each landform unit, and to identify linkages and transfers of regolith materials over space and time, two adjacent drainage basins were selected for detailed analysis. Each basin has an area of c. 200 km^2, one being underlain by granites and granodiorites, with a landscape of inselbergs, ferricrete mesas, glacis (pediments) and valley swamps; the other is underlain by granite–gneisses and migmatites, with near-planate interfluves and valley swamps (Figure 18.1 and Teeuw, 1991a).

Geomorphological maps, at 1:1250 scale, were compiled on surface mining plans from field surveys and the interpretation of large-scale aerial photographs, to show the distribution of landform types in both study areas. Regolith profiles were sampled using 1 m diameter pits positioned in each landform type that occurred along 10 transects from interfluve crest to valley floor. In each pit, samples for textural and petrographic analysis were collected at 30 cm intervals down to bedrock or saprolite. Samples were taken from the six most frequently occurring landform units (Figure 18.2) and used for micromorphological and mineralogical analyses. Thin sections were prepared using 9 × 4 × 2 cm *in situ* samples from each regolith layer, collected in aluminium Kubiena boxes, following the methodology of FitzPatrick (1980).

LANDFORM–REGOLITH UNITS

Regolith profile units can be grouped into five main layers: (1) soil, (2) soil/top gravel, (3) colluvial/alluvial sediments, (4) bottom gravel and (5) saprolite. The landform units are taken as the six slope facets shown in Figure 18.2. These are:

1. Interfluves (Figure 18.3) are near-planate, but tend to have pediment or glacis forms where they extend from residual hills: they cover c. 50% of the land surface.

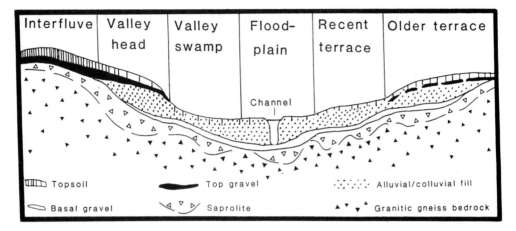

Figure 18.2   Diagrammatic summary of landform–regolith units

Their slope angles range from almost flat to 20° with a mean value of 3.5°. The soil has a mean depth of 0.28 m, and is underlain by a gravel dominated by iron sesquioxide accumulations (nodules and pisoliths) with smaller proportions of quartz pebbles. This top gravel is clearly polygenetic, reflecting not only contemporary pedogenic and cultivation processes, but including inherited pedogenic material, and, at some sites, clasts from former alluvial process domains. This gravel layer has a mean thickness of 0.39 m and overlies a mottled saprolite. Colluvial/alluvial sediments and a basal gravel dominated by rounded quartz

Figure 18.3   Near-planate interfluve: topsoil, top gravel and saprolite contact, exposed by diamond mining

pebbles are occasionally found as lenses: these appear to be remnants of former valley-floor deposits.

2. Valleyheads begin on the interfluves as shallow, elongate depressions, or 'swales', and extend down into valley swamps. Mean slopes increase from 2° to 7° and both the soil and top gravel layers are thinner than on the interfluves, averaging 0.16 m and 0.24 m respectively. Below the top gravel layer is a colluvial/alluvial fill, thickening downslope to *c*. 4 m, the upper part of which is frequently indurated with iron sesquioxides. A common occurrence between the saprolite and the fill layers, is a polymict basal gravel, averaging 0.14 m in thickness.

3. Valley swamps tend to be channelless, 5 m to 20 m wide, they cover *c*. 20% of the Koidu land surface and comprise about 80% of the drainage network. They contain a colluvial/alluvial fill, averaging 1.9 m in thickness, and consist, from the surface down, of an organic-rich clayey sand, an unstructured clayey sand and a basal gravel in the clayey-sand matrix, the latter averaging 0.4 m in thickness. This basal gravel layer is dominated by quartz clasts (21%, on average, being rounded pebbles) and it rests on a kaolinite-rich saprolite. The scarcity of alluvial sedimentary structures makes it difficult to regard the basal gravel and overlying clayey sands solely as fluvial deposits.

4. Floodplains have been formed by the axial streams which appear downstream of the swamps found in the headwater valleys of the two sampled drainage basins (Figure 18.4). They comprise a fining upwards sequence of alluvial sediments averaging 2.9 m in thickness, with structures indicative of channel, overbank and colluvial sedimentation. Despite the presence of permanent stream channels, the basal quartz gravel is less well rounded, thinner and contains a lower proportion of gravel clasts than that of the valley swamp.

5. Recent Terraces, depending on their location within the drainage basin, have regolith profiles that are similar to either the valley swamp or to the floodplain

Figure 18.4   Floodplain: diamond mining

deposits. Mean values for profile thickness and basal gravel are 1.62 m and 0.23 m respectively, whilst soil/top gravel layers are rare or absent. Radiocarbon dating of organic matter within these sediments indicates an age range of 7000–10 000 years BP (Thomas and Thorp, 1980).

6. Older Terraces (Figure 18.5) frequently underlie the convex valley-side slopes separating interfluve from swamp and are partially obscured by colluvium. Mean values for profile thickness and basal gravel are 2.55 m and 0.25 m respectively, with a moderately developed soil and pisolith-rich top gravel layer having a mean thickness of 0.17 m. Iron sesquioxide mottling with a distinctive morphology is present in the lower parts of the sedimentary profile, whilst pebbles in the basal gravels often display haloes indicating periods of impregnation with, or leaching of, iron sesquioxides. Radiocarbon dating of organic material in some of these terraces gives age ranges of 35 000 to 20 500 years BP (Thomas and Thorp, 1980) for floodplain formation. Incision to form the terrace features probably took place during the early Holocene 'Pluvial' (12 500–6 500 years BP).

Figure 18.5   Older Terrace: pit showing topsoil with incipient top gravel, alluvial/colluvial fill and basal gravel over saprolite

Periods of colluviation have partially buried many of the terrace sediments. On the Older Terraces these events are likely to have taken place during the dry climates of the last glacial maximum (20 500–13 000 years BP), while a mid-Holocene dry period may have contributed to colluviation on the Recent Terraces.

PARTICLE SIZE DISTRIBUTION

Previous studies of the textural and petrographic variations in the regolith of the Koidu Basin concentrated on the top gravels of the interfluves and valley-heads and on the basal gravels of the valley floors, with the purpose of identifying properties associated with concentrations of heavy minerals, particularly diamonds (Thomas *et al.*, 1985; Thomas and Thorp, 1985; Teeuw, 1989 and 1991a). These, and subsequent more detailed analyses, have confirmed a general decrease in the clay content of the gravel layers from interfluve to valley floor, together with a concomitant decrease in the numbers of iron sesquioxide clasts. A similar trend of decreasing clay content in the saprolite from interfluve zone to valley floor is indicated by variations in the saprolite textures from different parts of the landscape (Figure 18.6).

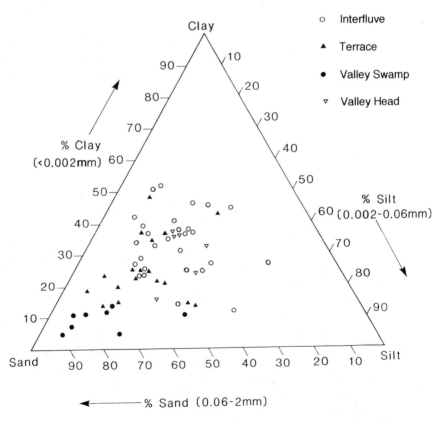

Figure 18.6   Variations in the texture of saprolite, by landform type

Table 18.1 Silt : clay ratios, minimum and maximum values

	Interfluve		Valley floor		Terrace	
	Near-planate	Valley-head	Valley swamp	Floodplain	Recent	Older
Soil or fill	0.3 0.9	0.4 2.0	0.3 2.8	0.2 1.8	0.2 1.1	0.1 1.0
Basal gravel	Not present	0.6 3.2	0.9 3.8	0.8 1.7	0.3 0.9	0.2 1.1
Saprolite	0.4 1.4	0.7 2.0	1.0 1.7	1.0 2.9	1.1 1.7	0.4 1.4
n:	28	6	4	4	10	7

Table 18.1 summarises silt, clay ratios for the various regolith layers sampled in each landform type. It should be noted that the size range of the silt used here was 2–62 μm, considerably coarser than the range used by Van Wambeke (1962) to classify tropical soils. Regolith layers showing the highest relative clay content are the soils on the interfluves and in the basal gravels of the Recent and Older Terraces. Layers that show the lowest relative clay contents are the basal gravel and saprolite of the valley-heads, swamps and floodplains, together with the saprolite of the Recent Terrace. Variations in the clay content of saprolite from valley-floor and interfluve settings are also seen in Figure 18.6. An investigation of regolith micromorphology and mineralogy was undertaken to ascertain whether or not these varying proportions of clay relative to silt and sand are due to transfers of clay, or to varying intensities and durations of weathering processes.

REGOLITH MICROMORPHOLOGY

Iron sesquioxide accumulations are clearly seen in the accompanying photomicrographs and can be grouped into three main types: mottles, nodules and pisoliths. Mottles (segregations) are soft, porous, irregularly shaped accumulations that are typically yellow–brown to red (Munsell colours: 7.5 YR 6/2 to 5 YR 7/8). Pisoliths are smooth, hard, dark brown/black, near spherical concretions. Nodules are concretions that are intermediate between mottles and pisoliths (colours: 7.5 YR 5/2 to 10 YR 6/2). Both near-spherical and irregularly shaped forms are found and usually show a hematite-rich aureole around a softer earthy interior.

The thin sections show that mottles are clearly autochthonous, indicating the input and 'fixing' of iron sesquioxides. The exact nature of the process is not clear, but it appears to be related to seasonal variations in redox conditions. The landforms with the greatest seasonal variations in watertable depth—valley-heads and Older Terraces—have the most mottles. Nodules appear to result from the induration of mottles, possibly due to watertable fall or during erosion or bioturbation of saprolite: the greater their exposure to reworking and induration, the greater their degree of rounding and hardness—eventually leading to the formation of pisoliths.

Table 18.2 Iron sesquioxide accumulations (rare occurrences in parentheses)

	Interfluve		Valley floor		Terrace	
	Near-planate	Valley-head	Valley swamp	Floodplain	Recent	Older
Soil or fill	Pisoliths Nodules (Mottles)		(Nodules) pisoliths	ABSENT		(Nodules)
Basal gravel	Mottles (Gritty nodules, pisoliths)		(Gritty nodules, pisoliths and broken nodules)		(Gritty nodules, pisoliths)	Mottles (Nodules)
Saprolite	(Mottles)		ABSENT		(Mottles)	

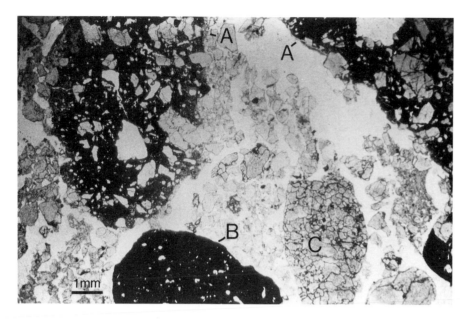

Figure 18.7 Thin section of a valley swamp basal gravel: (A) 'gritty nodules' showing dissolution rim, (b) pisolith, (C) disintegrating quartz pebble with probable relict impregnation by iron sesquioxides

Nodules with surfaces showing exposed sand grains ('gritty nodules'), indicating the etching of iron sesquioxides, occur in the basal gravel layer of the valley-head, swamp, floodplain and Recent Terraces units (cf. (A) in Figure 18.7). The absence of mottles, nodules and pisoliths from the saprolite of the swamps and floodplains is a further indication that iron sesquioxides are dissolved and removed in the valley-floor zone.

Coatings, or cutans, of clays and/or iron sesquioxides are important indicators of the routes followed by weathered material in the regolith. Clay linings along pores and passages in the soil layers suggest transfers from the interfluves to the valley-heads, and then down the valley-head regolith profiles to the colluvial fill layer, basal gravel and saprolite (cf. Table 18.3 and (A) in Figure 18.8). A thin layer within the saprolite (1–5 cm thick), immediately below the basal gravel layer in the swamps is conspicuously devoid of clays and displays enhanced macro-porosity, indicating enhanced clay eluviation below the fill.

Beaded cutans of clay and/or iron sesquioxides (Figure 18.9) along soil pores in the colluvial/alluvial sedimentary and basal gravel layers of the Older Terrace units indicate a major accumulation of weathered material in that part of the regolith as a result of illuviation. This is in agreement with field observations: basal gravels of the Early Holocene and Late Pleistocene ('Older') Terraces are characterised by clayey matrices, in contrast to the Recent Terraces of the later Holocene. An important observation concerns the absence of clay/iron sesquioxide coatings from all the regolith layers of the swamp and floodplain units. These sections of the regolith are also devoid of mottles and have high silt:clay ratios, indicating that they are sites with net losses of clay and iron sesquioxides.

Table 18.3 Distribution of clay along pores and/or iron sesquioxide coatings (cutans)

	Interfluve		Valley floor		Terrace	
	Near-planate	Valley-head	Valley swamp	Floodplain	Recent	Older
Soil or fill	Along pores Around nodules	Broken cutans Rare Mn coatings	ABSENT		Broken cutans Anisotropic domains	Goethite flecks Beaded cutans
Basal gravel		Along pores Around nodules	ABSENT			Anisotropic domains Beaded cutans
Saprolite	Isotropic domains	Along pores Around biotite	ABSENT		Rare: (around biotite)	Around biotite

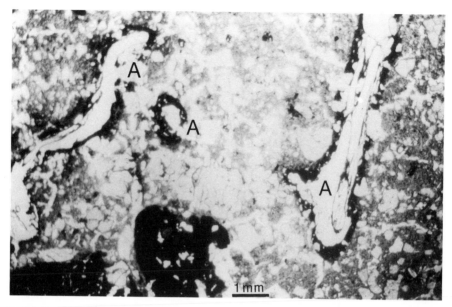

Figure 18.8   Thin section of interfluve topsoil; iron sesquioxide coatings along root passages and soil pores, (A); soil matrix dominated by iron-stained kaolinite

The hydromorphic layers of the valley-floor domain also appear to be sites for the disintegration of allochthonous quartz clasts. During their residence in regoliths of the interfluvial domains quartz grains are impregnated along micro-fissures by iron oxides, to produce 'runiform' iron staining, a process described by Eswaran *et al.* (1975). Upon arrival in the valley floors the grains disintegrate, as can be seen in Figure 18.7. This leads to the production of both sand-sized particles from the disintegration of quartz pebbles and of silt-sized particles from quartz grains. The dissolution of quartz may be enhanced by the creation of such large volumes of silt-sized quartz fragments with their greater relative surface areas. These processes may be important factors in the poorly understood process of quartz removal from tropical drainage basins.

REGOLITH MINERALOGY

The dominant clay mineral in the regoliths of the Koidu Basin is kaolinite (Table 18.4. In the regolith of the interfluve and terrace domains the kaolinite was frequently stained by iron sesquioxides (Figure 18.9). In contrast, iron staining was absent from the kaolinitic matrix of the floodplain regolith and was a rare occurrence in the swamp regolith. Comparison with the distributions of iron sesquioxide and clay coatings (Table 18.3) shows a similar pattern, indicating that the valley floor is a domain wherein there is a net removal of weathered material from the drainage basin.

Gibbsite, a rare occurrence in the regolith matrices, was not found in the interfluve samples, but was found in the valley-head units, probably because of the enhanced drainage there compared with the near-planate interfluves. The widespread distribu-

Table 18.4 Regolith minoeralogy (rare occurrences in parentheses; Fe = iron sesquioxides)

	Interfluve		Valley floor		Terrace	
	Near-planate	Valley-head	Valley swamp	Floodplain	Recent	Older
Soil or fill		Kaolinite + Fe (Gibbsite)	Kaolinite (Fe) (Gibbsite)	Kaolinite		Kaolinite (Fe) (Gibbsite)
Basal gravel		Kaolinite + Fe (Gibbsite)	Kaolinite (Fe)	Kaolinite (Gibbsite)	Kaolinite + Fe (Gibbsite)	
Saprolite		Kaolinite + Fe (Gibbsite)	Kaolinite (Gibbsite)		Kaolinite + Fe (Gibbsite)	

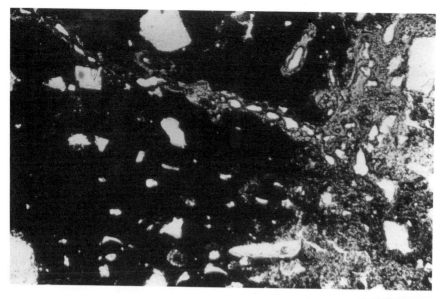

Figure 18.8 Beaded cutan along a crack in a pisolith (black); matrix (right) is dominated by iron-stained kaolinite: alluvial/colluvial fill layer, Older Terrace

tion of gibbsite in the valley-floor and terrace domains is consistent with the large throughputs of water in these sites and suggests active desilication.

DISCUSSION AND CONCLUSIONS

The following paths for iron and clay movement in the regolith are indicated:

1. pedogenic 'fixing' of Fe^{3+} in the B horizon of interfluve soils, producing a gravel layer dominated by iron sesquioxide nodules and pisoliths;
2. accumulation of iron sesquioxide mottles in the layer of seasonal watertable variation within valley-heads and terraces;
3. vertical eluvial/illuvial transfers of clays in interfluve and Older Terrace regoliths;
4. lateral transfers of clays and iron by eluviation in the soil water from interfluves, through valley-heads to valley floors;
5. colluvial transfers of iron sesquioxide accumulations, clay minerals and clasts from interfluves, valley-heads and valley sides into valley floors;
6. lateral transfers of iron and clays by groundwater eluviation, via the basal gravel layer, from valley-heads, Low Terraces and swamps;
7. removal of iron and clays from the drainage basins by groundwater eluviation, via the basal gravel and upper saprolite layers of floodplains, confirmed by analyses of seepages from mining faces;
8. periodic scouring and flushing-out of clastic sediments and upper saprolite along valley-heads and valley floors during major discharge events, possibly clustered during periods of rapid climate change.

The photomicrographs of the basal gravel and saprolite below the floodplain and swamps show clear evidence for:

1. the removal of clay and silica, and the dissolution of iron sesquioxides;
2. quartz clast disintegration and high silt:clay ratios;
3. the absence of autochthonous clay and iron sesquioxide accumulations, and;
4. the presence of gibbsite.

These observations indicate the important role played by the channelless valley swamps in the removal of weathered material from these drainage basins. The accumulation of complex gravel layers on interfluves is further evidence of the removal of fines and the collapse of the upper profile.

Fe^{3+} will be transported as oxides or with the kaolinite as clay and silt-sized particles and aggregates, but reduction to Fe^{2+} in the swamp environments leads to dissolution of the sesquioxides, and to the formation of highly dispersible iron-free kaolinite which moves freely in the low energy water flows. Hydromorphic environments in the tropics play a decisive role in the dissolution of rocks and are responsible for the formation of the 'white sand' deposits that occur extensively on lowland plains in Borneo, Amazonia and the Guyanas. The possible collapse of ferrallitic weathering residues under such conditions to form podzols, and a concomitant reduction in local relief by chemical denudation, have received much comment (Chauvel, 1977; Duchaufour, 1982; Fairbridge and Finkl, 1984; Lucas *et al.*, 1988); whilst the role of swamp conditions in long-distance transfers of solutes and dispersible clays has been emphasised by Sieffermann (1988).

The most effective environments for the operation of these processes occur in permeable substrates subject to high watertables, as in central parts of the Amazon Basin. However, the Koidu region is an erosional landscape, a plateau situated on the flank of a continental uplift, the Guinea Dome (Thomas, 1980; Thomas and Summerfield, 1987). The drainage is incised and interfluve areas commonly have less than two metres of saprolite over relatively impermeable crystalline rocks (Teeuw, 1991b). Whilst hydromorphic conditions in the study region are limited to the swamps and floodplains, there is ample evidence for episodes of erosion by surface water on interfluves and for mass movements down valley-sides, transferring large volumes of regolith into the valley-floor hydromorphic domains (Teeuw, 1989 and 1991a).

The upper slope profiles with their sesquioxide segregations are ferrallitic in character, and groundsurface lowering is likely to follow the progression proposed by McFarlane (1983), as lateral eluviation of fine sediment and solutes within the regolith profiles leads to increasing concentrations of nodules and pisoliths in the upper profile. Under forest, these processes may be mediated by chelating compounds and by micro-organisms (McFarlane and Heyderman, 1985).

These are all near-continuous processes, responding to low magnitude rainfall events throughout the wet season; their cumulative effects are likely to exceed the low frequency/high magnitude events that lead to surface flows on slopes and to sediment flushing from the headwater swamps. The ages of sediments within the headwater swamps indicate residence times for the sediment of from 10 000 to 100 000 years: this will be sufficient to permit the etching of subjacent rock (Thomas and Thorp, 1985). Unfortunately, the comparative amounts of material moved from the Koidu drainage systems by solution, suspension and as bedload are not known. Mass balance studies, similar to those carried out by Huggett (1976) and by Velbel (1985 and 1986), would not be possible, as much of the study region has been disturbed by surface mining; but, under controlled experimental conditions, such studies could provide clearer indications of how humid tropical denudation systems operate under conditions of low relief.

The chemical component of denudation, as detected in the regolith characteristics, is clearly very important in the Koidu landscape, and the processes at work appear to conform to concepts of etchplanation and to the notion of a cratonic regime (Fairbridge and Finkl, 1980; Thomas, 1989a and 1989b).

ACKNOWLEDGEMENTS

This study was funded by a Research Training Award from the UK Natural Environment Research Council; with field support from BP Minerals International and the National Diamond Mining Co., Sierra Leone; with institutional support from the Environmental Science Department, University of Stirling, Scotland, and the Department of Geography, University College, Dublin, Ireland. M. F. Thomas also wishes to thank the Carnegie Trust for the Universities of Scotland for fieldwork support.

REFERENCES

Aleva, C. J. J. (1987). Occurrence of stone-lines in tin-bearing areas in Belitung, Indonesia, and Rondonia, Brazil. *Geo-Eco-Trop*, **11**, Academie Royale des Sciences d'Outre-Mer, Brussels, pp. 197–203.

Avenard, J. M. (1973). Evolution géomorphologique au Quaternaire dans le centre-ouest de la Côte d'Ivoire. *Revue Géomorphologie Dynamique*, **22**, 145–160.

Bremer, H. (1981). Reliefformen und reliefbildende Prozesse in Sri Lanka. *Relief, Boden, Palaoclima*, **1**, 7–184.

Chauvel, A. (1977). Recherches sur la transformation des sols férralitiques dans la zone tropicale avec saisons contrastes. *Travaux et Documents de l'ORSTOM*, Paris, **62**.

Duchaufour, P. (1982). *Pedology, Pedogenesis and Classification*. (Translated by T. R. Paton), Allen and Unwin, London.

Eswaran, H., Sys, C. and Sousa, E. C. (1975). Plasma infusion—a pedological process of significance in the humid tropics. *Annales de Edofollogia y Agrobiologica*, **34**(9–10), 655–673.

Fairbridge, R. W. and Finkl, C. W. Jr (1980). Cratonic erosional unconformities and peneplains. *Journal of Geology*, **86**, 69–86.

Fairbridge, R. W. and Finkl, C. W. Jr (1984). Tropical stone lines and podzolised plains as palaeoclimatic indicators for weathered cratons. *Quaternary Science Reviews*, **3**, 41–72.

FitzPatrick, E. A. (1980). *The Preparation and Description of Thin Sections of Soils*. University of Aberdeen Press, Aberdeen.

Hawthorne, J. B. (1973). Model of a kimberlite pipe. *Earth Physics and Chemistry*, **9**, 1–16.

Huggett, R. J. (1976). Soil landscape systems, a model for soil genesis. *Geoderma*, **13**, 1–22.

Leveque, A. (1979). Pédogenèse sur le socle granito-gneissique du Togo—différenciation des sols et remaniements superficiels. *Travaux et Documents de l'ORSTOM*, Paris, 108.

Louis, H. (1964). Uber rumpfflachen und talbildung in den wechselfeuchten tropen besonders nach studien in Tanganyika. *Zeitschrift für Geomorphologie*, **8** (Sonderheft), 43–70.

Lucas, Y., Boulet, R. and Chauvel, A. (1988). Intervention simultane des phenomnes d'enforcement vertical et de transformation laterale dans la mise en place de systemes sols férralitiques–podzols de l'Amazonie Brasilienne. *Comptes Rendus de l'Academie des Sciences*, Paris, **306**, Serie II, 1395–1400.

McFarlane, M. J. (1983). Laterites. In Goudie, A. S. and Pye, K. (eds), *Chemical Sediments and Geomorphology*, Academic Press, London, pp. 7–58.

McFarlane, M. J. and Heydeman, M. T. (1985). Some aspects of kaolinite dissolution by a laterite-indigenous micro-organism. *Geo-Eco-Trop*, **8**, Academie Royale des Sciences d'Outre-Mer, Brussels, pp. 73–91.

Moeyersons, J. (1989). The concentration of stones into a stone-line, as a result from subsurface movements in fine and loose soils in the tropics. *Geo-Eco-Trop*, **11**, Academie Royale des Sciences d'Outre-Mer, Brussels, pp. 11–22.

Sieffermann, R. G. (1988). Le systeme des gandes tourbires equatoriales. *Annales de Géographie*, **544**, 642–666.

Teeuw, R. M. (1989). Variations in the composition of gravel layers across the landscape. Examples from Sierra Leone. *Geo-Eco-Trop*, **11**, Academie Royale des Sciences d'Outre-Mer, Brussels, pp. 151–169.

Teeuw, R. M. (1991a). A catenary approach to the study of gravel layers and tropical morphodynamics. *Catena*, **18**, 71–89.

Teeuw, R. M. (1991b). Comparative studies of two adjacent drainage basins in Sierra Leone: some insights into tropical landscape evolution. *Zeitschrift für Geomorphologie*, **NF35**(3), 257–267.

Teeuw, R. M., Thomas, M. F. and Thorp, M. B. (1991). Geomorphology applied to exploration for tropical placer deposits. In Sutherland, D. G. (ed.), *Alluvial Mining*, Institution of Mining and Metallurgy, London, pp. 458–480.

Thomas, M. F. (1980). Timescales of landform development on tropical shields—a study from Sierra Leone. In Cullingford, R. A., Davidson, D. A. and Lewin, J. (eds), *Timescales in Geomorphology*, Wiley, Chichester, pp. 333–354.

Thomas, M. F. (1989a). The role of etch processes in landform development I: etching concepts and their applications. *Zeitschrift für Geomorphologie*, **NF33**, 129–142.

Thomas, M. F. (1989b). The role of etch processes in landform development II: etching and the formation of relief. *Zeitschrift für Geomorphologie*, **NF33**, 257–274.

Thomas, M. F. and Goudie, A. S. (eds) (1985). Dambos: small channelless valleys in the tropics. *Zeitschrift für Geomorphologie*, NF, Supplementband, **52**.

Thomas, M. F. and Summerfield, M.J. (1987). Long-term landform evolution: key themes and research problems. In Gardiner, V. (ed.), *International Geomorphology 1986, Part II*, pp. 935–956.

Thomas, M. F. and Thorp, M. B. (1980). Some aspects of the geomorphological interpretation of Quaternary alluvial sediments in Sierra Leone. *Zeitschrift für Geomorphologie*, NF, Supplementband, **36**, 140–161.

Thomas, M. F. and Thorp, M. B. (1985). Environmental change and episodic etch-planation in the humid tropics of Sierra Leone. In Douglas, I. and Spencer, T. (eds), *Environmental Change in the Tropics*, Allen and Unwin, London, pp. 239–267.

Thomas, M. F. and Thorp, M. B. (1992). Landscape dynamics and surface deposits arising from late Quaternary fluctuations in the forest/savanna boundary of Sierra Leone. In Furley, P. A., Proctor, J. and Ratter, J. A. (eds), *The Nature and Dynamics of Forest/Savanna Boundaries*, Chapman and Hall, London, pp. 215–253.

Thomas, M. F., Thorp, M. B. and Teeuw, R. M. (1985). Palaeogeomorphology and the occurrence of diamondiferous placer deposits in the Koidu area, Sierra Leone. *Journal of the Geological Society of London*, **142**, 789–802.

Van Wambeke, A. R. (1962). Criteria for classifying tropical soils by age. *Journal of Soil Science*, **13**(1), 124–132.

Velbel, M. A. (1985). Geochemical mass balances and weathering rates in forested watersheds of the southern Blue Ridge. *American Journal of Science*, **285**, 904–930.

Velbel, M. A. (1986). The mathematical basis for determining rates of geochemical and geomorphic processes in small forested watersheds by mass balance: examples and implications. In Coleman, S. M. and Dethier, D. P. (eds), *Rates of Chemical Weathering in Rocks and Minerals*, Academic Press, New York, pp. 431–451.

19 The Behaviour of Chromium in Weathering Profiles Associated with the African Surface in parts of Malawi

M.J. McFARLANE
University of Botswana, Botswana

D.J. BOWDEN
Newman College, University of Birmingham, UK

and

L. GIUSTI
University of Lancaster, UK

ABSTRACT

Chromium, widely assumed to be geochemically immobile in lateritised profiles, has been used as a resistant index against which the leaching of other elements may be assessed. In a profile developed from granite/gneiss on a typical, flat interfluve of the African surface in the central plains of Malawi, the conversion of saprolite to lateritic residuum was calculated to result in the loss of some 76% of the Cr originally present.

Samples of groundwaters discharging into the seasonally waterlogged bottomlands were analysed by scanning proton microprobe (SPM) and atomic absorption spectrophotometry (AAS-graphite furnace). These showed that contemporary Cr mobilisation reaches high levels despite the Cr-poor parent material.

Microbial leaching experiments, using microorganisms indigenous to the interfluve profile, acting under aerobic conditions on a substrate of ilmenite, pseudorutile and rutile, mobilised substantial quantities of Cr into the leachates. No correlation was found between mobilised Cr and pH.

These results show that Cr cannot be used as a resistant index. It is mobile in this lateritic profile, mobility being microbially mediated. Further research is necessary on (1) the methodology for the determination of Cr, (2) seasonal and regional variations of levels, particularly in areas of Cr-rich rocks, and (3) the forms of the mobile Cr, their stability constraints and toxicology.

INTRODUCTION

Lateritic residual deposits are generally considered to be the products of differential leaching which removes relatively soluble elements from weathering profiles, allowing relatively resistant elements, notably Fe, to accumulate. Original rock textures and

Rock Weathering and Landform Evolution. Edited by D. A. Robinson and R. B. G. Williams
© 1994 John Wiley & Sons Ltd

structures are commonly destroyed in the Fe-enriched horizons, unlike in the underlying parent material, the kaolinised saprolite, where they survive intact. If such laterites have an Fe content many times that of the kaolinised saprolite, each metre of laterite represents the remains of many metres of saprolite. The scale of the requisite surface lowering has been the subject of geomorphological enquiry in areas where residual laterites occur on extensive interfluves. It was soon recognised that major element concentrations may not be the best indicators of this lowering. Although the Fe content of laterites may reach 40–60%, even very low quantities of some minor or trace elements may express enrichments or concentrations, compared with the parent saprolite, which are even higher than that of Fe (e.g. Hartman, 1955; Mulcahy, 1960; Esson and Surcan dos Santos, 1978; Widdowson, 1990).

Estimations of saprolite consumption that yielded residual mantles thus came to focus on resistant indices (RI). These are elements which would be expected to be immobile under the ambient conditions and which therefore give the highest concentration factors in the conversion of saprolite to residuum. Recognition of a RI element allows saprolite consumption and surface lowering to be calculated and the relative mobilities of other elements to be assessed against the RI.

Several elements have been considered as RIs, for example Al (Reiche, 1943), Ti (Gilkes and Suddhiprakaran, 1981; Leprun, 1981; Esson and Surcan dos Santos, 1978), Zr (Goldschmidt, 1937; Gordon and Murata, 1952; Wofenden, 1965; Rao and Krishnamurthy, 1981) and Cr (Schellmann, 1964; Zeissink, 1969; Mercado, 1981; Sahoo, 1981). This paper considers one of them: chromium.

CHROMIUM ABUNDANCE AND OCCURRENCE

Cr is the 21st most abundant element in the earth's crust (McGrath and Smith, 1990). The Cr content of rocks varies widely. The average is about 100 ppm, ranging from 5 to 5000 ppm (Rose et al., 1979, p. 31), the highest concentrations occurring in ultramafic rocks (Goldschmidt, 1954; Cannon, 1978). It exists in many oxidation states but is most commonly present as Cr(III) and Cr(VI). Trivalent Cr occurs in oxides such as chromite, $(Fe.Mg)Cr_2O_4$, and eskolaite, Cr_2O_3 and in sulphides (e.g. daubreelite, $FeCr_2S_4$ and brezianite, Cr_3S_4). Hexavalent chromium occurs as crocoite and chromatite (Matzat, 1978). Trivalent chromium also replaces other elements (especially Fe^{3+} and Al^{3+}) in numerous minerals such as garnets, tourmaline, micas, amphiboles, chlorite, magnetite, ilmenite, rutile, pyroxenes and olivines. Secondary minerals cited as hosts to Cr are kaolinite (McLaughlin, 1959; Rao and Krishnamurthy, 1981) and goethite (Sahoo, 1981), minerals which often dominate laterite assemblages.

CHROMIUM MOBILITY IN WEATHERING PROFILES AND SOILS

If lateritic weathering is a strictly geochemical differentiation process, as the abundance of publications on this aspect of lateritisation would appear to indicate, the chemical similarities of Al^{3+}, Fe^{3+} and Cr^{3+} would lead to the expectation of similar behaviour during lateritic weathering. Retention and accumulation of Fe in residua would therefore be expected to be paralleled by retention and accumulation of Cr. Based on charge and ionic radius, Cr is placed centrally in the field of elements considered to be geochemically immobile (Whittaker and Muntus, 1970; Rose et al., 1979, p. 24). Cr concentrations are found in various residual soils, including tropical residua (e.g. Rose et al., 1979, pp. 153, 268, 270 and 286; Murthy et al., 1981; Ogura et al., 1981; Sahoo, 1981; Zeegers et al., 1981; Schellmann, 1986), apparently in accordance with its assumed immobility. Values of up to several per cent of Cr_2O_3 are reported in laterites developed from ultramafic rocks (Hotz, 1964; Zeissink, 1969). The Cr content of residual overburden can, in many cases, be correlated with the underlying parent rock, for example in Sierra Leone (Webb, 1958), Upper Volta (Zeegers et al., 1981) and India (Murthy et al., 1981). This places it beyond question that this Cr is residual, deriving from the underlying parent rocks (Webb, 1958). Such patterns generally support the assumption that Cr may be regarded as an immobile element.

Nevertheless, Cr is not totally absent from ground and surface waters. Rose et al. (1979) reported a median Cr content of 1 ppb. Chromium in seawater is normally less than 2 ppb (Shiraki, 1978). In river water it ranges from less than 1 ppb to 134 ppb. The mean values of thousands of analyses from North America and Siberia range between 5.2 and 5.8 ppb. Values vary with parent rock. For example, very high values (>100 ppb) were reported in groundwater from the Hinokami chromite ore district in Japan (Yamagata et al., 1960). Similarly high values are found in river water polluted by industrial effluent (e.g. Marumo et al., 1970).

In natural leaching systems, the solubility of Cr(III) is limited primarily by the formation of insoluble $Cr(OH)_3$. Trivalent chromium (as Cr^{3+} and CrO^{2-}) is soluble at low pH, its solubility decreasing at pH >4, and at pH 5.5 it precipitates as $Cr(OH)_3$ (Bartlett and Kimble, 1976a and b; Rai and Szelmeczka, 1990). Since tropical leaching, which yields laterite, occurs under vadose conditions normally in the pH range 5–8, immobility could be expected in these profiles. Under reducing conditions Cr is generally considered to be immobile, the more mobile Cr(VI) being reduced to Cr(III), and, even under oxidising conditions within the pH range 5–8, Cr is regarded as immobile because the oxidisation of Cr(III) by dissolved O_2 is very slow and negligible at earth surface conditions of temperature and pressure (Schroeder and Lee, 1975; Eary and Rai, 1987).

The reported effects of organic matter vary with respect to chromium mobility. Thus, it has been established that, in the presence of soil organic matter, Cr(VI) is reduced to Cr(III), especially in acidic soils, forming insoluble hydroxides and oxides (Bartlett and Kimble, 1976a and b; Grove and Ellis, 1980; Bartlett and James, 1988). However, it has also been reported (Bartlett and Kimble, 1976b; Masscheleyn et al., 1992) that soil organic complexes of Cr(III), formed at low pH, may remain stable and soluble even at pH 7. Thus, organic compounds would appear to be capable of sustaining Cr(III) mobility even into the pH range of the lateritic profiles, following the formation of complexes at low pH, although the relevance of this to lateritic profiles would appear to be limited since they are generally very poor in organic compounds and have near neutral pH.

Data on the effects of the presence of other elements in solution are limited. Bartlett and Kimble (1976b) experimentally demonstrated that, in the presence of excess aluminium in solution, Cr(VI) precipitated as the pH was raised above 4–5 and became totally insoluble between pH 6 and pH 8. Its solubility increased again at pH >8. Masscheleyn *et al.* (1992) proposed that reduction of Fe and Mn under anaerobic conditions affects Cr(III) to Cr(VI) oxidation in natural soils and waters. This is regarded as an abiotic process, consistent with the findings of Bartlett and James (1979), Amacher and Baker (1982) and Eary and Rai (1987).

No direct microbial oxidation of Cr(III) to the more mobile Cr(VI) has been reported (Masscheleyn *et al.*, 1992). The effects of microorganisms have received attention largely with respect to their use in Cr retrieval. Karavaiko (1988) reports that three bacteria (*Pseudomonas dechromaticans*, *Pseudomonas chromatophila* and *Aeromonas dechromatica*) are known to be capable of reducing Cr(VI) to Cr(III), resulting in precipitation of Cr(OH)$_3$. All are capable of using chromates and bichromates as acceptors of electrons when grown anaerobically on organic media. This led to the development of a microbiological method of industrial effluents treatment for chrome removal, these microorganisms functioning at optimal pH conditions of 7.7–8.3, 7.0–8.5 and 5.0–9.0 respectively. Although the optimal activity is within the lateritic range, the effects are to fix chromium rather than mobilise it. Erlich (1983) showed, experimentally, that chromium in chromite ore can be mobilised into leachates by the action of sulphuric acid generated by *Thiobacillus thiooxidans* from sulphur. However, microbially-mediated Cr mobility in the natural environment appears to be unrecorded.

In short, the expectation that Cr should be essentially immobile in laterite profiles (e.g. Schellmann, 1964 and 1986; Zeissink, 1969; Dunham, 1978; Mercado, 1981; Sahoo, 1981) would appear to be well grounded both in geochemical and known biochemical terms. By the early 1970s, although chemical analyses of laterite profiles were sufficient to prompt reconsideration of the security of some assumed RIs in laterite profiles (McFarlane, 1976), Cr appeared to remain secure. Since then, however, with the growing abundance of trace element data it has become clear that patterns of Cr concentrations in laterites are remarkably varied. In some cases Cr concentration is very substantial, exceeding that of some or all other potential RIs (e.g. Zeegers *et al.*, 1981; Widdowson, 1990). In other profiles the Cr content in the laterite is no greater than that of the saprolite, despite large concentrations of other elements in the residual mantle (e.g. Rao and Krishnamurthy, 1981). In yet other cases Cr is reported to be depleted during lateritisation (e.g. Nambiar *et al.*, 1981), and Davy and El-Ansary (1986) presented data to show slight depletion of Cr from the saprolite underlying a lateritic bauxite, though not from the bauxite itself. This raises the question: 'If Cr is leached from some laterite and saprolite profiles how is this achieved?' If Cr is truly immobile in apparently depleted profiles, this would lead to the deduction that those elements with higher concentration factors must be allochthonous. This in turn raises the question of their provenance. Have they been transported from topographically higher sites? Where no higher-lying terrain occurs, should relief inversion be evoked? Or should selective aeolian input be considered? These are clearly important geomorphological issues and of considerable relevance to geochemical prospecting.

This paper addresses Cr behaviour in the African surface weathering profile of central Malawi, with a view to addressing these issues.

LOCATION

The study area is located in the central plains of Malawi (Figure 19.1), where the African surface is extensive. The surface is characterised by monotonously low relief, with wide flat interfluves dipping gently towards seasonally waterlogged bottomlands (dambos), usually streamless where the surface is well preserved. The parent rocks are Basement Complex granite-gneisses, with variably biotite-rich or quartzo-feldspathic facies.

Weathering is deep, averaging about 30 m (McFarlane *et al.*, 1992). Mean annual rainfall ranges from 800 to 1000 mm, falling dominantly from November to March. Mean monthly minimum temperature ranges from 8°C (August) to 18°C (January) and mean monthly maximum from 22°C (July) to 26°C (January) (National Atlas of Malawi, 1983). Fieldwork focussed on two localities, the environs of Chimimbe dambo (Figure 19.1, area 2) and of Linthembwe dambo (Figure 19.1, area 1).

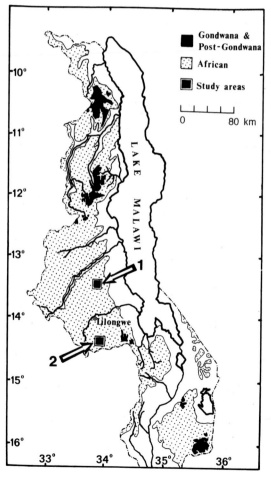

Figure 19.1 Location of the study areas. 1, Linthembwe; 2, Chimimbe. Major erosion surfaces of Malawi follow Lister (1967)

MASS BALANCE—THE CASE FOR Cr LEACHING FROM THE AFRICAN SURFACE PROFILE

The characteristic element chemistry of the African Surface profile was studied near Chimimbe dambo (Figure 19.1, area 2) where relief is very low and interfluves extensive and flat (Figure 19.2). Samples were analysed from a 22 m core drilled on the flat interfluve on the watershed to the southeast of the dambo (Figure 19.2). The profile components, described in more detail elsewhere (McFarlane, 1992), may be summarised as comprising 7–7.5 m of earthy reddish sandy colluvium or residuum, overlying some 13 m of saprolite, of which the upper 6 m was entirely kaolinised. This is the zone of 'superior alteration' (de Lapparent, 1941; Lelong and Millot, 1966). Kaolinisation extended below this level, but smectite was the dominant clay mineral (replacing feldspar) in the lower part of the zone of 'inferior alteration'. Selected samples in the residual mantle and saprolite were analysed for a range of major and trace elements by X-ray fluorescence (XRF) using standard procedures— fused glass discs for major and minor elements and pressed powders for trace elements.

The gross fabric change from saprolite to residuum makes isovolumetric mass balance calculations (Millot and Bonifas, 1955) inappropriate. However, relative mobilities of elements, during the conversion of saprolite to residuum, can be deduced

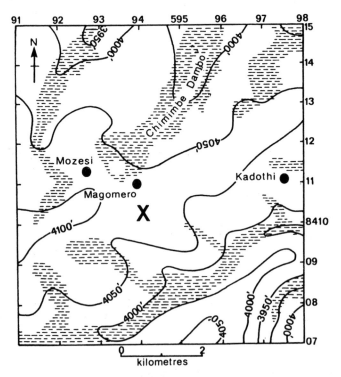

Figure 19.2 The Chimimbe area, showing relief (contours in feet), dambos (pecked shading) and the location of the core (X)

Table 19.1 Mean element concentrations in saprolite and residuum, concentration factors and percentage losses in the conversion of saprolite to residuum

	(a) Saprolite Mean of 6 (%)	(b) Residuum Mean of 5 (%)	(c) Concentration factor	(d) Saprolite consumed (m)	(e) Surface lowering (m)	(f) % retained cf MnO (MnO = 37.44 = 100%)	(g) % lost
			$\dfrac{51.035 \times 1.301^a}{54.370 \times 1.108^b}$ = × 1.102	7.5 × 1.102 = 8.27	8.27 − 7.5 = 0.77	$\dfrac{100}{37.44}$ × 8.27 = 22.1	
SiO_2	54.370	51.035					87.9
Fe_2O_3	10.960	21.440	× 2.297	17.23	9.73	46.0	54.0
Al_2O_3	28.148	23.354	× 0.970	7.28	–	19.4	80.6
TiO_2	1.117	2.559	× 2.689	20.17	12.67	53.9	46.1
MnO	0.118	0.499	× 4.992	37.44	29.94	100.0	–
CaO	1.429	0.064	× 0.052	0.39	–	1.0	99.0
MgO	1.730	0.148	× 0.100	0.75	–	2.0	98.0
Na_2O	0.595	0.032	× 0.064	0.48	–	1.3	98.7
K_2O	0.759	0.116	× 0.180	1.35	–	3.6	96.4
P_2O_5	0.098	0.140	× 1.685	12.64	5.14	33.8	66.2
	(ppm)	(ppm)					
Nb	4.0	14.7	× 4.277	32.08	24.60	85.7	14.3
Ni	105.5	64.2	× 0.715	5.36	–	14.3	85.7
Rb	34.5	15.4	× 0.524	3.93	–	10.6	89.4
Sr	73.2	52.8	× 0.847	6.35	–	17.0	83.0
Y	37.3	42.2	× 1.327	9.95	2.45	26.6	73.4
Zr	84.8	349.4	× 4.840	36.30	28.80	96.9	3.1
Ba	395.3	649.4	× 1.924	14.43	6.93	38.5	61.5
Cr	**253.5**	**259.6**	**× 1.200**	**9.00**	**1.50**	**24.0**	**76.0**
V	132.5	395.8	× 3.507	26.30	18.80	70.2	29.8

[a] residuum density
[b] saprolite density

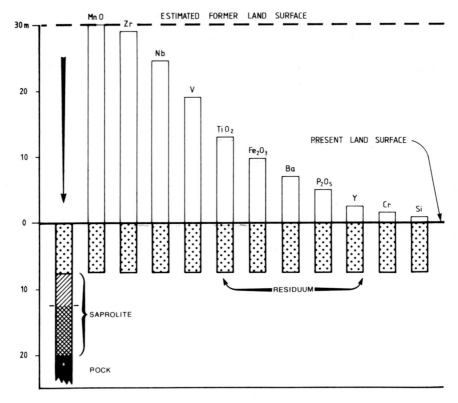

Figure 19.3 Surface lowering calculated for elements concentrated in the lateritic residuum of the interfluve core (X in Figure 19.2)

from the ratios of element concentrations, with reference to the element with the greatest concentration factor, in this case Mn, with Zr as a close second best (Table 19.1 columns a, b and c). In the conversion of saprolite to residuum, the order of mobilities is: $CaO > Na_2O > MgO > K_2O > Rb > Ni > Sr > Cr > Y > P_2O_5 > Ba > Fe_2O_3 > TiO_2 > V > Nb > Zr > Mn$ (least mobile).

From the concentration factors the column of saprolite which is necessary to yield 7.5 m of residuum can be calculated (Table 19.1 column d) and from this the surface lowering estimated (Table 19.1 column e, Figure 19.3). The percentage retention and loss of each element, with respect to MnO can also be calculated, as shown in Table 19.1 (columns f and g). These data would suggest that some 76% of the Cr in the column of parent material, which was consumed to yield the residuum, has been lost from the profile.

GROUNDWATER ANALYSES

Scanning Proton Microprobe (SPM) Analysis

Earlier work on shallow groundwater collected at the end of the dry season, from the same region, concerned Al mobilisation (McFarlane, 1991; McFarlane and Bowden,

1992). As a component of this research a sample (unfiltered) of Al-rich water was evaporated and multi-element mapping of the precipitates undertaken, using a scanning proton microprobe. The maps and chemical spectra showed the presence of Cr. Although this type of analysis cannot provide absolute values, it does give element ratios. Al:Cr was *c.* 55. Total Al, determined by AAS-graphite furnace, was 24 ppm. From this, the Cr level was estimated to be about 400 ppb.

AAS-Graphite Furnace Analysis

Subsequent to the Al study, a small batch of fresh samples was collected at the end of a poor and erratic wet season, for Cr determinations by AAS-graphite furnace. Collection was from the environs of Linthembwe dambo (Figures 19.1 and 19.4). This dambo is narrower than Chimimbe and in an area of slightly more pronounced relief, which favours greater and better sustained discharge at the dambo margins, at times when many shallow wells are dry in the Chimimbe area. Samples were collected

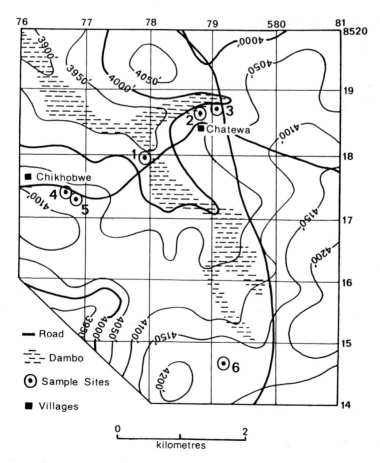

Figure 19.4 The Linthembwe area, showing relief (contours in feet), dambos (pecked shading) and the location of water sample sites

Table 19.2 Groundwater samples, their source, pH, EC and Al and Cr values, determined by AAS-graphite furnace. For locations see Figure 19.4

	Source	Level below surface (m)	Water depth (m)	pH	EC (ξmho)	Al (ppm)	(a) Al (ppm)	(a) Cr (ppb)	(b) Al (ppm)	(b) Cr (ppb)
1	Unlined well	0.45	1.64	7.0	280	0.20	0.06	8.6	0.08	8.2
2	Unlined well	0.35	1.26	5.6	32	24.00	13.43	15.5	27.10	14.2
3	Brick lines	0.46	2.13	5.7	49	0.27	26.14	30.2	41.09	41.0
4	Brick lined	0.85	>3.00	7.1	410	0.11	0.04	12.1	0.10	8.2
5	Unlined well	1.03	0.80	7.6	585	22.00	4.67	10.3	7.83	10.1
6	Tube-well	?	?	8.0	330	0.04	0.01	13.8	0.01	10.4

late dry season   late wet season

(a) 30 minutes sonication; (b) 70 minutes sonication.

unfiltered in polythene bottles, returned to the UK and cool stored before analysis. Immediately before analysis the samples were sonicated in order to resuspend any materials which could have precipitated in the interim. Two sets of determinations were made, the first following ½ hour sonication, the second following 1 hour and 10 minutes sonication, three determinations being made on each sample. The sample sites are described and analytical results given in Table 19.2.

MICROBIAL LEACHING EXPERIMENTS

The groundwater analyses having shown that Cr is mobilised under contemporary conditions and also having suggested a correlation between Al and Cr, a possible role for indigenous microorganisms in Cr mobilisation was explored, since previous work (McFarlane and Heydeman, 1984; McFarlane and Pollard, 1989) had implicated microbial agents in Al mobilisation. The core, upon which the estimation of element losses from the profile was based, was drilled with a triple core barrel. The cutting edge of the central corer protrudes beyond the others and thus cuts a core which is uncontaminated by drilling fluids. Thus the core was also suitably uncontaminated and appropriate for sampling of indigenous microorganisms.

A section of core in the lower part of the residuum was selected for microbial isolation. A small quantity of residuum was aseptically sampled directly onto a nutrient agar plate (Difco, pH 6.8) and incubated at 30°C. Morphologically distinct microbial colonies were isolated and purified on plates of the same nutrient.

The 20 microorganisms thus isolated were challenged with blacksand from Sri Lanka, composed of ilmenite (30%), pseudorutile (52–53%) and rutile (6%). These are highly stable minerals, in this case with a high Cr content. The results of chemical analysis of the ore, before leaching, are shown in Table 19.3.

Table 19.3 Analysis of blacksand (30% ilmenite, 52–53% pseudorutile, 6% rutile), before leaching. Values in %

TiO_2	Fe_2O_3	FeO	Cr_2O_3	V_2O_3	MnO	SiO_2	Al_2O_3	P_2O_5	MgO	CaO
53.4	23.0	17.9	0.15	0.35	0.85	0.60	1.0	0.06	0.66	0.10

The finely ground ore was sterilised by autoclaving at 121°C for 15 minutes. Leaching was undertaken, using separate isolates, on 5 g batches of ore in 500 ml conical flasks containing 200 ml nutrient broth (Difco, pH 7.4) to which glucose had been added (final glucose concentration 3%). Flasks were gently agitated on an oscillating platform at 30°C for 20 days. The samples were centrifuged at 4500 rpm for 20 minutes to bring down the solids, including microorganisms. One hundred millilitres were drawn from the top of the supernatant for analysis by AAS-graphite furnace, following sample dilution when necessary. Results are shown in Figures 19.5 and 19.6.

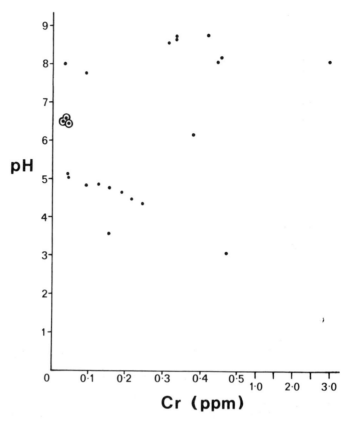

Figure 19.5 Scatter plot of chromium versus pH in the supernatant of a leaching experiment, using indigenous microorganisms acting on a substrate of ilmenite, pseudorutile and rutile. Circled points are the uninoculated controls

DISCUSSION

The mass balance calculation is empirically based on Mn because this has the highest concentration factor in the saprolite-to-residuum conversion. Of the 'traditional' RIs, Zr has the highest factor, very close to that of Mn (which may appear to strengthen the security of the calculation). However, despite having the highest concentration factor, Mn is certainly mobile in this and other African surface profiles. This is seen from analysis of groundwater discharging into dambos (McFarlane, 1992). It is also clear from Mn deposition in the saprolite of this profile, above the boundary between 'superior' and 'inferior' weathering (McFarlane and Bowden, 1992). Hence the calculation of the saprolite-to-residuum conversion is based on enriched saprolite, which means that the concentration factor is underestimated and surface lowering greater than that deduced. This, in turn, means that percentage losses of Cr, and other elements, with respect to Mn, are also greater than our calculation suggests. Regardless of the uncertainties which surround such calculations, there would seem

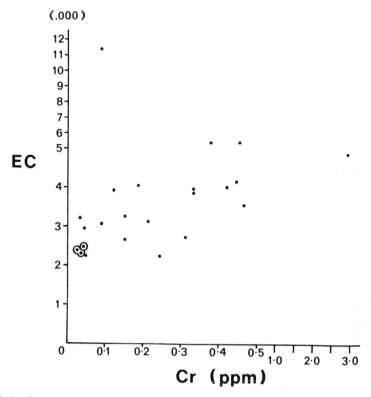

Figure 19.6 Scatter plot of chromium versus electrical conductivity (EC) of a leaching experiment, using indigenous microorganisms acting on a substrate of ilmenite, pseudorutile and rutile. Circled points are the uninoculated controls

little basis for questioning the deduction that Cr is lost from the profile during the formation of the residual mantle.

The Cr levels, determined by AAS for sample 2 at the end of the wet season, are much lower than that deduced from the same sample, analysed by SPM and collected at the end of the dry season. It is possible that this expresses a seasonal variation. This is well seen from the Al levels of samples 2 and 3 from closely adjacent wells; Al-bearing shallow throughflow at the end of the wet season overprints the deep water discharge of low Al which dominates (sample 3) at the end of the dry season. However, there are significant differences, in both Al and Cr results, between 30-minute and 70-minute sonicated samples. Both data sets had very low coefficients of variation on three determinations and it seems likely that the differences express substantial cohesion to the plastic surface of the sample bottles.

Such cohesion is known to occur in the case of organic materials (Allen, 1974). The disparity between samples sonicated for different periods indicates that even the higher value may still under-represent the true Cr levels. We can only say that Cr is present in at least the quantities indicated by the 70-minute sonicated data and these values are very substantially higher than the published median figure of 1 ppb (Rose *et al.*, 1979). It is possible that the published figures understate total Cr mobilisation, as in the case of Al (McFarlane, 1991; McFarlane and Bowden, 1992) if sample treatment involved filtration and acidification. Although procedural uncertainties exist, our data support the deduction of the mass balance calculation that Cr is leached from the African surface weathering profile despite Eh and pH conditions which are not conducive to Cr mobilisation in solution.

The ability of indigenous microorganisms to mobilise Cr under aerobic conditions is confirmed by the substantial amounts of this element in the leachates from inoculated flasks (compare with the three controls which were identically treated but uninoculated). The pH of the experiment was uncontrolled, final pH ranging from 3 to 8.7. There was no correlation between pH and Cr levels (Figure 19.5). A weak positive correlation ($r = 0.246$) exists between electrical conductivity (EC) and Cr levels (Figure 19.6). The favourably high nutrient levels used in this experiment facilitated microbial growth and hence Cr mobilisation at much more enhanced concentrations than in the natural environment where nutrient levels are low.

Although the form in which the Cr is leached from Malawian profiles is unknown, organic binding appears a logical conclusion. This opens the way to consideration of the variable behaviour of Cr in lateritic profiles in terms of variable microbial populations and activity. It can be said with confidence that Cr cannot be regarded as a resistant index for mass balance determinations, estimations of landsurface lowering or assessment of allochthonous contributions to residua.

In some profiles (e.g. Sahoo, 1981; Butt, 1985) the Cr maximum is in the saprolite immediately below the residuum. It is well known that heavy minerals gravitate to the base of residual mantles. However, Cr concentration in the underlying saprolite may suggest that a 'repeated solution and deposition' mechanism operates, a mechanism known to provide a cumulative effect with respect to other lateritic elements, for example Fe, Mn, Al and Ni. Thus, the accumulated elements are not enriched because they are immobile, but because they have limited mobility as compared with the protracted mobility of elements which are leached out of the profile. In the Chimimbe interfluve profile, the Cr maximum is at the very bottom of the residuum where saprolite clasts dominate the texture. Hence this profile may well provide another example of the distributional effects of a 'repeated solution and deposition' mechanism. It is possible that the mobilised Cr becomes adsorbed onto kaolinite, near the top of the saprolite, which is undergoing dissolution and depleted in Al, as is the case with K and with other metallic elements (McFarlane and Heydeman, 1984).

Although Cr is a biologically necessary element, essential for glucose metabolism (IPCS/WHO, 1988) the WHO recommended limit is 50 ppb in water; in excess of this level it is regarded as toxic. Its toxicity is reported to be almost exclusively confined to hexavalent compounds (IPCS/WHO, 1988). Research by Beyersmann (1989) showed that soluble hexavalent chromium is taken up rapidly and accumulated intracellularly, whereas compounds of trivalent chromium penetrate biomembranes about three orders of magnitude more slowly. Hexavalent Cr, after its uptake, is metabolised by electron donating compounds via $Cr(V)$ to $Cr(III)$ compounds. There appear to be two mechanisms of chromate genotoxicity; one with direct DNA damage caused by $Cr(V)$ species and one via DNA-protein crosslinks formed with $Cr(III)$, the final reduction state of chromate. Stackhouse and Benson (1989) researched the effect of humic acid on the toxicity and bioavailability of two forms of trivalent chromium and found that humic acid significantly decreased toxicity. The form in which Cr is mobilised by the microorganisms indigenous to Malawi profiles is not known and its toxicity remains unexplored. In terms of total Cr, it seems likely that, even in this area of Cr-poor rocks, levels of this element in the groundwater in some wells may exceed the WHO recommended limit at certain seasons. Levels in areas of Cr-rich basic and mafic rocks can be expected to be significantly higher. Further research is needed on the seasonal variability of Cr levels, the levels reached in areas of Cr-rich rocks, the forms in which Cr is mobilised and their stability constraints and toxicity.

The ability of the Malawian microorganisms to mobilise Cr from the lattice of such stable minerals as ilmenite, pseudorutile and rutile requires further exploration with respect to beneficiation of blacksands, a major ore for the TiO_2 industry. This is of particular relevance to Sri Lanka, with some of the world's largest deposits which suffer from unacceptably high Cr levels.

CONCLUSIONS

Cr is unsuitable as a resistant index for mass balance calculations and does not provide a basis for assessing either the relative mobilities of other elements or of the contingent landsurface lowering associated with the formation of lateritic residua. Nor can Cr concentrations be used as the threshold for identification of allochthonous enrichment of other elements, i.e. those with concentration factors greater than that of Cr. Chromium has been leached from the lateritic African Surface profile in Malawi and its presence in shallow groundwaters shows that Cr mobilisation is a contemporary process. As with the contemporary leaching of Al, following congruent kaolinite dissolution, this mobilisation of Cr raises questions concerning the antiquity of lateritic residual mantles on ancient landsurfaces.

The levels of Cr mobilised in Malawian groundwater are high in comparison with published global values, even in this area of Cr-poor rocks, and may seasonally exceed the recommended WHO limit. Further research is needed on the seasonal and regional distribution patterns of Cr in groundwater in lateritic areas, particularly those underlain by basic rocks where much higher levels can be expected.

Laterite-indigenous microorganisms are able to mobilise Cr under oxidising conditions within a wide range of pH, there being no correlation between pH and levels of Cr mobilised. The nature of the organically bound forms is not known. Further research on this, and on the stability constraints and toxicity of the mobile forms, is also needed. The ability of microorganisms to leach Cr from such stable minerals as ilmenite, pseudorutile and rutile has potential for the beneficiation of blacksand, an important ore for the TiO_2 extractive industry.

ACKNOWLEDGEMENTS

The Chimimbe core was drilled by British Geological Survey, as part of the ODA-sponsored Basement Aquifer Project. Chemical analysis of the core (XRF) was by Dr G. Hendry, Geology Department, Birmingham University. Financial support for the leaching experiments was provided by Tioxide, UK, who also provided the ground ore and its chemical analysis.

Dr L. Macaskie, Microbiology Department, Birmingham University, is thanked for help with the leaching experiments. Newman and Westhill College, Birmingham University, generously provided financial support for the collection of water samples. We thank Mr G. Duke, Luton Polytechnic, for helpful discussion of Cr toxicology and F. McFarlane for cartography.

REFERENCES

Allen, S. E. (1974). *Chemical Analysis of Ecological Materials*. Blackwell, Oxford.

Amacher, M. C. and Baker, D. E. (1982). Redox reactions involving chromium, plutonium and manganese in soils. *Report DE-AS02-77DPO4515, US Department of Energy*, US Government Printing Office, Washington, DC.

Bartlett, R. J. and James, B. R. (1979). Behaviour of chromium in soils: III. Oxidation. *Journal of Environmental Quality*, **8**, 31–35.

Bartlett, R. J. and James, B. R. (1988). Mobility and bioavailability of chromium in soils. In Nriagu, J. O. and Nieboer, E. (eds), *Chromium in the natural and human environment, Advances in Environmental Science and Technology*, **20**, 267–304, Wiley, Chichester.

Bartlett, R. J. and Kimble, J. M. (1976a). Behaviour of chromium in soils: I. Trivalent forms. *Journal of Environmental Quality*, **5**(4), 379–382.

Bartlett, R. J. and Kimble, J. M. (1976b). Behaviour of chromium in soils: II. Hexavalent forms. *Journal of Environmental Quality*, **5**(4), 383–386.

Beyersmann, D. (1989). Biochemical speciation of chromium genotoxicity. *Toxicological and Environmental Chemistry*, **22**, 61–67.

Butt, C. R. M. (1985). Granite weathering and silcrete formation on the Yilgarn Block, Western Australia. *Australian Journal of Earth Sciences*, **32**, 415–432.

Cannon, H. L. (1978). *Geochem. Environ.*, **3**, 17–31.

Davy, R. and El-Ansary, M. (1986). Geochemical patterns in the laterite profile at the Boddington gold deposit, Western Australia. *Journal of Geochemical Exploration*, **26**, 119–144.

de Lapparent, U. (1941). Logique des mineraux du granite. *Revue Scientifique*, pp. 248–294.

Dunham, Sir K. (1978). Contributed remarks. Chemistry and mineralogy of section through lateritic nickel deposit at Liberdade, Brazil. *Transactions Institute Mining and Metallurgy (Series B Applied Earth Sciences)*, **87**, B74.

Eary, L. E. and Rai, D. (1987). Kinetics of chromium (III) oxidation to chromium (VI) by reaction with manganese dioxide. *Environment Science Technology*, **21**, 1187–1193.

Erlich, H. L. (1983). Leaching of chromite ore and sulfide matte with dilute sulphuric acid generated by *Thiobacillus ferrooxidans* from sulphur. In Rossi, G. and Torma, E. A. (eds), *Recent Progress in Biohydrometallurgy*, Associazione Mineraria Sarda, Cagliari, Italy, pp. 19–42.

Esson, J. and Surcan dos Santos, L. C. (1978). Chemistry and mineralogy of section through lateritic nickel deposit at Liberdade, Brazil. *Transactions Institute Mining and Metallurgy (Series B Applied Earth Sciences)*, **87**, B53–B60.

Gilkes, R. J. and Suddhiprakaran, A. (1981). Mineralogy and chemical aspects of lateritisation in southwestern Australia. In *Lateritisation Processes*, Oxford and IBH Publishing Co., Calcutta, pp. 34–44.

Goldschmidt, V. M. (1937). The principles and distribution of chemical elements in minerals and rocks. *Journal of the Chemical Society of London*, **1**, 655–673.

Goldschmidt, V. M. (1954). *Geochemistry*. Clarendon, Oxford.

Gordon, M. and Murata, K. J. (1952). Minor elements in Arkansas bauxite. *Economic Geology*, **47**, 169–179.

Grove, J. H. and Ellis, B. G. (1980). Extractable chromium as related to soil pH and applied chromium. *Soil Science Society of America Journal*, **44**, 238–242.

Hartman, J. A. (1955). Origin of heavy minerals in Jamaican bauxite. *Economic Geology*, **50**(7), 738–747.

Hotz, P. E. (1964). Nickeliferous laterites in southwestern Oregon and northwest California. *Economic Geology*, **59**, 355.

IPCS/WHO (1988). *IPCS International Programme on Chemical Safety Environmental Health Criteria 61*, UN Environment Programme and WHO Publications.

Karavaiko, G. I. (1988). Micro-organisms and their significance for biogeotechnology of metals. In Karavaiko, G. I., Rossi, G., Agate, A. D., Groudev, S. N. and Avakyan, Z. A. (eds), *Biotechnology of Metals*, United Nations Environment Programme, Moscow.

Lelong, F. and Millot, G. (1966). Sur l'origine des mineraux micaces des alterations lateritiques. Diagenese Regressive-mineraux en transit. *Bulletin du Service Carte geologique Alsace Lorraine*, **19**, 271–287.

Leprun, J. C. (1981). Some principal features of ironcrusts in dry western Africa. In *Lateritisation Processes*, Oxford and IBH Publishing Co., Calcutta, pp. 144–153.

Lister, L. A. (1967). Erosion surfaces in Malawi. *Records of the Geological Survey of Malawi*, **7** (for 1965), 15–28.

Marumo, T., Matsuda, Y., Mizohata, A., Matsunami, T. and Takeuchi, T. (1970). Activation analysis of water in the river Yodo. *Annual Report Radiation Center Osaka Prefecture*, **11**, 23–??.

Masscheleyn, P. H., Pardue, J. H., DeLaune, R. D. and Patrick, W. H. Jr (1992). Chromium redox chemistry in a Lower Mississippi Valley Bottomland Hardwood Wetland. *Environment Science Technology*, **26** (6), 1217–1226.

Matzat, E. (1978). Chromium. In Wedepohl, K. H., (ed.), *Handbook of Geochemistry*, **II/3**, 24-A-1–24-A-7, Springer-Verlag, Berlin, Heidelberg.

McFarlane, M. J. (1976). *Laterite and Landscape*, Academic Press, London.

McFarlane, M. J. (1991). Aluminium menace in tropical wells. *New Scientist*, **1780**, 38–40.

McFarlane, M. J. (1992). Groundwater movement and water chemistry associated with weathering profiles of the African Surface in parts of Malawi. In Burgess, W. and Wright, E. P., (eds), *Hydrology of Basement Rocks with Particular Reference to Africa*, Geological Society of London Special Publication 65, pp. 101–129.

McFarlane, M. J. and Bowden, D. J. (1992). Mobilisation of aluminium in the weathering profiles of the African Surface in Malawi. *Earth Surface Processes and Landforms*, **17**, 789–806.

McFarlane, M. J. and Heydeman, M. T. (1984). Some aspects of kaolinite dissolution by a laterite-indigenous micro-organism. *Geo. Eco. Trop.* **8**, 73–91.

McFarlane, M. J. and Pollard, S. (1989). Some aspects of stone-lines and dissolution fronts in parts of Malawi and Zimbabwe. *Geo. Eco. Trop.* **11**, 23–35.

McFarlane, M. J., Chilton, P. J. and Lewis, M. A. (1992). Geomorphological controls on borehole yields: a statistical study in an area of basement rocks in central Malawi. In Burgess, W. and Wright, E. P., (eds), *Hydrogeology of Crystalline Basement Aquifers in Africa*, Geological Society of London Special Publication 65, pp. 131–154.

McGrath, S. P. and Smith, S. (190). Chromium and nickel. In Alloway, B. J. (ed.), *Heavy Metals in Soils*, Blackie, Glasgow, pp. 125–150.

McLaughlin, R. J. W. (1959). The geochemistry of some kaolinitic clays. *Geochimica et cosmochimica Acta*, **17**, 11–16.

Mercado, J. M. O. (1981). Geochemistry of the laterites in Nanoc Islands, Surigao Province, Philippines. In *Lateritisation Processes*, Oxford and IBH Publishing Co., Calcutta, pp. 45–57.

Millot, G. and Bonifas, M. (1955). Transformations isovolumetriques dans les phenomenes de lateritisation et de bauxitisation. *Bulletin du Service Carte geologique Alsace Lorraine*, **8**, 3–10.

Mulcahy, M. J. (1960). Laterites and lateritic soils in south-western Australia. *Journal of Soil Science*, **11**, 206–226.

Murthy, M. K., Kalsotra, M. R. and Prasad, S. (1981). In *Lateritisation Processes*, Oxford and IBH Publishing Co., Calcutta, pp. 193–201.

Nambiar, A. R., Sukumaran, P. V., Warrier, R., Nair, G. S. and Satyaseelan, P. (1981). Lateritisation of anorthosite, gabbro, granophyre and charnokite—a case study from Kerala. In *Lateritisation Processes*, Oxford and IBH Publishing Co., Calcutta, pp. 120–128.

National Atlas of Malawi (1983). Malawi Government Printer, Zomba, Malawi.

Ogura, Y., Ito, K., Koide, K. and Shimosaka, K. (1981). Geochemistry and mineralogy of nickel oxide ores in the southwestern Pacific area. In *Lateritisation Processes*, Oxford and IBH Publishing Co., Calcutta, pp. 58–67.

Rai, D. and Szelmeczka, R. W. (1990). Aqueous behaviour of chromium in coal fly ash. *Journal of Environmental Quality*, **19**, 378–382.

Rao, J. J. and Krishnamurthy, C. V. (1981). Some observations on the mineralogy and geochemistry of Hazaridadar and Raktidadar Plateaus, Amarkantak area, Madhya Pradesh, India. In *Lateritisation Processes*, Oxford and IBH Publishing Co., Calcutta, pp. 89–108.

Reiche, P. (1943). Graphic representation of chemical weathering. *Journal of Sedimentary Petrology*, **13**, 58–68.

Rose, A. W., Hawkes, H. E. and Webb, J. S. (1979). *Geochemistry in Mineral Exploration*, 2nd edition. Academic Press, London.

Sahoo, R. K. (1981). The mineralogy and geochemistry of nickeliferous laterites of Sukinda, Orissa, India. In *Lateritisation Processes*, Oxford and IBH Publishing Co., Calcutta, pp. 77–85.

Schellmann, W. (1964). Zur lateritischen Verwitterung von Serpentinit. *Geologisches Jahrbuch* **81**, 645–678.

Schellmann, W. (1986). On the geochemistry of laterites. *Chemi der Erde*, **45**, 39–52.

Schroeder, D. C. and Lee, G. F. (1975). Potential transformations of chromium in Natural Waters. *Water, Air, Soil Pollution*, **4**, 355–365.

Shiraki, K. (1978). Chromium. In Wedepohl, K. H. (ed.), *Handbook of Geochemistry*, **II/3**, 24-I-1, Springer-Verlag, Berlin.

Stackhouse, R. A. and Benson, W. H. (1989). The effect of humic acid on toxicity and bioavailability of trivalent chromium. *Ecotoxicology and Environmental Safety* **17**(1), 105–111.

Schellmann, W. (1986). On the geochemistry of laterites. *Chem. Erde*, **45**, 39–52.

Schroeder, D. C. and Lee, G. F. (1975). Potential transformations of chromium in Natural Waters. *Water, Air, Soil Pollution*, **4**, 355–365.

Shiraki, K. (1978). Chromium. In Wedepohl, K. H. (ed.), *Handbook of Geochemistry*, **II/3**, 24-I-1, Springer-Verlag, Berlin.

Stackhouse, R. A. and Benson, W. H. (1989). The effect of humic acid on toxicity and bioavailability of trivalent chromium. *Ecotoxicol. Environ. Saf.* **17**(1), 105–111.

Webb, J. S. (1958). Observations of geochemical exploration in tropical terrains. In *Symposium de Exploracion Geoquimica*, 20th International Geological Congress, Mexico City, 1956, Vol. 1, pp. 143–173.

Whittaker, E. J. W. and Muntus, R. (1970). Ionic radii for use in geochemistry. *Geochimica et Cosmochimica Acta*, **34**, 945–956.

Widdowson, M. (1990). The uplift history of the Western Ghats in India. Doctoral Thesis, Oxford University.

Wolfenden, E. B. (1965). Geochemical behaviour of trace elements during bauxite formation in Sarawak, Malaysia. *Geochimica et Cosmochimica Acta*, **29**, 1051–1062.

Yamagata, N., Murakami, Y. and Torii, T. (1960). Biogeochemical investigation in serpentine –chromite ore district. *Geochimica et Cosmochimica Acta*, **18**, 23–35.

Zeegers, H., Goni, J. and Wilhelm, E. (1981). Geochemistry of lateritic profiles over a disseminated Cu–Mo mineralisation in Upper Volta (West Africa)—preliminary results. In *Lateritisation Processes*, Oxford and IBH Publishing Co., Calcutta, pp. 359–368.

Zeissink, H. E. (1969). The mineralogy and geochemistry of a nickeliferous laterite profile (Greenvale, Queensland, Australia). *Mineral. Deposita*, **4**, 132.

20 The Effects of Tropical Weathering on the Characteristics of Argillaceous Rocks

CHEN-HUI FAN
University College London, UK

ROBERT J. ALLISON
University of Durham, UK

and

MERVYN E. JONES
University College London, UK

ABSTRACT

This paper presents the results of a study which examines the influence of weathering on the properties of argillaceous rocks. Data are presented for a heavily over-consolidated mudrock from Barbados, an environment characterised by tropical weathering. Weathering promotes substantial changes in clay fabric. This is accompanied by a large increase in pore volume and significant changes in material strength as weathering effectively converts the rock from an over-consolidated to a remoulded state. Weathering thus strongly influences material properties and hence material behaviour. In the case of the rock examined here, recorded changes in physical properties with depth in the weathering profile have implications for landform development.

INTRODUCTION

Most rocks and minerals exposed at and immediately beneath the surface of the Earth are in an environment quite unlike that under which they were formed. With this in mind, weathering can be defined as the process of rock and mineral alteration to more stable forms under various conditions of moisture, temperature and biological activity that prevail at the ground surface (Brunsden, 1979; Ollier, 1984).

Physical and chemical weathering are both important mechanisms of rock material alteration. With physical weathering the original rock disintegrates with no appreciable change in chemical or mineralogical composition. A mechanism common to all physical weathering is the establishment of sufficient stresses within the rock to cause material breakdown. Chemical weathering, on the other hand, is a process in which the chemical and mineralogical composition of the original rock material is changed.

Rock Weathering and Landform Evolution. Edited by D. A. Robinson and R. B. G. Williams
© 1994 John Wiley & Sons Ltd

Chemical weathering occurs because rocks and minerals are rarely in equilibrium with near-surface water, temperature and pressure conditions. Broadly speaking, chemical weathering is active in most areas where temperatures are high, since an increase in temperature often accelerates chemical reactions (Brunsden, 1979).

In tropical environments where tectonic activity promotes uplift, or serves to maintain elevation, weathering processes are highly active. Clays subject to such weathering often form steep, unvegetated slopes (Allison, 1991). Soil development is strongly restricted due to the continuous down-slope removal of the products of the weathering process. Badlands frequently develop. Furthermore, weathering promotes substantial changes in clay fabric. Usually a large increase in pore volume and significant changes in material strength occur, causing mudrocks to revert from an over-consolidated to an effectively remoulded state. Tropical weathering thus heavily influences material properties and the mechanical behaviour of argillaceous sediments.

STUDY AREA

In this study the effects of tropical weathering on the characteristics of a heavily over-consolidated argillaceous rock, named the Joe's River Formation, a scaly clay from Barbados, are investigated. The island of Barbados is part of the Lesser Antilles Fore-arc complex (Figure 20.1), a tectonically active area where the Atlantic Plate is being subducted beneath the Caribbean Plate at an average rate of 20 mm yr^{-1} (Pudsey and Reading, 1981). Seismic surveys and sonar images of the sea floor to the east of Barbados indicate the presence of mud diapirs associated with unconsolidated sediments at depth. It is these sediments which are thought to form the diapiric melange of which the Joe's River Formation is a part (Biju-Duval *et al.*, 1982; Stride *et al.*, 1982; Torrini *et al.*, 1985; Brown and Westbrook, 1987).

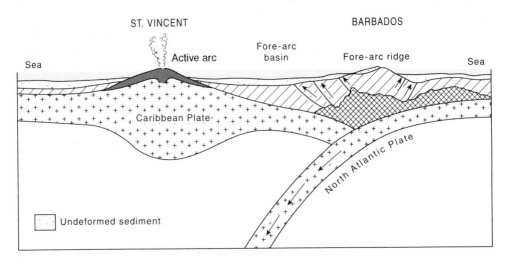

Figure 20.1 The tectonic setting of Barbados

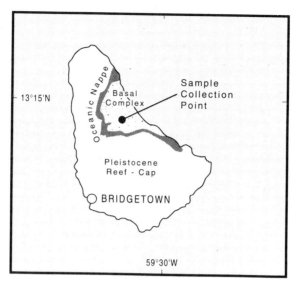

Figure 20.2 Location of the study area

The Joe's River Formation outcrops in the Scotland District of Barbados (Figure 20.2) and is thought to represent a diapir melange deposit, as opposed to an olistostrome melange (Senn, 1940; Pudsey and Reading, 1981). The Joe's River Formation has three components. First, a massive breccia of mudstone and sandstone clasts, with a foliated mudstone matrix. Secondly, a layered breccia of scattered, poorly to well-oriented blocks of quartzose turbidite, radiolarite and calcareous pelagic rocks in foliated, sandy, mud-pebbly mudstone. Thirdly, turbidite beds, one of which includes calcareous microfossils (Laure and Speed, 1984). The formation includes a number of large olistoliths embedded in a dark green to brown, silty clay matrix, which appears to be highly inhomogeneous, containing clay-rich and silt-rich areas. It also contains traces of uncemented, fine-grained sand and streaks of solid hydrocarbon (Enriquez-Reyes and Jones, 1991). In addition to the inhomogeneity, there are nodules of highly plastic, light green clay, which are of variable diameter.

The most distinctive feature of the melange clay is its pervasive fracture pattern (Yassir, 1989; Enriquez-Reyes and Jones, 1991). The material possesses numerous cleavage planes which vary in spacing between 2 mm and 15 mm (Enriquez-Reyes *et al.*, 1990). The cleavage surfaces are usually smooth, frequently curved and when exposed in outcrop display a polished, slickensided surface. Blocks of unweathered material break up easily, with the fabric acting as lines of weakness along which disintegration occurs.

FIELD SAMPLING

Samples were extracted from one of the major Joe's River Formation outcrops, identified in Figure 20.2. A trial pit was excavated to expose the weathering profile

Figure 20.3 Undisturbed sampling technique, showing a block of material ready for waxing

throughout the Joe's River Formation. Undisturbed samples were collected by digging to leave a free-standing column of material, which was then carefully shaved down to minimise disturbance (Figure 20.3). The block of material was then sectioned into samples of a suitable size for transport to the laboratory. Before removal each sample was covered in muslin, coated in paraffin wax and sealed within a protective air-tight tin to prevent moisture loss between extraction and testing and to minimise damage to the scaly fabric in transit.

By using the above technique, it was possible to collect a continuous column of material down through the entire weathering profile. Consequently, laboratory tests have been conducted on Joe's River Formation material taken from three different weathering zones. First, from the top of the column, material which is highly weathered (HW) in appearance, exhibiting a high void ratio. Secondly, material taken from 40–80 cm below the ground surface and partially weathered (PW) in appearance. Thirdly, material between 80 and 125 cm below the ground surface which was slightly weathered (SW). Additional material was extracted from a depth of 125 cm which displays little physical evidence of weathering, and, for the purposes of this study, is termed unweathered (UW). While to some extent these remain relative terms, field evidence in each of the excavated weathering profiles leaves little doubt about the differences used to make these subdivisions. Within a depth interval of little more than 1 m, the extracted samples varied from a soft, plastic mud at the ground surface which displayed little or no evidence of the scaly fabric, to a stiff, highly cleaved sediment, which sheared readily along its anastomosing cleavage planes.

SOIL MECHANICS THEORY

The physical properties of soil materials can be divided into two categories: inherent properties and mechanical properties (Terzaghi and Peck, 1967). Inherent properties are related to the constituents of the material and its structure, and include natural moisture content, particle size distribution, porosity and bulk density (Lambe and Whitman, 1979). The moisture content has a fundamental influence on the geotechnical characteristics of the material and, as a result, a number of empirical index property tests such as the plastic limit, liquid limit and plasticity index have been devised to permit the estimation of material strength directly from its moisture content. In addition to the above properties, the mineral composition of sediment is also an inherent property.

Mechanical properties are the response of material subject to change in stress. Shear strength is one of the most important mechanical properties. This parameter is dependent on properties such as particle size distribution, particle arrangement, mineralogy and degree of saturation (Ebuk et $al.$, 1990; El-Sohby et $al.$, 1990). Consequently, shear strength will alter if weathering causes any of the inherent properties to change.

The influence of weathering on the mechanical behaviour of the Joe's River Formation can be evaluated by examining the changes that an increase in weathering imposes on the material. A frequently used method for representing mechanical data in geomorphology is a simple stress–strain curve. A more rigorous description of mechanical behaviour can be attained by combining stress–strain curves with either Mohr Circles of Stress or a stress path diagram, plotted in terms of deviatoric stress (q) and mean effective stress (p').

The deviatoric stress (q) is the numerical difference between the effective vertical stress and the effective horizontal stress. In other words:

$$q = \sigma'_1 - \sigma'_3 \tag{1}$$

where q = deviatoric stress; σ'_1 = effective vertical stress; and σ'_3 = effective horizontal stress.

The mean effective stress (p') is defined as follows:

$$p' = \frac{(\sigma'_1 + 2 \times \sigma'_3)}{3} \tag{2}$$

Evolving states of volumetric strain in the deforming soil can be represented by plotting the mean effective stress with the volume. The specific volume (v) is the volume of a soil sample containing a unit volume of soil grains.

$$v = 1 + e \tag{3}$$

where v = specific volume and e = sample void ratio.

These parameters form the basis of the Cam-Clay model (Atkinson and Bransby, 1978). The approach used in this study has been to examine the relationship between weathering and mechanical behaviour using the Cam-Clay model. Cam-Clay is widely adopted in soil mechanics for the description of soil deformation and has been the subject of several major geotechnical publications (e.g. Schofield and Wroth, 1968; Atkinson and Bransby, 1978; Wood, 1990). The Cam-Clay model defines a series of state boundaries in q–p'–v space. A volume within the space is defined by these boundaries (Figure 20.4) and represents the only possible stress states that the material can attain. The boundaries are as follows.

1. The *normal consolidation line*, which is a stress path followed by a normally consolidated soil under isotropic compression. All materials, from normally consolidated to heavily over-consolidated, lie between this state boundary line and the p'–v axes.

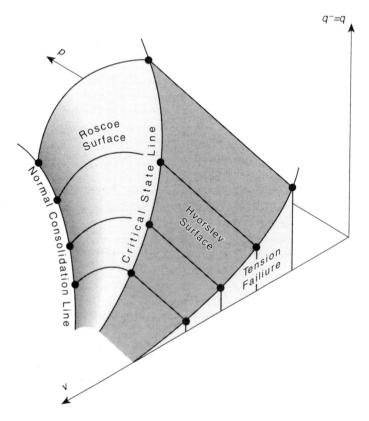

Figure 20.4 The Cam-Clay model as defined by four state boundaries

2. The *critical state line*, which can be described as an ultimate state line existing in stress-specific volume space on which all the stress paths from triaxial compression tests terminate. In other words, the critical state line defines yield conditions for a material under a given set of stress conditions. The crucial characteristic of the critical state line is that the failure of isotropically compressed samples will occur once the stress state of a sample reaches the line, irrespective of the test path followed. Failure will be manifested as a state at which large shear distortions occur with no change in stress or specific volume (Atkinson and Bransby, 1978; Farmer, 1983).

3. For normally consolidated to lightly over-consolidated soils, the stress paths for both drained and undrained tests will trace out a curved surface which links the normal consolidation line with the critical state line in stress–volume space. This curved surface is termed the *Roscoe surface*. On the other hand, for heavily over-consolidated soils, the stress path will trace out a curved surface known as the *Hvorslev surface*.

The Hvorslev surface and the Roscoe surface meet at the critical state line.

LABORATORY PROCEDURE

The inherent material properties noted above were all measured using standard techniques and following established experimental procedure (Head, 1984; BSI 1377, 1990). Material mechanical properties were measured using a conventional triaxial cell (Figure 20.5). Test specimens were prepared to a diameter of 38 mm and a length of approximately 76 mm. Each sample was enclosed in a thin rubber membrane sealed at the top and bottom on to loading platens using rubber O-rings. The membrane acts as a seal to separate the sample and its pore fluid from the confining fluid. Porous discs and filters were placed between the sample and end platens. Drainage connections pass through the top and the base of the sample to permit the monitoring of pore pressure and pore volume change.

Three sets of samples, each with different degrees of weathering, were tested using the triaxial cell (Table 20.1). To characterise the mechanical behaviour of the material, standard consolidation–undrained triaxial compression tests were used following the method proposed by Bishop and Henkel (1962). There are four reasons for adopting this procedure. First, under these constraints, test conditions most adequately reflect field conditions. Secondly, the test material is almost impermeable, making undrained tests an acceptable procedure. Thirdly, due to the excessive time which would be required to undertake drained tests, a programme of undrained experiments is the only feasible alternative. Finally, by conducting undrained tests, volume does not change during shear. In other words, the volume variable is held constant and does not have to be considered when interpreting results.

Each test has two distinct phases. The first was a consolidation phase, when each sample was subjected to three confining pressure increments. Following each increment, the confining pressure was held constant and pore fluid was allowed to drain out of the sample, thus permitting the dissipation of excess pore pressure. The volume of pore fluid draining out of the specimen and the gradual decrease in pore pressure

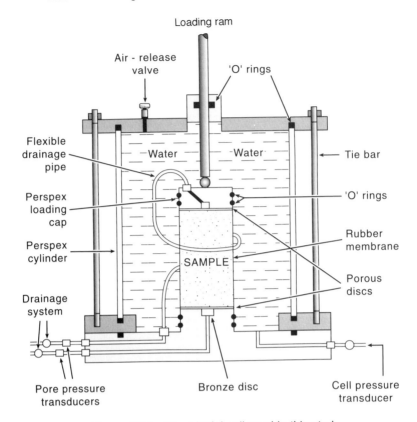

Figure 20.5   The triaxial cell used in this study

through time were both recorded at regular intervals, to allow the construction of consolidation curves (Figure 20.6). In the second phase of each test the samples were subject to undrained shear. The cell pressure was held constant while the specimen was sheared at a constant rate of axial deformation. Each test was terminated when the specimen failed. No drainage was permitted during the shear phase. Therefore the volume and the moisture content of the samples remained constant.

Table 20.1   Samples used in the triaxial cell tests and their confining conditions

Depth in profile (cm)	Sample number	Cell pressure (kPa)
0–40	HW 100	100
0–40	HW 200	200
0–40	HW 400	400
40–80	PW 100	100
40–80	PW 200	200
40–80	PW 400	400
80–125	SW 100	100
80–125	SW 200	200
80–125	SW 400	400

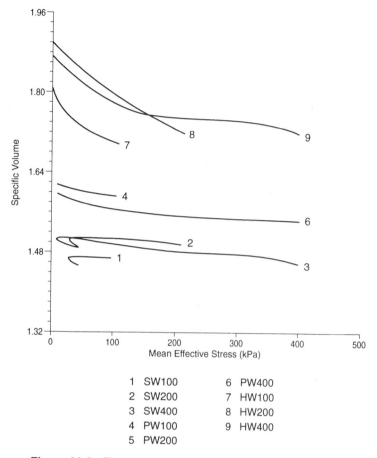

Figure 20.6 The consolidation curves of the test materials

RESULTS

The results of laboratory tests conducted to determine important material properties, including particle size distribution, moisture content, porosity, bulk density and index properties are presented in Table 20.2 and Figure 20.7. The particle size distribution curves show that the silt and clay fraction varies between different weathering grades but the sand fraction is similar in each case. The moisture content

Table 20.2 Inherent properties of the test materials

Depth in profile (cm)	Plastic limit (%)	Liquid limit (%)	Plasticity index	Bulk density (g cm^{-3})	Moisture content (%)	Porosity (%)
0–40	20.27	40.33	10.06	1.91	32.28	46.70
40–80	19.36	29.29	9.93	2.04	29.6	38.00
80–125	22.81	42.79	19.98	2.10	22.00	32.01

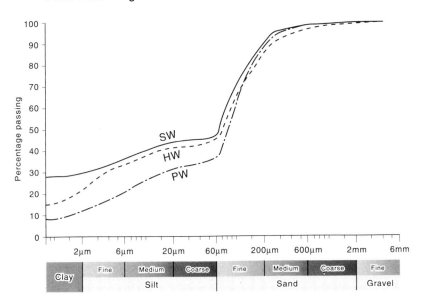

Figure 20.7   The particle size distribution curves of the test materials

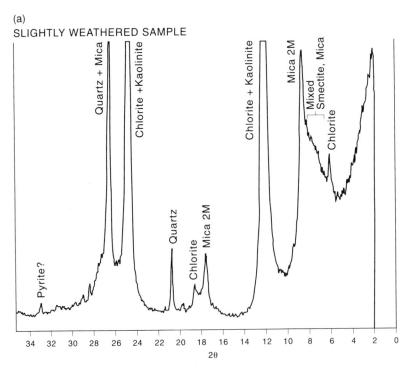

Figure 20.8   X-ray diffraction analysis of the Joe's River Formation scaly clay: (a) unweathered, (b) partially weathered, (c) highly weathered

(b)
PARTIALLY WEATHERED SAMPLE

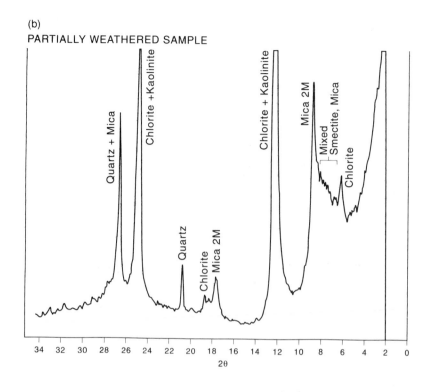

(c)
HIGHLY WEATHERED SAMPLE

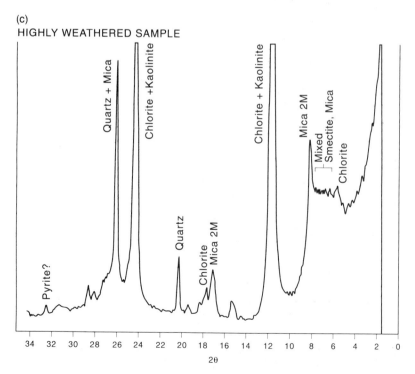

Table 20.3 Major mineral composition of the samples

Mineral	Unweathered	Slightly weathered	Partially weathered	Highly weathered
Mica	×	×	×	×
Smectite				
Chlorite	×	×	×	×
Kaolinite	×	×	×	×
Quartz	×	×	×	×
Mica 2M	×	×	×	×

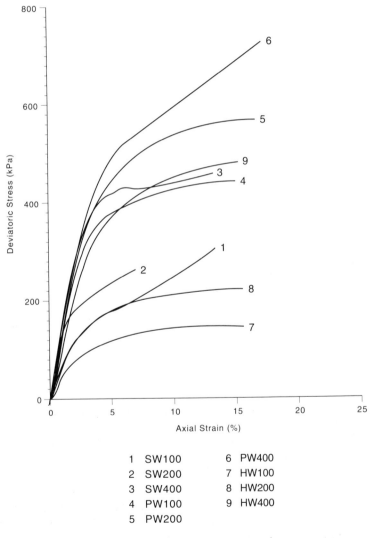

1 SW100	6 PW400
2 SW200	7 HW100
3 SW400	8 HW200
4 PW100	9 HW400
5 PW200	

Figure 20.9 Triaxial test stress–strain curves

and porosity increase from 22% to 32% and from 32% to 46% respectively between slightly weathered and highly weathered states. In contrast, the bulk density decreases from 2.1 g cm^{-3} for the slightly weathered sample to 1.91 g cm^{-3} for the highly weathered material. Index property data follow a similar trend to the particle size results.

Details of the main mineralogical constituents are presented in Figure 20.8 a–c. The major minerals found in all the samples are mica-smectite interlayer, chlorite, kaolinite, quartz and mica 2M. Weathering does not alter the mineral composition, although it is still possible that weathering may have an effect on the total quantity of clay present (Table 20.3).

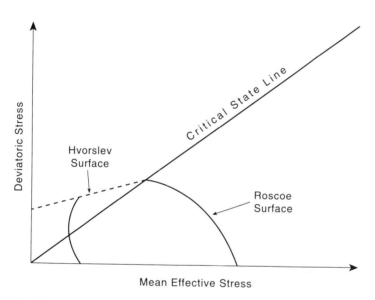

Figure 20.10 Stress paths for undrained, normally consolidated and over-consolidated soil samples

The mechanical tests conducted under different confining pressures were completed for each weathering class. The results were used to define stress paths and failure envelopes. The results are presented in Figures 20.9, 20.10 and 20.11. Typical stress paths for undrained tests on normally consolidated and over-consolidated soil samples are shown in Figure 20.10. Tests conducted on slightly weathered materials (SW1, SW2 and SW3) were terminated before they reached the critical state condition. In particular, the SW2 test, with a confining pressure of 200 kPa, was stopped shortly after failure occurred, at an axial strain of 7%. It is interesting to note that the shear resistance of sample SW1 at a confining pressure of 100 kPa is higher than for sample SW2.

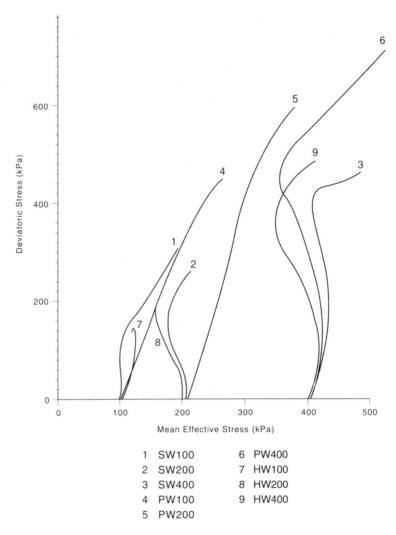

Figure 20.11   The stress paths ($q - p'$) of the test materials

1 SW100   6 PW400
2 SW200   7 HW100
3 SW400   8 HW200
4 PW100   9 HW400
5 PW200

DISCUSSION

The data confirm that weathering significantly affects material properties. For argilla-ceous materials under tropical weathering, the effects mainly manifest themselves as changes in the arrangement of particles, fabric, cement and pore space. The increasing porosity and moisture content, and the decreasing bulk density of the material as the weathering proceeds, indicates that weathering gradually promotes an increasing pore volume. This occurs at the expense of the scaly fabric which is progressively destroyed. X-ray diffraction analyses confirm that weathering does not significantly alter the mineralogy of the material. All the samples have similar XRD traces, suggesting that either the original mineralogy of the material is stable at the near-surface environment or the time for material to weather down and be stripped from the slopes is too short for considerable mineral alteration to occur.

The shear resistance of the least weathered sample is lower than that of the partially weathered sample at the same confining pressure. This is probably because the material in a relatively unweathered state exhibits a strong scaly clay fabric, which facilitates premature shear failure. In other words, the shear resistance and failure pattern of unweathered and slightly weathered scaly clay are strongly controlled by the scaly fabric. As the material weathers, the scaly fabric starts to break down and any preferred particle orientation is lost. The result of this loss of orientation is an increase in shear strength. This is an important point. The implication is that, for sedimentary materials with well-developed fabrics, the start of weathering leads to a strength increase rather than a strength reduction. The degeneration of the fabric and the associated polished slickensided surfaces as particle re-orientation starts, can perhaps be best described as a mechanism by which friction increases across fabric boundaries. It is this which is responsible for the initial strength increase. As weathering continues, the pore space is enlarged. The material becomes highly restructured, the fabric disappears and the sediment can be considered to be in a remoulded state. The result is a considerable reduction in shear strength.

The stress–strain plots indicate that strain hardening occurs in the least weathered sample and partially weathered sample. However, strain softening is completely absent. The inference to be drawn here is that within the range of confining pressures used in this study, when materials are subject to a deviatoric stress, particle rearrangement does not occur but the frictional resistance between individual grains increases, leading to a gradual increase in strength.

CONCLUSIONS

There are three main conclusions to be drawn from this study. First, tropical weathering leads to an increase in the pore volume of argillaceous materials. This in turn is responsible for increases in moisture content and porosity and a decrease in the bulk density. Secondly, the scaly clay fabric is progressively destroyed by weathering. The result of the disappearance of the scaly clay fabric is an initial increase in material shear strength with the start of weathering. Further weathering results in the material becoming remoulded, and the shear strength drops considerably. Finally, materials with a well-developed fabric or cleavage pattern do not obey the critical state model of soil behaviour. However, as the shear planes are destroyed during weathering, the properties of the clay become increasingly similar to those of an equivalent highly porous, remoulded, critical-state-type material.

ACKNOWLEDGEMENTS

The generous assistance of Dr M.J. Leddra, University of Sunderland, with the laboratory testing is warmly appreciated. The X-ray diffraction samples were kindly prepared by Mr A. Osborn, Department of Geological Sciences, University College London. The diagrams were drawn with dexterity by staff in the Drawing Office, Department of Geography, University of Durham.

REFERENCES

Allison, R. J. (1991). Slopes and slope processes. *Progress in Physical Geography*, **15**, 412–437.

Atkinson, J. H. and Bransby, P. L. (1978). *The Mechanics of Soils—an Introduction to Critical State Soil Mechanics*. McGraw-Hill, London.

Biju-Duval, B., Le Quelleck, P., Mascle, A., Renard, V. and Valery, P. (1982). Multibeam bathometric survey and high resolution seismic investigation on the Barbados Ridge complex (eastern Caribbean): a key to the knowledge and interpretation of an accretionary wedge. *Tectonophysics*, **86.1**, 275–304.

Bishop, A. W. and Henkel, D. J. (1962). *The Measurement of Soil Properties in the Triaxial Test*, 2nd edition. Edward Arnold, London.

British Standards Institution 1377. (1990). *Methods of Test for Soils for Civil Engineering Purposes*. HMSO, London.

Brown, K. and Westbrook, G. (1987). The tectonic fabric of the Barbados Ridge accretionary complex. *Marine and Petroleum Geology*, **4**, 71–81.

Brunsden, D. (1979). Weathering. In Embleton, C. and Thornes, J. (eds), *Process in Geomorphology*, Edward Arnold, London, pp. 73–129.

Ebuk, E. J., Hencher, S. R. and Lumsden, A. C. (1990). The influence of structure on the shearing mechanism of weakly bonded soils. In Cripps, J. C. and Moon, C. F. (eds), *The Engineering Geology of Weak Rock*, Geological Society of London Engineering Group, London, pp. 242–253.

El-Sohby, M. A., Mazen, S. O. and Aboushook, M. I. (1990). Deformational characteristics of some weakly cemented and over-consolidated soil formations. In Cripps, J. C. and Moon, C. F. (eds), *The Engineering Geology of Weak Rock*, Geological Society of London Engineering Group, London, pp. 155–163.

Enriquez-Reyes, M. P. and Jones, M. E. (1991). On the nature of the scaly texture developed in melange deposits. *Proceedings 32nd International Conference on Rock Mechanics as a Multidisciplinary Science*, pp. 713–722.

Enriquez-Reyes, M. P., Allison, R. J. and Jones, M. E. (1990). Slope stability in scaly clay terrains. In Cripps, J. C. and Moon, C. F. (eds), *The Engineering Geology of Weak Rock*, Geological Society of London Engineering Group, London, pp. 422–432.

Farmer, I. W. (1983). *Engineering Behaviour of Rock*, 2nd edition. Chapman and Hall, London.

Head, K. H. (1984). *Manual of Soil Laboratory Testing Volumes 1 and 2*. Pentech Press, London.

Lambe, T. W. and Whitman, R. V. (1979). *Soil Mechanics, SI Version*, Wiley, New York.

Laure, D. K. and Speed, R. C. (1984). Structure of the accretionary complex of Barbados. II: Bissex Hill. *Geological Society of America Bulletin*, **95**, 1360–1372.

Ollier, C. D. (1984). *Weathering*. Longman, New York.

Pudsey, J. C. and Reading, H. G. (1981). Sedimentology of the Scotland Group. *Geological Society of London Special Publication*, **10**, 291–308.

Schofield, A. and Wroth, C. P. (1968). *Critical State Soil Mechanics*. McGraw-Hill,London.

Senn, A. (1940). Palaeogene of Barbados and its bearing on the history and structure of the Antillean Caribbean region. *Bulletin American Association of Petroleum Geologists*, **24**, 548–610.

Stride, A., Belderson, R. and Kenyon, N. (1982). Structural grain, mud volcanoes and other features on the Barbados Ridge complex revealed by GLORIA side-scan sonar. *Marine Geology*, **49**, 189–196.

Terzaghi, K. and Peck, R. B. (1967). *Soil Mechanics in Engineering Practice*. Wiley, New York.

Torrini, R., Speed, R. and Mattioli, A. (1985). Tectonic relationships between forearc-basin strata and the accretionary complex at Bath, Barbados. *Geological Society of America Bulletin*, **96**, 861–874.

Wood, D. M. (1990). *Soil Behaviour and Critical State Soil Mechanics*. Cambridge University Press, Cambridge.

Yassir, N. A. (1989). Mud volcanoes and the behaviour of over-pressured clays and silts. Unpublished Ph.D. thesis, University of London.

21 The Groundplan of Cuesta Scarps in Dry Regions as Controlled by Lithology and Structure

KARL-HEINZ SCHMIDT
Freie Universität Berlin, Germany

ABSTRACT

More than 100 structural landforms (cuesta scarps, hogbacks) on the Colorado Plateau in the semiarid southwestern United States were investigated in an analysis of the groundplan attributes of scarps in dry regions. The shape of the groundplan is character- ised by the embayment index (BI), which is calculated by dividing the actual length of the scarp by the corresponding straight line distance. The BI-value increases with the degree of embayment. Data on independent variables in the system of structural landforms were collected for individual cuesta scarps. This was followed by a statistical analysis to examine the influence of these variables on the embayment of the groundplan. Independent system variables controlling the variation of the embayment index are the lithological and structural attributes of the caprock, its thickness, the structural dip, the direction of the dip, the configuration of the drainage net, and the distance to base level. Under arid and semiarid conditions straight scarps are found as well as highly embayed ones. Cuesta scarps are less embayed with decreasing resistance and increasing thickness of the caprock and with increasing dip. Backscarps (dip out of the scarp slope) are more embayed than frontscarps (dip into the scarp). When a scarp is accompanied by a river parallel to its foot or breached by an allogenic channel, the influence of the other independent variables is masked. The lithological and structural characteristics of the caprock, its thickness and the dip account for more than 60% of the total variance in the groundplan embayment of cuesta scarps.

INTRODUCTION

Form and process relationships for structural landforms such as cuestas, hogbacks or horizontal plateaus were investigated on the central Colorado Plateau, USA (Figure 21.1). Structural landforms generally consist of (1) a bipartite scarp slope with a steep or cliff-like upper slope in the resistant caprock and a moderately inclined lower slope in the underlying less competent rock and (2) a backslope (dipslope) supported by the caprock (Figure 21.2). On the central Colorado Plateau mean annual precipitation values range from 150 to 350 mm with most of the study area lying in the range between 150 and 250 mm in the Canyonlands and Navajo sections of the Colorado Plateau according to the physiographic subdivision proposed by Hunt (1974). Areas

Rock Weathering and Landform Evolution. Edited by D. A. Robinson and R. B. G. Williams
© 1994 John Wiley & Sons Ltd

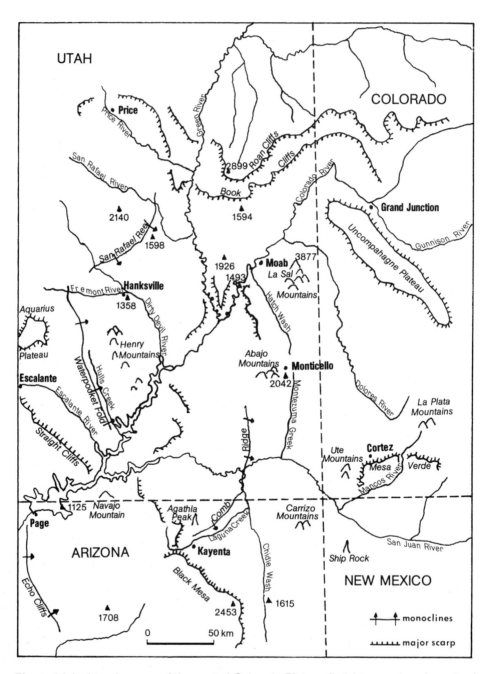

Figure 21.1 Location map of the central Colorado Plateau (heights are given in metres)

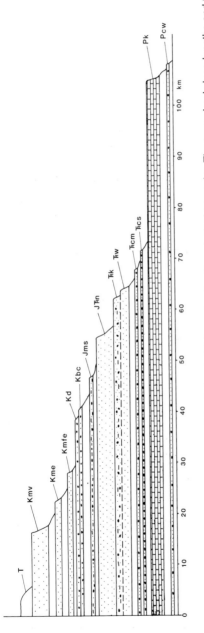

Figure 21.2 Generalized stratigraphic column of the Colorado Plateau showing the principal caprocks. The mean backslope lengths and the mean thickness of the caprocks are drawn to scale. For thickness information the Kaibab Limestone (Pk) may serve as a reference unit with its average thickness of 100 m.

T	Tertiary cover
Kmv	Mesaverde Sandstone, moderately cemented, porous (3)
Kme	Emery Sandstone, often only moderately cemented (3)
Kmfe	Ferron Sandstone, often only moderately cemented (3)
Kd	Dakota Sandstone, strongly cemented, conglomeratic (12.5)
Kbc	Burro Canyon Formation, strongly cemented, conglomeratic (12.5)
Jms	Salt Wash Sandstone, strongly cemented, with mudstone lenses (6)
JTrn	Navajo Sandstone, poorly cemented, porous, with intra-formational partings (1)
Trk	Kayenta Formation, sandstone with siltstones and conglomerates (8)
Trw	Wingate Sandstone, only moderately cemented, with vertical joints (5)
Trkw	Kayenta and Wingate Sandstones as joint caprocks (7)
Trcm	Moss Back Sandstone, silica cemented, conglomeratic (9)
Trcs	Shinarump Conglomerate, silica cemented (11)
Pk	Kaibab Limestone, very resistant (14)
Pcw	White Rim Sandstone, strongly cemented (10)

The numbers in parentheses indicate the rank of resistance of the caprocks, ordered from least to most resistant. A detailed description of the lithological attributes of the caprocks and the deduction of the ranking is given in Schmidt (1988, 1989 and 1991). Note: the stratigraphic column is generalized. Due to lateral intertonguing, thickness and facies changes and erosional unconformities, there is no location on the Colorado Plateau where the complete caprock sequence is found in a single profile (therefore no vertical scale is included). The combination Trkw is omitted from the figure for clarity

Table 21.1   Independent control variables

(1)   lithology and structure (resistance) of caprock
(2)   lithology and structure of underlying soft rock
(3)   thickness of caprock
(4)   thickness of soft rock
(5)   thickness ratio
(6)   structural dip
(7)   direction of dip
(8)   position of drainage lines
(9)   distance to base level

with precipitation amounts exceeding 350 mm were not included in the analysis. As there are no major spatial variations of the semiarid climate in the study area, the control of the lithological and structural variables on landform characteristics and geomorphological processes can be studied under conditions similar to an experimental setup (cf. Schmidt, 1988).

There are a number of independent non-climatic control variables which influence the dependent form and process variables of structural landforms (Schumm and Chorley, 1966; Small, 1970; Blume, 1971; Schmidt, 1987, 1988 and 1991) (Table 21.1). Data on these variables were collected for about 120 structural landforms on the Colorado Plateau. The influence of the controlling factors was statistically analysed for the set of dependent form and process variables and a great number of systematic functional relationships were found (Schmidt, 1988). In this paper the control of the variation of groundplan embayment as an example of a dependent variable will be discussed.

VARIATIONS OF GROUNDPLAN EMBAYMENT AND CONTROLLING FACTORS

The groundplan pattern is quantitatively described by the embayment index or ratio (BI), which is calculated by dividing the actual length of the scarp by the corresponding straight line distance (Schmidt, 1980). Cuesta scarp groundplans were measured for scarps formed in the major caprock units of the Colorado Plateau (Table 21.2, Figure 21.2), and also for scarps in caprocks of only local importance such as the Cedar Mountain, Bluff, Entrada and Cedar Mesa sandstones (Schmidt, 1988). The following conditions were set for inclusion in the analysis. Scarps must have caprock thicknesses of more than 5 m. They must have a separate backslope, and not be just ledges on a complex scarp slope. Scarp segments must be at least 4 km long (straight line distance) without changes in independent system variables. Outliers were not considered in the calculation of the embayment index for scarp segments.

Curvimeters were applied for length measurements. The maps used were 1:62500 (or larger scale) geological maps. Most of the study area was geologically mapped during prospecting for coal and later uranium-bearing rocks. Where geological maps were not available topographic maps and aerial photographs of similar scales were used.

Table 21.2 Means and standard deviations of the embayment index (BI) for individual caprocks

Caprocks	BI	SL	n
Mesaverde Sandstone	1.51 ± 0.44	95%	7
Emery Sandstone	1.38 ± 0.40	99%	6
Ferron Sandstone	1.44 ± 0.42	95%	6
Dakota Sandstone	2.42 ± 0.77	–	11
Burro Canyon Formation	1.90 ± 0.78	n.s.	9
Salt Wash Sandstone	2.18 ± 0.78	n.s.	5
Navajo Sandstone	1.46 ± 0.33	99.9%	12
Kayenta Sandstone	1.86 ± 0.30	95%	10
Kayenta/Wingate Sandstone	1.59 ± 0.56	99%	14
Wingate Sandstone	1.37 ± 0.11	99%	9
Moss Back Sandstone	2.30 ± 0.53	n.s.	4
Shinarump Conglomerate	2.30 ± 0.36	n.s.	3
Kaibab Limestone	1.46 ± 0.24	95%	5
White Rim Sandstone	3.00 ± 0.10	n.s.	3

Only caprock units with three or more measured scarps are shown. The mean embayment index for the total sample is 1.8.
SL = level of significance: The t-test was applied to test if there is a significant difference between the means of the embayment ratios of the individual caprocks and the reference caprock (Dakota Sandstone).
n.s. = not significant.
n = number of scarps included in the calculations.

The BI-value increases with the degree of embayment. Not too long ago less embayed groundplans were thought to be typical features of dryland cuesta scarps (Louis and Fischer, 1979), because spring sapping, mainly responsible for scarp re-entrants in humid areas, is only of minor importance. In dry areas, however, embayments are mainly a consequence of concentrated linear surface runoff after torrential rainfall. On the Colorado Plateau we find straight scarps as well as highly embayed ones (Barth and Blume, 1973; Schmidt, 1988). Embayment ratios range from unity (straight) to values of more than 10.

Control by Resistance and Thickness of the Caprock

A great number of major caprock units are involved in the formation of structural landforms on the Colorado Plateau (Figure 21.2). A ranking of rock resistance is indicated in this figure. This ordinal scale, with the most resistant caprock getting the highest number, is based on variable caprock properties such as porosity, density, cementing material and jointing (Schmidt, 1988, 1989 and 1991). When the embayment ratios of the major caprock units are compared, there are significant differences in the mean values (Table 21.2). The Dakota Sandstone was chosen as a reference unit, because it is representative of a group of caprocks with similar attributes of lithology and thickness [Burro Canyon Formation (Kbc), Salt Wash Sandstone (Jms), Moss Back Sandstone (Trcm), Shinarump Conglomerate (Trcs), White Rim Sandstone (Pcw)]. The Dakota Sandstone itself is a highly resistant silica-cemented

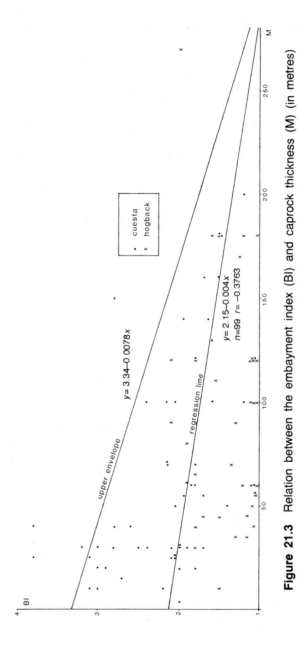

Figure 21.3 Relation between the embayment index (BI) and caprock thickness (M) (in metres)

caprock with a widely spaced joint structure. The *t*-test demonstrated that the caprocks with high resistance ranks [rank 6 (Jms), 9 (Trcm), 10 (Pcw), 11 (Trcs),12.5 (Kbc)] (Figure 21.2) do not have significantly different groundplan patterns. This group of caprocks is at the same time characterised by small to moderate thicknesses (less than 50 m). The other group of caprocks with significantly smaller ratios consists of less resistant caprocks, some of them with a much greater thickness (e.g. Mesaverde Group) and thick caprocks of high resistance (e.g. Kaibab Limestone).

The combined influence of thickness and resistance already becomes obvious in this simple data set. The high embayment ratios of the resistant and relatively thin caprocks may be explained by two circumstances: (1) the competent caprocks are stable even on long and narrow spurs because of their higher shearing resistance, and (2) thin caprocks are more easily dissected by headward erosion of slope rills and drainage lines on the dipslope. Rock collapse, on the other hand, is more common in less resistant caprocks, and caprocks of great thickness promote slope failures because of increased overburden pressure with resulting higher shear stresses. This causes instability of narrow elongated spurs. Moreover, thick caprocks are more efficient obstacles to dissection.

Figure 21.3 shows the relation between caprock thickness and the embayment index, regardless of the caprocks involved. Supporting the interpretations given above, increasing thickness leads to a decreasing groundplan embayment, but there is a very large amount of scatter in the diagram. The upper envelope gives an impression of the limit of maximum possible embayment index values for a given caprock thickness. Five points plot above the upper envelope, because of particular circumstances. The three points on the left side (small thickness) of the scatter plot belong to scarps in the Dakota Sandstone in the Four Corners area in close proximity to the San Juan River, where short horizontal distances to base level, low dip and high resistance combine to produce these extreme embayment ratios. The same applies for the point in the central part of the plot, a scarp in the Kayenta and Wingate Sandstones in the central Canyonlands close to the Colorado River. The fifth outlier on the plot is a much dissected hogback in the Navajo Sandstone. In this diagram cuestas and hogbacks are distinguished to give some impression of the structural dip of the scarps involved. Hogbacks develop when the structural dip exceeds 6° (Barth and Blume, 1973; Schmidt, 1988). Embayment index values of the hogbacks (high structural dip) are most frequently found below the regression line (relatively low embayment ratios), which is indicative of the controlling effect of dip.

When embayment and thickness variations are compared for individual caprocks, the relationships become clearer. Here, those caprocks with marked regional variations in thickness (Mesaverde Sandstones, Salt Wash Sandstone, Navajo Sandstone) show highly significant correlations (Schmidt, 1980 and 1988).

Control by Lithology, Structure and Thickness of the Underlying Less Competent Rock and by the Thickness Ratio

The lithological attributes of the underlying less resistant rocks have no significant influence on groundplan embayment. The Dakota Sandstone, for instance, shows the same embayment characteristics when overlying varying lithologies of the Morrison

Combined Control by Caprock Thickness and Structural Dip

When the combined influence of caprock thickness and structural dip is evaluated in a multiple regression for the total sample, about one third of the variance of groundplan embayment can be explained (Table 21.3). A major part of the variance not explained by these controlling factors is caused by differences in the response of individual caprocks to varying structural dip and caprock thickness. When the control exerted by structural dip, and the combined influence of dip and caprock thickness, are statistically analysed for individual caprocks, the coefficients of determination become considerably higher (Table 21.3). The correlations were calculated only for caprocks in which a sufficient number of cuestas and hogbacks were found to make statistical tests applicable and which have formed structural landforms with sufficient variation of the respective controlling factors. Backscarps and scarps with special positions in relation to the drainage lines (parallel to scarp foot, scarp breached by allogenic river) are not included in the calculations. For the majority of caprocks, either dip alone, or dip and caprock thickness, explain 60% or more of the variance of the embayment ratio.

Some peculiarities in the response of individual caprocks are evident from Table 21.3. The Emery, Ferron and Dakota Sandstones show too little variation in their thickness to make a multiple correlation worth calculating. The coefficient of determination of the multiple correlation is particularly high for the Salt Wash Sandstone. This is due to the high thickness variations of this unit. For the Navajo Sandstone there is no apparent correlation between dip and groundplan embayment, but with the inclusion of thickness the multiple correlation coefficient becomes significant. The Navajo Sandstone is one of the most prominent caprocks on the Colorado Plateau, in some places exceeding 300 m in thickness. It is a weakly cemented sandstone and the uppermost member of the Glen Canyon Group (lower units Kayenta and Wingate

Table 21.3 Correlation coefficients for the relationships between the embayment index (BI) and controlling variables

Caprocks	n	Structural dip		Multiple correlation (dip and thickness)		
		r	SL	R	R^2	SL
Emery Sandstone	5	−0.7197	n.s.	–	–	–
Ferron Sandstone	6	−0.9726	99%	–	–	–
Dakota Sandstone	11	−0.7740	99%	–	–	–
Burro Canyon Formation	8	−0.7282	95%	0.7714	59.5%	95%
Salt Wash Sandstone	5	−0.6368	n.s.	0.9960	99.2%	99%
Navajo Sandstone	9	0.2752	n.s.	0.7853	61.7%	95%
Kayenta Sandstone	10	−0.6294	95%	0.8295	68.8%	99%
Kayenta/Wingate Sandstone	12	−0.6772	95%	0.6773	45.9%	95%
Wingate Sandstone	8	−0.0381	n.s.	–	–	–
Undifferentiated	99	−0.5364	99.9%	0.5737	32.9%	99.9%

Embayment ratios of backscarps and scarps with special positions of the drainage lines (parallel to scarp foot, scarp breached by allogenic river) are not included.
r = correlation coefficient; R = multiple correlation coefficient; R^2 = coefficient of determination (in %); SL = significance level; n.s. = not significant; n = number of values

Sandstones). In many cases it is the top unit of the scarp slopes in the Glen Canyon Group, or it covers the Kayenta Sandstone backslope with residual hills (see southern part of Mancos Mesa in Figure 21.6). As a separate scarp-former it only appears on the steeply dipping eastern flanks (e.g. Comb Ridge, Waterpocket Fold, San Rafael Reef, see Figure 21.1) of the large anticlines, where structural dip mostly exceeds 10°. Here it forms conspicuous hogbacks. In contrast to the general trend (Figure 21.3) there is a significant positive correlation between thickness and groundplan embayment for the Navajo Sandstone. Being confined to hogbacks, there is only a limited variation of structural dip as a controlling variable. Thus the differences in thickness exert major control on embayment variations. Increasing thicknesses of the Navajo Sandstone result in significantly increased widths of the outcrop areas, which, even in the strongly dipping hogback positions, allow the formation of deeper re-entrants.

The embayment index of the Wingate Sandstone, which covers only a very narrow range between 1.3 and 1.4, is independent of structural dip and thickness. The

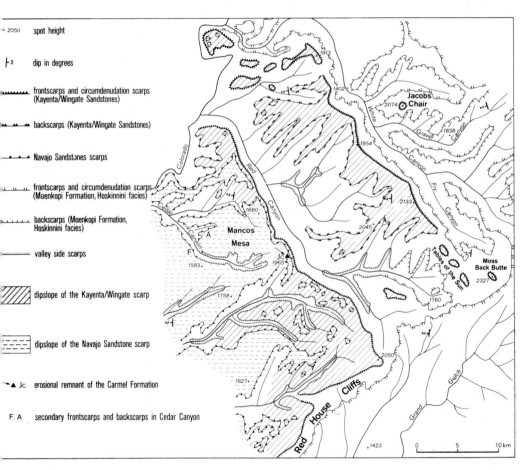

Figure 21.6 Frontscarps and backscarps in the Red House Cliffs area, southeastern Utah. Dip is to the west at about 2°. Note the straight east-facing frontscarps and the deeply embayed backscarps on the downdip sides. Spot heights are given in metres

Wingate Sandstone is much less resistant than the overlying Kayenta Sandstone and is characterised by regular sets of vertical joints. It forms detached scarps only on the flanks of anticlines when the dip exceeds 6°. Without the protecting cover of the Kayenta Sandstone, blocks and slaps disintegrate more easily along the joints, and frequent rockfalls result in a rapid recession of the cliff. Narrow spurs become unstable, deep re-entrants do not develop, and the embayment ratio reaches its upper limit at a value of 1.4. Thus, the restricted range of the embayment index of the Wingate Sandstone is a consequence of the control of its own lithological and structural attributes.

Control by the Direction of the Dip

Mortensen (1953) was the first to differentiate between frontscarps (dip into the scarp slope) and backscarps (dip out of the scarp slope) and to draw attention to some of the important morphometric and process differences between these scarp types. The preceding presentation only referred to frontscarps, because backscarps reveal a significantly different groundplan configuration and process control. Earlier investigations already described the more embayed groundplan of backscarps (Mortensen, 1953; Ahnert, 1960; Blume, 1971; Barth and Blume, 1973). These descriptions are supported by the quantitative analysis of the embayment ratios. When the mean values of the ratio for the backscarps and frontscarps in the study area are compared, a highly significant difference is visible. Separate tests for individual caprocks with a sufficient number ($n > 3$) of cuestas with both backscarps and frontscarps also demonstrate that the backscarps always have a significantly higher mean embayment index (Table 21.4). A typical example of the embayment differences is shown in Figure 21.6 with an almost straight groundplan on the frontscarp (northeast) and a deeply embayed groundplan on the opposite, downdip side.

As the drainage divide on a cuesta is generally located close to the crest of the frontscarp, runoff on the dipslope is directed towards the backscarp, which is therefore much more dissected than the corresponding frontscarp (Figure 21.6). In addition to the headward erosion of autochthonous slope rills, which is mainly

Table 21.4 Embayment ratios (BI) for frontscarps and backscarps

Caprocks	(BI) frontscarps	(BI) backscarps	SL	n
Mesaverde Sandstone	1.30	8.40	99%	5
Dakota Sandstone	1.48	3.43	99.5%	4
Kayenta Sandstone	1.17	2.90	99.5%	4
Kayenta/Wingate Sandstones	1.34	2.61	99.5%	8
Undifferentiated	1.33	3.90	99.9%	25

SL = significance level (t-test)
n = number of paired values
The data used for this table refer to cuestas with frontscarps on the updip and backscarps on the downdip side. There are not many cuestas which comply with this requirement, so there are only few paired data. The means for the frontscarps differ from those in Table 21.2, because only selected scarps could be included in this comparison.

responsible for the re-entrants on frontscarps, the embayment of backscarps is also caused by runoff generated in the catchment area of the dipslope and concentrated in consequent drainage lines. Linear erosion in cuesta scarp development is of particular importance on the backscarps. Sapping processes are also much more effective on backscarps (Laity, 1988). Backwearing of backscarps, therefore, runs at a much faster pace (Schmidt, 1987). Headward erosion of the backscarp valleys may come close to the position of the frontscarp and may ultimately lead, together with recession processes on the frontscarp, to the separation of parts of the cuesta and to the formation of outliers (see Tables of the Sun or Moss Back Butte in Figure 21.6).

Control by the Position of the Drainage Lines

There are three different scarp categories with respect to the position of drainage lines: (1) a river runs parallel to the scarp foot, (2) the scarp is breached by an allogenic river, and (3) the scarp is not affected by these special conditions. The groundplan embayments of these categories differ significantly. Scarps with parallel (subsequent) drainage lines have very low embayment ratios, because the basal erosion keeps the scarp straight (mean value of $BI = 1.07$; $n = 9$). When scarps are breached by allogenic rivers, these form deep re-entrants or superimposed canyons, and the linear dissection of the scarp slopes is intensified by tributaries to the major rivers. Scarps of this category have a mean embayment index of 2.24 ($n = 14$). The difference of means of these two categories is significant at the 99.9% level. The mean embayment ratio of the scarps not affected by these special circumstances lies close to 1.8 ($n = 99$) and differs significantly (99%) from both of the other categories. When scarps belong to one of the first two groups, the influence of the other independent variables is obscured. These scarps were, therefore, not included in the preceding calculations and comparisons.

Control by Distance to Base Level

This independent variable includes both vertical and horizontal distances to base level. Great vertical distance to base level in structurally elevated areas leads to high embayment ratios, because the great differences in altitude supply the washes and slope rills with higher erosional energy resulting in enhanced dissection. When the scarps lie horizontally close to the regional base level, the groundplan is very irregular. This is particularly emphasised by the outliers on the left side in Figures 21.3 and 21.4, scarps in the Dakota Sandstone and Burro Canyon Formation in the Four Corners area, where they are extremely embayed and dissected by the tributaries of the San Juan River. High embayment can also be observed in the north-central section of Figure 21.6, which lies close to the Colorado River.

CONCLUSION

Independent system variables significantly controlling the variation of the embayment index are the lithological and structural attributes (resistance) of the caprock, the thickness of the caprock, the structural dip, the direction of the dip, the configuration of the drainage net, and the vertical and horizontal distance to base level. We find straight scarp groundplans as well as highly embayed ones in close spatial proximity. The variations of groundplan embayment are explained at high levels of determination. A number of functional relationships between dependent and independent system variables may be formulated. The strong morphometric diversity of groundplan embayment and the diversity of the morphological attributes of the structural landforms on the Colorado Plateau as a whole (Schmidt, 1988) are indicative of the great variation in processes active on the scarp slopes and backslopes of cuestas, hogbacks and horizontal plateaus. This diversity is caused by the great differences in the attributes of the non-climatic control factors. The results demonstrate that, within a given climatic regime, a specific landform element may reveal a high degree of form and process variation, which is exclusively effected by changes in internal system conditions.

REFERENCES

Ahnert, F. (1960). The influence of Pleistocene climates upon the geomorphology of cuesta scarps on the Colorado Plateau. *Annals of the Association of American Geographers*, **50**, 139–156.

Barth, H. K. and Blume, H. (1973). Zur Morphodynamik und Morphogenese von Schichtkamm- und Schichtstufenrelief in den Trockengebieten der Vereinigten Staaten. *Tübinger Geographische Studien*, **53**, 1–102.

Blume, H. (1971). *Probleme der Schichtstufenlandschaft*. Erträge der Forschung 5, pp. 1–117, Darmstadt.

Howard, A. D. and Kochel, R. C. (1988). Introduction to cuesta landforms and sapping processes on the Colorado Plateau. In Howard, A. D., Kochel, R. C. and Holt, H. E. (eds), *Sapping Features on the Colorado Plateau, A Comparative Planetary Geology Field Guide*, National Aeronautic and Space Administration, Special Publication 491, pp. 6–56.

Hunt, C. B. (1974). *Natural Regions of the United States and Canada*. Freeman, San Francisco.

Laity, J. E. (1988). The role of groundwater sapping in valley evolution on the Colorado Plateau. In Howard, A. D., Kochel, R. C. and Holt, H. E. (eds), *Sapping Features on the Colorado Plateau, A Comparative Planetary Geology Field Guide*, National Aeronautic and Space Administration, Special Publication 491, pp. 63–70.

Laity, J. E. and Malin, M. C. (1985). Sapping processes and the development of theater-headed valley networks on the Colorado Plateau. *Geological Society of America Bulletin*, **96**, 203–217.

Louis, H. and Fischer, H. (1979). *Allgemeine Geomorphologie*. De Gruyter, Berlin.

Mortensen, H. (1953). Neues zum Problem der Schichtstufenlandschaft. Einige Ergebnisse einer Reise durch den Südwesten der USA. *Nachrichten ver Akadamie der Wissenschaften Göttingen*, **2**, 3–22.

Schmidt, K.-H. (1980). Der Grundriss von Schichtstufen in Trockengebieten. *Die Erde*, **111**, 231–246.

Schmidt, K.-H. (1987). Factors influencing structural landform dynamics on the Colorado Plateau—about the necessity of calibrating theoretical models by empirical data. *Catena Supplementband* **10**, 51–66.

Schmidt, K.-H. (1988). Die Reliefentwicklung des Colorado Plateaus. *Berliner Geographische Abhandlungen*, **49**, 1–183.

Schmidt, K.-H. (1989). The significance of scarp retreat for Cenozoic landform evolution on the Colorado Plateau, USA. *Earth Surface Processes and Landforms*, **14**, 93–105.

Schmidt, K.-H. (1991). Lithological differentiation of structural landforms on the Colorado Plateau, USA. *Zeitschrift für Geomorphologie*, NF Supplementband **82**, 153–161.

Schumm, S. A. and Chorley, R. J. (1966). Talus weathering and scarp recession in the Colorado Plateau. *Zeitschrift für Geomorphologie*, NF**10**, 11–36.

Small, R. J. (1970). *The Study of Landforms*. Cambridge University Press, Cambridge.

Section 5

WEATHERING AND LANDFORM DEVELOPMENT IN TEMPERATE ENVIRONMENTS

22 Sandstone Weathering and Landforms in Britain and Europe

D. A. ROBINSON and R. B. G. WILLIAMS
University of Sussex, Brighton, UK

ABSTRACT

Although sandstones are estimated to occupy 15% of the world's land surface, the systematic study of their weathering and landform features has until recently been largely overlooked, especially in temperate environments. By reference to outcrops from a number of sites across Britain and mainland Europe, this paper describes the distinctive suite of landforms and weathering features that are characteristic of sandstone terrains within the region, discusses their origin, and examines why they are particularly characteristic of such terrains.

Sandstones in Europe are very variable in both their composition and structure. Weakly cemented or thinly bedded sandstones generally form areas of low relief with no rock exposures. In contrast, stronger, more massive sandstones frequently give rise to outcrops of valley-side cliffs or crags, isolated rock towers, buttes and mesas. Traditionally sandstones were considered to be chemically inert because of the low solubility of crystalline quartz, and weathering was believed to be largely restricted to physical processes. However, it has become increasingly apparent that quartz is far more soluble in near-surface environments than was originally thought. Sandstones undergo a variety of surface and sub-surface weathering processes that sometimes lead to a weakening and disintegration of the rock, but may also lead to a hardening and toughening of the outer surface of the rock. It is the interplay of this combination of surface and sub-surface processes that explains many of the weathering features that are developed and it is suggested that chemical processes are more important in the weathering of sandstones than has hitherto been suggested.

INTRODUCTION

Although limestones outcrop over only about 7% of the Earth's land surface (White *et al.*, 1984), the landforms that they produce have been intensively studied by geomorphologists. In contrast, granite landforms have been investigated in much less detail, and sandstone landforms have been remarkably neglected, even though each of these rock types occupies about 15% of the Earth's land surface.

There is no obvious reason why sandstone terrains should have been so conspicuously ignored by geomorphologists. Although sandstones do not give rise to such distinctive landforms as limestones and granites, they are responsible for some of the most spectacular scenery in the world, whose origins are far from being fully

Rock Weathering and Landform Evolution. Edited by D. A. Robinson and R. B. G. Williams.
© 1994 John Wiley & Sons Ltd

understood and clearly deserve much closer study. Early French and German geomorphologists, such as Hettner (1887) and de Martonne (1909), gave due prominence to
sandstone landforms in their writings, but interest then waned. In 1972, Mainguet
tried to rekindle research into sandstone landforms with a masterly study of the
sandstone scenery of parts of the Sahara and contrasting areas such as the Vosges of
eastern France. In spite of this timely intervention, most geomorphological textbooks
continued to give scant attention to sandstone scenery. More recently, Young and
Young (1992) have offered an up-to-date account of the geomorphology of sandstones, drawing mainly on Australian and American examples. In the present paper,
we shall focus primarily on the sandstone terrains of temperate Europe, and examine
the extent to which these are characterised by a distinctive suite of weathering
processes and landforms.

THE NATURE AND PROPERTIES OF SANDSTONES

Sandstones are indurated sands. Sand is a size grade, not a mineralogical classification, and sandstones need not necessarily be dominated by any single mineral species
(Pettijohn et al., 1972). Nevertheless, most sandstones consist mainly or entirely of
grains of quartz cemented together, and it is the geomorphological evolution of rocks
of this type that form the subject of this paper.

Quartz-rich sandstones vary considerably in texture and composition. They show
wide differences in grain size characteristics, density, strength and porosity, as well
as in the kinds of minerals that are present in addition to quartz, as either grains or
cement.

Sand is normally considered to comprise grains of material in the size range
0.05–2.0 mm in diameter. Although by definition sandstones are mainly composed of
material within this size range, they may contain appreciable amounts of coarser or
finer material, sometimes both. In some sandstones the grains have been well sorted
and are all of similar size; other sandstones consist of poorly sorted grains of variable
size. The grains can vary in form from very angular to perfectly spherical. The size
and shape of the sand grains determines how tightly the grains can be packed
together. The porosity of sands without interstitial cement can vary from 27 to 47%
(Blyth and de Freitas, 1984, p. 119).

Cementation or lithification of sand grains to form sandstone can occur either by
pressure melting and recrystallisation of the grains, by compaction and consolidation
of included clay material or by the precipitation of authigenic minerals from water
passing through the pores. The most common authigenic minerals are silica, oxides,
hydroxides or carbonates of iron, and calcite, although other cements are known.
Silica may be deposited either as an interstitial cement or as a crystal coating in optical
continuity with the crystal structure of the original grains. Pressure melting and
recrystallisation of quartz during diagenesis tends to produce interlocking grains that
fit together like a three-dimensional jigsaw. One of the effects of lithification is to
reduce the pore space and therefore the porosity of the cemented sandstones, usually
to within the range 5–25%, but in extreme cases to practically zero. However, as
Sparks (1971) points out, the reduction in porosity is often accompanied by increased
brittleness and jointing so that losses in mass permeability through the pores of the

rock tend to be offset by fracture permeability through flaws and along joints and bedding planes.

Quartz-rich sandstones are frequently referred to as orthoquartzites, and are the hardest and strongest of the sandstones. The major grain components of these sandstones other than quartz are feldspars and clay minerals, including micas. Micaceous sandstones frequently form flagstones with closely spaced bedding planes lined with mica crystals. Sandstones rich in feldspars are termed arkoses. Fresh grains of feldspar are almost as resistant as quartz to physical breakdown, but they are more susceptible to chemical alteration, principally to kaolinite, which softens and weakens the rock. Clay-rich sandstones are frequently termed wackes (Pettijohn, 1957, and Pettijohn *et al.*, 1972). They generally have little strength and relatively low porosity, although they do absorb moisture. They seem to be particularly vulnerable to frost action, and to wetting and drying and, as a result, break down much more rapidly than other sandstones.

The compressive strength of sandstones depends on their porosity, the amount and type of bonding material and the composition of the grains. Sandstones in which the grains are bonded together by cement precipitated from percolating solutions are much stronger than those in which clay and other fine detritus bonds the grains together. The amount of cement appears to be more important than its composition. Poorly cemented, porous sandstones have low strength while dense, well-cemented sandstones have great strength (Table 22.1). For sandstones with the same level of cementation, silica cement is stronger than a calcareous cement (Bell, 1983). The strength of sandstones is reduced considerably when the pores are saturated with water. Bell (1983) gives examples of a reduction of 30% in the compressive strength of Fell Sandstone from northern Britain after saturation, and a reduction of 60% for

Table 22.1 Properties of some British sandstones

Stone	Density $(kg\ m^{-3})$	Crushing strength $(kgf\ cm^{-2})$	Porosity (% vol)
Fell Sandstone (Carboniferous)	2201	287	–
Birchover Gritstone (Millstone Grit)	2365	540	15.2
Springwell Sandstone (Coal Measures)	2174	595	16.6
Blue Pennant Sandstone (Coal Measures)	2815	1643	2.6
Penrith Sandstone (Permian)	2240	400	11.5
St Bees Sandstone (Bunter)	2160	343	25.8
Grinshill Sandstone (Keuper)	1952	266	23–26
Ardingly Sandstone (Cretaceous)	–	321–529	25–26

Data compiled from: Leary, 1986; *Natural Stone Directory*, 1990.

Bunter Sandstone from the English Midlands. The strength properties of many sandstones are similar to those of limestones of equivalent age, but are markedly lower than most granites and other igneous rocks. Caenozoic sandstones tend to be markedly weaker than Mesozoic or older sandstones. Notable exceptions are provided by the sarsen stones of southern England and northern France, and some of the more quartzitic sandstones from Fontainebleau near Paris, which possess exceptional strength for their (Tertiary) age.

Sandstones vary considerably in the frequency of bedding planes and joints. Parts of the Millstone Grit of northern England, and the Ardingly Sandstone of southeast England, are very massive with joints and bedding planes spaced five or more metres apart. Others, such as the micaceous flagstones already referred to, are very closely bedded. Beds of quartz-rich sandstone are often underlain or succeeded by bands of weaker, clay-rich sandstone, shales or mudstones, along which weathering is often concentrated. Even where such argillaceous bands are absent and the rock is uniformly quartz-rich, lithification is often variable with beds of well-cemented rock alternating with beds of more loosely cemented sands, which are much more easily weathered and eroded. The effects of unequal lithification and differential weathering are well displayed by the sandstone outcrops at Fontainebleau (Williams and Robinson, 1982).

MACRO LANDFORM FEATURES

Cliffs and Pillars

Clay-rich or silty sandstones, such as the Sandgate Beds of the Lower Greensand of southeast England, are generally only weakly cemented and have a low infiltration capacity because the clay and silt particles tend to clog the pores. In temperate Europe they are on the whole very susceptible to weathering and erosion and form low relief, in contrast to arid parts of the world where they can give rise to distinct cuestas. Because they readily support surface runoff they generally support quite dense stream networks.

Weakly cemented but more quartzose sands and sandstones such as the Folkestone Beds frequently form areas of higher ground in temperate Europe, seemingly because their high porosity greatly restricts runoff and erosion. First order valleys are generally streamless or carry runoff only in very wet weather.

Hillslopes on the above types of sandstone are rarely steep and almost always soil-covered. Surface exposures of bare rock are very unusual. In contrast, massive and more strongly cemented sandstones generally form steep slopes, and often bare cliffs or isolated pillars of rock, especially on the sides of deeply incised valleys or around the edges of plateaus. Such cliffs and pillars are a characteristic feature of sandstone terrains throughout the world from Antarctica to the humid tropics, but are especially prominent in arid areas, as Young and Young (1992) have pointed out. In Europe their widespread occurrence is often obscured by woodland or other vegetation. Many of the cliffs developed in the Ardingly Sandstone of southeast England, for example, are completely hidden in dense woodland (Robinson and Williams, 1976 and 1981). Woodlands also mask many cliffs on the Triassic sandstone outcrops in the Vosges of France and the Pfazerwald of Germany. Even the spectacularly high cliffs of Cretaceous sandstone in Saxony (Germany) and Bohemia (Czech Republic) are partly obscured by forest.

The height and scale of many sandstone cliffs is impressive. In Britain, cliffs of Ardingly Sandstone reach 15 m in height, some of those developed in Millstone Grit top 30 m in the Pennines, while some of those developed in New Red Sandstones are of a similar height. In mainland Europe, cliffs of 20 m or more occur in both the Vosges and the Pfazerwald, whilst the sandstone cliffs of Saxony and Bohemia have individual faces that sometimes exceed 60 m, and the total height of the cliffs in some cases is more than 200 m.

The height of sandstone cliffs in Europe seems more related to the thickness of the individual beds than to the strength of the rock. The sandstones of Saxony and Bohemia, for example, are surprisingly weakly cemented. It is possible to detach grains of sand from the surface of some cliffs merely by gentle rubbing, and one can quickly bore a deep hole in the surface by rotating a sharp metal object, such as a screwdriver. The Ardingly Sandstone of southeast England is also extremely soft (Robinson and Williams, 1976 and 1981).

Sandstone cliffs in Britain and Europe vary greatly in profile. Some have a smooth, vertical face, which is broken only by tight angular joints, and have a flat summit plane or platform, which makes a sharp, almost right-angled junction with the vertical face below. Others are much more rounded and bulbous in form, especially on top, and have widened joints and bedding planes that are hollowed out and weathered back at the edges. It is tempting to relate this contrast to differences in the toughness of the rock and the dominant processes that have shaped the cliffs. Compare, for example, some of the cliffs of Millstone Grit, such as Stannage Edge to the west of Sheffield, which are sharply angular, and have a great jumble of mostly angular blocks at their base, with the more bulbous forms of the Ardingly Sandstone cliffs of southeast England. The contrast in form could be due merely to the difference in toughness between the hard, relatively impervious gritstone and the much weaker Ardingly Sandstone. Also, the frostier climate of the Pennines compared with the southeast, both today and perhaps also during the Quaternary cold stages, may account for some of the difference. However, explanation of the contrast is complicated by the fact that both sandstones are remarkably resistant to frost action (Robinson and Williams, 1981; Sampson, undated), and some gritstone cliffs are unusually rounded, for example Plumpton Rocks and Brimham Rocks in Yorkshire.

Sandstone cliffs also differ greatly in plan. Although some cliffs form near-vertical walls, broken only by lines of tight vertical joints, others are more varied in form with deep embayments and protruding buttresses. The floors of some embayments have reverse slopes and appear to have been formed by massive rotational landslips involving the collapse and outward movement of whole sections of cliff face. Several examples of this phenomenon can be seen on the Millstone Grit in northern Britain, notably in Longdendale (Johnson, 1980) and at Alport Castle (Goudie and Gardener, 1985) in Derbyshire. In the Czech Republic there are excellent examples near the famed sandstone arch of Pravická brána. Others have no reverse slope and may result from the sliding forward, downslope, of large blocks or sections of cliff at some time in the past (Young and Young, 1992). In part at least, this difference in the form of embayments is probably related to the dip of the sandstone beds with respect to the cliff face (Schmidt, this volume, Chapter 21), but this has not been systematically studied.

Many cliffs are dissected by widened vertical joints forming open clefts or passageways (Figure 22.1) that extend a short distance back into the cliff face or run parallel to the face a short distance behind. On some outcrops a complex labyrinth of passageways has developed through which it is possible to walk (Robinson and Williams, 1976 and 1981). In Britain these widened joints are often called gulls or wents, which are traditional terms used by quarrymen.

In some areas the joint widening has been caused by downslope movement and separation of the intervening joint blocks. Such movement is greatly facilitated if the dip of the rock is valleywards and if the underlying strata are relatively impermeable and yielding when wet, or are composed of loose sand or silt.

Downslope movement is particularly evident in the case of the Ardingly Sandstone outcrops of southeast England which overlie the Wadhurst Clay. Often, hollows in the faces of the joint blocks forming one side of each widened joint match protrusions

Figure 22.1 A widened joint forming an open gull through an outcrop of Ardingly Sandstone near West Hoathly, southeast England

in the faces of the blocks on the other side, and vice versa, thus demonstrating that the blocks once fitted together and have maintained vertical alignment as they have moved. In other cases, however, it is clear that the blocks have rotated and become backtilted or else have tilted forward. Sometimes blocks have moved so far and undergone so much backtilting that they now rest on the footslope beneath the cliffs with their bedding lying at right angles to the angle of slope. Likewise, blocks which have overtoppled forward can be seen in front of a number of cliffs. In a few cases, individual joint-bounded blocks of rock stand in total isolation separated by at least a metre and sometimes several metres from any other sandstone block.

Dating of the movement is problematical, but most authors suggest that it is not occurring at the present day, and that it probably developed under periglacial conditions, most likely as permafrost was melting (Hollingworth et al., 1944; Gallois, 1965). This is supported by infillings of loess and the recovery of the remains of cold-loving animals from within similar gulls developed in the Hythe Beds of southeast England (Dines et al., 1969; Worssam, 1981). Others have suggested that movement might have occurred at some time in the past when rainfall was higher (Bird, 1964).

At Fontainebleau, some blocks appear to be still moving apart. Here, the sandstone forms a cap rock on linear ridges and the opening up of the joints seems to result from a loss of loose unconsolidated sand from beneath the sandstone causing under-mining and downslope movement (Williams and Robinson, 1982). Where the vegetation cover of the slopes is broken, erosion of the loose sand and undermining of the summit sandstone still occurs.

On many outcrops, it appears that the joint faces have weathered back, rather than moved apart. Where the joints intersect the cliff faces, the weathered debris, mostly loose sand, is washed away during heavy rains, thus forming the open passageways. Further into the cliffs, the widened joints are infilled with debris and are impassable.

The weathering could have occurred either before or during the formation of the cliffs, and could be due to enhanced chemical activity, frost action, or a combination of the two. Joints act as channels for downward percolating groundwater, which doubtless penetrated into the adjacent rock, and thus could be expected to have accelerated chemical weathering, particularly of any included feldspars, micas or soluble cementing agents. At the same time, the flow of water would have tended to flush the soluble products of weathering away, thereby ensuring that reactions did not stop due to the establishment of chemical equilibria. During periods of extreme cold, the greater amounts of water in the joints could also make them potential centres for intense frost action.

Although this explanation has the virtue of simplicity, it encounters many prob-lems. The walls of many passageways, for example, have iron-rich casings and have evidently been foci for the deposition of solutes passing through the sandstone rather than sites of weathering. Also, many sandstones are remarkably resistant to frost weathering even when saturated at normal temperature and pressure.

A variant explanation holds that the passageways have been opened up by weather-ing along narrow zones of closely spaced joints. It is suggested that these zones intersect with one another, separating 'islands' of more massive rock with widely spaced joints. Differential weathering along such closely spaced joints appears to be a major factor in the development of the maze of passageways which isolate massive towers of relatively unjointed sandstone in the so-called 'rock cities' in Saxony and the Czech Republic, at, for example, Tiské Stěny, Malá Skála and Prachovské Skály (Figures 22.2 and 22.3). The far ends of the passageways, in the interior of the rock outcrops, often expose unusually close jointed, somewhat broken sandstone.

Differential weathering of sandstone bedrock with variable joint spacing was used by Palmer and Radley (1961) to explain the development of gritstone tors in the English Pennines under intense periglacial frost action. They envisage that hillsides developed in zones of closely spaced joints suffered more rapid erosion, due to intense frost riving of the bedrock and gelifluction of the resulting debris, than areas with more widely spaced joints where the bedrock was less susceptible to frost action, and survived to give rise to impressive valley-side crags or 'tors'. Linton (1964) agreed that differential spacing of joints was important in the differential survival of the gritstone, but argued, by analogy with his ideas about the evolution of granite tors (Linton, 1955), that the closely jointed areas of gritstone were first rotted by deep chemical weathering and then stripped of their weathered regolith later.

Despite Linton's assertions, it seems likely that many valley-side cliffs and crags have a one- rather than two-stage origin. Obviously, there has to be a period of incision when the sandstone is breached and dissected by streams. The rate of incision has to be greater than the rate at which weathering and mass transport processes can wear back the sandstone in order that a free face can develop without any soil cover. The sandstone clearly has to have sufficient strength to maintain a free face. The high permeability of most sandstones, either through pores or down joints and along bedding planes, and the relative inertness of quartz probably assists in this process.

Figure 22.2   Isolated towers of sandstone at the Bastei, a famous viewpoint overlooking the River Elbe in eastern Germany

Weathering produces relatively few clasts of coherent debris, mostly just grains of sand that fall away from the exposed face. The greater the incision, the drier the sandstone tends to become, and moisture drawn to the surface case-hardens the surface which protects rather than weakens the rock (Robinson and Williams, 1987). Thus, once developed, cliffs and crags may well increase their resistance to weathering and erosion, and become relatively persistent landscape features.

In some areas of the Pennines weathering and erosion have very largely removed what were once continuous sheets of jointed sandstone, leaving only a few isolated pillars of rock, which may be remnants of the largest and most resistant of the original joint blocks. Parts of Brimham Rocks in Yorkshire may represent this stage of destruction, as may isolated blocks such as Mother Cap on Hathersage Moor and the Eagle Stone on the backslope of the Baslow Edge cuesta in Derbyshire.

The extent to which a sandstone outcrop is broken up into isolated pillars depends upon the joint pattern of the rock and other factors. If the sandstone is massive and

Figure 22.3 Milenci (the lovers), a spectacular, 60 m high tower of cross bedded sandstone at Adrpach 'rock city' near Broumov in northern Bohemia. Weathering along subsidiary joints within the tower has created a deep 'window'

coherent, with only a few widely spaced joints, the formation of pillars or other detached blocks may be limited to sites where blocks slip or tilt forward due to undermining. However, if valley incision has dissected the sandstone layer into several isolated units, retreat of cliffs from all sides where incision has occurred will create buttes and mesas, and eventually pillars. Also, if weaknesses exist in the sandstone due to close jointing patterns, or if the sandstone sits on a very plastic or yielding clay over which it can slide, extensive opening up of the joints will tend to occur, leading to the isolation of numerous pinnacles.

In temperate regions, the evolution of sandstone scenery cannot ignore Quaternary climatic change. It might be envisaged that the development and later melting of permafrost greatly assisted the opening up of many joints due to cambering of the sandstone over yielding underlying clays (Worssam, 1963; Gallois, 1958). Increased incidence and severity of frost may have been responsible for enhancing the weathering of some of the closely jointed sandstones. Alternatively, rotting of closely jointed sandstone may have been enhanced by warm, moist, inter-glacial conditions. Freezing of pore water within the sandstones may have facilitated incision by rivers charged with meltwater. Solifluction of weak strata above or below sandstone beds may have accentuated exposure of the sandstone from beneath a soil cover. However, in making such suggestions, it is essential to remember that sandstone cliffs are widespread in environments never affected by intense frost action or periglaciation and, even within the temperate zone, are well developed in areas such as the Sydney Basin, Australia, which have never suffered from a periglacial climate (Young and Young, 1992).

Basal Undercutting

Many sandstone cliffs are strikingly undercut towards their base (Figure 22.4), sometimes with caves. Where undercutting affects isolated boulders or pillars, so-called mushroom or pedestal rocks are formed (Figure 22.5). Explanations of the undercutting include: the presence of softer, more easily weathered bands of rock; abrasion by wind-blown sand; salt weathering, possibly by salts drawn up by capillary action from the soil-covered talus forming the footslope at the base of the cliffs, and enhanced rotting of the sandstone due to rising damp from the soil or talus. A further possibility is that the undercutting is the result of sub-surface, rather than sub-aerial, weathering followed by footslope lowering (Robinson and Williams, 1976; Twidale and Campbell, 1992). Whatever the process, it would seem that increased moisture is almost certainly involved for it is very noticeable that wherever water seeps out of bedding planes on cliff faces, deep hollows and linear cavities tend to develop across the faces of the cliffs. Moreover, churches built of sandstone blocks invariably show enhanced weathering where the blocks are in contact with damp soil.

Sandstone Pavements

Sandstone surfaces that are approximately horizontal and devoid of soil cover are known as pavements, or 'platières'. They are particularly well developed at Fontainebleau (Figure 22.6) where they occupy large areas of the sandstone-capped ridge crests (Williams and Robinson, 1982). Like limestone pavements, they are broken by vertical joints into rectangular, subangular or rhomboidal blocks depending

Figure 22.4 Cliffs of Ardingly Sandstone at Bowles Rocks near Groombridge, southeast England, showing basal undercutting The cliffs are some 12 m high

Figure 22.5 Le Champignon, a pedestal rock in the Fontainebleau Forest near Barbizon. The rock is approximately 2 m in height and the upper surface is patterned by polygonal cracking

Figure 22.6 A sandstone pavement (platière) capping a ridge top in the Fontainebleau Forest. Widened joints cross the pavement, forming open clefts 1 to 2 m deep. A shallow rock basin filled with rainwater is visible in the foreground

upon the jointing pattern of the rock. If the bedding of the sandstone is dipping down a hillside, the surface of the pavements will normally mirror the angle of dip (Robinson and Williams, 1992). The joints vary in width from hair-line cracks to gaping fissures a metre or more across. These fissures may be entirely infilled with weathered debris and soil, or gape open to a depth of a metre or more. The upper surfaces of the joint-bound blocks may be case-hardened and exhibit a variety of micro-weathering features.

The origin of the pavements remains uncertain. Unlike in limestone pavements, the joints cannot have been opened up by solution nor can their exposure be a product of glaciation for they are well developed on outcrops well beyond the limits of glacial influence. Twidale (1980) suggested that sandstone pavements developed in the Drakensberg Mountains of South Africa began forming beneath the ground surface as a result of moisture attack on the bedrock and that the weathered debris was later stripped off by erosion. In many European examples, the joints appear to have opened up as a result of the sandstone blocks moving apart as a result of cambering.

MICRO-WEATHERING FEATURES

Surface Crusting

One of the most important characteristics of many porous sandstones is their tendency to case-harden owing to the development of a surface crust or rind (Robinson and

Williams, 1987). Laboratory experiments suggest that on soft sandstones the crusting results from the deposition of solutes by pore water that has been drawn to the surface of the rock and has then evaporated (Paraguassu, 1972; Whalley, 1978).

The crusts vary in colour, but are invariably darker than the fresh rock. The colouring results from the inclusion of carbon compounds and the deposition of iron and, in some cases, manganese or titanium compounds. Some crusts are black or dark brown, others are grey. On iron-rich sandstones the crusts may be reddish or even pink (Robinson and Williams, 1992). However, the most important component of many, if not all, crusts is redeposited silica. This may form coatings around the surface grains of the sandstone, develop bridges spanning the pores between grains, or be precipitated as small irregular particles on the grain surfaces. The ingress of water into the rock is reduced by the blocking of the surface pores, and the increased cementation between the grains increases surface strength.

The formation of crusts and the importance of their strengthening properties is particularly marked in the case of the more friable, porous sandstones. When fresh, these sandstones have low strength due to weak bonding between the grains, yet when crusted they are much tougher and can form impressive cliffs. The crusts exercise profound control over the morphology of these cliffs and their micro-weathering features.

Crusts also form on harder sandstones, but are less obvious. In Britain, for example, they exist on Millstone Grit, Pennant and Old Red Sandstones. These rocks are much less porous, and it remains uncertain whether the crusts form primarily by the evaporation of solute-rich waters from within the rock, or whether they are composed largely of extraneous materials deposited on the surface of the sandstones by wind and rain. They appear to play a much less protective role than crusts on weaker sandstones, and may indeed be destructive.

Polygonal Cracking

Some sandstone outcrops are patterned by networks of shallow cracks that penetrate the rock surfaces perpendicularly irrespective of whether the surfaces are flat or sloping (Williams and Robinson, 1989). The cracks often extend less than 5 mm into the rock, and are seldom more than 10 mm deep. Most join at angles of about 120°, producing polygons that are either five or six-sided. However, in some localities, mainly on steeply sloping or vertical faces, the cracks intersect at right angles, forming square or rectangular polygons. The majority of polygons are between 100 and 200 mm in diameter, although miniature polygons as small as 10 mm and coarser patterns up to 1 m in diameter also exist. Initially, the cracks are narrow, but, with time, weathering tends to wear back the edges, causing them to widen.

Polygonal cracking is most often recorded and is most extensively developed on sandstones, but it also occurs quite commonly on granites, and occasionally on other rocks such as basalts, andesites and limestones (Williams and Robinson, 1989). It is most typical of hot, arid or seasonally arid regions, but it is also found in areas of humid temperate climate, for example in Britain and Western Europe. It is a strangely localised phenomenon, being common on some sandstone outcrops and yet entirely absent from others with seemingly similar characteristics. In Britain, good examples of polygonal cracking can be found on the Fell Sandstone cliffs in Northumberland. Cracking is rare on Millstone Grit outcrops, but is well developed at 'Whipsnade' on

Kinderscout. It is quite common, but never extensive, on the Ardingly Sandstone cliffs in Kent and Sussex (Robinson and Williams, 1976 and 1981) and on the New Red Sandstone cliffs in Cheshire. In temperate Europe, spectacular cracking occurs on the sandstone outcrops of Fontainebleau, France (Robinson and Williams, 1989), and good examples can be seen on the Externsteine cliff in north Germany. For some reason, cracking is extremely rare on the extensive sandstone outcrops of Saxony and Bohemia.

The causes of polygonal cracking remain uncertain. Although a few investigators have suggested that it is associated with pre-existing structures in the rock surface, most writers agree that it is a surface weathering phenomenon. It has been suggested that it is a product of insolation weathering (Mainguet, 1972), frost weathering (Franzle, 1971), and wetting and drying, particularly of sandstones containing clay (Netoff, 1971). Twidale and Bourne (1975) have suggested that it is a sub-surface weathering phenomenon associated with deep weathering.

In a detailed review, Williams and Robinson (1989) concluded that none of these theories satisfactorily explains either the world distribution of polygonal cracking or its spatial distribution on the rock outcrops on which it occurs. They argued that the only factor common to all polygonised surfaces is the presence of a case-hardened crust, and suggested that the cracking might be a direct result of case-hardening, possibly due to the drying and shrinkage of silica deposited in the crust by evaporating pore water as a gel or opal. Alternatively, they suggested that it could be an indirect result of crusting and the enhanced stresses that develop because of the different physical and mechanical properties of the crust and the underlying rock.

The association between case-hardening of sandstones and the development of polygonal cracking is acknowledged by Young and Young (1992). They, however, favour an idea of Branagan (1983), who likened the cracking to the crazing that affects glazed pottery as it ages. They argue that the cracking results from surface strain overcoming the strength of ageing crust due to fatigue effects. They note, however, that the crazing seen in pottery lacks the regular polygonal form that is so characteristic of sandstones and other rock surfaces.

Alveole and cavernous weathering

Honeycombing

Many vertical and sloping faces of sandstone become pitted by masses of roughly circular hollows or alveoles, commonly referred to as honeycomb weathering (Figure 22.7). The hollows vary in size, but rarely exceed 100 mm in diameter, and are often so closely spaced that only a thin wall of rock separates one hollow from another. Within the hollows the rock tends to be rather soft and friable, whereas the walls are harder and often crusted. The hollows are generally U-shaped in cross section, although some widen a little inwards behind the hardened outer rim of the surrounding wall.

Honeycomb weathering attacks many types of rock, but is most frequently and extensively found on sandstones, especially the weaker and more porous varieties, such as the Ardingly Sandstone and the Cretaceous sandstones of Saxony and

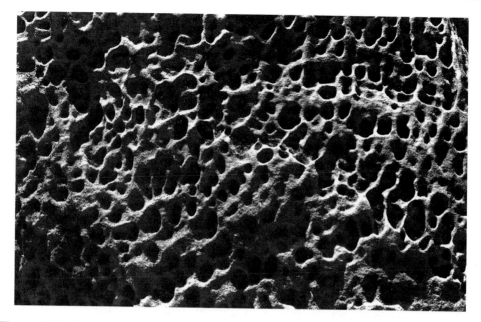

Figure 22.7 Honeycomb weathering on a cliff face in Chiddingly Wood, near West Hoathly, southeast England. The hollows vary between 100 and 300 mm in diameter

Bohemia. It is much less common on tougher harder sandstones such as Millstone Grit.

Honeycomb weathering is often attributed to salt action, but there is some uncertainty over whether the salts induce physical or chemical weathering of the stone. According to most investigators (e.g. Mustoe, 1982) the growing salt crystals prize surface grains of sand apart, causing physical destruction of the sandstone, but Young (1987) has suggested that the salts increase the dissolution of silica by modifying the pH and therefore the solubility of silica. Both processes may act together.

Gypsum and halite are commonly found in coastal rocks undergoing honeycombing and doubtless assist the weathering (see, for example, the studies of Gill, 1981; McGreevy, 1985; and Mottershead, this volume, Chapter 10). In many inland locations in Britain and Europe, however, little or no salt can be found in the hollows created by honeycomb weathering. An important exception is provided by the sandstone outcrops of Saxony and Bohemia where conspicuous efflorescences of gypsum and alum salts often line the hollows (Lentschig-Sommer, 1961). There can be little doubt that much, if not all, of the honeycombing on these particular outcrops is due to salt action. The problem is to explain the honeycombing in the many inland locations where salts are largely, if not entirely, absent, for example in southeast England. Many of the Ardingly Sandstone outcrops show conspicuous honeycombing, which in some localities is actively forming and in others is completely crusted over and seemingly fossil. Why there should be a change from active excavation to senescence at such sites remains unknown. Churches built of Ardingly Sandstone during the last century have often developed quite deep honeycombing, especially close to the soil surface, which suggests that rising damp and irregular surface crusting may be key factors.

Tafoni

Sandstones also exhibit larger forms of alveolar and cavernous weathering that are usually termed tafoni, a term originally introduced by Penck (1894) to describe hollows excavated in granites on the island of Corsica. Tafoni are well-rounded hollows or cavities with domed or scalloped roofs that develop in vertical or near-vertical faces of rock (Figure 22.8). They range in size from as little as 0.1 m to several metres in depth and diameter. The openings are often smaller than the cavity behind and are sometimes protected by a projecting 'hood' or 'visor', especially if they are developed in a rock that is case-hardened. Some tafoni penetrate right through boulders or slabs of rock, creating rounded shafts or windows. Many tafoni develop

Figure 22.8 Tafoni developed in a cliff face at Hrubá Skála in the Český ráj ('Bohemian Paradise'), Czech Republic

at the base of cliffs or on the face of cliffs where bedding planes emerge. Where they occur at the base of cliffs, they often grade into basal undercutting, and the two features may not be readily distinguishable.

Although common in sandstones, tafoni are especially plentiful in granites (Twidale, 1982), and also occur in trachyte, diorite, gneiss, schist and conglomerate. They seem to be rare and poorly developed in limestones, except in parts of North Africa (Smith, 1978). Thus, they are a less characteristic feature of sandstones than the smaller-scale honeycombing. They are very poorly developed in British sandstones, especially the harder varieties, although on the Millstone Grit what appear today to be inactive tafoni can be seen in exposures at a few locations, such as at Birchover in Derbyshire. On mainland Europe, they are quite common at Fontainebleau, and on some of the central European sandstone outcrops, such as at Tiské Stěny in the Czech Republic. Nowhere in Europe, however, do they seem as common as on exposures in some of the more arid areas of the world (Mainguet, 1972).

The origins of tafoni are uncertain, although, as with honeycombing, many writers attribute their development to salt weathering, and salt efflorescences combined with weathered debris have been found on tafoni surfaces in a number of localities (see, for example, Rondeau, 1961; Bradley *et al.*, 1978 and 1980; Mustoe, 1983). However, other writers are sceptical of the role of salts because not all tafoni contain significant quantities of salt (Dragovich, 1969; Conca and Rossman, 1985).

In sandstones and granites, several writers have related the development of tafoni to uneven case-hardening or crusting (Wilhelmy, 1964; Mabbutt, 1977; Winkler, 1975). Areas where the crust is well developed are able to resist erosion whilst areas where the crust is thin or absent are attacked, and develop into hollows. With time, the hollows extend laterally into the softer interior rock behind the areas protected by better developed crust.

Some sandstone tafoni-like forms may not be weathering features at all. From observations in quarries at Fontainebleau, Williams and Robinson (1982) described how some curved and hollowed-out forms, very similar to tafoni, were a consequence of differential cementation of sand during diagenesis.

Rock-basins (Weather-pits or Gnammas)

Rock-basins are closed depressions developed on flat or gently sloping rock surfaces that are found in every climatic environment from the tropics to the poles. They occur on a great variety of rock types, including limestones, granites, basalts and gneiss, but are particularly well developed on some sandstone outcrops. Most basins are circular or oval in plan and less than 500 mm deep. Very often they are flat-floored and steep sided, but some are more saucer shaped with gently sloping sides. Steep-sided basins frequently exhibit overhanging rims and undercut sides. Rainwater collects in the basins, and in some cases overflows, carving a distinct spillway across the lowest point of the rim. Over time, incision of the spillways allows less water to collect in the basins before overflow commences. In a few cases, the spillways are incised to the level of the basin floor and the basins are permanently drained. Between rains, any water in the basins evaporates, and even in the wetter parts of temperate Europe few basins are permanently filled with water. As weathering proceeds, basins may widen and coalesce with neighbouring basins to produce compound forms. In extreme cases, Twidale (1982) has described how all that may remain is a confused fretwork of rounded or angular rims and intersecting basins which he termed a 'meringue' surface, but this has not been observed by the authors on any of the sandstone outcrops in Europe.

Basins developed on sandstones tend to be most abundant on the highest and most exposed surfaces, especially on the summit edges of cliffs and pillars, although some are also found on sandstone pavements away from cliff edges. They are most widespread on sandstones of low porosity, such as the Millstone Grit, and are rare on more porous sandstones such as the Ardingly Sandstone.

On limestones, basins are usually attributed to solutional processes, and similar processes may be important in calcareous cemented sandstones. In granites and arkosic sandstones, hydrolysis of the feldspars is probably important. On the more quartzitic sandstones, however, the origin of the basins is problematical. Most workers agree that the dominant processes are chemical or biochemical dissolution of either the cement or the grains of the sandstones, but the precise mechanism remains unclear. Uncertainty also surrounds the role of mechanical processes. It has been suggested, for example, that frosts (Matthes, 1930; Dahl, 1966), or lichen (Franzle, 1971) help to excavate basins by prizing off flakes and grains of rock. On some sandstones it is clear that basins develop as a result of the breaching of the case-hardened outer layer and excavation of the weakened rock beneath.

Fluting (Pseudokarren)

Sub-parallel flutes and runnels resembling those developed on limestone are found on many of the more massively jointed sandstones in a wide variety of environments. They most frequently develop on steeply sloping rock surfaces, usually descending for a short way down the upper convex slopes of the most exposed edges of cliffs, buttresses and pinnacles. Some descend from rock basins and carry overflow water, but others descend from knife-edge ridges. Like rock-basins, they are rare on very porous sandstones, and most frequent on relatively impervious sandstones.

Fluting on sandstones has been little studied, and its origin remains something of a mystery (Williams and Robinson, this volume Chapter 24). During rains the flutes undoubtedly act as channels for water flow, but the extent to which the water aids weathering or simply helps to remove pre-weathered debris is uncertain. They are, however, a very characteristic feature of many sandstone outcrops, but one of the most difficult to study because of their development in very inaccessible locations.

CONCLUSION

Sandstone outcrops do not possess any unique morphological characteristics, but exhibit several features that are better, and more commonly, developed, than they are on any other rock type. As a result, many sandstone terrains exhibit a suite of features that combine to produce a distinct and characteristic morphology. As with the development of limestone karst, the relative development of the various features that characterise this morphology seems to depend on the mineralogical composition and porosity of the rock, the persistence of bedding planes and joints, the climate, both past and present, and the denudational history. Case-hardening of exposed surfaces seems to play a particularly important role in determining the detailed morphology and range of weathering forms that develop. The most characteristic sandstone landforms are found on silica-rich, massive sandstones; the sandstones giving least character to the areas where they outcrop are those containing appreciable quantities of clay or silt. Feldspathic sandstones (arkoses) tend to produce landform assemblages very similar to those of many granites to which their composition is closely related. There is increasing evidence that the dissolution and mobility of silica and other chemical constituents of sandstones is very important in the evolution of sandstone landforms, and that the role of physical weathering has possibly been overemphasised.

REFERENCES

Bell, F. G. (1983). *Engineering Properties of Soils and Rocks*. Butterworths, London.

Bird, E. C. F. (1964). Tor-like sandstone features in the central Weald. *Abstract 1156, 20th International Geographical Congress*, London.

Blyth, F. G. H. and de Freitas, M. H. (1984). *A Geology for Engineers*, 7th edition. Arnold, London.

Bradley, W. C., Hutton, J. T. and Twidale, C. R. (1978). Role of salts in the development of granite tafoni, South Australia. *Journal of Geology*, **86**, 647–654.

Bradley, W. C., Hutton, J. T. and Twidale, C. R. (1980). Role of salts in granite tafoni, South Australia, a reply. *Journal of Geology*, **88**, 121–122.

Branagan, D. F. (1983). Tessalated pavements. In Young, R. W. and Nanson, G. C. (eds), *Aspects of Australian Sandstone Landscapes*, Australian and New Zealand Geomorphology Group, Wollongong, pp. 11–20.

Conca, J. C. and Rossman, R. R. (1985). Core softening in cavernously weathered tonalite. *Journal of Geology*, **93**, 59–73.

Dahl, R. (1966). Block fields, weathering pits and tor-like forms in the Narvik Mountains, Nordland, Norway. *Geografiska Annaler*, **45A**, 55–85.

Dines, H. G., Buchan, S., Holmes, S. C. A. and Bristow, C. R. (1969). *Geology of the Country Around Sevenoaks and Tonbridge*. Memoir Geological Survey Great Britain, HMSO, London.

Dragovich, D. J. (1969). The origin of cavernous surfaces (tafoni) in granitic rocks of southern South Australia. *Zeitschrift für Geomorphologie*, NF13, 163–181.

Franzle, O. (1971). Die opferkessel im quartzitischen sandstein von Fontainebleau. *Zeitschrift für Geomorphologie*, NF15, 212–235.

Gallois, R. W. (1965). *The Wealden District*. British Regional Geology, HMSO, London.

Gill, E. D. (1981). Rapid honeycomb weathering (tafoni formation) in greywacke, S.E. Australia. *Earth Surface Processes and Landforms*, **6**, 81–83.

Goudie, A. and Gardener, R. (1985). *Discovering Landscape in England and Wales*. George Allen and Unwin, London.

Hettner, A. (1887). Gebirgsbau und oberflächengestaltung der Sächsischen Schweiz. *Forsch. zur D. Landes-und Volkskunde*, **2**, 245–355.

Hollingworth, S. E., Taylor, J. H. and Kellaway, G. A. (1944). Large-scale superficial structures in the Northamptonshire Ironstone field. *Quarterly Journal Geological Society*, **100**, 1–35.

Johnson, J. H. (1980). Hillslope stability and landslide hazard—a case study from Longendale, north Derbyshire, England. *Proceedings Geological Society, London*, **91**(4), 315–326.

Leary, E. (1986). *The Building Sandstones of the British Isles*. Building Research Establishment Report, Watford.

Lentschig-Sommer, S. (1961). Ein vorkommen von Alunogen (Keramohalit) im Elbsandsteing-
ebeit. *Jb. Staatl. Mus. Mineral Geol.*, 109–110.

Linton, D. L. (1955). The problem of tors. *Geographical Journal*, **121**, 470–487.

Linton, D. L. (1964). The origin of Pennine tors; an essay in analysis. *Zeitschrift für Geomorphologie*, NF**8**, 5–24.

Mabbutt, J. A. (1977). *Desert Landforms*. MIT Press, Cambridge, Massachusetts.

Mainguet, M. (1972). *Le modèle des gres*. IGN, Paris.

Martonne, E. de (1909). *Traité de Géographie physique*. Armand Colin, Paris.

Matthes, F. E. (1930). Geologic history of the Yosemite Valley. *United States Geological
Survey, Professional Paper*, **160**.

McGreevy, J. P. (1985). A preliminary scanning electron microscope study of honeycomb
weathering of sandstone in a coastal environment. *Earth Surface Processes and Landforms*,
10, 509–518.

Mustoe, G. E. (1982). The origin of honeycomb weathering. *Bulletin Geological Society of
America*, **93**, 108–115.

Mustoe, G. E. (1983). Cavernous weathering in the Capitol Reef desert, Utah. *Earth Surface
Processes and Landforms*, **8**, 517–526.

Natural Stone Directory–1991, 8th edition (1990). Stone Industries, Ealing Publications,
Maidenhead.

Netoff, D. I. (1971). Polygonal jointing in sandstone near Boulder, Colorado. *The Mountain
Geologist*, **8**, 17–24.

Palmer, J. and Radley, J. (1961). Gritstone tors of the English Pennines, *Zeitschrift für
Geomorphologie*, NF**5**, 37–52.

Paraguassu, A. B. (1972). Experimental silicification of sandstone. *Bulletin Geological Society
America*, **83**, 2853–2858.

Penck, A. (1894). *Morphologie der Erdoberflache*. Stuttgart (Vol. 1, p. 214).

Pettijohn, F. J. (1957). *Sedimentary Rocks*. Harper, New York.

Pettijohn, F. J., Potter, P. E. and Siever, R. (1972). *Sand and Sandstone*. Springer-Verlag,
Berlin, Heidelberg.

Robinson, D. A. and Williams, R. B. G. (1976). Aspects of the geomorphology of the
sandstone cliffs of the central Weald. *Proceedings Geological Association*, **87**, 93–100.

Robinson, D. A. and Williams, R. B. G. (1981). Sandstone cliffs on the High Weald landscape.
Geographical Magazine, **53**, 587–592.

Robinson, D. A. and Williams, R. B. G. (1987). Surface crusting of sandstones in southern
England and northern France. In Gardiner, V. (ed.), *International Geomorphology 1986*,
part 2, Wiley, Chichester.

Robinson, D. A. and Williams, R. B. G. (1989). Polygonal cracking of sandstone at Fontainebleau, France. *Zeitschrift für Geomorphologie*, NF**33**, 59–72.

Robinson, D. A. and Williams, R. B. G. (1992). Sandstone weathering in the High Atlas,
Morocco. *Zeitschrift für Geomorphologie*, NF**36**, 413–429.

Rondeau, A. (1961). *Recherches géomorphologique en Corse*. Armand Colin, Paris.

Sampson, D. N. (undated). The geology of the Edale–Kinderscout area. In Nunn, P. J. (ed.),
Rock Climbs in the Peak: Volume 7—The Kinder Area, British Mountaineering Council, pp.
22–25.

Smith, B. J. (1978). The origin and geomorphic implications of cliff foot recesses and tafoni
on limestone hamadas in the northern Sahara. *Zeitschrift für Geomorphologie*, NF**22**, 21–43.

Sparks, B. W. (1971). *Rocks and Relief*. Longman, London.

Twidale, C. R. (1980). Origin of some minor sandstone landforms. *Erdkunde*, **34**, 219–224.

Twidale, C. R. (1982). *Granite Landforms*. Elsevier, Amsterdam.

Twidale, C. R. and Bourne, J. A. (1975). The subsurface initiation of some minor granite
landforms. *Journal Geological Society Australia*, **22**, 477–484.

Twidale, C. R. and Campbell, E. M. (1992). On the origin of pedestal rocks. *Zeitschrift für
Geomorphologie*, NF**36**, 1–14.

Whalley, W. B. (1978). Scanning electron microscope examination of a laboratory simulated
silcrete. In Whalley, W. B. (ed.), *Scanning Electron Microscopy in the Study of Sediments*,
Geo-Abstracts, Norwich, pp. 399–405.

Wilhelmy, H. (1964) Cavernous rock surfaces (tafoni) in semi-arid and arid climates. *Pakistan Geographical Review*, **19**, 9–13.

White, I. D., Mottershead, D. and Harrison, S. H. (1984). *Environmental Systems: An Introductory Text*. George Allen and Unwin, London.

Williams, R. B. G. and Robinson, D. A. (1982). Sandstone sculptures in the Fontainebleau woods. *Geographical Magazine*, **54**, 572–579.

Williams, R. B. G. and Robinson, D. A. (1989). Origin and distribution of polygonal cracking of rock surfaces. *Geografisker Annaler*, **71A**, 145–159.

Winkler, E. M. (1975). *Stone Properties, Durability in Man's Environment*. Springer-Verlag, Wien.

Worssam, B. C. (1963). *Geology of the Country around Maidstone*. Memoir Geological Survey Great Britain, HMSO, London.

Worssam, B. C. (1981). Pleistocene deposits and superficial structures, Allington Quarry, Maidstone, Kent. In Neale, J. and Fenley, J. (eds), *The Quaternary in Britain*, Pergamon, Oxford.

Young, A. R. M. (1987). Salt as an agent in the development of cavernous weathering. *Geology*, **15**, 962–966.

Young, R. and Young, A. (1992). *Sandstone Landforms*. Springer-Verlag, Berlin, Heidelberg.

23 Classification of Dartmoor Tors

JUDY EHLEN
US Army Topographic Engineering Center, Alexandria, Virginia, USA

ABSTRACT

Forty-eight Dartmoor tors were classified using field and laboratory measurements of various geomorphic, petrographic and structural factors. The data were evaluated using: (1) non-parametric correlations, (2) joint spacing frequency distributions, (3) variable spatial distributions, and (4) multivariate statistics. Analyses of correlations, joint spacing frequency distributions and variable spatial patterns allowed definitions by landform type (summit tors, spur tors, and valleyside tors), and multivariate analysis indicated the relative importance of individual variables in defining the different tor types. The information provided by the joint spacing frequency distributions and interpretation of the variable spatial patterns is semi-quantitative, thus allowing a semi-quantitative classification.

INTRODUCTION

There are many examples in the literature on granite landforms that relate structural, petrographic and geomorphic characteristics to landform. The relation between joints and granite landforms, for instance, is frequently mentioned. Joints are usually described as controlling the general outlines of the landforms. Water enters the rock through the joints, which become locations for more intense chemical weathering, which in turn produces the characteristic rounding of the landforms (e.g. Linton, 1955; Waters, 1957; Brunsden, 1964; Thomas, 1974). Certain landforms are defined in terms of joint spacing, e.g. domes develop only where joints are very widely spaced (e.g. Ormerod, 1859 and 1869; Mabbutt, 1952; Twidale, 1964 and 1982). Ranges for joint spacing are occasionally given, but appear to be based on casual observation, not measurement (e.g. de la Beche, 1839; Caine, 1967; Brunsden and Gerrard, 1970). Most studies refer to vertical or steeply dipping joints, and there is little reference to the systematic horizonal joints that are an essential part of the characteristic ortho-gonal joint pattern in granitic rocks. Mention of sheeting joints, however, is common. Spacing between horizontal joints is typically described as increasing with depth (e.g. Jahns, 1943; Oen, 1965; Hawkes, 1982).

Chapman and Rioux (1958), Dumanowski (1964) and Snow (1968) presented quantitative data on joint spacing, but only in passing. Thorpe (1979), however, measured joint spacing and distribution in an underground cavity in the Stripa granite

Rock Weathering and Landform Evolution. Edited by D. A. Robinson and R. B. G. Williams.
© 1994 John Wiley & Sons Ltd

in Sweden, as did Bourke *et al.* (1981 and 1982) in bore-holes in the Carnmenellis granite in Cornwall. Whittle (1989) presented more recent and comprehensive data, both surface and subsurface, collected on the Carnmenellis granite in conjunction with the Hot Dry Rock geothermal project. Only Gerrard (1974 and 1978) and Ehlen and Zen (1990) have reported detailed fracture spacing data. Gerrard (1974 and 1978) described the joint patterns on Dartmoor and Bodmin Moor and evaluated the relations between jointing and various geomorphic factors. Ehlen and Zen (1990) reported fracture spacings and modal analyses for several types of granitic rocks in the United States and on Dartmoor.

The mineralogical and textural characteristics of granitic rocks also affect landforms. Lithologic or petrographic boundaries often coincide with landform boundaries (e.g. Demek, 1964; Jeje, 1973). Dumanowski (1964 and 1968), Brook (1978), Robb (1979) and Pye *et al.* (1986) determined that inselbergs tend to contain abundant potassium feldspar whereas rocks in areas without inselbergs tend to be low in potassium feldspar. Gibbons (1981) and Pye *et al.* (1984) found porphyritic rocks are more resistant than non-porphyritic rocks. Gibbons (1981) also found that tor densities increased in coarser-grained rocks.

With respect to classification using these factors, Linton (1955) referred to tors according to topographic position, and offered a series of hypotheses for the formation of certain types of landforms, some of which are topographically dependent. He believed these landforms were formed by a two-stage process. Linton did not explain how the landform types differ with respect to composition, grain size, or texture, but stated that 'There is every indication that the main factor at work on Dartmoor . . . is the spacing of joints.' (p. 474).

Gerrard (1974) used a classification similar to Linton's in his evaluation of the Dartmoor tors. He classified the tors as summit tors, valleyside and spur tors, and emergent tors. Including tors on Bodmin Moor, Gerrard (1978) determined that there are two groups: (1) summit and valleyside tors and (2) emergent tors. Joints are more closely spaced and spacing is more variable in summit and valleyside tors. Gerrard included joint spacing, tor height, relative relief and landform type in his analysis.

Twidale (1980, 1981 and 1982) has also classified granite landforms. He separated inselbergs into three, genetically related types: bornhardts (domes), castle koppies and nubbins. Castle koppies develop in massive bedrock on what will be large-radius domes. Nubbins are block- or boulder-strewn residuals that develop on what will be small-radius domes. Jointing is important in this classification in that sheeting structure is dominant in nubbins whereas, although fractures are few, orthogonal joints are dominant in castle koppies. Castle koppies and nubbins develop subsurface, but under slightly different climatic conditions. The Dartmoor tors are classified as castle koppies.

This paper presents a summary of the results of four different approaches to sorting out the relations among this multiplicity of factors with respect to classifying granite landforms. All methods incorporate a landform classification similar to those presented by Linton (1955) and Gerrard (1974 and 1978). The results from the different approaches are combined to form a semi-quantitative classification of the Dartmoor tors.

VARIABLE DEFINITIONS AND PROCEDURES

The twenty-two variables comprise three groups: geomorphic variables, petrographic variables and structural variables. Table 23.1 lists the variables and their abbreviations. All variables were defined in Ehlen (1991 and 1992a), but the definitions are repeated here for ease of reference. The procedures used to obtain the data are briefly described for the same reason.

Table 23.1 Variables and their abbreviations

Geomorphic variables:	*Petrographic variables:*
Relative relief (REL)	Grain size:
Landform type (LF)	Rock (MRGS)
Joint control of landform (JC)	Quartz (MQUA)
	Potassium feldspar (MKSP)
Structural variables:	Plagioclase (MPLG)
Mean primary spacing:	
Vertical joints (MVPS)	Composition:
Horizontal joints (MHPS)	Per cent quartz (PQUA)
Mean secondary spacing:	Per cent potassium feldspar (PKSP)
Vertical joints (MVSS)	Per cent plagioclase (PPLG)
Horizontal joints (MHSS)	Per cent tourmaline (PTM)
Secondary horizontal/vertical joint spacing	Presence/absence schorl (SCHR)
ratio (SHVR)	Presence/absence tourmaline veins (VEIN)
	Presence/absence clay (CLAY)
	Rock texture:
	Grain size distribution (GSD)
	Number of megacrysts (PHEN)
	Per cent volume megacrysts (MEGA)

Geomorphic Variables

Three geomorphic variables were addressed: relative relief, landform and joint control of tor shape. Relative relief, determined from 1:25 000 scale topographic maps, is defined as the vertical distance between the outcrop and the nearest main stream within a horizontal distance of 800 m (Gerrard, 1974).

Three landform types, summit tors, spur tors and valleyside tors, were identified in the field and from 1:25 000 scale topographic maps. This classification does not directly correspond to those presented by Linton (1955) or Gerrard (1974 and 1978), but contains aspects of both and is based upon them. The landform types are defined in Table 23.2. Figure 23.1 shows the typical topographic position of each type.

Tor shape is usually controlled by one of three joint sets—a horizontal set that forms the top of the tor, and vertical sets (1) perpendicular to the face of the tor forming its sides, and (2) parallel to and forming the face of the tor. The variable 'joint control' identifies which type of joint, i.e. vertical or horizontal (see below), appears most important to the shape of the tor. If neither is dominant, joint control is equal or combined. Joint control was determined visually.

Table 23.2 Types of landforms

Tor type	Tor size	Tor location	Nearby slope
Summit	Large	On hill and ridge crests	Gentle
Spur	Typically small	Ends of ridges or spurs	Gentle
Valleyside	Large and massive to small ledges	Along valley sides, below the break in slope, usually on upper slopes	Steep above and below

Figure 23.1   Typical topographic positions of Summit (1), Valleyside (2) and Spur Tors (3)

Petrographic Variables

The petrographic variables include rock grain size, composition and texture. Grain size and most composition variables were determined microscopically on stained, cut slabs. Grain size was measured for quartz, potassium feldspar and plagioclase feldspar. The modal analyses also included tourmaline. Potassium feldspar grain size and modal per cent refer to groundmass; the large megacrysts are considered separately with respect to texture.

Additional composition variables are the presence or absence of clay, tourmaline veins and schorl. Clay, probably kaolinite, was identified in thin section. The presence of clay may be an indicator of degree of weathering and its presence may affect outcrop size and/or joint spacing. Tourmaline veins, which occur in closely spaced secondary vertical joints, typically form zones that are more weathered than the surrounding rock. Presence or absence was determined in the field. Schorl is an intergrowth of quartz and tourmaline. It occurs as small, usually rounded blebs that are very hard, and that are distributed unevenly throughout the rocks. Presence or absence, which might affect joint spacing, was determined in the field and visually from cut slabs.

Three variables address textural characteristics of the rocks. Megacryst counts for the variable 'number of megacrysts' were done in the field. The number of megacrysts longer than 2.5 cm within a standard botanist's quadrat ($0.25 \, m^{-1}$) on a typical surface was recorded. The variable 'per cent megacrysts' was determined microscopically and is thus on a volume basis. The variable 'grain size distribution' was determined visually from the stained, cut slabs. The rocks were classified as equigranular or megacrystic. Only potassium feldspar megacrysts are included in these variables; large quartz and plagioclase crystals are present, but are less than 2.5 cm in length.

Structural Variables

Joint spacing is the distance between successive joints in a given joint set, measured normal to the planes of the joints along a linear traverse of continuous outcrop. A joint set is a collection of individual joints with similar inclination that are essentially parallel. Primary joints are long, usually open, tor-shape-controlling joints that typically cross other joint traces. Secondary joints are shorter joints, local in extent, that rarely cross other joint traces. No genetic connotation is implied by these terms. Vertical joints are defined as dipping 70° or more, whereas horizontal joints dip 25° or less. Joints in the intermediate category are rare on Dartmoor.

Joint spacings were measured between approximately 7000 joints in 185 joint sets at 58 sample sites on 48 tors (Figure 23.2). The spacings, for both primary and secondary vertical joints and primary and secondary horizontal joints, are reported in Ehlen and Zen (1990). Mean joint spacing for a sample site was determined by averaging the spacings for all joints of one type regardless of joint set. Ratios between horizontal and vertical secondary joints are also included as a variable. The ratio was calculated using mean joint spacings.

DATA ANALYSIS

Introduction

Four approaches were used to evaluate the data: (1) statistical correlations, (2) joint spacing frequency distributions, (3) spatial distributions of the individual variables, and (4) multivariate statistical analyses. Each section below includes a brief description of the procedure, a note as to which variables were used and in what form, and a summary of the results of the analysis. Some of these results were published previously; sources are cited where appropriate. Because the variables differ slightly among procedures, the results described below for each procedure do not necessarily agree. For example, the actual grain size measurements were used in the calculation of correlation coefficients, but on the joint spacing frequency histograms, grain size is addressed in ranges which were coded. The way the results can be interpreted thus

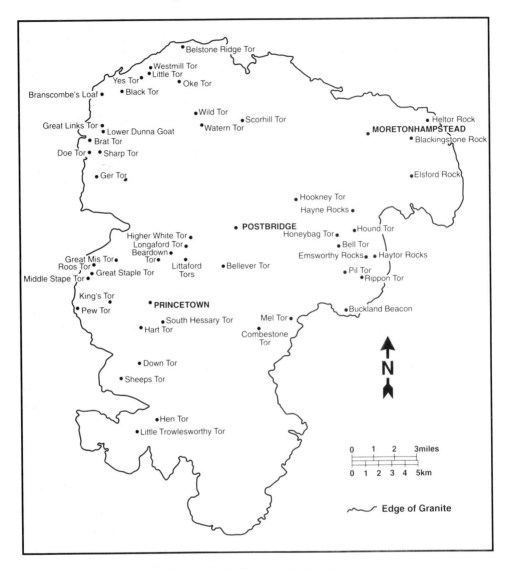

Figure 23.2   Sample site locations

differs slightly. In addition, not all variables were used in all procedures; exceptions are noted below. The data are shown in Table 23.3; the abbreviations used are in Table 23.1; and the codes used in Table 23.3 are explained in Table 23.4.

Correlations

Analysis of statistical correlations comprised the initial evaluation of the multitude of possible interrelationships between these variables, and is described in detail in Ehlen (1993). The non-parametric procedure Spearman's rank correlation coefficient was used because (1) it allows inclusion of binary and nominal variables, and (2) ranks are not necessarily affected by closed or constant-sum data, i.e. percentages (Rock, 1988). The following discussion refers to significant correlations at the 95% confidence level unless otherwise noted. The correlation matrix comprises Figure 23.3; Table 23.1 explains the variable abbreviations used in Figure 23.3.

In order to understand how the relations described below were determined, the coding system used for the nominal and ordinal variables must be explained. As stated above, Table 23.4 shows the codes assigned to these variables. For the binary variables, a positive correlation is with a higher number; a positive correlation indicates absence. Regarding the nominal variables, if a given variable is positively correlated with landform, for example, the landform type involved would be summit tors, which are coded with the highest number. High positive correlations with the nominal variables thus refer to summit tors, megacrystic rocks, and combined horizontal and vertical joint control of tor shape. Unfortunately, valleyside tors could not be evaluated using this procedure: the coding system was such that valleyside tors always appear to have characteristics intermediate between those of spur and summit tors. All variables except the secondary joint spacing ratio were used in the analysis of correlations.

Two types of landform characterisation are possible: (1) direct characterisation based on significant correlations with landform and (2) indirect characterisation based on correlations between two other variables, one of which is significantly correlated with landform.

Analysis of significant correlations indicates that summit tors have high relative relief and wide vertical joint spacing. Their shapes are usually controlled by vertical joints alone or horizontal and vertical joints in combination. The rocks are strongly megacrystic, schorl is typically absent, and tourmaline veins are rare. Indirect relations indicate summit tors are likely to have widely spaced horizontal joints and to be composed of coarse-grained rock that is high in plagioclase and tourmaline, but low in quartz and potassium feldspar. Clay is likely to be absent.

Spur tors have lower relative relief and narrower vertical joint spacing. They are controlled by horizontal joints. The rocks are feebly megacrystic or equigranular in texture, and contain both schorl and tourmaline veins. Indirect relations indicate horizontal joints in spur tors are likely to be closely spaced and that the rocks are likely to be finer grained, to contain clay, and to have low tourmaline and plagioclase abundances and high quartz and potassium feldspar abundances.

Joint Spacing Frequency Distributions

The analysis of the relations defined by the joint spacing frequency distributions can be found in Ehlen (1991). Frequency distributions were used to determine how the variables affected joint spacing (e.g. is there a difference in joint spacing between rocks with grain size <1 mm and rocks with 1–2 mm grain size?). The variables used are: joint type, grain size, grain size distribution, number of megacrysts, relative relief and landform. The variables were plotted in combination; Figure 23.4 shows one set of frequency histograms as an example, all histograms can be found in Ehlen (1989). In addition, all frequency distributions were compared using chi-square so that

Table 23.3 The data

Tor	Geomorphic variables			Structural variables					Petrographic variables Grain size			
	REL (m)	LF	JC	MVPS (cm)	MVSS (cm)	MHPS (cm)	MHSS (cm)	SHVR	MRGS (mm)	MQUA (mm)	MKSP (mm)	MPLG (mm)
Honeybag	140	3	1	126.8	69.3	30.4	15.7	0.23	2.68	2.91	2.90	3.15
Bell	40	1	2	205.1	56.8	39.5	12.1	0.21	2.55	2.72	2.61	2.79
Yes	70	3	3	146.4	41.8	43.2	16.5	0.40	1.42	1.26	1.81	1.18
Little[a]	100	3	1	126.3	56.3	26.4	8.7	0.15	1.45	1.55	1.62	1.23
West Mill	70	3	2	116.3	45.4	82.4	25.6	0.56	2.00	1.98	2.40	1.99
Great Staple	110	3	2	256.7	71.0	35.8	9.1	0.13	2.34	1.83	3.05	3.69
Hound	160	3	2	175.7	56.2	14.2	14.2	0.25	1.64	1.39	1.65	2.43
Hound	160	3	3	290.0	76.1	96.5	13.2	0.17	1.28	1.24	1.42	1.60
Hound	160	3	1	213.1	40.7	81.6	12.8	0.31	1.06	0.93	1.06	1.72
Blackingstone	40	2	1	585.6	192.3	106.5	15.3	0.08	1.61	1.93	1.53	2.12
Heltor	80	3	1	560.5	120.4	102.1	14.1	0.12	1.46	1.45	1.71	1.25
Elsford	159	2	1	392.1	137.4	55.2	10.2	0.07	1.87	2.00	2.08	2.32
Emsworthy[a]	60	1	1	958.5	125.0	123.7	15.0	0.12	1.83	1.87	2.42	1.94
South Hessary	50	3	3	227.2	45.6	82.0	12.2	0.27	1.67	1.67	2.03	1.35
Combestone	130	3	3	214.4	60.9	70.3	11.7	0.19	1.56	1.78	1.53	1.71
Scorhill	40	2	1	221.8	32.2	48.4	8.1	0.25	1.48	1.31	1.78	1.97
Hayne Rocks	120	3	3	250.7	67.2	91.0	16.9	0.25	1.60	1.61	1.63	2.28
Pil	190	3	1	245.0	69.9	67.9	10.0	0.14	1.99	2.31	2.26	2.33
Pil	190	3	1	300.1	78.5	87.0	13.5	0.17	1.99	2.31	2.26	2.33
Hookney	120	3	1	202.8	92.6	43.8	7.8	0.08	1.35	1.14	1.47	2.00
Bellever	90	3	1	462.3	105.1	71.3	10.6	0.10	1.84	1.81	2.23	1.70
Belstone Ridge[b]	120	1	3	78.0	31.0	80.0	20.8	0.67	1.49	1.64	1.55	1.62
Roos	140	3	2	327.5	91.0	79.3	24.2	0.27	2.27	2.21	2.70	3.17
Middle Staple	120	3	2	167.1	45.1	75.9	14.1	0.31	1.53	1.46	1.85	1.53
Littaford	100	1	3	187.5	68.8	57.3	12.7	0.18	1.53	1.32	1.49	2.57
Branscombe's Loaf	180	3	1	385.6	96.4	35.5	5.3	0.05	1.36	1.37	1.61	1.27
Lower Dunna Goat	30	2	1	293.1	33.5	102.0	8.3	0.25	1.48	1.63	1.81	1.16
West Mill	70	3	2	160.0	37.2	39.5	8.8	0.24	0.85	0.74	0.99	0.84
Hen	110	2	3	75.0	18.1	25.1	10.1	0.56	1.56	1.59	2.09	1.23
Ltl. Trowlesworthy	80	1	2	155.4	45.1	68.9	20.2	0.45	2.40	3.16	2.72	1.87
Higher White	120	3	1	361.8	78.5	44.8	8.2	0.10	1.33	1.26	1.55	1.39
Longaford	120	3	3	187.1	54.5	66.1	10.2	0.19	1.95	2.07	2.83	1.61
Haytor	100	3	3	303.0	43.0	81.0	10.7	0.25	1.96	2.36	1.89	2.54
Haytor	110	3	2	202.1	45.4	98.0	17.1	0.38	0.67	0.77	0.64	0.63
Pew	130	3	3	334.9	63.5	135.0	25.6	0.40	1.71	1.83	1.98	1.83
Hart	40	2	1	289.5	121.6	71.9	15.5	0.13	2.04	2.21	2.78	1.98
Sheepstor	60	3	3	316.8	54.0	113.8	28.4	0.53	1.08	0.71	1.50	1.44
Sheepstor	130	3	3	244.3	67.8	104.3	14.6	0.22	1.46	1.64	1.47	1.92
Great Links	230	3	1	441.7	128.0	53.1	9.4	0.07	1.36	1.26	1.39	2.14
Brat	150	1	1	238.9	86.9	84.8	10.2	0.12	1.67	1.69	1.78	2.21
Wild	80	1	3	312.3	96.4	55.6	10.8	0.11	1.19	0.96	1.37	1.47
Watern	120	1	1	270.4	80.7	49.9	7.8	0.09	1.13	1.18	1.27	1.05
Beardown	80	3	3	412.5	121.2	82.0	12.5	0.10	1.41	1.52	1.44	2.26
Beardown	110	3	3	255.3	85.5	63.6	12.3	0.14	1.69	1.60	2.17	1.55
Great Mis	200	3	2	297.1	55.0	86.9	8.3	0.15	1.38	1.37	1.42	1.92
Great Mis	160	3	2	410.9	62.5	160.4	20.2	0.32	1.61	1.77	1.77	1.67
Black	100	2	1	334.0	51.9	117.8	14.1	0.27	1.36	1.36	1.28	1.87
Mel	100	1	3	245.9	38.1	78.0	9.7	0.25	2.05	2.28	2.64	1.69
Ger	110	1	3	209.3	66.6	71.7	16.7	0.25	1.31	1.34	1.50	1.25
King's	190	3	3	220.4	79.5	67.0	9.4	0.12	1.42	1.48	1.67	1.50
Oke	110	3	3	235.8	69.3	109.4	13.4	0.19	1.98	2.40	2.56	1.60
Oke	110	3	3	326.1	107.4	97.3	23.5	0.22	1.98	2.40	2.56	1.60
Down	130	3	3	247.9	54.8	97.8	14.6	0.27	2.19	2.62	2.51	2.40
Rippon	110	3	3	364.1	38.8	69.0	10.6	0.27	1.19	1.10	1.25	1.57
Buckland Beacon	290	1	1	225.3	43.8	113.0	18.5	0.42	1.41	1.42	1.70	1.58
Sharp	130	1	1	157.5	33.7	54.0	12.7	0.38	1.33	1.30	1.65	1.15
Doe	110	1	1	198.4	44.6	56.9	8.5	0.19	1.15	1.06	1.38	0.95

nd = no data available.
[a]These tors are not named on the Ordnance Survey map. The names are used by Hemory (1983).
[b]This tor is not named on the Ordnance Survey map. I invented this name.

Table 23.3 (Continued)

Composition							Rock texture		
PQUA (%)	PKSP (%)	PPLG (%)	PTM (%)	SCHR	VEIN	CLAY	GSD	PHEN (No.)	MEGA (%)
31.2	26.6	25.6	11.3	1	2	nd	2	22	15.3
35.3	29.6	24.5	4.5	1	2	nd	2	23	17.3
35.2	35.8	23.1	4.7	1	2	2	2	11	17.0
32.7	36.5	21.8	6.0	1	2	nd	2	0	nd
33.4	34.3	19.1	8.1	1	1	1	2	5	9.9
34.1	41.7	7.9	8.4	1	1	2	2	23	14.0
24.0	25.1	27.0	4.6	1	2	1	2	22	11.5
35.1	27.5	20.5	10.5	1	2	1	2	15	11.5
34.7	34.5	14.0	8.0	2	2	2	2	10	11.5
27.9	27.7	21.1	8.1	1	2	2	2	22	16.1
36.7	39.6	11.2	7.8	1	2	2	2	32	22.1
31.2	31.7	17.5	12.3	1	2	1	2	14	13.0
32.5	25.5	21.3	10.1	2	2	1	2	17	nd
38.0	38.2	9.4	9.4	1	1	1	2	4	6.3
33.3	34.5	19.6	9.1	2	1	2	2	4	13.3
33.3	32.2	13.6	12.9	1	2	2	2	4	3.7
31.8	29.6	18.7	7.6	1	2	2	2	24	17.6
29.2	30.8	18.1	14.7	1	2	1	2	16	11.6
29.2	30.8	18.1	14.7	1	2	1	2	16	11.6
33.1	26.3	20.8	8.8	1	1	2	2	19	12.0
36.4	36.8	13.8	9.2	1	1	1	2	1	5.7
30.6	29.0	20.7	15.4	1	2	nd	2	4	5.0
32.0	30.9	16.0	14.3	2	1	2	2	13	15.8
35.9	38.3	10.5	9.5	1	1	nd	2	4	14.0
35.6	28.9	17.5	14.2	1	1	2	2	2	5.0
34.1	32.1	17.6	13.6	1	2	2	2	11	14.8
31.3	33.6	20.6	11.6	1	1	2	1	0	0
37.9	39.6	18.1	1.3	1	2	nd	1	0	0
38.1	29.4	16.6	9.5	1	1	nd	1	0	5.0
30.4	30.0	25.7	5.5	2	1	2	1	0	2.0
34.0	31.1	20.2	11.1	2	1	2	2	12	18.0
32.3	27.1	18.2	17.6	1	1	2	2	6	11.9
29.3	23.2	25.1	14.2	2	1	1	2	44	17.6
39.1	38.0	16.5	2.9	2	2	1	1	0	0
33.7	28.8	20.4	11.7	2	1	1	2	3	6.0
35.7	25.9	15.6	11.5	1	1	nd	2	3	13.5
34.4	38.9	11.6	9.7	2	1	nd	2	33	20.5
30.7	28.5	17.6	13.7	1	1	2	2	30	20.5
33.1	33.2	15.5	12.6	1	2	2	2	16	8.5
34.0	31.3	16.1	14.4	2	1	2	2	12	10.4
36.2	32.1	21.4	6.0	2	1	2	2	4	5.0
31.1	34.9	23.5	7.0	1	1	2	1	0	1.0
31.4	33.8	13.4	15.4	2	1	1	2	3	5.0
36.2	37.4	11.2	10.0	1	1	2	2	2	5.0
35.8	30.2	15.0	12.9	1	1	2	2	18	20.0
33.5	33.5	16.1	12.5	2	1	1	2	8	20.0
35.6	34.9	16.5	8.9	1	2	2	2	4	6.3
34.3	29.1	20.9	9.4	2	1	1	.2	7	11.1
34.1	34.1	18.0	7.8	2	1	2	2	5	8.0
37.0	30.5	16.6	10.8	2	1	2	2	5	7.6
25.8	28.9	28.3	9.0	1	2	2	1	1	4.6
25.8	28.9	28.3	9.0	1	2	2	1	1	5.0
30.7	29.7	17.7	10.4	1	2	2	2	22	15.4
37.2	31.5	22.3	9.3	1	1	1	2	41	16.4
30.2	25.8	26.5	12.9	2	1	2	2	21	12.8
37.2	38.8	14.1	4.3	1	1	1	2	0	7.0
36.0	34.4	17.5	7.2	1	1	nd	2	0	3.2

	Geomorphic Variables:			Structural Variables:				Petrographic Variables: Grain Size:				Composition:							Rock Texture:		
	REL	LF	JC	MVPS	MVSS	MHPS	MHSS	MRGS	MQUA	MKSP	MPLG	PQUA	PKSP	PPLG	PTM	SCHR	VEIN	CLAY	GSD	PHEN	MEGA
REL	1.000																				
LF	.363	1.000																			
JC	.068	.198	1.000																		
MVPS	-.020	-.270	-.169	1.000																	
MVSS	.142	-.278	.264	.622	1.000																
MHPS	-.012	-.202	.184	.422	.055	1.000															
MHSS	.058	-.183	.333	-.021	-.002	.565	1.000														
MRGS	-.068	.101	.044	-.040	.120	.031	.233	1.000													
MQUA	.024	.116	.050	-.044	.168	.172	.263	.940	1.000												
MKSP	.112	.097	.074	-.096	-.040	-.041	.186	.918	.842	1.000											
MPLG	.173	.140	.073	.135	.262	.027	.141	.655	.559	.420	1.000										
PQUA	-.211	.102	.167	-.139	-.153	-.223	-.193	-.355	-.471	-.216	-.440	1.000									
PKSP	.201	.218	.003	-.047	-.190	-.179	.181	.315	-.397	-.207	-.010	.528	1.000								
PPLG	-.009	.023	.143	-.141	-.024	-.004	.133	-.107	-.225	-.043	-.018	-.575	-.603	1.000							
PTM	-.326	.215	.028	-.309	.219	.219	-.034	.246	.301	.211	.363	-.286	-.412	-.185	1.000						
SCHR	.284	.237	.352	.297	.172	.413	.390	.114	.162	.141	.124	.229	.121	.267	.212	1.000					
VEIN	.193	-.387	.062	.166	.248	.140	.234	-.147	.184	.088	.230	.072	.050	-.343	-.006	.116	1.000				
CLAY	-.266	.199	.184	.258	.265	.163	-.016	.017	.017	.027	.144	.018	.130	.057	.076	.357	.160	1.000			
GSD	.452	.503	.399	.429	.430	.305	.344	.366	.318	.330	.563	.363	.288	.152	.505	.529	.422	.408	1.000		
PHEN	.241	.308	.041	.295	.160	.134	.119	.092	.045	.104	.549	.300	.264	.033	.187	.218	.271	.161	.653	1.000	
MEGA	.267	.417	.091	.249	.111	.086	.164	.161	.141	.118	.354	-.101	-.116	-.112	.201	.212	.637	.186	.637	.805	1.000

Significant correlations are shown in bold.

Figure 23.3 Correlation matrix (from Ehlen, 1993)

Table 23.4   Nominal and ordinal variable codes

Variable	Code
Binary variables	
Schorl	1 = presence; 2 = absence
Tourmaline veins	1 = presence; 2 = absence
Clay	1 = presence; 2 = absence
Landform	
Spur tors	1
Valleyside tors	2
Summit tors	3
Grain size distribution	
Equigranular	1
Megacrystic	2
Joint control by	
Horizontal joints	1
Vertical joints	2
Vertical + horizontal joints	3

significant differences could be identified between pairs of spacing distributions. All three tor types are significantly different from each other.

Joint spacing is widest in summit tors. Mean spacing for primary horizontal joints is 73 cm; for secondary horizontal joints, 13 cm; and for primary vertical joints, 260 cm. Horizontal joint spacing is similar to that in spur tors and both vertical and horizontal joint spacing are more like that in spur tors than in valleyside tors. Summit tors with coarser grain have intermediate joint spacing. They contain the largest numbers of megacrysts, and summit tors with high relative relief contain the most abundant megacrysts.

Joint spacing is second widest in spur tors. Mean primary vertical joint spacing is 258 cm; mean primary horizontal joint spacing, 73 cm; and mean secondary horizontal joint spacing, 12 cm. Mean secondary vertical joint spacing is the widest among the three types of tors (70 cm). Horizontal joint spacing in spur tors is very similar to that in summit tors, and spur tors are more like summit tors than valleyside tors with respect to vertical joint spacing. In coarser-grained spur tors, joint spacing is narrower than it is in either coarser-grained summit tors or valleyside tors, but becomes wider with increasing numbers of megacrysts. In this respect, spur tors are similar to valleyside tors.

Valleyside tors, which are controlled by secondary vertical joints, have the narrowest joint spacing of the three types. Mean spacing for primary vertical joints is 257 cm; for secondary vertical joints, 48 cm; for primary horizontal joints, 57 cm; and for secondary horizontal joints, 11 cm. Valleyside tors are very different from spur and summit tors with respect to both kinds of joints. In addition, valleyside tors always have the narrowest joint spacing regardless of relative relief. Joint spacing becomes wider, however, as the rock becomes coarser grained. Finally, joint spacing in valleyside tors becomes wider with increasing numbers of megacrysts, similar to spur tors.

DARTMOOR, ENGLAND VERTICAL JOINTS NO. MEGACRYSTS: 0-2	MIN	MAX	MEAN	MODE
P	6	1575	229.3	700 - 705
S	2	690	54.1	5 - 10

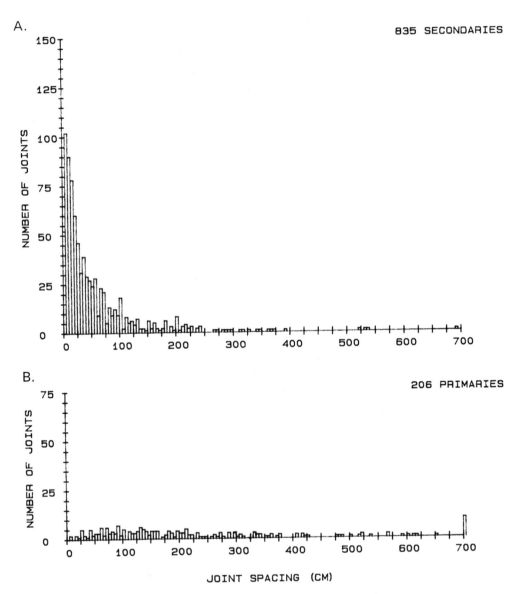

Figure 23.4 Frequency histogram showing the distribution of vertical joint spacings in rocks with <2 megacrysts for Secondary Joints (A) and Primary Joints (B)

Spatial Patterns

Spatial patterns were identified by visual analysis of contour maps showing the distribution of each variable (Ehlen, 1992a). These maps were generated using the TIN module of ARC/INFO. Similarities between variable patterns were determined by overlaying the contour maps on a light table. As with the correlation analysis, two types of characteristics with respect to landform are possible: (1) those determined by overlaying a variable map on to the landform map, and (2) those determined by overlaying two variable maps, only one of which exhibits a definable pattern when compared directly to the landform map. Figure 23.5 shows the distribution of plagioclase feldspar as an example; additional maps can be found in Ehlen (1989 and 1992a). All variables except the joint spacing ratio were used in this analysis.

Analysis of spatial patterns indicates that primary vertical joints in summit tors are widely spaced (usually >300 cm). Horizontal joints have intermediate to wide

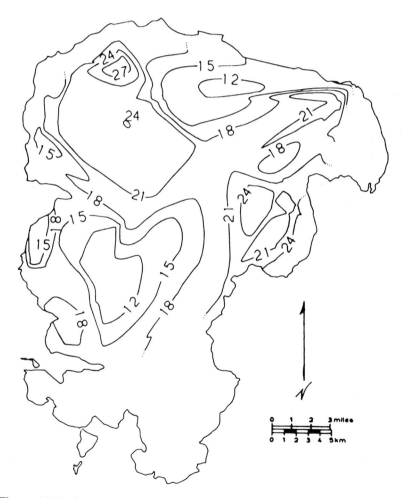

Figure 23.5   Spatial distribution of plagioclase feldspar. Contour interval 3%

spacing: primary spacing ranges from 60 to 80 cm and secondary spacing is >10 cm. Summit tors occur where relative relief is high (mean: 126 m). The rocks contain abundant feldspar, usually >30% potassium feldspar and >18% plagioclase. They are strongly megacrystic (usually >15 per quadrat and >15% megacrysts) as well as coarse grained (>2 mm). Schorl is often present. Spatial relations between other variables suggest that summit tors may be controlled by vertical joints, and are likely to contain abundant quartz and no tourmaline veins.

Joint spacing is narrower in spur tors. Primary vertical joint spacing is <200 cm; secondary vertical spacing ranges from 50 to 75 cm; primary horizontal joint spacing is usually <60 cm; and secondary horizontal spacing, <10 cm. Relative relief is typically intermediate (mean: 116 m). The rocks in spur tors are fine grained (<1 mm) and often contain tourmaline veins. Texture is feebly megacrystic, generally with <5% megacrysts. Spatial relations between other variables suggest that quartz abundance may be high and tourmaline abundance is likely to be low. Tor shape may be controlled by vertical joints and rock texture is likely to be equigranular.

In valleyside tors, primary vertical joint spacing is narrow to intermediate (<300 cm), but horizontal joint spacing is wide. Primary horizontal joint spacing ranges from 60 to 200 cm, and secondary spacing is usually >10 cm. Valleyside tors are controlled by horizontal joints, and relative relief is typically low (mean: 73 m). Although potassium feldspar is coarse grained (>2 mm), the rocks typically have fine to intermediate grain size (<2 mm). Potassium feldspar is not abundant (<31%) and quartz abundance is intermediate (30–33%). Spatial relations between other variables suggest that plagioclase feldspar may be abundant. Tourmaline abundance is likely to be low, and tourmaline veins are likely to be absent. Rock texture ranges from equigranular to very feebly megacrystic.

Multivariate Analysis

Ordination and classification procedures were used to group similar tors and to identify the most important variables. Principal coordinates analysis, a Q-mode procedure, was chosen for ordination because (1) it accepts nominal and ordinal variables and (2) it is distribution free. The non-hierarchical classification also allowed inclusion of nominal and ordinal variables. Five clusters were identified (Figure 23.6). All variables except clay were used. Details of these procedures and the results presented below are reported in Ehlen (1993).

Tors in the first cluster occur mainly south of a line connecting Great Mis Tor and Bell Tor (see Figure 23.2 for the locations of specific tors). They are characterised by medium to high numbers of megacrysts, medium- to coarse-grained feldspar, narrow to intermediate vertical joint spacing, medium to high secondary joint spacing ratios, and low to intermediate quartz abundances. Tourmaline veins are present, but there is generally no schorl. Most of the tors are summit tors (e.g. Roos Tor), but spur tors are present as well (e.g. Mel Tor).

Members of the second cluster are present throughout Dartmoor, except in the south. Many of them are lamellar, e.g. Great Links Tor. They are characterised by fine- to medium-grained feldspar, widely spaced vertical joints, low secondary joint spacing ratios, and low to intermediate quartz abundances. Tourmaline veins are absent.

The two sites in Cluster 3 have no megacrysts, and plagioclase feldspar is fine grained. Vertical joint spacing is narrow. The tors occur in the northwest and east.

Most of the tors in the fourth cluster occur in the east. They have medium to high numbers of megacrysts, medium- to coarse-grained feldspar, intermediate vertical joint spacing, low quartz abundances, moderately to highly abundant plagioclase, and form summit tors (e.g. Hound Tor). Schorl is typically present and tourmaline veins are absent.

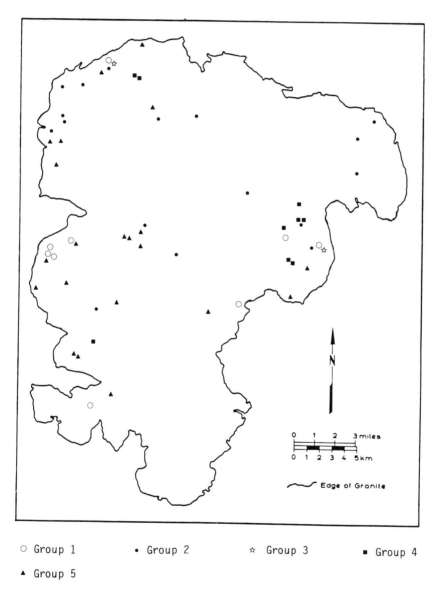

○ Group 1   • Group 2   ☆ Group 3   ■ Group 4

▲ Group 5

Figure 23.6 Spatial distribution of the five clusters identified using multivariate analysis (from Ehlen, 1992b)

The fifth cluster is the largest and is present throughout Dartmoor except in the northeast. These tors are often located near the granite boundary (e.g. Pew Tor) and many are altered or reddened (e.g. Doe Tor). They are characterised by few megacrysts, fine- to medium-grained feldspar, narrow to intermediate vertical joint spacing, medium to high secondary joint spacing ratios, low to intermediate plagioclase abundances, and form summit and valleyside tors (e.g. Rippon Tor and Hen Tor respectively). Tourmaline veins are typically present.

The results of the multivariate analyses do not allow the characteristics of each landform type to be defined as do the other procedures discussed above, but they do provide significant insight into which variables are important in defining groups of tors. These variables can then be used to define better each landform type. The important variables, in coordinate order, are vertical joint spacing, the secondary joint spacing ratio, the presence or absence of tourmaline veins, number of megacrysts, plagioclase grain size, quartz and plagioclase abundances, landform, the presence or absence of schorl, and potassium feldspar grain size.

Joint spacing frequency distributions were used to evaluate the statistical significance of the results. Joint spacings for each joint set in each tor in each cluster were tabulated and the frequency distributions were determined. These were compared using chi-square to ascertain whether the clusters were significantly different with respect to joint spacing. The groups along each of the important coordinates were also compared in this manner.

There are significant differences in joint spacing between six of the ten possible combinations of the five clusters. Each cluster is significantly different from at least two others, and one cluster, Cluster 3, is significantly different from all others. In addition, statistical comparison of the joint spacing distributions for the five tor clusters indicates that the most important variables distinguishing the five groups are number of megacrysts, feldspar grain size, quartz abundance, the presence or absence of tourmaline veins and schorl, and joint spacing, excluding primary horizontal joint spacing. With respect to joint spacing, secondary joint spacing is more important than primary joint spacing regardless of joint type, and vertical joints are more important than horizontal joints.

CONCLUSIONS

Although the descriptions of each tor type using the different procedures are not identical, the similarities among the results allow the three tor types to be described semi-quantitatively. The results of the multivariate procedures indicated which variables to use in this classification. Only those characteristics that are common to the results of several procedures and are not contradicted by the results of others are used. Wherever possible, general terms are quantified. The classification is:

Summit Tors generally have high relative relief (mean 126 m), are megacrystic (>15 per quadrat and/or 15% megacrysts), and have the widest joint spacing. For primary vertical joints, spacing is usually >300 cm; for primary horizontal joints, the mean is 73 cm; and for secondary horizontal joints, the mean is 13 cm. Summit tors are usually controlled by vertical joints or by vertical and horizontal joints

Figure 23.7   Summit Landform—Rippon Tor

combined. Feldspar is usually abundant (>30% potassium feldspar; >18% plagioclase). Example: Rippon Tor (Figure 23.7).

Spur Tors generally have narrower vertical joint spacing: primary joint spacing
is usually <200 cm; and secondary joint spacing is 50–75 cm (mean 70 cm).
Horizontal joint spacing is intermediate: mean primary spacing is 73 cm; mean
secondary spacing, 12 cm. Spur tors occur where relative relief is intermediate
(mean 115 m). The rocks are fine grained (<1 mm) and feebly megacrystic (<5%
megacrysts) or equigranular. Potassium feldspar abundance is low. Example:
Littaford Tors (Figure 23.8).

Valleyside Tors occur where relative relief is low (mean 73 m) and have narrow
joint spacing. Primary vertical joint spacing is typically <300 cm (mean 257 cm);

Figure 23.8   Spur Landform—Littaford Tors (from Ehlen, 1992a). Reproduced by permission

Figure 23.9 Valleyside Landform—Black Tor (from Ehlen, 1992a). Reproduced by permission

mean primary horizontal joint spacing is 57 cm; and mean secondary horizontal joint spacing is 11 cm. Secondary vertical joint spacing is widest in valleyside tors: the mean is 48 cm. Horizontal joints control tor shape. The rocks are finer grained (<2 mm), feebly megacrystic and quartz abundance is low. Example: Black Tor (Figure 23.9).

This classification refers only to the true granites on Dartmoor, and only further testing will show whether these characteristics are typical of other granitic rocks (e.g. granodiorite, quartz diorite) or of granites in other areas. Similarly, the quantitative descriptors determined by analysis of spatial patterns and joint spacing frequency distributions refer only to the Dartmoor tors. Until the same procedures are applied to granite landforms in other areas, it is not known whether these characteristics and values are universal or are peculiar to Dartmoor.

ACKNOWLEDGEMENTS

I wish to thank J.R. Hawkes, British Geological Survey, for providing the per cent megacryst data; Margaret Oliver, University of Birmingham, and Richard Morgan, University of Otago, New Zealand, for their assistance with the multivariate analyses; and E-an Zen, University of Maryland, USA, for field assistance and reviewing an early version of this paper. John Gerrard, University of Birmingham, has provided advice, support and assistance throughout the period in which this work was done.

REFERENCES

Bourke, P. J., Bromley, A., Rae, J. and Sincock, K. (1981). A multi-packer technique for investigating resistance to flow through fractured rocks and illustrative results: Siting of radioactive waste repositories in geological formations. In *Proceedings of the Nuclear Energy Agency Workshop (Paris)*, Organization for Economic Co-operation and Development, pp. 173–187.

Bourke, P. J., Evans, G. V., Hodgkinson, D. P. and Ivanovich, M. (1982). An approach to prediction of water flow and radionuclide transport through fractured rock: Geophysical investigations in connection with geological disposal of radioactive waste. *Proceedings of a Nuclear Energy Agency Workshop (Ottawa, Canada)*, Organization for Economic Co-operation and Development, pp. 189–198.

Brook, G. A. (1978). A new approach to the study of inselberg landscapes. *Zeitschrift für Geomorphologie*, Supplementband **31**, 138–160.

Brunsden, D. (1964). The origin of decomposed granite on Dartmoor. In Simmons, I. G. (ed.), *Dartmoor Essays*, Devonshire Association for the Advancement of Science, Literature and Art, Exeter, pp. 97–116.

Brunsden, D. and Gerrard, J. (1970). The physical environment of Dartmoor. In Gill, C. (ed.), *Dartmoor, A New Study*, David and Charles, Newton Abbot, pp. 21–53.

Caine, N. (1967). The tors of Ben Lomond, Tasmania. *Zeitschrift für Geomorphologie*, **11**, 418–429.

Chapman, C. A. and Rioux, R. L. (1958). Statistical study of topography, sheeting and jointing in granite, Acadia National Park, Maine. *American Journal of Science*, **256**, 111–127.

de la Beche, H. T. (1839). *Report on the Geology of Cornwall, Devon, and West Somerset*. Longmans, London, pp. 156–192, 270–282.

Demek, J. (1964). Castle koppies and tors in the Bohemian Highland (Czechoslovakia). *Biulteyn Peryglacjalny*, **14**, 195–216.

Dumanowski, B. (1964). Problem of the development of slopes in granitoids. *Zeitschrift für Geomorphologie*, Supplementband **5**, 30–40.

Dumanowski, B. (1968). Influence of petrographical differentiation of granitoids on landforms. *Geographica Polonica*, **14**, 93–98.

Ehlen, J. (1989). Geomorphic, petrographic and structural relations in the Dartmoor Granite, Southwest England. Unpublished Ph.D. Thesis, University of Birmingham.

Ehlen, J. (1991). Significant geomorphic and petrographic relations with joint spacing in the Dartmoor Granite, southwest England. *Zeitschrift für Geomorphologie*, NF35, 425–438.

Ehlen, J. (1992a). Analysis of spatial relationships among geomorphic, petrographic and structural characteristics of the Dartmoor Tors. *Earth Surface Processes and Landforms*, **17**, 53–67.

Ehlen, J. (1992b). *Comparison of Air Photo Landform Units to Statistically Defined Groups of Tors in Dartmoor, Southwest England*. US Army Topographic Engineering Center, TEC-0001, Fort Belvoir, Virginia.

Ehlen, J. (1993). *Statistical Analysis of Geomorphic, Structural and Petrographic Characteristics of the Dartmoor Tors, Southwest England*. US Army Topographic Engineering Center, TEC-0027, Fort Belvoir, Virginia.

Ehlen, J. and Zen, E. (1990). Joint spacings, mineral modes and grain size measurements for selected granitic rocks in the Northern Rockies and in Southwest England. *US Geological Survey Open-File Report* 90–48.

Gerrard, A. J. W. (1974). The geomorphological importance of jointing in the Dartmoor granite. In Brown, E. H. and Waters, R. S. (eds), *Progress in Geomorphology*, Institute of British Geographers Special Publication No. 7, pp. 39–50.

Gerrard, A. J. W. (1978). Tors and granite landforms of Dartmoor and eastern Bodmin Moor. *Proceedings of the Ussher Society*, **4**, 204–210.

Gibbons, C. L. M. H. (1981). Tors in Swaziland. *Geographical Journal*, **147**, 72–78.

Hawkes, J. R. (1982). The Dartmoor granite and later volcanic rocks. In Durrence, E. M. and Laming, D. J. C. (eds), *Geology of Devon*, University of Exeter, Exeter, pp. 85–116.

Hemery, E. (1983). *High Dartmoor, Land and People*. Robert Hale, London.

Jahns, R. H. (1943). Sheet structure in granites: its origin and use as a measure of glacial erosion in New England. *Journal of Geology*, **51**, 71–98.

Jeje, L. K. (1973). Inselbergs' evolution in a humid tropical environment: the example of South Western Nigeria. *Zeitschrift für Geomorphologie*, NF**17**, 194–225.

Linton, D. L. (1955). The problem of tors. *Geographic Journal*, **121**, 470–487.

Mabbutt, J. A. (1952). A study of granite relief from South-West Africa. *Geological Magazine*, **89**, 87–96.

Oen, I. S. (1965). Sheeting and exfoliation in the granites of Sermersoq, South Greenland. *Meddelelser om Gronland, Kommissionen for Videnskabelige Undersogelser I. Gronland*, **179**, 1–40.

Ormerod, G. W. (1859). On the rock basins in the granite of the Dartmoor district, Devonshire. *Quarterly Journal of the Geological Society of London*, **15**, 16–29.

Ormerod, G. W. (1869). On some of the results arising from the bedding, joints and spheroidal structure of the granite on the eastern side of Dartmoor, Devonshire. *Quarterly Journal of the Geological Society of London*, **25**, 273–280.

Pye, K., Goudie, A. S. and Thomas, D. S. G. (1984). A test of petrological control in the development of bornhardts and koppies on the Matapos Batholith, Zimbabwe. *Earth Surface Processes and Landforms*, **9**, 455–467.

Pye, K., Goudie, A. S. and Watson, A. (1986). Petrological influence on differential weathering and inselberg development in the Kora area of central Kenya. *Earth Surface Processes and Landforms*, **11**, 41–52.

Robb, L.J. (1979). *The Distribution of Granitophile Elements in Archean Granites of the Eastern Transvaal, and their Bearing on Geomorphological and Geological Features of the Area*. University of the Witwatersrand, Economic Geology Research Unit, Information Circular 129, Johannesburg, South Africa.

Rock, N. M. S. (1988). *Numerical Geology*. Springer-Verlag, New York.

Snow, D. T. (1968). Rock fracture spacings, openings, and porosities. *Journal of the Soil Mechanics and Foundations Division, Proceedings of the American Society of Civil Engineers*, **94**(SM-1), 73–91.

Thomas, M. F. (1974). Granite landforms: a review of some recurrent problems in interpretation. In Brown, E. H. and Waters, R. S. (eds), *Progress in Geomorphology*, Institute of British Geographers Special Publication No. 7, pp. 13–35.

Thorpe, R. (1979). *Characterization of Discontinuities in the Stripa Granite Timescale Heater Experiment*. Lawrence Berkeley Laboratory, for the US Department of Energy, LBL-7083.

Twidale, C. R. (1964). A contribution to the general theory of domed inselbergs, conclusions derived from observations in South Australia. *Transactions of the Institute of British Geographers*, **34**, 91–113.

Twidale, C. R. (1980). The origin of bornhardts. *Journal of the Geological Society of Australia*, **27**, 195–208.

Twidale, C. R. (1981). Granitic inselbergs: domed, block-strewn and castellated. *Geographic Journal*, **147**, 54–71.

Twidale, C. R. (1982). *Granite Landforms*. Elsevier, New York.

Waters, R. S. (1957). Differential weathering and erosion on oldlands. *Geographical Journal*, **123**, 503–513.

Whittle, R. (1989). The granites of Southwest England, Section 2, Jointing in the Carnmenellis Granite. In Parker, R.H. (ed.), *Hot Dry Rock Geothermal Energy, Phase 2B Final Report of the Camborne School of Mines Project, Vol. I*, Pergamon Press, Oxford, pp. 121–182.

24 Weathering Flutes on Siliceous Rocks in Britain and Europe

R. B. G. WILLIAMS and D. A. ROBINSON
University of Sussex, Brighton, UK

ABSTRACT

Weathering flutes or runnels resembling limestone karren are much more widely developed on siliceous rock outcrops in Britain and Europe than is generally recognised. The flutes, which are sometimes as much as 5 m long and 0.5 m deep, are particularly characteristic of steeply sloping rock faces, and run directly down the faces, channelling water away during rains. They are normally spaced about 0.15 to 0.2 m apart.

In Britain, well-developed flutes occur quite commonly on Millstone Grit crags in the Pennines and also on crags of Fell Sandstone in Northumberland. A few examples are present on Triassic Sandstone crags in Cheshire and Shropshire. In eastern Germany, the Czech Republic and southern Poland, fluting is common on outcrops of Upper Cretaceous sandstone. It also occurs on granite outcrops in France, Spain, the Czech Republic and Poland.

The origin of flutes is problematic. Although streamlets of water running down the flutes may have an erosive effect, weathering processes are probably more important in flute formation. Solutional weathering plays a key role in the development of limestone karren, but is apparently much less important in forming flutes on siliceous rocks, despite some suggestions to the contrary. Flutes tend to develop on the most exposed rock surfaces that rapidly become soaked during rains but dry out quickly afterwards, which suggests that frequent wetting and drying may be an important aid to their development. Differential growth of algae and lichens, and the uneven development of surface crusting may also be contributory factors. Some flutes have developed as overflows from cliff-top weathering pits or rock basins.

Several writers have suggested that the flutes that they have studied in Europe, especially on granite outcrops, are fossil features, dating in some cases from the Tertiary. The view advanced here, however, is that the majority of flutes are likely to be recent in age as is demonstrated by their occurrence on many prehistoric standing stones and on some historic buildings.

INTRODUCTION

While undergoing weathering some sloping or vertical rock surfaces develop more or less regularly spaced flutes or grooves, which are aligned down the surfaces and carry runoff away during rains. The flutes are most frequent on limestone outcrops, where they vary considerably in size and form, even though most researchers attribute them all to solutional weathering (Ford and Williams, 1989).

Rock Weathering and Landform Evolution. Edited by D. A. Robinson and R. B. G. Williams.
© 1994 John Wiley & Sons Ltd

Flutes also develop on siliceous rocks, such as granites and sandstones, which are far less soluble than limestone, except at unusually high temperatures or in very alkaline conditions. The origin of these flutes, which are sometimes called pseudo-karren or pseudo-lapiés, is still not firmly established, although a number of theories exist, which have been usefully summarised by Twidale (1982).

Fluting on siliceous rocks is most common in the tropics and subtropics, more especially in rain forest areas, but also in the savanna zone and on desert margins. As well as occurring inland, it is a conspicuous feature of many outcrops on tropical coasts, for example on granite in the Seychelles and on Pulau Ubin island off Singapore (Hsi-lin, 1962).

There have been relatively few reports of fluting on siliceous rocks in Britain and Europe, although, as this paper will show, it is quite common and well developed in certain areas. After a review of the distribution of flutes in relation to geological and other environmental factors, the paper will go on to discuss their morphology, possible mode of origin and date of formation.

FLUTE DISTRIBUTION ON SILICEOUS ROCKS IN THE BRITISH ISLES

Fluting occurs quite widely on cliffs of Upper Carboniferous (Namurian) Millstone Grit in the Pennines (Figure 24.1 and Table 24.1). In Derbyshire particularly good examples can be seen at Robin Hood's Stride and Cratcliffe Tor, near Birchover (Figures 24.2 and 24.3). Well-developed fluting also occurs on the cliffs of Hen Cloud and The Roaches in Staffordshire. In the Yorkshire Pennines the best examples are found at Almscliffe (Figure 24.4) and Plumpton Rocks near Harrogate.

Flutes also occur on many of the crags of Lower Carboniferous (Dinantian) Fell Sandstone in Northumberland, notably at Kyloe Crag, Bowden Doors, Back Bowden Doors and St Cuthbert's Cave near Belford. Figure 24.1 probably greatly under-estimates the true frequency of fluting in Northumberland because the crags are widely scattered, and many of the less accessible examples have not been inspected.

Fluting occurs rather sparingly on outcrops of Triassic (or New Red) sandstone in Cheshire and Shropshire. It is perhaps best developed on the former river cliffs at Helsby on Merseyside, but it also occurs at Beeston Crag near Nantwich, and, further south, on crags at Lee Brockhurst and Grinshill near Wem. No flutes have been found on the well-known cliffs of Triassic sandstone at Hawkstone Park, near Wem, or on the many cliffs of Triassic sandstone and conglomerate in the Churnet Valley in Staffordshire.

No flutes occur on the Lower Cretaceous sandstone cliffs in Kent and Sussex, although these display a wealth of other weathering features. Nor have any examples been found as yet on cliffs of Devonian (Old Red) and Torridonian Sandstone.

No fluting has been reported on granite outcrops anywhere in the British Isles. Water draining from rock basins or weather pits has, in some instances, carved overflow channels, but they do not form parallel groupings in the same way as flutes. Fine flutes do occur, however, on Roche Rock, an isolated crag of quartz–tourmaline schorl near St Austell in Cornwall.

In Britain, fluting occurs not only on rock outcrops but also on many prehistoric standing stones. Some of the more pillar-like stones have clearly been shaped by man, and some archaeologists (e.g. O'Toole, 1939, and Burl, 1976) have assumed that the fluting is also artificial, but most consider it to be due to weathering that has occurred after the stones were erected. Since the fluting is similar in all respects to the natural fluting found on rock outcrops, the suggestion that it was carved by the people who raised the stones is unconvincing. Nor, in the majority of cases, is there any reason to suppose that it was created by weathering before the stones were erected. Nearly all the stones have vertically oriented bedding planes, and the fluting is invariably

Figure 24.1 Rock outcrops in England and Wales with well-developed fluting

vertical. The prehistoric people who erected the stones must have obtained them from natural rock outcrops or from glacial deposits. On the rock outcrops the bedding planes are horizontal or dip at a small angle, except in a few cases where blocks or boulders have become detached and have tipped over so that the bedding is vertical or near vertical. Prehistoric people may well have selected detached blocks to use as standing stones in order to avoid having to quarry them, but the orientation of the blocks at this stage would not have interested them. On the rock outcrops the vertical fluting normally crosses the bedding planes more or less at right angles, but on the standing stones both the fluting and the bedding planes are vertical, and the obvious inference is that it developed in all, or nearly all, cases after the stones were erected.

Table 24.1 Rock outcrops in the British Isles with well-developed fluting

---

Peak District and the Yorkshire Pennines (Millstone Grit)
Hen Cloud. NGR: 40083616
The Roaches. NGR: 40023630
Black Rocks, Cromford. NGR: 42933558
Robin Hood's Stride, Harthill Moor. NGR: 42243623
Cratcliffe Tor, Harthill Moor. NGR: 42263624
Rowtor Rocks, Birchover. NGR: 42383624
Stone Edge, near Chesterfield. NGR: 43433673
Curbar Edge, Baslow. NGR: 42553755
Froggatt Edge, near Baslow. NGR: 42473765
Stanage Edge, near Hathersage. NGR: 42453834
Almscliffe Crag, between Harrogate and Otley. NGR: 42684490
Plumpton Rocks, near Harrogate. NGR: 43554536
Earl Crag, between Colne and Keighley. NGR: 39884430
Brimham Rocks, near Pateley Bridge. NGR: 42064650

Northumberland (Fell Sandstone)
Great Wanney, near West Woodburn. NGR: 39335835
Callerhues, near Bellingham. NGR: 38525863
Rothley Crags, near Elsdon. NGR: 40435885
Simonside, near Rothbury. NGR: 40305985
Ravensheugh Crag, near Rothbury. NGR: 40125988
Drake Stone, near Harbottle. NGR: 39216044
Corby's Crag, near Alnwick. NGR: 41276101
Bowden Doors, between Belford and Wooler. NGR: 40706325
Back Bowden Doors, between Belford and Wooler. NGR: 40656336
St Cuthbert's Cave, between Belford and Wooler. NGR: 405966351
Kyloe-in-the-Wood, between Lowick and Belford. NGR: 40456388
Kyloe Crag, between Lowick and Belford. NGR: 40406395
Berryhill Crag, between Wooler and Berwick. NGR: 39386403

Cheshire and Shropshire (New Red Sandstone)
Helsby, between Runcorn and Ellesmere Port. NGR: 34943755
Beeston Crag, near Nantwich. NGR: 35383592
Lee Brockhurst, near Wem. NGR: 35503269
Grinshill, near Wem. NGR: 35183237

Cornwall (schorl)
Roche Rock, near St Austell. NGR: 19910596

Figure 24.2 Deeply cut flutes on Millstone Grit. Weasel Pinnacle at Robin Hood's Stride, near Birchover, Derbyshire

Figure 24.3 Fluting on the exposed upper surfaces of Cratcliffe Tor, near Birchover

Figure 24.4 Fluted blocks of Millstone Grit in a field beside Almscliffe Crag, near Harrogate, Yorkshire

Table 24.2 Examples of strongly fluted standing stones in the British Isles

The Queen Stone. Standing stone beside River Wye, Symonds Yat, Hereford and Worcester (NGR: 35622182). Rock type: Old Red Sandstone.

Nine Stone Close stone circle, Harthill Moor, Birchover, Derbyshire (NGR: 42253625). Two of the four surviving stones are fluted. Rock type: Millstone Grit.

Devil's Arrows, Boroughbridge, North Yorkshire (NGR: 43914666). These three standing stones, which are arranged roughly in line, are impressively fluted. Rock type: Millstone Grit, supposedly quarried at Knaresborough, 10 km to the south.

Rudston Monolith, Rudston churchyard, near Bridlington, North Humberside (NGR: 50974677). This 7.8 m high, 26 tonne menhir is the tallest in the British Isles. Rock type: Jurassic (?) sandstone thought to have been brought from Cayton Bay, 16 km to the north.

Duddo Four Stones stone circle, Duddo, near Berwick, Northumberland (NGR: 39316437). Contrary to the name, there are five standing stones; fluting is very well developed. Rock type: Fell Sandstone.

Swinburne Castle standing stone, near Hexham (NGR: 39375745;. Rock type: Fell Sandstone.

Machrie Moor stone circles, Blackwaterfoot, Isle of Arran (NGR: 19106324). Several of the standing stones are strikingly fluted. Rock type: Old Red Sandstone.

Standing stone, Brodick Castle, Isle of Arran (NGR: 20046375). Rock type: Old Red Sandstone.

Stone of Setter, Eday, Orkney (NGR: 356410371). Rock type: Old Red Sandstone.

Ardistran, Aghade and other standing stones, near Tullow, North Carlow, Eire. Rock type: Granite.

As Table 24.2 and Figure 24.5 demonstrate, flutes occur on standing stones not only in England but also in Scotland and Eire where they are unrecorded from natural rock outcrops. Another interesting difference is that they are found on a wider range of rock types than their counterparts on the outcrops.

Flutes in the British Isles are commonest at low elevations and are absent from mountain tops and cirques. On the Millstone Grit outcrop all the best sites are situated below the 300 m contour, with the exception of Hen Cloud and The Roaches,

Figure 24.5 Examples of standing stones with well-developed flutes in the British Isles

which have an unusually exposed and elevated situation, reaching 450–500 m OD. The many cliffs and crags at higher elevations, up to about 600 m on Kinderscout and 720 m on Whernside, have few or no flutes. On the Fell Sandstone flutes become progressively scarcer with increasing altitude. There are, for example, very few flutes on Simonside (420 m OD) and Ravensheugh (390 m) compared with such low-lying sites as Kyloe Crag (108 m) and Back Bowden Doors (170 m).

It is not entirely clear why flutes should be most frequent at low-lying sites. There is less frost, the rock outcrops are more strongly heated in summer, and probably remain drier for longer periods, although the number of cycles of wetting and drying may be greater than at higher elevations, where the rocks often fail to dry out for long periods. On the Millstone Grit outcrop, differences in rock type may also be important. Thus in the Peak District the cliffs with the most conspicuous fluting are formed of the Ashover and Roaches Grits, which being massively jointed and strongly cemented weather relatively slowly. Although these grits outcrop at lower elevations, they were laid down later in the Millstone Grit succession than the Kinderscout Grit, which forms the cliffs high on Kinderscout. The notoriously severe weather on this mountain, and the fact that the grit is quite closely jointed, ensures that the cliffs are subject to frequent rockfalls (Sampson, undated), which doubtless helps inhibit flute formation. The more massive joint blocks on the cliffs of the Ashover and Roaches Grits provide flutes with large uninterrupted surfaces on which they can form, and this may explain why the cliffs show much more fluting. The Chatsworth or Rivelin Grit, which forms many of the gritstone edges on the eastern side of the Peak District, is another massive sandstone, but it is somewhat coarser grained and less feldspathic than the underlying Ashover and Roaches Grits (Aitkenhead et al., 1985), and is not as frequently fluted.

FLUTE DISTRIBUTION ON SILICEOUS ROCKS IN MAINLAND EUROPE

It is likely that flutes occur on siliceous rocks in many parts of mainland Europe, but details of their distribution remain somewhat sketchy. They are conspicuous features of some granite outcrops in Corsica (Klaer, 1956 and 1957; Rondeau, 1961) and the Massif Central (Godard et al., 1972). They have also been described from granite in Galicia in northwest Spain, mainly from coastal localities (Carlé, 1941; Bulow, 1942; Vidal Romani, 1989). On the Brittany coast well-developed flutes can be seen in granite at Ploumanac and Trégastel. Allegedly fossil examples are found on granite outcrops near Prague and also further east in the Czech Republic (Demek, 1964a and b). Czerwinski and Migon (1993) describe other instances of fluting on the Sudetan granite outcrop in southern Poland.

For some reason in Britain and Europe flutes are much more common on sandstones than on granites, whereas the opposite is true in the tropics and subtropics. In northern Germany, for example, fine fluting occurs on the pinnacles of Cretaceous sandstone that form the Externsteine in the Teutoburger Wald near Bielefeld. Numerous flutes are to be found on the cliffs of Cretaceous sandstone (Elbsandstein) in Sächsische Schweiz (Saxon Switzerland) in eastern Germany, and across the frontier in the Czech Republic. Particularly good examples can be seen on the Schrammsteine near Bad Schandau, the Pfaffenstein at Königstein, and near the natural bridge known as Pravcická brána on the Czechoslovakian side of the border. Fluting also occurs further east in Bohemia on the cliffs and pinnacles of Upper Cretaceous sandstone which constitute the so-called rock cities, for example at Prachovské Skály near Jicín. Further examples are found in southern Poland on the same sandstone (Alexandrowicz, 1990).

Surprisingly, fluting is not found on the extensive outcrops of Oligocene sandstone at Fontainebleau, near Paris, nor apparently on the cliffs of Triassic sandstone and conglomerate in the Vosges and the Black Forest.

Little information is available about fluting on standing stones in Europe, but some of the granite menhirs in Brittany (Finistère), such as the Kerscavan monolith near Brest and the Kerreneur menhir at Penmarch (Burl, 1985), are strikingly fluted, as is the Gollenstein menhir in the Saarland, which is the tallest standing stone in central Europe.

FLUTE DIMENSIONS, MORPHOLOGY AND FIELD RELATIONS

Flutes on siliceous rocks in the British Isles and the temperate zone of Europe are generally shorter and less deeply incised than their tropical or subtropical counterparts. The maximum depth of incision is about 0.5 m, whereas in the tropics some examples attain a depth of 2 m (Twidale, 1982). It is not possible to compare mean depths because flutes in their initial stages of development are so shallowly indented as not to be identifiable with any certainty. It is only as they deepen that they become easily recognisable.

Most flutes in Britain are between 0.5 and 2 m long; the longest example recorded to date is 5.7 m (at Back Bowden Doors). In central Europe some flutes in very inaccessible locations may attain a length of 10 m, but no accurate measurements are available. By contrast, in the tropics flutes sometimes extend for many tens of metres down the sides of granite domes.

On Millstone Grit and Fell Sandstone outcrops most flutes are spaced between 0.05 and 0.25 m apart, with a mean of 0.15 m (140 measurements). On the Elbsandstein outcrops in eastern Germany preliminary data suggest a mean of about 0.20 m. It is difficult to give useful figures for the width of flutes because their sides generally merge imperceptibly with the intervening rock surfaces, but most British and European examples are between 10 and 70 mm wide as measured between points half-way up their sides.

The shallower flutes have gently sloping sides and rounded floors, but deeper examples tend to be U or V-shaped in cross-section. The ribs of rock between the flutes are flat-topped or convex; the crests are never as sharp as in limestone rillenkarren.

Flutes are most frequently developed on steeply sloping rock surfaces. They are rarely found on vertical surfaces except as downward extensions of flutes that originate above on steeply sloping surfaces. The reason why flutes are seldom initiated on vertical surfaces would seem to be that the rainwater can descend quite freely, with very little frictional resistance. The force of the flowing water perpendicular to the rock surfaces may be too low to allow any significant entrainment of rock particles.

Alternatively, the water may descend not in separate trickles but as a continuous film or sheet so that loosened material is removed from the whole face rather than in discrete channels.

Most flat or gently sloping surfaces on rock outcrops where flutes have developed are either smooth or pitted by rock basins. A few, however, develop drainage channels as they weather down, which tend to wander across the surfaces, forming irregular or dendritic networks. It is only where the surfaces become steeper that the channels become sufficiently parallel to be described as flutes. The parallelism develops best on planar surfaces; surfaces that are convex in plan, such as rounded cornices, have flutes that tend to fork or splay out in a downward direction, allowing new flutes to develop in between, thus preserving approximate constancy of spacing as the area of rock surface increases.

Many flutes start at the top edges of cliffs or fallen blocks and descend only short distances. In the same way, flutes on standing stones invariably start at the tops of the stones and fade away downwards, only rarely reaching the ground surface.

Although a few flutes on the rock outcrops are located on fallen blocks that are readily accessible, the majority tend to be located high up on exposed cornices, projecting buttresses, isolated pinnacles or towers. Their frequent inaccessibility, except to rock climbers, makes them difficult to study, which is probably the main reason why they have not received the same attention as other weathering phenomena.

PROPERTIES OF ROCKS WITH FLUTES

Flutes are found only on rocks that are fairly massively jointed. Closely jointed or flaggy bedrock, even if it forms cliffs, is evidently unsuited to flute development, partly because surface runoff becomes diverted into the joints, but also because rockfalls are more likely to occur, which destroys the flutes.

It seems likely that flutes are unable to form on very pebbly sandstones and conglomerates or on sandstones that are highly porous, weakly cemented and very friable. These restrictions apart, the most obvious characteristic of flutes is their seeming indifference to the mineralogical composition and texture of the bedrock. Few cliff-forming sandstones, for example, are more dissimilar than Millstone Grit and Fell Sandstone, yet flutes occur commonly on both. The Millstone Grit is usually highly feldspathic; analyses of the Roaches and Ashover Grits, for example, indicate feldspar contents of between 11 and 27% (Aitkenhead *et al.*, 1985). Most varieties are strongly cemented and have a high compressive strength, normally 50–75 MN m^{-2} (*Natural Stone Directory*, 1990). The effective porosity is low, about 2–6%, and water absorption is correspondingly limited, normally about 2.5–3.5%. Fell Sandstone, by contrast, is composed almost entirely of quartz. Scanning electron microscope examination reveals that the grains fit loosely together, and only a little cement is present. As a result, the grains can be very easily separated by gentle rubbing using a pestle and mortar. The compressive strength of the sandstone is only 20–30 MN m^{-2} while the effective porosity averages about 10%, and water absorption is normally about 5–6%, about double that for Millstone Grit (Bell, 1978; *Natural Stone Directory*, 1990). The Elbsandstein is another highly quartzose sandstone that is rather weakly bonded.

This seeming indifference of flutes to many rock properties that could have been expected to be important for flute development is in sharp contrast to the strong preferences exhibited by several other weathering phenomena. Honeycombing, for example, frequently occurs on outcrops of the more porous and friable sandstones, but is rarely developed on outcrops of harder varieties such as Millstone Grit. Weather pits or rock basins have the opposite distribution, being common on Millstone Grit and other hard sandstones, and rarely developed on outcrops of very soft and porous sandstone, such as the Ardingly Sandstone of Kent and Sussex, which has an effective porosity as high as 27%, and absorbs as much as 9–12% water (Williams and Robinson, 1981). Sloping surfaces on the Ardingly Sandstone shed runoff only in very heavy rains; in gentle rains the water drains straight into the rock. The few weather pits that have managed to form fill with runoff during storms, but are unable to retain the water for long after the storms finish. The general dryness of the rock surfaces, and the infrequency of runoff, which greatly hinders the development of weather pits, is presumably also the reason why there are no flutes on the outcrops.

Schmidt hammer readings have been made on fluted surfaces of Millstone Grit in Derbyshire and Elbsandstein in eastern Germany. Unfortunately, the hammer head is too wide to insert into the deepest and narrowest flutes, which cannot, therefore, be tested. Furthermore, the readings that have been made may have been affected by differences in surface roughness. It is normal practice to use a carborundum stone to grind rock surfaces smooth prior to testing, but no surface preparation (other than lichen removal) was undertaken in the tests on the flutes as it would have tended to remove any weathering rind or loose weathered material from the surfaces, thus creating artificial test conditions. Despite these drawbacks, the tests provided no evidence that the rock underneath the flutes is generally weaker or more strongly weathered than the rock between the flutes. The flutes do not appear to be aligned along pre-existing lines of weakness in the rocks, except in a very few cases.

Hand-lens examination has revealed no obvious signs of weathering on the floors of the flutes, apart from the loosening of a few individual grains. There is no evidence that flute formation on the Millstone Grit is assisted by preferential decay of the feldspars. The feldspars exposed at the base of the flutes at sites such as Robin Hood's Stride appear to be no more weathered than the feldspars exposed on the ribs of rocks between the flutes. They have not been etched out leaving the quartz grains projecting, as often happens on the rock outcrops and on blocks of gritstone buried in weathered regolith. On the contrary, the larger grains of feldspar often appear quite fresh and sometimes project into the flutes, which suggests that they are actually quite resistant. Mitchell (in O'Toole, 1939) also recorded feldspar crystals projecting into flutes on the granite standing stones in Eire.

Samples of rock from the base of flutes on Millstone Grit and Fell Sandstone have been examined using a scanning electron microscope, but no evidence of weathering has been found, other than minor solutional etching of grain surfaces. No chemical alteration of the feldspars is visible.

KENDALL AND WROOT'S EXPLANATION OF THE ORIGIN OF FLUTES

Kendall and Wroot (1924, p. 121) suggested that flutes on Millstone Grit outcrops in Yorkshire are caused by 'differential erosion by rain'. They postulated that rainwater running down the rock faces tends initially to follow slight inequalities of the surface. Because the rock along these tracks is more frequently and more thoroughly wetted than the surrounding rock it tends to weather more rapidly, and with the passage of time the tracks become progressively incised and develop into recognisable flutes. The process is self-reinforcing in that the deeper the flutes become, the damper they tend to remain, so increasing the rate of weathering.

Although the Kendall and Wroot theory is convincing as far as it goes, it does not explain why the flutes tend to be regularly spaced. Initial slight inequalities of the surface could be expected to be randomly distributed. A second failing is that the theory does not identify the process of weathering that helps rainwash carve out the flutes. Kendall and Wroot emphasised that the feldspar in Millstone Grit readily decays in contrast to the quartz which is much more resistant to chemical attack. However, they stopped short of identifying chemical decay as the main weathering process creating the flutes.

Although Kendall and Wroot did not give any reason for the quasi-regular spacing of flutes, a process of capture seems likely. Flutes that are spaced some distance apart can secure more rainwater and become more rapidly incised than closely spaced flutes which compete with one another for rainwater and therefore become incised more gradually. Shallow flutes that are near to deep flutes stand a good chance of being captured as the rock surfaces are weathered down. However, deep flutes that are too far apart will not be able to drain the surfaces efficiently and new flutes will form in between, thus evening out the spacing.

CUNNINGHAM'S EXPLANATION

A very different theory was advanced by Cunningham (1964), who drew attention to the small enclosed hollows that often crowd the edges of crag tops immediately above flutes (Figure 24.6). He introduced the term crenellations to describe the hollows, and also referred to them as moulins, but neither term has found much favour with later researchers. It seems likely that the hollows are just youthful weather pits, but their close spacing is very unusual.

Cunningham supposed that the hollows slowly enlarge as a result of continued weathering, and begin to coalesce, creating narrow, winding depressions that act as drainage channels. The process is aided by rainwater, which collects in the hollows in wet weather and overflows, creating additional lengths of channel. Eventually, each group of hollows become completely replaced by an intricate network of channels, which Cunningham rather confusingly also termed crenellations.

According to Cunningham, rainwater from the hollows or the channels carves out the flutes as it spills over the edges of the cliffs and gains erosive energy. If this is correct, then the hollows develop first and are responsible for the formation of the flutes. The flutes are fairly regularly spaced because the hollows are crowded together and roughly equal sized.

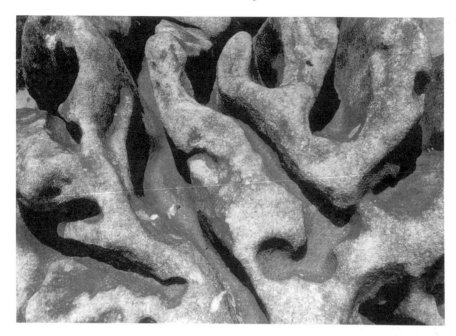

Figure 24.6 Vertical view of the edge of a joint block at Cratcliffe Tor. Weather pits or crenellations on the flat, upper surface of the block (bottom of photograph) merge with flutes (top of photograph) that run down the face of the block away from the camera

Cunningham's theory is attractive because it explains why so many flutes end headwards in small weather pits. This close association between what might be thought at first glance to be two quite separate phenomena cannot be accidental. It is observable not only on the Millstone Grit outcrops in the Peak District, where Cunningham carried out his studies, but also in other areas, for example on the Fell Sandstone outcrops in Northumberland and even on the Corsican granites (Klaer, 1956 and 1957). Nevertheless, on many sandstone crags, even in the Peak District, groups of flutes can be found that drain unpitted, smoothly sculptured surfaces, and not all groups of weather pits on the edges of crag tops discharge into flutes. Even at sites where many of the flutes start from breaches in the walls of weather pits, there are usually some that start from the tops of the walls or from ribs of rock separating the pits and, therefore, never carry overflowing rainwater from the pits. Further evidence is provided by the fluted standing stones. Several have such a narrow crest that there is little or no room for weather pits to develop. Flutes descending from the crest down opposing sides of the stones give it a characteristically jagged profile. This jaggedness is not the result of weather pitting as Cunningham appears to have supposed.

Cunningham suggested that the small, densely packed hollows or weather pits on the crag tops are initiated when a hard casing of iron-enriched sandstone, possibly a weathering rind, becomes breached as quartz pebbles or larger-than-average fragments of gritstone are weathered out. The holes vacated by the pebbles become points of attack for a variety of weathering processes.

Further study of the sites mentioned by Cunningham casts some doubt on this explanation. Hard casings are present on the sides of many cliffs, but are generally missing from the weathered tops where the weather pits occur. It seems clear that weather pits can be created at any stage during the lowering of the cliff tops by weathering, and the presence of a hard casing (and a coincidental layer of quartz pebbles) is not a necessary prerequisite. Nor is there any good reason to believe that all the networks of drainage channels on the cliff tops have evolved from small, closely spaced weather pits. Both types of crenellated surface can develop seemingly independently.

Extensive crenellated surfaces occur on top of some of the Elbsandstein cliffs in eastern Germany. They differ from British examples in their generally larger scale and in having very broad, ill-defined channels, which enclose rounded knobs of rock and gently curving ridges. The channels tend to have abrupt heads, like miniature corries, and the side channels often hang as they enter the main channels. The illusion of a glaciated terrain in miniature is furthered by channels that pass directly through ridges, as if created by glacial transfluence or diffluence. It is unclear how these crenellated surfaces were created, but like British examples, they often connect with flutes where they reach the cliff edges.

RAINWATER EROSION

Although rainwater flows down the flutes in wet weather, often at a considerable speed, there is no evidence that it has much erosive power. The flutes normally accumulate such small quantities of debris between rains that there is almost nothing for the water to pick up to use as an erosive tool. Moreover, the flutes reach their maximum depth only short distances below their heads, before the water could be expected to pick up much debris even supposing it existed.

The obvious conclusion is that flute formation is due largely to one or more processes of weathering, and that the periodic discharges of water flush away the weathering products, but have little power to abrade the rock. It is clear that the weathering processes, whatever they are, operate directly on the flute surfaces, releasing individual grains or crystal fragments, and have little effect on the rock at depth.

PROCESSES OF MECHANICAL WEATHERING

To be acceptable any theory of flute formation must explain why flutes are present on the tops of many standing stones and why on rock outcrops they often occur in the most exposed and inaccessible positions on the tops of pinnacles and narrow buttresses. This distribution would seem to rule out frost weathering as a major cause of fluting because cold air could be expected to drain off the flutes during all but the most prolonged spells of frosty weather. Moreover, the rocks on which flutes occur are not noted for their frost susceptibility. Although in the case of Millstone Grit it is claimed that 'in severe winters frost may lift off small flakes and also produce granular disintegration of patches of less well cemented rock' (Brown, 1976, p. 10), it is doubtful if frost is capable of much weathering of granite and some of the other types of sandstone on which flutes occur.

Salt weathering may explain occurrences of flutes on rocky coasts, for example in Brittany (Godard *et al.*, 1972), but it would seem to be of very little importance at inland sites in Britain and Europe. Fluting at these sites occurs on surfaces that are repeatedly rainwashed and are generally free of salts. Salts tend to accumulate on rock surfaces that are protected from the rain and are slow to dry out, for example in cavities and under overhangs at the base of cliffs. Numerous efflorescences of gypsum and 'alum' salts, which Lentschig-Sommer (1961) identifies as alunogen, $Al_2(SO_4)_3 \cdot 16H_2O$, and alaun, $K(NH_4)Al(SO_4)_2 \cdot 12H_2O$, occur on the lower parts of the sandstone cliffs in eastern Germany and the Czech Republic, but never on the exposed upper surfaces where the flutes develop.

Insolation weathering can also be ruled out because the flutes are developed on rock surfaces of every possible orientation. They are as common on north-facing surfaces as on south-facing ones which receive greatest insolation.

Field observations confirm that the sites where flutes occur tend to be the first to become thoroughly wetted during rains because they are upstanding and relatively exposed. They also quickly dry off after rains. Not all sites face the sun, but they are open to the wind and get the benefit of the slightest breeze. Wind blast not only assists drying but may also aid rain penetration. In addition, it seems quite likely that rain draining off the tops of cliffs is blown back up the flutes during sudden gusts and spattered as spray on the cliff edges, thus initiating the weather pits. If this conjecture is true, then the weather pits may form after the flutes, and not before as proposed by Cunningham.

The frequent cycles of wetting and drying to which the flutes are exposed may increase the intensity of mechanical weathering. Many types of sandstone, including Millstone Grit, expand appreciably on wetting and contract again on drying (Schaffer, 1932). This moisture movement, as it has been termed, may be sufficient to cause surface grains to become detached, although no evidence has yet been collected to prove that this is the case. It is difficult to believe, however, that moisture movement is a significant process in the formation of flutes on granite.

PROCESSES OF CHEMICAL WEATHERING

Fluting on siliceous rocks in the tropics has often been attributed to chemical weathering, particularly solution and hydrolysis (Twidale, 1982). As has already been noted, many flutes in Britain and Europe are developed in sandstones that consist almost entirely of quartz, such as Fell Sandstone and Elbsandstein, and it might seem perverse to propose a solutional origin for the flutes since bulk quartz has only very limited solubility in pure water within the range of temperatures normally experienced in this area. However, grains of quartz in sandstone may have coatings of amorphous silica that are very much more soluble than their interiors, and the presence of organic acids may increase the solubility of quartz still further (Young and Young, 1992). In addition, it is important to note that the cement between the grains may be more soluble than the grains themselves. In the temperate zone sandstones in which flutes occur, solutional weathering is probably concentrated along the grain boundaries and it may well loosen the grains without materially affecting their interiors, allowing rainwash to carry the grains away. On granites solutional weathering may have a similar action, preferentially following the crystal boundaries, thus releasing the crystals in a relatively intact state.

Confirmation that appreciable solution of quartz actually occurs in temperate zone sandstones is provided by the silica-cemented surface rinds or crusts that often develop on their outcrops, especially on their lower surfaces (Robinson and Williams, 1987). The cement is deposited by evaporating pore water, which enters the sandstones as rain or as groundwater seepage from adjacent strata. In coastal environments the presence of salt and the weak alkalinity of seawater may enhance rates of quartz solution.

The main obstacle to believing that chemical attack plays an important part in flute formation on siliceous rocks in Britain and Europe is the well-preserved state of the feldspars in granite flutes and flutes on the feldspathic Millstone Grit. Feldspar is much more susceptible to chemical weathering than quartz, and in the presence of water decomposes to clay by hydrolysis. As with solution, hydrolysis probably penetrates faster along the boundaries of the crystals or grains than through their interiors, thus causing granular weathering. Given that feldspar is very liable to decay, the fact that crystals or grains of feldspar often project into flutes on granite and Millstone Grit may seem odd. But the projecting feldspars tend to be larger than the average crystals or grains that make up the rocks, and it would seem probable that they are less easily removed by granular weathering. The real puzzle is that the rocks show few signs of preferential removal of feldspars even when the quartz and feldspar crystals are the same size. As has already been mentioned, Millstone Grit outcrops often have very pitted surfaces because the grains of feldspar have been selectively removed leaving the more resistant quartz grains upstanding. This characteristic surface 'etching' is absent or, at the most, only weakly developed on the floors of the flutes, which suggests that chemical weathering may not be as important in flute formation as is generally supposed.

It must be emphasised that solution and hydrolysis operate only while the rock surfaces are wet. Although fluted surfaces soon start to carry runoff after rain begins to fall, they dry out quickly when the rain finishes. One can readily understand why rock basins that retain water between rains should become enlarged as a result of chemical weathering, but it is less obvious why such weathering should be important in creating self-draining structures such as flutes. Because of the long residence time of the water in rock basins it probably becomes saturated between rains, whereas the water that descends the flutes is constantly changing and remains aggressive, but this advantage that the flutes possess persists only as long as the rains last, and the inconstancy of the flow must greatly limit the mobilisation of silica in the flutes.

Some flutes on sandstone outcrops have become incised through a surface crust which helps waterproof and strengthen the rock. Once the flutes have breached the protective crust, weathering can proceed more rapidly on their floors than on the rock surfaces in between which still retain the crust. However, many flutes occur on surfaces that are uncrusted or only lightly crusted, and sometimes there is a thicker crust within the flutes than on the surfaces in between. It would appear, therefore, that crust formation can both help and hinder flute development.

Twidale (1982) has argued that flutes on granites are initiated by chemical weathering processes and groundwater flows operating at the weathering front, which separates fresh bedrock from overlying weathered granite or saprolite. The flutes become exposed at the surface when the overlying weathered material is removed by erosion. This theory does not provide a satisfactory explanation for British and European flutes. There is no evidence that the flutes on the standing stones have developed below the ground surface and then been exhumed. In the case of the rock outcrops, the flutes are often found on the highest pinnacles and rock towers, which would be the first parts of the outcrops to be exposed if an overlying weathered mantle had been removed. They would thus have had a relatively long time to have become remodelled or totally destroyed by subaerial weathering processes, and their existence indicates that it is actually subaerial, and not subsurface, weathering that is responsible for their formation.

BIOLOGICAL WEATHERING

There remains the possibility of biological weathering. At most sites lichens and surface algae grow more successfully within the flutes than on the rock surfaces in between because there is more moisture and shelter. However, at other sites the lichens and algae avoid the flutes and colonise the interflute areas or show no clear preferences.

As pointed out by Twidale (1982) and other writers, it is unclear whether lichens and algal coatings have a protective role or enhance weathering. Because they partially shield the rock, they could be expected to reduce rates of weathering, and there can be little doubt that while each lichen is alive it hinders the removal of weathered debris. On the other hand, there is a possibility that lichens and surface algae increase weathering by retaining moisture on the rock surface and by secreting acid substances that help dissolve the rock. In addition, lichens secure themselves to the rock with hyphae that penetrate between the sand grains or crystals and may prize them apart. Lichens that grow on siliceous rocks are generally thought to be less penetrative and destructive than lichens on calcareous rocks (Schaffer, 1932), but nevertheless they may significantly aid flute formation at some sites.

The majority of sites where flutes occur are regularly visited by birds and become conspicuously coated with faecal splats. Standing stones, in particular, are often used as perches, and in dry weather accumulate large numbers of splats, which wash either into or off the rock during rains. Observations at The Roaches and at Kyloe Crag suggest that cliff edges and projections are also favoured perching places and receive unusual numbers of splats. Doubtless this continuing avian bombardment accelerates rock weathering at sites where flutes occur.

AGE OF THE FLUTES

Many flutes are apparently still being excavated by rainwash and weathering, but others are probably relict forms, although not necessarily very old. On some outcrops, for example, the growth of trees has resulted in groups of flutes becoming infilled with soil and plant litter, which has apparently arrested their development. One can also find flutes that seem to be undergoing slow destruction. For instance, at Robin Hood's Stride in Derbyshire, the thick, black weathering rind that has formed within some of the deeper flutes on the Weasel Pinnacle suggests that they are no longer being actively excavated. The rind is absent from much of the rock surface in between the flutes, which is continuing to undergo weathering. The end result could be the elimination of the flutes.

Demek (1964a and b) and Czerwinski and Migon (1993) considered that the flutes on granite outcrops in the Czech Republic and Poland are relict features because they are generally covered with lichens and mosses, and are not present on blocks created by frost wedging during periglacial periods. They suggested that the flutes originated in the late Tertiary under a very different climatic regime from the present. However, they did not explain how the flutes have survived weathering and erosion during the Quaternary Era, and even if it is accepted that the flutes are no longer forming, there would appear to be no good reason to suppose that they are of any great age.

Archaeological evidence indicates that in Britain most standing stones were erected from Middle Neolithic times onwards until the Middle Bronze Age, between about 3370 and 1200 BC (Burl, 1988). The highly weathered Stone of Setter on Eday in the Orkney Isles, which has been likened to a huge hand reaching out of the ground (Burgher, 1991), is believed to date from the 2nd millenium BC (Ritchie, 1985). Similarly, the deeply fluted stones at Machrie Moor (Isle of Arran), Nine Stone Close (Derbyshire) and the Devil's Arrows (Yorkshire) are thought to have been erected in the Early or Middle Bronze Age, between 1975 and 1200 BC. Since the fluting can be presumed to have formed after these stones were erected it cannot be more than 3200–4000 years old. Evidently, fluting is a comparatively rapid process of rock sculpture where conditions are favourable. Since the climate has remained broadly the same during the last 4000 years, there is no need to invoke major changes in climate in order to explain flute formation, although it is important to recognise that forest clearances may have been important in facilitating the growth of flutes.

There is clear evidence that flutes in Britain can develop even more rapidly than the 3200 to 4000 years suggested by the prehistoric standing stones. Sueno's Stone at Forres in Moray is Scotland's tallest surviving cross-slab (Richie and Breeze, 1989). Standing 6.5 m high, it is covered in remarkable Pictish carvings depicting a violent battle, probably with the Scots in 966. The top of the stone is quite strongly fluted, so much so that a metal cap has been fitted to try to protect its top from further weathering. Yet when the stone was discovered in the early 18th century it lay buried in the ground. The fluting has formed only while the stone has been in an upright position, before it either fell or was cast down, and after its re-erection in 1726. It has, therefore, had less than a thousand years in which to develop.

Many of the coping stones of Fell Sandstone on the battlements of Alnwick Castle in Northumberland are badly weathered, and in some cases have developed quite deep, regularly spaced flutes. The flutes are about the same distance apart as the flutes on the Fell Sandstone outcrops, but are only about 0.3 m long, being limited in length by the size of the coping stones. Construction of the castle began in the 12th century, but many of the fluted stones cap walls and towers that date from the 14th century. Warkworth Castle, a few kilometres south of Alnwick on the River Coquet, is also built of Fell Sandstone, and has similarly fluted battlements and towers, dating from the late 14th and early 15th centuries. In Derbyshire many cottages and field barns built in the last two centuries have end-walls capped by slabs of sandstone. Where these slabs meet at the roof crest they are very exposed to the weather and quickly wear down. In some cases, miniature channels have developed down their surfaces, which appear to represent the first stages in flute formation. The evidence from Northumberland and Derbyshire suggests, therefore, that quite deeply incised flutes are able to develop in a mere 500 years in susceptible types of sandstone in highly exposed situations.

ACKNOWLEDGEMENTS

The authors thank Dr Karl-Heinz Schmidt (Freie Universität Berlin) and Dr Stephen Harrison (Middlesex University) for providing useful information about flute localities.

REFERENCES

Aitkenhead, N., Chisholm, J. I. and Stevenson, I. P. (1985). *Geology of the Country around Buxton, Leek and Bakewell.* British Geological Survey Memoir, HMSO, London.

Alexandrowicz, Z. (1990). The optimum system of tors protection in Poland. *Polska Akademia Nauk, Zaklad Ochrony Przyrody i Zasobów Naturalnych,* **47**, 277–308.

Bell, F. G. (1978). Petrographical factors relating to porosity and permeability in the Fell Sandstone, *Quarterly Journal of Engineering Geology,* **11**, 113–126.

Brown, R. D. (1976). Geology. In Griffiths, B. and Wright, A. (eds), *Rock Climbs in The Peak: Volume 1—Stanage Area,* British Mountaineering Council, pp. 7–10.

Bulow, K. (1942). Karrenbildung in kristallen Gesteinen? *Zeitschrift für Deutsch. Geol. Geselt.,* **94**, 44–46.

Burgher, L. (1991). *Orkney—An Illustrated Architectural Guide.* Royal Incorporation of Architects in Scotland, Edinburgh. RIAS/Landmark Trust Guide No. 11.

Burl, A. (1976). *The Stone Circles of the British Isles.* Yale University Press, New Haven.

Burl, A. (1985). *Megalithic Brittany.* Thames and Hudson, London.

Burl, A. (1988). *Prehistoric Stone Circles,* 3rd edition. Shire Publications, Princes Risborough.

Carlé, W. (1941). Karrenbildung im granit der galischen Kuste bei Vigo (Norwestspanien). *Geol. Meere Binnengewasser,* **5**, 55–63.

Cunningham, F. (1964). A detail of process on scarp edges of Millstone Grit. *East Midland Geographer,* **3**, 322–326.

Czerwinski, J. and Migon, P. (1993). Mikroformy wietrzenia granitow w masywie Karkonosko-Izerskim. *Czasopismo Geograficzne,* **64** (3–4), 265–284.

Demek, J. (1964a). Castle koppies and tors in the Bohemian Highland (Czechoslovakia). *Biuletyn Peryglacjalny,* **14**, 195–216.

Demek, J. (1964b). Slope development in granite areas of Bohemian Massif (Czechslovakia). *Zeitschrift für Geomorphologie,* Supplementband **5**, 82–106.

Ford, D. C. and Williams, P. J. (1989). *Karst Geomorphology and Hydrology.* Unwin Hyman, London.

Godard, A. *et al.* (1972). Quelques enseignements apportés par le Massif Central français dans l'étude géomorphologique des socles cristallins. *Revue de Géographie Physique et de Géologie Dynamique,* **14**(3), 265–296.

Hsi-lin, T. (1962). The pseudokarren and exfoliation forms of granite on Pulau Ubin, Singapore, *Zeitschrift für Geomorphologie,* **NF5**, 302–312.

Kendall, P. F. and Wroot, H. E. (1924). *The Geology of Yorkshire.* Privately printed, Vienna.

Klaer, W. (1956). Verwitterungsformen in granit auf Korsika. *Petermanns Geogr. Mitt. Erganzungsheft.,* **261**, 1–146.

Klaer, W. (1957). 'Verkarstungserscheinungen' in silikatgesteinen. *Abhandlungen Geographischer Institute der Freien Universitate Berlin,* **5**, 21–27.

Lentschig-Sommer, S. (1961). Ein vorkommen von Alunogen (Keramohalit) im Elbsand-steingebiet. *Jb. Staatl. Mus. Mineral Geol.*, 109–110.

Natural Stone Directory—1991, 8th edition (1990). Stone Industries, Ealing Publications, Maidenhead.

O'Toole, E. (1939). A group of grooved standing stones in North Carlow. *Journal of the Royal Society of Antiquaries of Ireland*, **69**, 99–111.

Ritchie, A. (1985). *Exploring Scotland's Heritage—Orkney and Shetland*. Royal Commission on the Ancient and Historical Monuments of Scotland, HMSO, Edinburgh.

Ritchie, A. and Breeze, D. J. (1989). *Invaders of Scotland*. Historic Buildings and Monuments, HMSO, Edinburgh.

Robinson, D. A. and Williams, R. B. G. (1987). Surface crusting of sandstones in southern England and northern France. In Gardiner, V. (ed.), *International Geomorphology, 1986, Part II*, Wiley, Chichester, pp. 623–635.

Rondeau, A. (1961). *Recherches géomorphologique en Corse*. Armand Colin, Paris.

Sampson, D. N. (undated). The geology of the Edale–Kinderscout area. In Nunn, P. J. (ed.). *Rock Climbs in The Peak: Volume 7—The Kinder Area*, British Mountaineering Council, pp. 22–25.

Schaffer, R. J. (1932). *The Weathering of Natural Building Stones*. Department of Scientific and Industrial Research, Building Research Special Report No. 18, HMSO, London.

Twidale, C. R. (1982). *Granite Landforms*. Elsevier, Amsterdam.

Vidal Romani, J. R. (1989). Granite geomorphology in Galicia (NW Spain). *Cuarderno Lab. Xeoloxico de Luxe. Coruna*, **13**, 89–163.

Williams, R. B. G. and Robinson, D. A. (1981). Weathering of sandstone by the combined action of frost and salt. *Earth Surface Processes and Landforms*, **6**, 1–9.

Young, R. and Young A. (1992). *Sandstone Landforms*. Springer-Verlag, Berlin.

25 Limestone Weathering in the Supra-tidal Zone: An Example from Mallorca

C. A. MOSES and B. J. SMITH
The Queen's University of Belfast, UK

ABSTRACT

Coastal weathering gradients are investigated through a study of the Miocene limestone platform on the southern coast of Mallorca. Two broad supra-tidal divisions are identified —a spray zone, close to the edge of the platform, in which salt weathering dominates to produce alveoli and fretted hollows, and a landward zone in which solution dominates to produce solution pits and pans. This apparent simplicity, compared with complex zonations identified in other areas, is largely a response to the micro-tidal regime of the area. Similar morphological zonation is observed on coastal defence works and break-waters in southern Mallorca, and suggests that the Miocene platform provides a model for the future behaviour of these structures.

INTRODUCTION

Limestone terrains are generally considered to be dominated by both inorganic and organic/biological dissolution. Discussions on limestone coastal platforms are no different and emphasis is placed on dissolution and bioerosion simply because they are more noticeable on this rock type. This view, as Trudgill (1985, p. 156) points out, neglects two things: first, these processes also occur on other rock types, and secondly, limestone coastal platforms are subject to processes common to all coasts. Furthermore, most studies have concentrated on the role of one particular process rather than evaluating the relative importance of a range of processes. For example, Lundberg (1977), Schneider (1977), Jones (1987 and 1989) and Viles (1987b and 1988) all stress the role of biological processes on limestone in a coastal environment. However it seems probable that a range of mechanisms, which are not mutually exclusive, operate in these areas. These include abrasion, wave and spray impact, wetting and drying, and salt action, as well as dissolution and organic action. The interaction of these mechanisms with rock properties is reflected in the morphology of coastal platforms. Morphological features frequently vary with distance inland from the platform edge, as certain weathering mechanisms become dominant and others cease to operate effectively. Working in the Bristol Channel area, Ley (1979) regarded such morphological assemblages as the response to a varying 'erosive energy input' over the platform foreshore. Because of this, it is to be expected that

Rock Weathering and Landform Evolution. Edited by D. A. Robinson and R. B. G. Williams.
© 1994 John Wiley & Sons Ltd

weathering processes, products and their morphological expression will vary with coastal energy conditions as well as broader climatic controls of weathering. The purpose of this paper is, through a study of the Miocene platform on the southern coast of Mallorca, to investigate coastal morphological zonation in a Mediterranean environment and to identify possible weathering gradients which reflect the unique character of this area. A number of complementary analytical techniques are used to elucidate the role of various processes and their mechanisms of operation. The implications of these results are then examined in the context of the durability of coastal engineering structures in a Mediterranean environment.

LOCATION

The southern coast of Mallorca consists of a series of benches formed as the Pleistocene sea-level fell. The oldest has been assigned to Tyrrhenian I at 15–25 m above present sea-level, and the youngest to Tyrrhenian III at 2–5 m above present sea-level (Butzer and Cuerda, 1962).

   The field area, at Cala Pi (Figure 25.1), has been identified as representative of the oldest sea-level Tyrrhenian I. It forms a prominent rocky coastal bench which

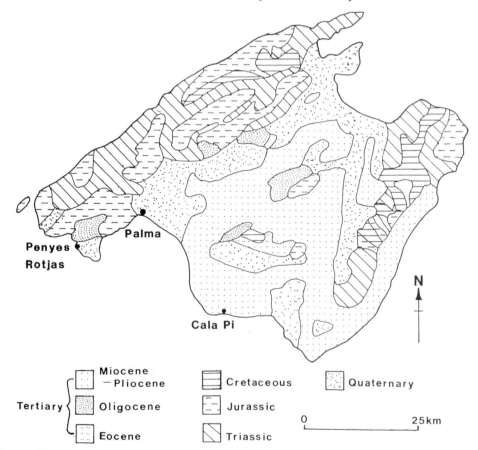

Figure 25.1 Geology of Mallorca and location of field area (from Crabtree *et al.*, 1978). Reproduced by permission

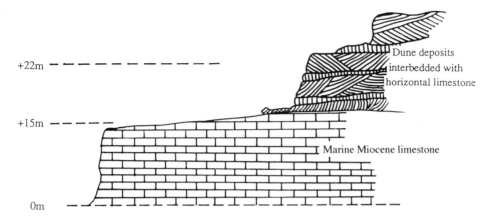

Figure 25.2   Geology of the coastal platform at Cala Pi (adapted from Cuerda, 1975)

faces east and is exposed to strong onshore winds. The present-day tidal range is only 0.2–0.4 m. The geology of the platform is predominantly a marine Miocene limestone backed by calcite-cemented calcareous dune deposits interbedded with thin layers of horizontal limestone (Figure 25.2).

Mallorca's climate is Mediterranean: winters are typically windy, mild and wet and summers are relatively calm, hot and dry. This regime is slightly modified in the south of the island where the dry season is accentuated and mean annual rainfall at Palma is 415.2 mm and mean annual temperature 17.1°C (Crabtree *et al.*, 1978).

MORPHOLOGICAL ZONATION OF THE COASTAL PLATFORM

The cliff face, or splash zone, is characterised by small, closely spaced, roughly circular depressions of a few centimetres in diameter and depth. Such features are variously termed honeycomb weathering or alveoli and are attributed to a number of origins including wind erosion, exfoliation, frost shattering and salt weathering (Mustoe, 1982; Trenhaile, 1987). The most favoured explanation, according to Sunamura (1992), is that of salt crystallisation as evidenced by scanning electron microscopy and/or X-ray diffraction analysis (McGreevy, 1985; Matsukura and Matsuoka, 1991).

Close to the platform edge, in the zone subjected to salt-laden spray, the morphology is dominated by alveoli and fretted hollows. Alveoli of diameter 10–50 mm and depth 10–20 mm occur on both horizontal and vertical surfaces. Many have a basal coating of granular rock meal which is attributed to the disintegration of the rock by the mechanical process of salt weathering (Mustoe, 1982; McGreevy, 1985). Solution pans are also present close to the platform edge. The origin of these features, as their name implies, is widely attributed to dissolution (Sweeting, 1972). Biochemical processes are also important in their formation both directly, causing water in basins to become undersaturated with respect to $CaCO_3$ at night (Emery, 1946; Trudgill, 1976), and directly as a result of biological corrosion of the substrate by algae (Schneider, 1977). Many of the solution pans have overhanging side walls, some of

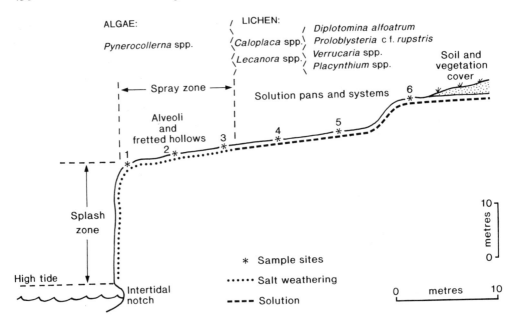

Figure 25.3 Weathering zones differentiated on the basis of morphology. Location of sample sites and microflora species zonation are also shown

which are 'fretted' by alveoli. All have basal coverings of blue-green algae (species unidentified) while some also contain a layer of salt crystals. This zone of alveoli and fretted hollows extends some 7 m inland from the cliff edge. Salt spray is also concentrated here. In general, the rock surface between solution pans has a rough texture and much of it is colonised by algae (see Figure 25.3). Samples fractured perpendicular to the weathered surface show chasmolithic organic material as a green line which extends some 5 mm below the surface.

Beyond the cliff edge zone, solution pans become larger and have a greater degree of connectivity. Alveoli are no longer present and rock surfaces are smooth. Algae are concentrated in the bottoms of the solution pans which provide a moister microenvironment with the presence of occasional standing water. The more arid intervening areas between pans are colonised mainly by lichens (see Figure 25.3 for species distribution). Such biological zonation, from algae in the splash and spray zones through to lichens in the supra-tidal back platform area, is common. It is attributed to the degree and frequency of moistening by tides, waves, splash, spray and rainfall, with organisms arranged according to their ecological tolerances (Schneider, 1977).

Surface morphology, therefore, allows the subdivision of this coastal platform into two broad process zones (Figure 25.3). However, on the basis of morphology alone, it is difficult to state unequivocally which processes operate since there have been many hypotheses put forward as to the genesis of alveoli and solution pans. Mechanical salt weathering appears, though, to be the dominant process within the splash and spray zone, some 7 m from the cliff edge. Beyond this zone of concentrated spray, dissolution seems to become the dominant shaping force with salt weathering of little or no importance. However, although it is possible to infer which processes are operative from morphology alone, the actual mechanisms are not clear. For example, it is not known which salts are responsible for the formation of alveoli in the spray and splash zones. Nor can any comment be made on which salt weathering mechanisms operate. Furthermore, although the surface of the 'solution zone' is smoother than that of the 'salt weathering' zones, indicating chemical as opposed to mechanical weathering, the actual mechanism of surface recession is not clear. A number of complementary analytical techniques have been used to investigate these problems.

SAMPLING AND RESULTS

Water Chemistry Analysis

The aim of the water chemistry analysis was simply to illustrate any spatial variations in ion concentration associated with distance from the platform edge at any one time. All samples were collected on the morning of the same day in summer. Had the intention been to investigate temporal variations it would have been necessary to have collected samples over a longer time period.

Water samples were collected from standing water in hollows and pans in a series of four parallel profiles normal to the cliff edge (Figure 25.3). These were analysed for the major cations present in sea water, Na^+, K^+ and Ca^{2+}, and for the anion SO_4^{2-} using atomic absorption spectrophotometry. Sample sites (1–3) were located within 7 m of the cliff edge, in the spray zone. The remaining sample sites (4–6) were located between 7 and 22 m up to the line where vegetation begins. For comparative purposes a sample of sea water was also analysed for the same ions (Figure 25.4).

Within the spray zone ion concentration is evident. Hollows are fed with a fairly constant supply of salts in the form of spray. Evaporation of the standing water concentrates these ions to levels greater than those of sea water itself. Beyond the spray zone these ions are still found, though in much lower quantities. Small amounts of spray carried by strong onshore winds and concentrated by runoff into the pans are possibly responsible for this.

Mineralogical Analysis: X-ray diffraction

Specimens of bedrock and weathered surface were collected from both morphological zones to establish mineralogy. The surface weathered layer was removed to a depth of no more than 5 mm which, from analysis of hand specimens, was the depth to which the weathered zone extended. Samples from the solution zone were ground to <63 μm for analysis of the powdered sample. Samples from the spray zone were ground to <63 μm and the water soluble components extracted. Then 0.5 g of each were heated in deionised water for 60 minutes, particulate material removed by centrifuging and the remaining solution concentrated by rapid boiling in a microwave oven. Solutions were then pipetted onto glass slides and dried beneath an infra-red lamp. Samples of pure rock from beneath the weathered zone were also analysed to determine mineralogy. Samples were leached in 100 ml of deionised water for 24

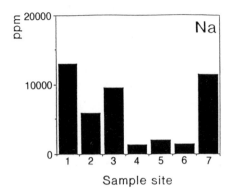

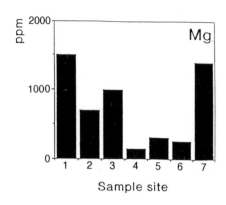

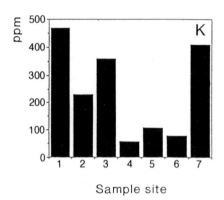

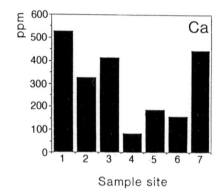

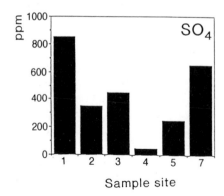

Figure 25.4 Ion concentration (ppm) of standing water. (1, 2, 3—spray zone; 4, 5, 6—solution zone; 7—sea water)

hours, to remove any soluble components which may have penetrated beyond the general depth of the weathered zone, and dried at a temperature of 105°C for a further 24 hours. They were then ground to <63 μm and the powdered sample analysed. All samples prepared for X-ray diffraction were analysed using a Siemens D5000 diffractometer (CuK).

Analysis of the parent rock identified major peaks for calcite with some minor peaks for dolomite. This compares with the results for 25 samples of weathered rock shown in Figure 25.5. In the spray zone appreciable amounts of gypsum, epsomite and halite were detected. Five samples from the cliff face showed the presence of gypsum and epsomite, three showed halite and epsomite. Seven samples from approximately 3 m inland from the cliff edge in the spray zone showed gypsum. Beyond this zone only trace amounts of gypsum were detected and very small amounts of quartz. Seven samples from approximately 15 m inland showed mainly calcite with trace amounts of gypsum and quartz. It is likely that both the gypsum and quartz are windblown—the former from spray, the latter from the dune deposits backing the platform. The calcite is simply derived from the limestone substrate as shown by analysis of unweathered samples.

From the data presented so far it is possible only to infer the causes of weathering in each zone, even though it is now known which salts are concentrated in the spray zone. The evidence is indirect, with the presence or absence of particular salts merely implying the mechanisms responsible for the formation of particular morphological features. Two techniques which may give a clear indication of the causes of weathering are scanning electron microscopy (SEM) and energy dispersive X-ray analysis (EDS). These give direct evidence of how the rock is weathered at the scale of individual crystals (Lewin and Charole, 1978; McGreevy, 1985).

Scanning Electron Microscope (SEM) and Energy Dispersive X-ray (EDS) Examination

Rock samples collected at sites adjacent to water sampling sites shown in Figure 25.3 were analysed by optical microscope to select samples for more detailed electron microscope analysis. Approximately 50 samples were analysed in this way. Duplicate samples of approximately 100 mm² were then mounted on 10 mm aluminium stubs for examination, one set using a Jeol 35CF Scanning Electron Microscope and the other using a Jeol 733 Superprobe. A summary of the results is shown in Table 25.1. This table presents the relative abundance of a range of features found at microscopic level and identified as representative of a range of process mechanisms by a number of authors. Crystal boundary widening, cleavage widening, V-in-V and blocky etching have all been used as evidence for inorganic dissolution (Gillot, 1978; Viles, 1990). Circular etch pits and tunnels formed at this scale are evidence of biological corrosion by algae and/or fungi (le Campion-Alsumard, 1979; Jones, 1987 and 1989; Viles, 1987a and b). In this study no attempt is made to suggest whether these features are formed by algae or fungi unless the species is known. They are simply used as evidence for the occurrence of biological corrosion.

Examination by SEM and EDS confirms the presence of gypsum and halite, and also shows crystals of mirabilite on samples from the spray zone. All three salt types occupy micro-depressions on the rock surface, but are also present as coatings and

(a)

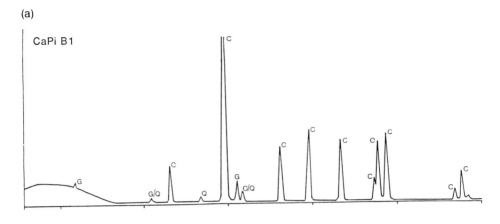

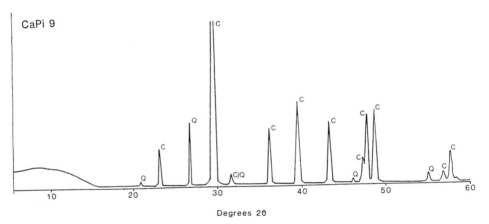

Degrees 2θ

Figure 25.5 X-ray diffraction patterns of weathered rock surface. (a) Solution zone (C—calcite, $CaCO_3$; G—gypsum, $CaSO_4 \cdot 2H_2O$; Q—quartz, SiO_2). (b) Spray zone (E—epsomite, $MgSO_4 \cdot 2H_2O$; G—gypsum, $CaSO_4 \cdot 2H_2O$; H—halite, $NaCl$)

as individual crystals in rock interstices (Figures 25.6 and 25.7). Surface coatings of salts such as halite make little contribution to mechanical breakdown of the rocks. Where, however, salts have crystallised within the rock there is evidence that this is associated with fracturing of the confining crystals and grains (Figure 25.7).

Examination of samples from the solution or landward zone detected no obvious evidence of salt weathering; instead solutional etching patterns are widespread (Figure 25.8). These take two forms: first, micrite cement around larger sparite crystals is preferentially dissolved and individual crystals are loosened. These are readily removed either by wind action, rainfall or runoff water leading essentially to solutional disaggregation of the rock surface (Trudgill, 1976). Secondly, there is preferential dissolution along lines of cleavage within calcite crystals. Again, this results in a weakening of the rock fabric and, in this case, fragments of crystals are readily removed by wind or water action, resulting in rock disintegration.

(b)

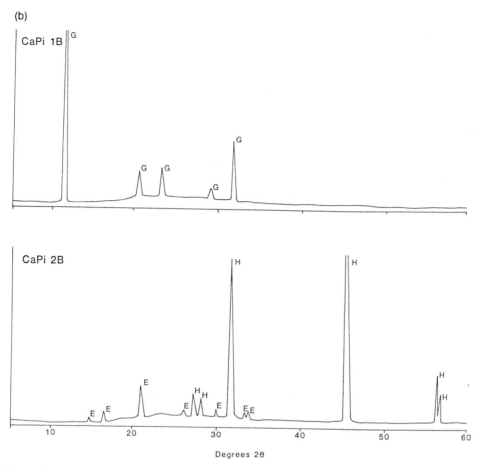

Figure 25.5 (Continued)

Biological corrosion is clearly evident across the platform (Table 25.1, Figure 25.9). Circular etch pits and tunnels occur in association with algal cover in the spray zone. Adjacent rock surfaces with no algal cover are, however, subject to salt crystallisation. In the solution zone, where a lichen cover is present, circular etch pits and tunnels are also evident with uncolonised surfaces exhibiting features associated with inorganic etching.

DISCUSSION

The morphological zonation on the coastal platform is clearly a response to changes in weathering characteristics. Examination of platform morphology shows a clearly defined zone fretted by alveoli in which mechanical weathering predominates. This is the region of the platform in which salts are in constant and abundant supply in the form of both direct splash and windblown spray. However, for morphology to be used as evidence for weathering processes, links between process and form must be

Table 25.1 Summary of weathering features identified by SEM analysis

Site code	Crystal boundary widening	Cleavage widening	V-in-V etching	Blocky etching	Microfractures	Circular etch pits and tunnels	Salt crystallisation
1	0	0	0	*	*	**	**
2	0	0	0	*	**	*	**
3	*	0	*	*	*	*	**
4	*	0	*	**	0	0	0
5	**	**	*	*	0	*	0
6	*	0	0	*	0	**	0

Key: 0 = not evident
* = evident
** = particularly well developed

(a)

(b)

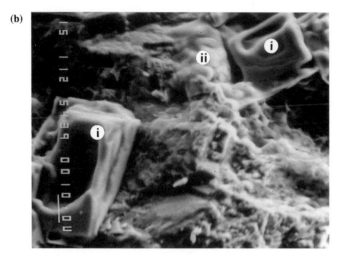

Figure 25.6   (a) Mirabilite (Na$_2$SO$_4$) crystals occupying, and growing against the sidewalls of depressions in the limestone surface. (b) Halite present as cubic crystals (i) in the interstices between calcite crystals, and as an amorphous coating (ii) on calcite crystal surfaces

demonstrated (Viles, 1987a and b). The alveoli and irregular microtopography of the zone close to the platform edge would appear to be the direct result of mechanical salt weathering (Figures 25.6 and 25.7). This zone not only receives a constant supply of salts, but is likely to experience frequent wetting and drying cycles as the result of wave splash interspersed with calm periods. Hence salts are constantly being deposited both on the rock surface and in pore spaces, with crystallisation due to evaporation and phase changes from anhydrous to hydrated forms in the case of gypsum. Salt efflorescences are visible evidence of the concentration and crystallisation of salts here. Yet these salts, crystallised in unconfined space, are not responsible for the limestone disintegration into a rock meal. Rather it is those salts occurring in

(a)

(b)

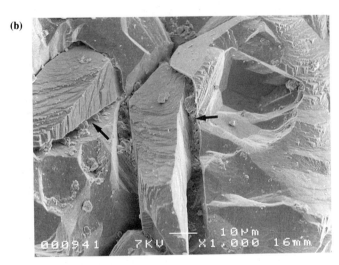

Figure 25.7 (a) Disaggregation of individual calcite crystals with halite present within microfractures (shown by arrows). (b) Microfracture development across calcite crystals (shown by arrows). Once initiated, microfractures could be propagated as the result of forces exerted by the presence of halite crystals

confined spaces, such as rock interstices or pore spaces, which cause mechanical breakdown. Individual crystals are prised from their position by halite crystals growing in pore spaces between crystal boundaries and by subsequent differential expansion on heating (Figure 25.7a). Not only are complete crystals removed intact, but fragments of crystals may be removed as the result of microfracture propagation across crystals (Figure 25.7b). Such fracturing could be achieved by the pressure of crystal growth (Cooke, 1981) in response to concentration of a saline solution through evaporation, or a decrease in temperature which can be sufficient to initiate salt

(a)

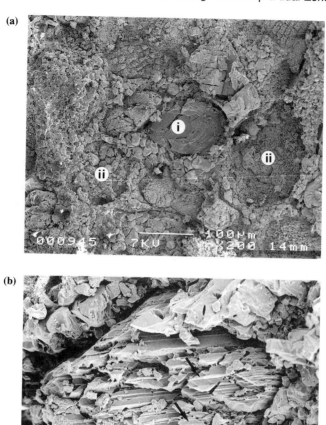

(b)

Figure 25.8 (a) Irregular micromorphology produced by solutional disaggregation. Preferential dissolution of cement material around larger calcite crystals (i) leads to their loosening and release, as evidenced by surface micropits (ii). (b) Preferential dissolution along lines of cleavage within individual calcite crystals causes dissection and eventual disintegration by the release of tiny fragments. Arrows show fragments almost ready for release

crystal growth within rock pores and cracks. Where salts can exist in a number of hydrated phases, volume change can be brought about by changes in air temperature and/or humidity. For example, when anhydrite ($CaSO_4$) takes up moisture to become gypsum ($CaSO_4 \cdot 10H_2O$) at 100% relative humidity and 0°C, it can exert a pressure of 2190 atmospheres (Goudie, 1985). This is likely to be an important mechanism within the spray zone where the rock surface experiences continual wetting and drying by salt spray. Halite does not, however, have a hydrated phase above 0°C so this type of volumetric expansion cannot be invoked to explain any associated weathering

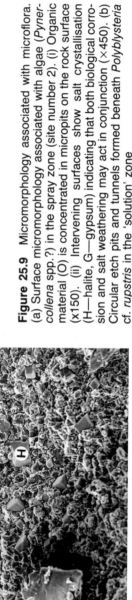

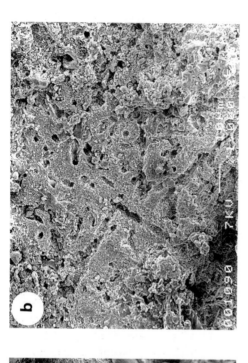

Figure 25.9 Micromorphology associated with microflora. (a) Surface micromorphology associated with algae (*Pynercollena* spp.?) in the spray zone (site number 2). (i) Organic material (O) is concentrated in micropits on the rock surface (×150). (ii) Intervening surfaces show salt crystallisation (H—halite, G—gypsum) indicating that both biological corrosion and salt weathering may act in conjunction (×450). (b) Circular etch pits and tunnels formed beneath *Polyblysteria* cf. *rupstris* in the 'solution' zone

effects. A third mechanism of salt weathering, namely differential thermal expansion may, however, operate (Cooke and Smalley, 1968). Many salt crystals have a higher coefficient of thermal expansion than the host rock minerals. Thus on heating they expand at a greater rate and exert pressures upon confining host minerals. The coefficient of volumetric expansion for halite, between the temperatures of 20 and 100°C is 0.936, while that for calcite is only 0.105. Similarly, gypsum has a coefficient of 0.58 which suggests that thermal expansion could play a role, as well as crystal growth and hydration pressures, in initiating and exploiting microfractures. Microfracture development is possibly controlled by lines of cleavage within the crystal, although a more detailed study would need to be carried out to confirm this.

Direct observation has also been made of mirabilite crystals in confined spaces on the rock surface (Figure 25.6a) while X-ray diffraction analysis confirms the additional presence of gypsum. Mirabilite may be responsible for similar micromorphological changes as is halite, simply as a result of crystal growth and differential thermal expansion. The needle-shaped crystal habit of mirabilite (Figure 25.6a) has been cited as an important factor in increasing its disruptive capability (Goudie, 1985). Furthermore, hydration of mirabilite to form thenardite causes a volume change of more than 300% and occurs at 32.4°C for a pure solution. In a NaCl saturated environment this temperature drops to 17.9°C (Sperling and Cooke, 1980). These temperatures frequently occur in Mallorca and it is possible that the phase change from mirabilite to thenardite in the type of confined space shown in Figure 25.6a could cause mechanical disintegration of the rock. Anhydrite ($CaSO_4$) will also effect similar damage by volume expansion on hydration to gypsum ($CaSO_4 \cdot 10H_2O$) due to wetting by wave splash or rainfall. Not only does wetting contribute to mechanical disintegration by hydrating salts, but it is a contributing factor in removal of the rock meal thus produced. Wave splash and raindrop impact may provide sufficient force finally to dislodge crystals and crystal fragments loosened by mechanical salt weathering. Strong onshore breezes, which contribute to salt crystallisation by aiding evaporation, may also help remove rock meal.

The presence of discrete solution pans in this zone indicates that some dissolution also occurs. This is probably assisted by the presence of algae in the base of the pans as indicated by their overhanging sides. But alveoli, fretting the sides of these pans as well as dissecting the intervening rock surface, indicate the predominant role of salt weathering. The consistently limited spatial extent of the salt weathering zone along the platform is probably a function of the micro-tidal regime. A tidal range of only 0.2–0.4 m means that the spatial extension of wave splash and spray inland is limited, and the morphological features associated with salt weathering are consequently concentrated in a discrete zone close to the platform edge. Beyond this distinct zone alveoli are absent and solution pans predominate. The major ions found in sea water are also found in the standing water of these pans (Figure 25.4), though in much lower concentrations than in either sea water itself, or standing water from pans in the salt weathering zone. Chemical analysis of the intervening rock surface indicates the presence of only trace amounts of gypsum (Figure 25.5a). This could have been deposited as an aerosol by onshore breezes and subsequently concentrated into solution pans by rainfall and runoff, where it is further concentrated by evaporation. It is unlikely that these trace amounts of salt have any significant weathering effect and no evidence was found for such weathering. It is possible to infer from this that the predominant weathering force in this zone is solution-induced disaggregation and disintegration. Trudgill (1976 and 1985) reports that dissolution on the coastal platform of Aldabra Atoll proceeds by a similar mechanism whereby the selective dissolution of high magnesium calcite or aragonite cements releases low magnesium calcite grains. Although in this study individual grains were not analysed for magnesium content, it is evident that selective dissolution of cement material ultimately has led to the disaggregation of the rock surface. Grains more resistant to solution are loosened by selective dissolution of cement material and are eventually released (Figure 25.8a). Not only is the surface disaggregated by this preferential dissolution process, but individual crystals also disintegrate through a similar mechanism. Cleavage lines within the crystals offer zones of weakness which are preferentially dissolved. This causes the dissection and eventual disintegration of whole crystals (Figure 25.8b). Again, it is most probably rainfall impact and runoff, or strong breezes, which finally dislodge and remove these particles.

Given the limited nature of the evidence provided in this study for biological processes, it is difficult to establish their role in the range of process mechanisms responsible for platform morphology. Similar problems have been encountered by Dalongeville (1977) in a study of morphological zonation on limestone coastal platforms in Libya. It is unlikely that the algae play a major role in the formation of alveoli as suggested by Mustoe (1982) since their occurrence is not limited to intervening walls. They are, however, found concentrated within solution pans where it is possible that they enhance the dissolving power of the water by production of CO_2 and organic acids (Trudgill,1985). Solution pans, with overhanging sidewalls as the result of aggressive solution, seem to substantiate this. It is also evident that by dissecting the rock surface with etch pits and tunnels the surface area exposed to other process mechanisms is increased. It is possible, for example, that in the spray zone salts may more readily crystallise in such dissected surfaces (Figure 25.6a and 25.9a(ii)) so that biological corrosion and mechanical salt weathering act in conjunction. Biological corrosion and inorganic dissolution may also act synergistically in the solution zone. Epi- and endolithic lichens have been found to be responsible for breaking the rock surface down into 'splinters' which are more readily dissolved inorganically (Schneider, 1977).

Finally, algae and lichen appear to have colonised all morphological features. As Viles (1987b) points out for Aldabra Atoll, it is difficult to establish a clear link between microscale features such as boreholes, formed in this case by cyanobacteria, and larger topographic features. This would require a detailed ecological study of microflora in relation to platform surface morphology which is beyond the scope of this particular study. A number of other studies on Mediterranean coastal platforms do, however, indicate the importance of biological processes in the production of platform micromorphology, e.g. Schneider, 1977; Focke, 1978; Le Campion-Alsumard, 1979; Higgins, 1980; Kelletat, 1980; and Schneider and Torunski, 1983. The relatively minor role attributed in this study to biological processes is probably a function of the study site location in the supra-tidal zone. Conditions for biological colonisation are much harsher here than in the intertidal and splash zones which experience biological abrasion as well as biological corrosion.

IMPLICATIONS AND CONCLUSIONS

Despite the fact that this is a site-specific study, it has implications for studies on the durability of coastal engineering structures such as breakwaters and coastal defence units. More than 95% of all breakwaters constructed are made of natural rock or concrete, or a mixture of the two (Fookes and Poole, 1981). In numerous locations around the coast of Mallorca coastal defence systems and breakwaters have been constructed. For example, natural limestone blocks have been used on a coastal defence wall at Palma and concrete blocks on a major breakwater at Penyes Rotges and numerous marinas around Palma Bay. The durability of these materials is defined as their ability to withstand the forces of weathering (Fookes and Poole, 1981, p. 97)—the disruptive forces of the decay processes versus the cohesive forces of the rock. The latter is influenced by the mineral species present; the degree of alteration or weathering state of the rock; grain size, shape and degree of interlock; the nature of the intergranular cement; lengths of joints, microfractures and any other discontinuities; and porosity and permeability (Dibb *et al.*, 1983a and b). These are the same factors which control the surface morphology of the Miocene limestone platform at Cala Pi.

When suitable materials are chosen for coastal defence construction a number of factors besides durability must be taken into account. For example, local availability of suitably sized material as well as an acceptable cost of acquisition and transport both govern the type of materials used. Where coastal areas are used as a social amenity, such as in resorts like Palma or Penyes Rotges, further consideration must be given to details such as the aesthetic and environmental impact of the finished product (Clark, 1988). Indeed, visual assessment of the performance of in-service structures in particular environments has been put forward as a useful estimate of the potential performance of new structures built from similar materials (Fookes *et al.*, 1988). The weathering and degradation that occurs over geological time works to reduce material durability within an engineering time scale of 10s of years. The coastal defence works at Palma and at Penyes Rotges are subject to the same process zonation as is the Miocene limestone platform at Cala Pi. This is reflected in particular in the coastal defence unit at Palma constructed of quarried blocks of natural limestone and in position for some 20 years. A visual inspection indicates a similar morphological zonation to that observed at Cala Pi, although not as clearly developed on the coastal defence unit because of its relative youth. Analysis by X-ray diffraction confirms the presence of halite within the spray and splash zones in association with incipient alveoli and fretting. Similarly, the breakwater at Penyes Rotges, although in place for less than five years, is already beginning to show some signs of deterioration just above high tide level. In both cases two discrete morphological zones can be identified similar to those present on the Miocene platform.

ACKNOWLEDGEMENTS

We would like to thank H. Fox for lichen identification, the cartographic staff of the School of Geosciences and staff of the Electron Microscope Unit at Queen's University. CAM is in receipt of a research studentship from the Department of Education, Northern Ireland.

REFERENCES

Butzer, K. W. (1962). Coastal geomorphology of Majorca. *Annals of the Association of American Geographers*, **52**, 191–212.

Butzer, K. W. and Cuerda, J. (1962). Coastal stratigraphy of Southern Mallorca and its implications for the Pleistocene chronology of the Mediterranean Sea. *Journal of Geology*, **70**, 398–416.

Clark, A. R. (1988). The use of Portland Stone rock armour in coastal protection and sea defence works. *Quarterly Journal of Engineering Geology, London*, **21**, 113–136.

Cooke, R. U. (1981). Salt weathering in deserts. *Proceedings Geologists Association*, **92**, 1–16.

Cooke, R. U. and Smalley, I. J. (1968). Salt weathering in deserts. *Nature*, **220**, 1226–1227.

Crabtree, K., Cuerda, J., Osmaston, H. A. and Rose, J. (1978). The Quaternary of Mallorca. *Quaternary Research Association, Field Meeting Guide*, December 1978.

Cuerda, J. (1975). *Los tiempos Cuaternarios en Baleares*. Diputacion Provincial de Baleares, Palma.

Dalongeville, M. (1977). Formes littorales de corrosion dans les roches carbonatees au Liban: etude morphologique. *Mediterranee*, **3**, 21–33.

Dibb, T. E., Hughes, D. W. and Poole, A. B. (1983a). Controls of size and shape of natural armourstone. *Quarterly Journal of Engineering Geology, London*, **16**, 31–42.

Dibb, T. E., Hughes, D. W. and Poole, A. B. (1983b). The identification of critical factors affecting rock durability in marine environments. *Quarterly Journal of Engineering Geology, London*, **16**, 149–161.

Emery, K. O.(1946). Marine solution basins. *Journal of Geology*, **54**, 209–228.

Focke, J. W. (1978). Limestone cliff morphology on Curacao (Netherlands Antilles) with special attention to the origin of notches and vermitide/coralline algal surf benches. *Zeitschrift für Geomorphologie*, NF22, 329–349.

Fookes, P. G. and Poole, A. B. (1981). Some preliminary considerations on the selection and durability of rock concrete materials for breakwaters and coastal protection works. *Quarterly Journal of Engineering Geology, London*, **14**, 97–128.

Fookes, P. G., Gourley, C. S. and Ohikere, C. (1988). Rock weathering in engineering time. *Quarterly Journal of Engineering Geology, London*, **21**, 33–57.

Gillot, J. E. (1978). Effect of deicing agents and sulphate solutions on concrete aggregate. *Quarterly Journal Engineering Geology, London*, **11**, 177–192.

Goudie, A. S. (1985). *Salt Weathering*. Research Paper 33, School of Geography, University of Oxford.

Higgins, C. G. (1980). Nips, notches and the solution of coastal limestone: an overview of the problem with examples from Greece. *Estuarine Sand Coastal Marine Science*, **10**, 15–30.

Jones, B. (1987). The alteration of sparry calcite crystals in a vadose setting, Grand Cayman Island. *Canadian Journal of Earth Science*, **24**, 2292–2304.

Jones, B. (1989). The role of microorganisms in phytokarst development on dolostones and limestones, Grand Cayman, British West Indies. *Canadian Journal of Earth Science*, **26**, 2204–2213.

Kelletat, D. (1980). Formenschatz und Prozessgefuge des 'Biokarstes' an der Kuste von Nordost-Mallorca (Cala Guya). *Berliner Geographisches Studien*, **7**, 99–113.

Le Campion-Alsumard, T. (1979). Le biokarst marin: role des organismes perforants. *Actes du Symposium International sur l'erosion karstique U.I.S., Aix en Provence–Marseilles–Nimes*, pp. 133–140.

Lewin, S. Z. and Charola, A. E. (1978). Scanning electron microscopy in the diagnosis of 'diseased' stone. *Scanning Electron Microscopy*, **1**, 695–703.

Ley, R. G. (1979). The development of marine karren along the Bristol channel coast. *Zeitschrift für Geomorphologie Supplementband*, NF 32, 75–89.

Lundberg, J. (1977). Karren of the littoral zone. Burren District, Co. Clare, Ireland. In *Proceedings of the 7th International Speleological Congress, Sheffield 1977*, British Cave Research Association, pp. 291–293.

Matsukura, Y. and Matsuoka, N. (1991). Rates of tafoni weathering on uplifted shore platforms in Nojima-Zaki, Boso Peninsula, Japan. *Earth Surface Processes and Landforms*, **16**, 51–56.

McGreevy, J. P. (1985). A preliminary scanning electron microscope study of honeycomb weathering of sandstone in a coastal environment. *Earth Surface Processes and Landforms*, **10**, 509–518.

Mustoe, G. E. (1982). The origin of honeycomb weathering. *Geological Society of America Bulletin*, **93**, 108–115.

Schneider, J. (1977). Carbonate construction and decomposition by epilithic and endolithic micro-organisms in salt and freshwater. In Flugel, E. (ed.), *Fossil Algae: Recent Results and Developments*, Springer-Verlag, Berlin.

Schneider, J. and Torunski, H. (1983). Biokarst on limestone coasts, morphologenesis and sediment production. *Marine Ecology*, **4**, 45–63.

Sperling, C. H. B. and Cooke, R. U. (1980). *Salt weathering in Arid Environments: Theoretical Considerations*. Papers in Geography No. 8, Department of Geography, Bedford College, University of London.

Sunamura, T. (1992). Geomorphology of Rock Coasts. Wiley, Chichester.

Sweeting, M. M. (1972). *Karst Landforms*. Macmillan Press, London.

Trenhaile, A. (1987). *The Geomorphology of Rock Coasts*. Oxford University Press, Oxford.

Trudgill, S. T. (1976). The marine erosion of limestones on Aldabra Atoll, Indian Ocean. *Zeitschrift für Geomorphologie Supplementband*, **26**, 164–200.

Trudgill, S. T. (1985). *Limestone Geomorphology*. Longman, London.

Viles, H. A. (1987a). A quantitative scanning electron microscope study of evidence for lichen weathering of limestone, Mendip Hills, Somerset. *Earth Surface Processes and Landforms*, **12**, 467–473.

Viles, H. A. (1987b). Blue-green algae and terrestrial limestone weathering on Aldabra Atoll: an SEM and light microscope study. *Earth Surface Processes and Landforms*, **12**, 319–330.

Viles, H. A. (1988). Cyanobacterial and other biological influences on terrestrial limestone weathering on Aldabra: implications for landform development. *Biological Society of Washington Bulletin*, No. 8, 5–13.

Viles, H. A. (1990). The early stages of building stone decay in an urban environment. *Atmospheric Environment*, **24A**(1), 229–232.

Section 6

WEATHERING AND LANDFORM DEVELOPMENT IN HIGH LATITUDE AND HIGH ALTITUDE ENVIRONMENTS

26 Lithological and Structural Effects on Forms of Glacial Erosion: Cirques and Lake Basins

IAN S. EVANS
University of Durham, UK

ABSTRACT

Forms of glacial erosion have been said (a) to be indifferent to geological structure and lithological variations; or (b) to be closely controlled by them. This contradiction may be resolved by the generalisation that where rock structure is either homogeneous or complex, with tight folds and frequent lithological changes, glaciers have eroded indifferently; whereas simple structures with strong variations in rock hardness, jointing and bedding have great influence on the progress of erosion. Some of the simplest glacial cirques are formed in flat-bedded, well-jointed lavas or sandstone/shale successions.

In the English Lake District, there is less contrast in cirque form between the Skiddaw Slates and Borrowdale Volcanics than is commonly supposed: however, cirques on mixed volcanic rocks (tuff and lava) are larger, better developed and account for most of the large cirque lakes. In the Cayoosh Range of the British Columbia Coast Mountains, the largest and best developed cirques are on granodiorite, and these are much more likely to contain lake basins.

Cirque floors are often controlled by highly resistant layers. Inward-dipping joints or bedding planes favour development of rock basins and steep headwalls. The presence of closed rock basins in cirques is rare on schists and weak sedimentary rocks, but common on well-jointed granitic rocks and massively bedded sandstones. Evidence for the expected relationship between headwall gradient and structure is weak. Glacial troughs and roches moutonnées are closely influenced by structure. Like other erosive agents, glaciers are selective in exploiting differentiated structures.

INTRODUCTION

The aim of this paper is to demonstrate the magnitude of lithological and structural effects in some forms of glacial erosion. This is attempted by providing new information from two large data sets on cirques and rock basins and reviewing some of the more relevant papers.

All erosional landforms reflect the action of forces on resistances, and it seems obvious that variations in lithology or rock structure may often produce variations in resistance sufficient to affect surface form. Nevertheless the importance of these geological effects may vary between different geomorphological processes. There is

Rock Weathering and Landform Evolution. Edited by D. A. Robinson and R. B. G. Williams
© 1994 John Wiley & Sons Ltd

a notion deeply ingrained in the geomorphological literature that geologically differentiated effects are a characteristic of fluvial erosion and weathering, while glacial erosion is relatively insensitive to geology. This may go back at least to W.M. Davis (1900; Chorley et al., 1973, chapters 14 and 15). Davis did not deny that lithology and structure could influence glacial action, but his emphasis was elsewhere and he interpreted their effects as inherited from previous fluvial action.

Thus von Engeln (1938, p. 426) proposed a process law: 'The importance of structure in the degradational development of landforms decreases as the grossness of the attacking processes increases'. By this he followed Flückiger in claiming that glacial erosion (and marine abrasion) were unaffected by structural differences. A fully glacial landscape, independent of pre-existing relief, would reflect structure only in detail. Derbyshire (1968, p. 122) claimed that 'Large dynamic glaciers tend to produce symmetrical cirques regardless of the bedrock structure'.

On the other hand, in the same paper (pp. 120–121) Derbyshire stated that 'The influence of faults, joint and bedding planes can be traced in the shape of many cirques'. Derbyshire and Evans (1976) reviewed the climatic factor in cirque variation, but concluded that geological effects were more easily established. Climate clearly has a dominating effect on cirque distribution, followed by topography rather than geology (Evans, 1977), but this does not necessarily apply to cirque form.

Modern research papers tend to the view that glacial erosion varies in rate and even in type between different rock types, and that structures such as joints, faults and bedding planes have a strong influence on glacial plucking. Textbook accounts, however, are often notable for dealing with glacial erosion in one chapter, with little mention of geological effects; and structural control in another, with little mention of glacial erosion. While geological effects on glacial erosion are not explicitly denied, this separation carries the implicit message that they are unimportant compared with effects on weathering and on fluvial landscapes.

PREVIOUS LITERATURE ON GLACIATED LANDSCAPES, TROUGHS AND ROCHES MOUTONNÉES

For Monjuvent (1974), glacial morphogenesis is simply one aspect of differential erosion, strongly controlled like fluvial erosion by lithology and structure. Peulvast (1986) emphasised strong differential erosion in the mountain glaciation of the Lofoten Islands. Some 'crag and tail' features are entirely produced by lithologically differentiated glacial erosion (Linton, 1963). For the Alps, Galibert (1962, p.13) claimed that 'Always the elementary forms of glacial sculpture are exactly adapted to the net of principal joints . . .' (more than to faults and thrusts); overall forms as well as details are related to structure. Zumberge (1955), in disagreement with von Engeln, emphasised the interaction of force and resistance in glacial erosion. He showed how rock strike, rather than ice flow direction, controlled plucking (quarrying) around Lake Superior. Similar cuesta landscapes were found on Isle Royale, where ice flowed along the strike of lavas, conglomerates and breccias, and in the Rove area of northeastern Minnesota where ice flowed across the strike of diabase sills and slates. In both, the Laurentide Ice Sheet enhanced asymmetry by excavating elongated rock basins in the strike valleys, deepest beneath scarp slopes. Zumberge's further suggestion that abrasion is controlled by 'active' (glaciological) factors requires qualification.

Rudberg (1973) showed strong structural influence in glacial erosion forms of medium size in four areas of Sweden; ice did not simply remodel fluvial valleys, because many valleys cross each other or cross previous divides. Sissons (1967, chapter 3) emphasised structural effects on glacial troughs and rock basins, with the best U forms along faults, or on rocks such as the Arran granites in Scotland which show no marked erodibility contrasts. Pippan (1967) summarised the effect of rock type on glacial troughs; these are narrowest on intrusive or quartzose rocks and limestones, and gentlest on soft rocks. Metamorphism makes resistance more uniform and favours 'typical' troughs as in the Jotunheimen, Norway.

In Dauphiné, France (Pelvoux, Belledonne and Dévoluy massifs), Monjuvent (1974) found that breadth and depth of glacial troughs varied inversely with rock resistance, rather than being calibrated to ice discharge. Although he suggested that this implied limited glacial transformation of the valleys, he accepted rock basins as unequivocal evidence of glacial erosion and gave many examples of lithological control on the basins and on the rock bars (verrous) which separate them. He found (1974, p. 471) that long profiles were regular on homogeneous rocks whether valleys were glaciated or not. Matthes (1930) found that the configuration of the Yosemite and neighbouring glacial troughs in California was closely related to rock type, being broad on the coarsely jointed Sentinel granodiorite and well-jointed gabbro and diorite, but constricted on massive granite.

Matthes (1930) maintained that 'the quarrying action of glaciers is inherently selective, especially in regions where the rocks are hard and tough'; its effectiveness depends specifically upon the spacing of joints. Where joint spacing is too great for quarrying, steps are formed and lowered by the slower process of abrasion. Many authors since have exemplified the effects of jointing, for example on the form and orientation of roches moutonnées (Gordon, 1981; Rastas and Seppälä, 1981). The effect is such that roches moutonnées are not reliable indicators of ice flow orientation, though they do show the sense of flow and if striations are present the direction is determined. There may be a distinction between medium-relief areas, of moderate erosion guided by structure (Zumberge, 1955), and mountain areas of deep erosion with stronger control by ice flow. Nevertheless, the best-developed 'glacially stream-lined' rock forms are where valleys trend parallel to rock structure.

In general, abrasion is a function of rock hardness. Gerrard (1988, p. 263) states that rock lithology and structure have a major influence on abrasion processes and the resultant landforms; limestone lacks resistance to abrasion, and forms long, regular grooves and ridges. Shale is especially susceptible to quarrying which, despite the possible creation of new joints by glacier-induced stresses, is inevitably sensitive to the degree of jointing (Linton, 1963, p. 2; Sissons, 1967, p. 33).

GEOLOGY AND CIRQUE FORM

Geological contrasts might be expected to affect rates of abrasion, quarrying, rock weathering and mass movement, and hence distort the shape and vary the size of erosional forms such as glacial cirques. Sugden (1969) suggested that the regularity of cirques in the Cairngorm Mountains, Scotland, was due partly to homogeneity of the granite, and partly to the smoothness of the plateau into which they cut. However, if geological contrasts are repeated sufficiently frequently, the net effect is of having homogeneous rocks, and simple, well-developed cirques may still form. Charlesworth (1957, p. 295) proposed that cirques are regular and simple in homogeneous or horizontal strata.

For example, the Cenozoic lava successions of East Greenland, Iceland, and the Faeroes support large numbers of easily defined cirques with semi-circular headwalls. Headwall crests are sharp, with cirques intersecting either each other or rather flat plateaux. Long- and cross-profiles are smoothly concave, and floors are flat, though rarely with rock basins. Thresholds are often lacking where valley-head cirques merge directly into glacial troughs. The lavas remain near-horizontal, with dips of a few degrees or less. Zones within each flow are eroded differentially to give steps in the headwalls (though often the 'treads' slope over 30 degrees), but this detail does not disturb the overall simplicity of the cirque form because there are many lava flows per headwall. In British Columbia, cirques in Quaternary lava flows are also well-developed, and headwall recession has sometimes gone so far as to break through intervening ridges.

We may generalise that on near-horizontal lava sequences, headwall recession is rapid but floor deepening is not marked. Comparable cirques are found in well-bedded, near-horizontal sandstones or sandstone–shale alternations, for example in South Wales, or in West Spitsbergen on flat-lying sandstones, limestones, mudstones and gypsiferous rocks (McCabe, 1939). But if the headwall contains a massive sandstone, it may be steeper and give an 'L-shaped' cirque as in Applecross, Scotland (Haynes, 1968 and 1971). Thicker ice and enhanced accumulation beneath an escarpment encourage rotational ice flow and hence the erosion of a rock basin, e.g. at Craig-y-Llyn on the Pennant Sandstone 'north crop' of the South Wales coalfield, or Coire na Poite in Applecross. These cannot be included in Embleton and King's (1975, p. 208) generalisation that 'In areas of sedimentary rock, cirques are often only poorly developed, as in southern and central Wales . . .'.

On the Cader Idris escarpment of Ordovician igneous rocks in mid Wales, Llyn y Gader exemplifies a similar 'escarpment-foot' rock basin cirque (Lewis, 1938). (Literature on rock basins in cirques is discussed further below.) Inclined massive layers may produce asymmetrical shapes (Peulvast, 1986, p. 147). In Scotland, Haynes (1968) attributed the 'schrundline' break in cirque headwall slope to medium-dip joints, rather than to the glaciological interpretation proposed in California by Gilbert (1904).

It is sometimes suggested that cirques do not develop on weak rocks. In parts of the Alps, for example south of the Pelvoux massif, shales are caught up among more resistant rocks and do form cirque headwalls, but this is rare. Monjuvent (1974) regarded cirques in Dauphiné as less influenced by geology than are other glacial forms. Lithology plays an important role in cirque development, but is secondary to altitude and initial topography. Cirques are favoured by faulted or broken rock, and by juxtaposition of hard and relatively soft rock (see also Peulvast, 1986, p. 148): in Belledonne in the French Alps they are preferentially incised in gabbro and serpentine rather than the very hard amphibolite. In central Sweden, Vilborg (1984, pp. 53–57) found many more cirques on high-relief ranges of highly metamorphosed rocks (massive greenstones, amphibolites, mica-schists and gneisses) than on ranges of comparable altitude on steeply dipping or low-grade metamorphic rocks (phyllites and spotted schists).

In detail, particular sets of joints influence cirque form as discussed by McCabe (1939) for West Spitsbergen, Schwan (1974) for the Vosges, Godard (1969) for the Isle of Arran, Clough (1977) for the English Lake District, and Addison (1981) for Snowdonia. Addison exemplified the importance of rock mass discontinuities for plane, wedge, toppling and compound failures of headwalls: the resulting erosional forms are structurally controlled at all scales.

Charlesworth (1957) maintained that jointing, cleavage and dip direction control cirque size and shape. However, their variation between lithologies seems to have less influence on measurable aspects of overall cirque form. Embleton and Hamann (1988) studied selected well-developed cirques on nine rock types in Austria and Britain, and found little difference in plan and profile closure other than surprisingly low values for granites.

Nevertheless, it is desirable to support generalisations by measurements on large populations of cirques, and not to rely solely on detailed consideration of case studies which may not be broadly representative. To do this, the use of generalised indices is unavoidable. We may hypothesise that cirque size and headwall gradient are particularly likely to vary between lithologies, reflecting both rock properties and the jointing characteristics en masse of different rock types.

CIRQUE FORM IN THE LAKE DISTRICT AND CAYOOSH RANGE

An attempt is now made to provide some quantitative generalisations about the effect of geology on cirque form, using two large data sets. The main limitation is reliance on mapped geological units which are often internally heterogeneous, but both data sets do cover considerable geological contrasts.

Data for the English Lake District, in Cumbria around 3°W, 54.5°N, are based on fieldwork over a considerable period, air photo interpretation, and measurement from 1:10 000 scale maps with a 10 m contour interval. This area is noted for the contrast between rugged, complex topography on Borrowdale Volcanic rocks and smoother slopes on the Skiddaw Slates, both of Ordovician age. In reality this is an oversimplification; smooth slopes are found on the Volcanics of the northern Helvellyn Range and rugged areas around Coledale and Blencathra are on the Skiddaw Slates, albeit largely where thermally metamorphosed (Evans and Clough, 1977).

Hence the distinction between thermally and merely pressure metamorphosed slates is recognised. A few cirques on the western fringe are on a granitic rock, the Ennerdale Granophyre. The great majority are on Borrowdale Volcanics and it is desirable to subdivide these. This was done from the small-scale maps of Moseley and Millward (1982) distinguishing tuffs from lavas, with which are grouped ignimbrite, rhyolite, dacite and sills. Even so the most common cirque lithology is a mixture of these two.

The 158 cirques recognised are spread over an area extending 48 km north–south and 40 km east–west. Some have been modified by overriding ice during ice sheet phases. Table 26.1 gives average values of cirque size, development and altitude

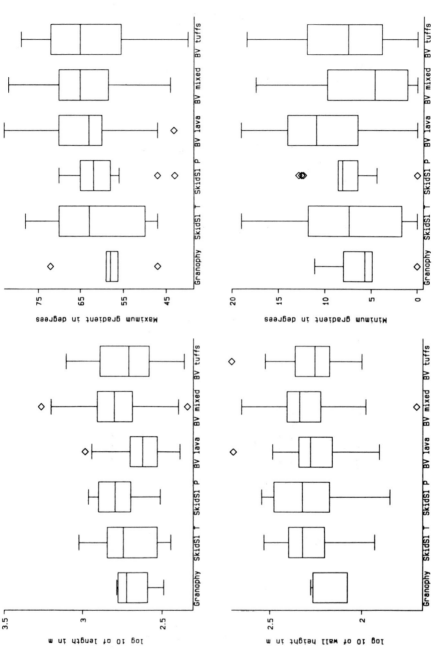

Figure 26.1 Box plots of (left) logarithms of length and wall height and (right) maximum and minimum gradient of English Lake District cirques. For each rock type, the central box shows upper and lower quartiles, with a line for the median. Lines are drawn outward to the last value within one 'step' of each quartile: the 'step' is defined as 1.5× interquartile range. Any values beyond those lines are plotted individually by diamonds, and are regarded as unusual and perhaps as outliers. The six rock types are in the same order as in Table 26.1, with 5, 15, 13, 29, 64 and 32 cirques respectively.

Table 26.1 Mean values of cirque size, development and altitude for six classes of geology in the English Lake District, Cumbria

No. of cirques: Geology	5 Ennerdale Granophyre	Skiddaw Slates		Borrowdale Volcanics			x158 Overall
		15 thermally-metamorphosed	13 pressure-metamorphosed	29 lavas etc.	64 mixed	32 tuffs	
Length (m)	488	572	639	467	715	602	620
Width (m)	666	573	648	556	780	659	680
Wall height (m)	147	204	218	192	215	200	205
Max. gradient (°)	58	62	60	65	64	63	63
Min. gradient (°)	6	8	8	11	6	8	8
Plan closure (°)	66	97	117	110	133	122	120
Cols (over 30 m)	0	0.2	0.231	0.24	0.594	0.406	0.405
Floor altitude	452	471	384	550	490	519	495
Max. alt. above	660	729	723	803	793	788	778
Lake status (ordinal)	4.40	4.47	4.92	5.17	3.98	4.63	4.47

Table 26.2 Effects of geology (six classes) on the size, development and altitude of cirques in the English Lake District, Cumbria. The adjustment of R^2 is to allow for degrees of freedom lost in fitting the model constants. $p*$, the probability of no relationship, is given only for comparison between effects, owing to violation of the independence assumption

	R^2		ANOVA	
	crude	adjusted	F	$p*$
Length (log)	0.118	0.089	4.1	0.0017
Width (log)	0.107	0.078	3.6	0.0039
Wall height (log)	0.035	0.004	1.1	0.3570
Max. gradient	0.028	−0.004	0.9	0.5060
Min. gradient	0.117	0.088	4.0	0.0018
Plan closure	0.089	0.059	3.0	0.0134
Cols (sq. root)	0.101	0.072	3.4	0.0058
Floor altitude	0.106	0.077	3.6	0.0040
Max. alt. above	0.113	0.084	3.9	0.0024
	chi-square	df	$p*$	
Lake status	40.5	25	0.026	
Cols	32.1	15	0.006	

on each of the six rock types mapped. Cirques on mixed volcanics are wider and longer, with gentler floors (lower minimum gradient; Figure 26.1) and greater closure in plan. Cirques on the resistant lavas (etc.) are narrower and shorter, their floors are steeper and they are at higher altitudes. Cirques on Skiddaw Slates and Ennerdale Granophyre are at lower altitudes; those on the Granophyre have poorer closure and lower wall heights, but this may be because of their surrounding topography rather than being intrinsic to the rock type.

Variation around these mean values is considerable (Figure 26.1 shows extremes, medians and quartiles), and may be assessed in terms of analysis of variance (Table 26.2). The positive skewness of the size variables is first remedied by logarithmic transformation, that of 'number of cols in cirque rim' by taking square roots. It is clear that 'wall height' and 'maximum gradient' (Figure 26.1) show no more variation between classes than within. Probability values show that the between-geology differences for the other variables are 'significant' in conventional terms, but it must be remembered that the cirque attributes exhibit a complex form of autocorrelation through space, and that the complete population of present-day Lake District cirques is included.

A better measure of the relevance of geology for each variable is R^2, the proportion of total variance attributed to division into these six geologic classes; this is low in each case. After adjusting for the five degrees of freedom lost in fitting the model, no R^2 is above 9%. (These calculations are performed using the program 'Stata'.) In other words, geology as represented by these classes is a minor factor in cirque variation. So too, however, are altitude, aspect and relief.

In various multivariate combinations (analyses of covariance) 'geology' likewise has little predictive power to add. Its relation to a qualitative attribute and to 'cols' (which takes only the values 0, 1, 2 and 3 in the Lake District) is shown by the chi-square values in Table 26.2. That for cols is consistent with the analysis of variance result; that for lake status is discussed below.

A second large data set has been measured from two-fold enlargements of 1:50 000 scale maps of part of the Coast Mountains in southern British Columbia. The Cayoosh Range lies around 122.3°W, 50.6°N, between Seton, Anderson and Duffey Lakes, Cayoosh and Haylmore Creeks, southwest of Lillooet on the dry side of the Coast Mountains. Cirques were identified on the basis of air photo interpretation, with limited field checking. The 198 cirques are concentrated in an area 29 km × 19 km.

The Cayoosh Range consists mainly of Triassic metamorphic rocks, with intrusive stocks of granodiorite in the centre and east and of quartz-diorite in the northwest and east (Roddick and Hutchison, 1973) providing the desired contrast in rock type. Most of the metamorphic rocks belong to the Fergusson (Bridge River) Group, a mixture of chert, phyllite, argillite and 'greenstone', with minor limestone and schist. Strike varies around northwest, with recorded dips of 40 to 85 degrees. A small area in the southwest is eroded into the Hurley Formation, which has a greater volcanic component; it contains thin-bedded limy argillite, phyllite, limestone, conglomerate, tuff, agglomerate and andesite, with minor chert. Dips of 30, 30 and 70 degrees have been recorded (Roddick and Hutchison, 1973).

Despite the complexity of these metamorphic sequences, most of the cirques developed on them are simple in form (with the usual complexities at major valley heads). On both the metamorphic and the igneous rocks, the cirques show strong asymmetry of climatic origin, as discussed by Evans (1977) for the adjacent Bridge River District.

Table 26.3 shows that in the Cayoosh Range, cirques on granodiorite are on average considerably larger than others, with gentler floors, steeper headwalls and more cols. Figure 26.2 confirms these differences in size and gradient in terms of quantiles. Granodiorite cirques are at higher altitudes than cirques on other lithologies, but their floors are at the same mean level as those on Fergusson metamorphics. The few cirques on quartz-diorite are smaller than those on granodiorite or on Fergusson, but the smallest and most poorly developed cirques are on Hurley metamorphics. The Cayoosh cirques as a whole are larger and steeper than those of the Lake District and are a little more closed in plan, but the biggest difference is the much greater number of cols around their crests. This reflects the higher relief and greater dissection of the Coast Mountains.

Again there is considerable scatter around these mean values (Figure 26.2), and the differences between geologies are assessed by analysis of variance or by chi-square (Table 26.4). Differences in wall height, in maximum and minimum gradient and in cols are considerable and probably significant, while those in length and width (with or without transformation to logarithms) are not. Differences in plan closure between the four classes of geology are so slight, in relation to within-class variation, that the F (variance) ratio of 0.09 is suspiciously low. Differences in altitude between the geologic classes are more significant than differences in size or development, where the highest adjusted R^2 is 8.2%.

Since each geologic type, other than Fergusson metamorphics, is grouped into one or two areas, there is a nagging possibility that the effects of geology are confounded with those of altitude and position, with larger, better-developed cirques in the

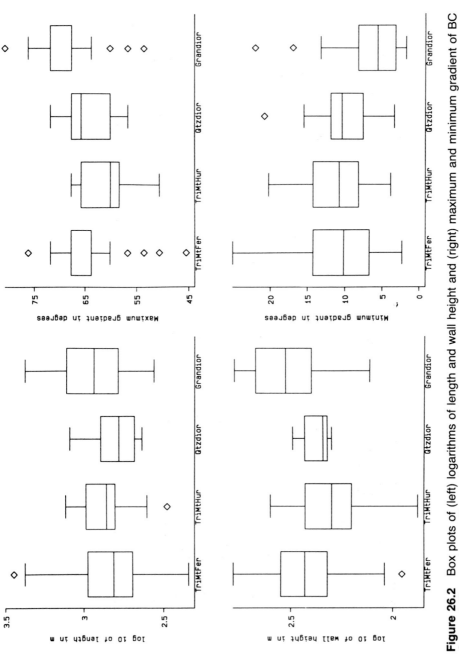

Figure 26.2 Box plots of (left) logarithms of length and wall height and (right) maximum and minimum gradient of BC Cayoosh Range cirques. For each rock type, the central box shows upper and lower quartiles, with a line for the median. Lines are drawn outward to the last value within one 'step' of each quartile: the 'step' is defined as 1.5× interquartile range. Any values beyond those lines are plotted individually by diamonds, and are regarded as unusual and perhaps as outliers. The four rock types are in the same order as in Table 26.3, with 140, 16, 10, and 32 cirques respectively, from left to right.

465

Table 26.3 Mean values of cirque size, development and altitude for four classes of geology in the Cayoosh Range, southern Coast Mountains, British Columbia. The Fergusson Group metamorphics are mainly chert, phyllite, argillite and greenstone. The Hurley Formation metamorphics are mainly limy argillite, phyllite, limestone, conglomerate, agglomerate, tuff and greenstone (Roddick and Hutchison, 1973)

No of cirques:	140	16	10	32	198
	Triassic metamorphics		Intrusive igneous		
Geology:	Fergusson Group	Hurley Formation	quartz-diorite	granodiorite	Overall
Length (m)	772	764	688	965	798
Width (m)	738	584	692	898	749
Wall height (m)	284	215	242	355	288
Max. gradient (°)	65.4	61.3	64.8	68.9	65.6
Min. gradient (°)	11.1	11.3	10.6	6.5	10.3
Plan closure (°)	135	127	133	135	134
Cols (over 30 m)	1.14	0.69	1.60	1.88	1.25
Floor altitude (m)	2094	1917	2006	2094	2060
Max. alt. above (m)	2451	2198	2344	2545	2425
Lake status (ordinal)	4.52	4.69	4.2	2.94	4.26

Table 26.4 Effects of geology (four classes) on the size, development and altitude of cirques in the Cayoosh Range, southern Coast Mountains, British Columbia

	R^2		ANOVA	
	crude	adjusted	F	p^*
Length (log)	0.030	0.015	2.0	0.1200
Width (log)	0.033	0.018	2.2	0.0900
Wall height (log)	0.086	0.072	6.1	0.0006
Max. gradient	0.085	0.071	6.0	0.0006
Min. gradient	0.096	0.082	6.9	0.0002
Plan closure	0.001	−0.014	0.1	0.9680
Cols (sq. root)	0.057	0.042	3.9	0.0099
Floor altitude	0.111	0.097	8.1	0.0000
Max. alt. above	0.057	0.042	3.9	0.0000
	chi-square	df	p^*	
Lake status	65.8	12	0.000	
Cols	18.9	9	0.026	
(pooling '3 or more')				

higher, central part of the Range, on granodiorite or Fergusson metamorphics. This was tested by analysis of covariance. The size and shape variables in Table 26.4 are correlated very weakly (<0.14) with cirque floor altitude, but moderately (0.32 to 0.59) with maximum altitude above; might the latter, then, account for the differences between rock types?

The combined analyses (of covariance) show strong relationships between each variable and maximum altitude above, with geology accounting for 5.2% of variance in log(length) and 6.7% of minimum gradient, but for insignificant parts of the other variables. In combinations of three controls (log(length), maximum altitude above, and geology), geology accounts for 4.7% of minimum gradient and 2.8% of (square root of) cols. Hence even allowing for cirque length, and altitude above, geology does have small effects on minimum gradient and number of cols, and it has a further small effect on length once altitude above is allowed for.

LAKE BASINS IN LAKE DISTRICT AND CAYOOSH CIRQUES

The development of rock basins preferentially on granitic rocks has often been noted. In Norway, Strøm (1938) noted lakes, some very deep, in almost all the cirques on the plutonic rocks of Moskenesøy in the Lofoten Islands. Later (1945) he contrasted these with long though flat-floored cirques on the easily frost-shattered felspathic quartzite of the Rondane mountains. In the central Pyrénées, Chevalier (1954, pp. 109 and 112) found rock basins incised into the lower parts of 'cirques en van' in granite, but absent from the shallow cirques on schists and most sedimentary rocks.

In the French Alps, rock basins are common in cirques on granite or gneiss in the Belledonne and Aiguilles Rouges massifs.

Granitic or other intrusive rocks are not, however, essential for the production of rock basins in cirques. Haynes (1968 and 1971) demonstrated the importance of joints dipping into a cirque. Vilborg (1977, p. 145) and Clough (1977) also found that these helped cirque development in profile. Conversely, outdipping joints, if dominant, may inhibit deepening of a cirque floor. Even more effective is the presence of a much more resistant lithology forming the floor, for example the Lewisian Gneiss beneath the frost-susceptible Cambrian quartzite of northwest Scotland (Thompson, 1950).

To provide further quantitative evidence on this topic, a six-fold classification of 'lake status' was developed, partly following Zienert (1967, p. 53), to reflect the probability of a rock basin being present. Class 1 is a major lake (occupying a large proportion of the cirque floor), believed to be in a rock basin because of its size, rounded shape, or the exposure of rock in the cirque threshold. Class 2 lakes are major, but may be due solely to a morainic dam. Class 3, not detected in the Cayoosh Range, is a major peat bog or infilled lake, and Class 4 is a cirque floor with minor lakes or peat bogs. Most cirque floors fall into Class 5, where the presence of drift deposits, talus or glaciers on an outsloping floor may be hiding the presence of a rock basin. Finally, in Class 6, exposures of bedrock are sufficiently well distributed over a completely outsloping floor to provide some confidence that a rock basin of any size is absent.

The chi-square values in Tables 26.2 and 26.4 show considerable differences in 'lake status' between geologic classes; these are clearly significant at least in the Cayoosh Range. These are based on the data cross-tabulated in Tables 26.5 and 26.6, where counts are given in bold type, followed by percentages (with decimal points) of the number of cirques on that rock type.

For the Lake District the Skiddaw Slates have many drift-covered cirque floors, the lavas etc. of the Borrowdale Volcanics have an excess of outsloping bedrock floors and the tuffs have a lot of minor lakes (often structurally determined). But the most notable deviation from 'equal shares' is the occurrence of 10 of the 13 major rock basins on mixed Borrowdale Volcanics, which form 64 out of the 158 cirques. For Clough (1977), the most important factor favouring overdeepening of cirque floors is inward or transverse dip, facilitating joint block removal in the Borrowdale Volcanics. He further related three lake basins to intersecting faults.

The concentration of major rock basins on Borrowdale Volcanics, especially where tuffs and harder rocks are mixed, is a real effect which must be related to the factors mentioned by Clough, or to lithological contrasts. Again, however, a cautionary note may be in order. The 'mixed volcanics' cirques are larger than others (Table 26.1). But the larger a cirque, the more likely it is to contain different lithologies, i.e. to be 'mixed'. Hence cirques may be 'mixed' because they are large, rather than large because they are 'mixed'! Large cirques are likely to be better developed, with larger, gentler floors clearly providing better opportunities for the excavation of rock basins.

Even clearer contrasts are seen in the Cayoosh Range (Table 26.6). Although there are 15 major rock basins on Fergusson metamorphic rocks, this is low in relation to the large number (140) of cirques on these rocks. The greatest concentrations of both rock and possibly moraine-dammed basins are on granodiorite, where 18 out of 32 cirques have major rock basins. There are none (out of 16 cirques) on Hurley

Table 26.5 Number (in bold face) and percentage of Lake District cirques with different lake basin status, on (six) different lithologies. Overall chi-square is 40.5, with 25 degrees of freedom; $p^* = 0.026$

Lake status	Ennerdale Granophyre	Skiddaw Slates		Borrowdale Volcanics			Total
		thermally-metamorphosed	pressure-metamorphosed	lavas etc.	mixed	tuffs	
Major, rock basin	**1** 20.0	**0** 0.0	**0** 0.0	**1** 3.4	**10** 15.6	**1** 3.1	**13** 8.2
Major, possibly solely moraine-dammed	**0** 0.0	**2** 13.3	**1** 7.7	**0** 0.0	**2** 3.1	**1** 3.1	**6** 3.8
Major peat bog	**0** 0.0	**1** 6.7	**0** 0.0	**1** 3.4	**11** 17.2	**2** 6.3	**15** 9.5
Minor peat or lake(s)	**0** 0.0	**2** 13.3	**1** 7.7	**1** 3.4	**5** 7.8	**8** 25.0	**17** 10.8
Drift-covered floor	**3** 60.0	**8** 53.3	**8** 61.5	**14** 48.3	**28** 43.8	**13** 40.6	**74** 46.8
Bedrock floor, no basin	**1** 20.0	**2** 13.3	**3** 23.1	**12** 41.4	**8** 12.5	**7** 21.9	**33** 20.9
Total	**5** 100	**15** 100	**13** 100	**29** 100	**64** 100	**32** 100	**158** 100

Table 26.6 Number (in bold face) and percentage of Cayoosh Range cirques with different lake basin status, on (four) different lithologies. Overall chi-square is 65.8, with 12 degrees of freedom; $p^* < 0.001$

Lake status	Triassic metamorphics		Intrusive igneous		Total
	Fergusson Group	Hurley Formation	quartz-diorite	granodiorite	
Major, rock basin	**15** 10.7	**0** 0.0	**1** 10.0	**10** 31.3	**26** 13.1
Major, possibly solely moraine-dammed	**1** 0.7	**0** 0.0	**0** 0.0	**8** 25.0	**9** 4.6
Minor lake(s)	**22** 15.7	**7** 43.8	**4** 40.0	**2** 6.3	**35** 17.7
Drift-covered floor	**84** 60.0	**7** 43.8	**5** 50.0	**12** 37.5	**108** 54.6
Bedrock floor, no basin	**18** 12.9	**2** 12.5	**0** 0.0	**0** 0.0	**20** 10.1
Total	**140** 100	**16** 100	**10** 100	**32** 100	**198** 100

Table 26.7 Condensed versions of Tables 25.5 and 25.6, contrasting cirques with a major lake (or bog) to those without. Numbers of cirques are in bold type, followed by percentages of the column total. For the Lake District, chi-square is 14.6 with 5 degrees of freedom; for the Cayoosh Range, 40.3 with 3. p^* as in Table 26.2

Lake District	Ennerdale Granophyre	Skiddaw Slates		Borrowdale Volcanics			Total
		thermally-metamorphosed	pressure-metamorphosed	lavas etc.	mixed	tuffs	
Major lake or bog	**1** 20.0	**3** 20.0	**1** 7.7	**2** 6.9	**23** 35.9	**4** 12.5	**34** 21.5
Minor or no lake	**4** 80.0	**12** 80.0	**12** 92.3	**27** 93.1	**41** 64.1	**28** 87.5	**124** 78.5
$p^* = 0.012$ Total	**5** 100	**15** 100	**13** 100	**29** 100	**64** 100	**32** 100	**158** 100

Cayoosh Range	Triassic metamorphics		Intrusive igneous		Total
	Fergusson Group	Hurley Formation	quartz-diorite	granodiorite	
Major lake or bog	**16** 11.4	**0** 0.0	**1** 10.0	**18** 56.2	**35** 17.7
Minor or no lake	**124** 88.6	**16** 100.0	**9** 90.0	**14** 43.8	**163** 82.3
$p^* = 0.000$ Total	**140** 100	**16** 100	**10** 100	**32** 100	**198** 100

metamorphics, and only one (out of 10) on quartz-diorite, though both these litholo-gies have more than their share of minor lakes. Thus it can be said that major lake basins are most likely to form in granodiorite cirques, but they do occasionally form on metamorphics and quartz-diorite.

Table 26.7 summarises these effects for both areas by collapsing the 'lake status' classification into a binary one; major lake basins or bogs (Classes 1 to 3) versus minor or (more usually) no lakes or bogs (Classes 4 to 6). For the Lake District, this improves the apparent significance of geological effects ($p = 0.012$ rather than 0.026) and confirms the dominance of major lakes (23 out of the 34 in total) on the mixed Borrowdale Volcanics. The excess is balanced by deficits on the other volcanics and on pressure-metamorphosed Skiddaw Slates. The overall 'Borrowdale/Skiddaw' con-trast is weak: Borrowdale Volcanics have 29 out of the 34 major lake basins, whereas 'equal shares' would given them 26.9.

The already very clear result for the Cayoosh Range is strengthened further by the aggregation. The granodiorite has 18 of the 35 major lakes and the intrusives together have 19, compared with 5.7 and 7.4 expected from equal shares. Yet without an excess on the peripheral plutons, we cannot be confident that the geological effect has been separated from others.

CONCLUSIONS

The large-sample statistical results are ambivalent. On the one hand, they confirm the reality of geological effects on cirque form, especially on major lake basins. On the other hand, these effects are small in relation to variability within each of the mapped rock types. Within the crystalline rocks studied here, geological effects on cirque size and gross cirque morphology are small. There are probably greater differences in size and morphology between crystalline, sedimentary and young volcanic rocks. These are not tested here, and their testing requires different rock types with comparable relief, climate and tectonic environment, that is, they probably need to be closely juxtaposed.

The greatest geological effect demonstrated here is in the development of major rock basins. These are much more frequent on granodiorite in the Cayoosh Range, and on mixed volcanic rocks in the Lake District.

More generally, it can be stated that simple, rounded cirques form, not only where rocks (such as the Cairngorm granite) are homogeneous, but also where frequent alternations of resistance (such as lava flow sequences or well-bedded sedimentary rocks) occur. The complex structures of metamorphic rocks also lend themselves to the development of simple cirques. The main 'distortions' of cirque form arise from single major contrasts in lithology, as between quartzite and gneiss or shale and massive limestone. Many studies in the literature demonstrate the great effects of jointing on details within cirques, on both the wall and the floor, or upon roches moutonnées.

In general, glacial erosion (especially plucking) is sensitive to geological variations. Structural control is more obvious in ice-scoured lowlands than in mountain troughs, but the best glacially streamlined rocks are where valleys parallel rock structure. Since structure affects glacial erosion, the final implication is that structural influence on glaciated landscapes is not necessarily inherited from non-glacial processes. All erosional landscapes reflect the interplay between process and form, which includes geology as well as geometry.

REFERENCES

Addison, K. (1981). The contribution of discontinuous rock-mass failure to glacial erosion. *Annals of Glaciology*, **2**, 3–10.

Charlesworth, J. K. (1957). *The Quaternary Era*. Arnold, London.

Chevalier, M. (1954). Le relief glaciaire des Pyrénées du Couserans. I—Les cirques. *Revue Géographique des Pyrénées et du Sud-ouest*, **25**, 97–124.

Chorley, R. J., Beckinsale, R. P. and Dunn, A. J. (1973). *The History of the Study of Landforms*, Vol. 2. Methuen, London.

Clough, R. McK. (1977). Some aspects of corrie initiation and evolution in the English Lake District. *Proceedings of the Cumberland Geological Society*, **3**, 209–232.

Davis, W. M. (1900). Glacial erosion in France, Switzerland, and Norway. *Proceedings of the Boston Society of Natural History*, **29**, 273–322. Reprinted 1909 in Johnson, D. W. (ed.), *Geographical Essays*, Ginn, Boston, pp. 635–689.

Derbyshire, E. (1968). Cirques. In Fairbridge, R. W. (ed.), *The Encyclopedia of Geomorphology*, Reinhold, New York, pp. 119–123.

Derbyshire, E. and Evans, I. S. (1976). The climatic factor in cirque variation. In Derbyshire, E. (ed.), *Geomorphology and Climate*, Wiley, Chichester, pp. 447–494.

Embleton, C. and Hamann, C. (1988). A comparison of cirque forms between the Austrian Alps and the Highlands of Britain. *Zeitschrift für Geomorphologie, NF Supplementband*, **70**, 75–93.

Embleton, C. and King, C. A. M. (1975). *Glacial Geomorphology*. Wiley, New York.

Evans, I. S. (1977). World-wide variations in the direction and concentration of cirque and glacier aspects. *Geografiska Annaler*, **59A**(3–4), 151–175.

Evans, I. S. and Clough, R. McK. (1977). Cirques of Mungrisdale: Bowscale and Bannerdale. In Tooley, M. J. (ed.), INQUA X Congress Guidebook for excursion A4, *The Isle of Man, Lancashire Coast and Lake District*, Geo Abstracts, Norwich, pp. 52–55.

Galibert, G. (1962). Recherches sur les processus d'érosion glaciaire de la Haute Montagne. *Bulletin Association Géographes Français*, **303–304**, 8–46.

Gerrard, A. J. W. (1988). *Rocks and Landforms*. Unwin Hyman, Maidstone.

Gilbert, G. K. (1904). Systematic asymmetry of crest lines in the High Sierra of California. *Journal of Geology*, **12**, 579–588.

Godard, A. (1969). L'Ile d'Arran (Écosse). *Revue de Géographie physique et de Géologie dynamique* (2), **11**, 3–30.

Gordon, J. E. (1981). Ice-scoured topography and its relationships to bedrock structure and ice movement in parts of northern Scotland and West Greenland. *Geografiska Annaler*, **63A**, 55–65.

Haynes, V. M. (1968). The influence of glacial erosion and rock structure on corries in Scotland. *Geografiska Annaler*, **50A**, 221–234.

Haynes, V. M. (1971). The relative influence of rock properties and erosion processes in the production of glaciated landforms with especial reference to corries in Scotland. Ph.D. Thesis, University of Cambridge, England.

Lewis, W. V. (1938). A meltwater hypothesis of cirque formation. *Geological Magazine*, **75**, 249–265.

Linton, D. L. (1963). The forms of glacial erosion. *Transactions Institute of British Geographers*, **33**, 1–27.

Matthes, F. E. (1930). Geologic history of the Yosemite Valley. *US Geological Survey, Professional Paper*, **160**, 54–103.

McCabe, L. H. (1939). Nivation and corrie erosion in West Spitzbergen. *Geographical Journal*, **94**, 447–465.

Monjuvent, G. (1974). Considérations sur le relief glaciaire à propos des Alpes du Dauphiné. *Revue de Géographie physique et de Géologie dynamique* (2), **16**, 465–502.

Moseley, F. and Millward, D. (1982). Ordovician vulcanicity in the English Lake District. In Sutherland, D. (ed.), *Igneous Rocks of the British Isles*, Wiley, Chichester, pp. 93–111.

Peulvast, J. P. (1986). Structural geomorphology and morphological development in the Lofoten–Vesterålen area, Norway. *Norsk Geografisk Tidsskrift*, **40**, 135–161.

Pippan, Th. (1967). On slope development in cirques and trough valleys. *Revue Géomorphologie dynamique*, **17**, 187–188.

Rastas, J. and Seppälä, M. (1981). Rock jointing and abrasion forms on roches moutonnées, southwest Finland. *Annals of Glaciology*, **2**, 159–163.

Roddick J. A. and Hutchison, W. W. (1973). Pemberton (east half) map area, B.C. *Geological Survey of Canada*, Paper 73–17.

Rudberg, S. (1973). Glacial erosion forms of medium size—a discussion based on four Swedish case studies. *Zeitschrift für Geomorphologie, Supplementband*, **17**, 33–48.

Schwan, J. (1974). Joint patterns in cirque walls of the Hautes Vosges, France. *Revue Géographie de l'Est*, **14**(1–2), 95–110.

Sissons, J. B. (1967). *The Evolution of Scotland's Scenery*. Oliver and Boyd, Edinburgh.

Strøm, K. M. (1938). Moskenesøy: a study in high latitude cirque lakes. *Skrifterutgitt av Det Norske Videnskaps-Akademi i Oslo, Math.-Nat. Klasse.*

Strøm, K. M. (1945). Geomorphology of the Rondane area. *Norsk Geologisk Tidsskrift*, **25**, 360–378.

Sugden, D. E. (1969). The age and form of corries in the Cairngorms. *Scottish Geographical Magazine*, **85**, 34–46.

Thompson, H. R. (1950). Some corries of northwest Sutherland. *Proceedings Geologists Association*, **61**, 145–155.

Vilborg, L. (1977). The cirque forms of Swedish Lapland. *Geografiska Annaler*, **59A**, 89–150.

Vilborg, L. (1984). The cirque forms of central Sweden. *Geografiska Annaler*, **66A**, 41–77.

Von Engeln, O. D. (1938). Glacial geomorphology and glacier motion. *American Journal of Science*, 5th Series, **35**(210), 426–440.

Zienert, A. (1967). Vogesen- und Schwarzwald-Kare. *Eiszeitalter und Gegenwart*, **18**, 51–75.

Zumberge, J. H. (1955). Glacial erosion in tilted rock layers. *Journal of Geology*, **63**, 149–158.

27 Joint Control in the Formation of Rock Steps in the Subglacial Environment

B. R. REA

The Queen's University of Belfast, UK

ABSTRACT

The recently deglaciated foreland of a small outlet glacier from Øksfjordjøkelen, north Norway, contains a set of rock bars (vertical lee faces 10–12 m high), and a smaller set of rock steps, both characterised by striated stoss surfaces and steep (almost vertical) lee sides. Outside this area the remainder of the plateau is covered extensively with autochthonous blockfield which shows similar jointing to that on the recently deglaciated foreland. A study of the rock jointing pattern of the deglaciated foreland and the blockfield has been carried out. Reconstruction of the former ice dynamics is undertaken using a number of techniques, allowing former conditions, including the velocity and ice thickness, to be estimated. Modelling of the stresses induced in the bed due to cryostatic pressures and basal shear stress has been undertaken for the inferred ice dynamics; stresses were found to be insufficient to 'quarry' the bedrock. Field observations showed that lee side faces of rock steps, however, did appear joint-controlled, and joint-controlled blocks were found in the moraines. For inferred basal shear stresses it was found that the glacier could have plucked blocks of approximately 1 m^3.

INTRODUCTION

This paper investigates the control which the bedrock joints have upon the subglacial formation of rock steps for the inferred ice dynamics of the Little Ice Age (LIA) advance on the foreland of Camp Glacier, Øksfjordjøkelen, north Norway. The subglacial environment is, in geomorphic terms, frequently a very destructive region as shown by the smoothed and fractured rock bed often exposed as glaciers retreat. In the subglacial environment the area of most importance is the ice/rock interface. It is here that sliding ice, sliding rock clasts, pressurised water and the bedrock are all in intimate contact. Under great ice depths extremely large cryostatic loads will be imposed on the bed, and these alone may be sufficient to fracture intact bedrock (Drewry, 1986). The presence of joints in the bedrock will obviously be important in relation to the process of rock step formation associated with subglacial bed quarrying, as they represent lines of weakness in the bed, and will thus facilitate the fracturing and removal of bed material at lower stress values. Here, quarrying is taken to encompass both bed failure, defined as crushing, fracture, cyclic loading and fatigue, and plucking, defined as the removal of pre-loosened bed material and/or

Rock Weathering and Landform Evolution. Edited by D. A. Robinson and R. B. G. Williams
© 1994 John Wiley & Sons Ltd

material resulting from bed failure (Drewry, 1986). This was noted as early as 1930 by Matthes who concluded from his work in Yosemite Valley that, 'joint structure plays a very important role in glacial quarrying'. The process of subglacial cavity formation also influences the fracturing of bedrock beneath sliding glaciers as their presence effectively reduces the strength of the bed which lies inside the cavity. There are a number of important aspects of ice dynamics which are relevant to the erosion processes active at the ice/rock interface in the formation of a rock step. Generally, it is only after glacier retreat that rock steps can be identified and so there is a problem in understanding the nature of their development, i.e. at what stage of the glacier coverage were they formed, and what ice dynamics or statics formed them. Beneath Camp Glacier, subglacial access is possible, which allows observation of the actual process, and measurement of some of the relevant ice dynamics.

FIELD AREA

The glacier foreland studied is that of 'Camp Glacier', a small outlet from Øksfjord-jøkelen (Figure 27.1). The ice field covers approximately 40 km^2 and has larger outlets than Camp Glacier, but this site was selected because it provides access to a large subglacial cavity which enabled measurement of basal sliding velocities and direct observation of subglacial processes. The geology of the plateau comprises mainly Siluarian banded gabbros (Krauskopf, 1954; Meier, 1987) which are dissected by three joint sets, two of which are represented on Figure 27.2. The third joint set is not displayed on the joint rosette as it is almost horizontal (see Brown, 1986).

The foreland is enclosed by two 'recent' moraines which are believed to be of Little Ice Age origin (W.B. Whalley, personal communication). The foreland is characterised by a series of large rock bar steps with vertical or near-vertical lee side faces (average height 10–12 m) which trend 110°–290° (see Figure 27.3). They are assumed to be relict from an earlier phase of ice coverage because they are seen to run across the foreland, below the moraines and out into the blockfield area. It is on the stoss faces of these large rock bar steps that smaller sets of rock steps have formed. The crests of these smaller steps trend at varying directions between 230° and 270° (Figure 27.3) and many of them have lee side faces which appear to be joint-controlled. They have well-striated stoss sides, and striation directions indicate an ice flow direction of 170° for the LIA advance (Figure 27.3), which would seem favourable for exploitation of the majority of the sub-vertical joints indicated on Figure 27.2.

INFERRED ICE DYNAMICS

The existence of cavities at the bed is important in the subglacial quarrying process. Inside a cavity the imposed normal load is zero and so the frictional component of the rock strength is not mobilised. (This assumes that the cavity is not water-filled as the pressurised water may exert a similar compressive stress on the rock). It should be noted that fluctuations in water pressure have been shown under certain circum-

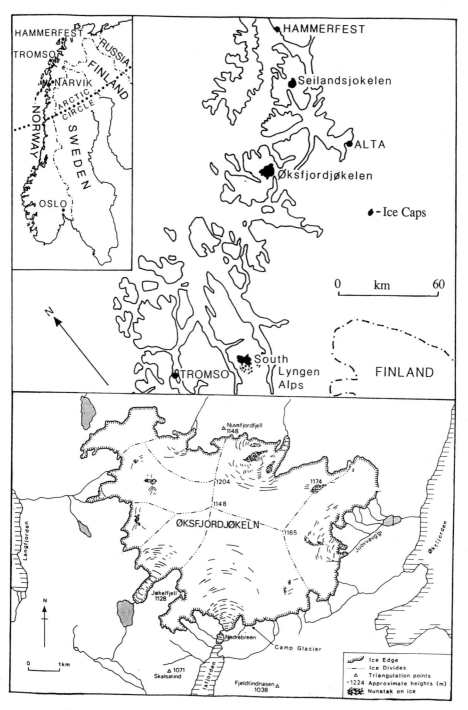

Figure 27.1 Location map of Øksfjordjøkelen and Camp Glacier

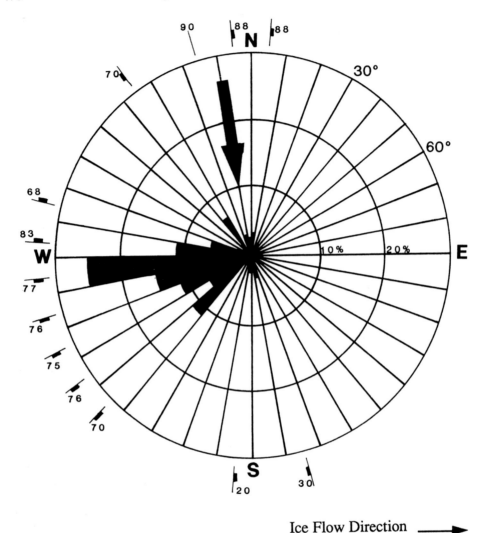

Figure 27.2 Joint rosette to show strike and dip of the two sub-vertical joint sets on the foreland and the ice flow direction associated with the LIA advance

stances to enhance the quarrying process (Röthlisberger and Iken, 1981; Iverson, 1991).

Cavities form in the lee side of rock steps provided that the velocity and the ice thickness are favourable (Boulton, 1974), i.e. when the pressure fluctuation associated with the flow (δP) is greater than the cryostatic load (ice density $\{\rho_i\}$ × gravity $\{g\}$ × ice thickness $\{h\}$):

$$\delta P/(\rho_i \, g \, h) > 1 \qquad (1)$$

An examination of the foreland showed small unstriated and unsmoothed areas on stoss surfaces which are taken to indicate regions where the ice had lost contact with its bed. Cavities are thus assumed to have been present during the LIA advance.

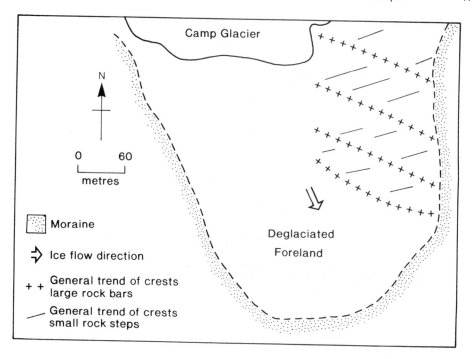

Figure 27.3 Map indicating the crests of large rock bars and small rock steps and ice flow direction (LIA advance), on the foreland of Camp Glacier

The existence of cavities associated with the fact that the rock bars are close to the snout suggests that the water present at the bed would have been free draining, at or close to atmospheric pressure, and water pressures in cavities at the bed would be accordingly low.

The basal velocity was measured over a five-week period during the summer of 1992. An ice screw embedded in the basal ice was attached to a transducer (LVDT type), and, as the ice flowed, the transducer core was pulled out and a trace given on an analogue recorder. The velocities were found to vary between 1.5 and 4 cm day^{-1}.

Reconstructions for the maximum ice thicknesses during the LIA advance were obtained by overlaying a series of down-glacier profiles of the present snout on to longitudinal profiles of the foreland. The ice depths were then measured, and the snout drawn as a contour map. From this reconstruction, transverse glacier surface profiles were constructed. These were then 'smoothed' with the use of the present cross glacier surface snout profiles. Ice depths were again calculated and found to reach a maximum thickness of 100 m.

The minimum ice thicknesses were constructed on a simpler basis. The foreland is, in effect, a topographic basin in which the minimum height of ice needed to form a moraine around it would be level with its top. Using the cross valley profiles previously mentioned and the assumption that ice would have been level with the top of the basin, minimum ice depth was calculated to be approximately 50 m.

MODELLING OF FLOW-INDUCED STRESSES

The resultant stress pattern in the bedrock produced by ice flow across it has been analysed by Boulton (1974), and from this the likelihood of failure can be measured by the ratio of rock strength to induced stress referred to as the safety factor, i.e. from equations (2) and (3).

Safety factor with cavitation =

$$\frac{\tau_c}{\dfrac{1.25 \, \eta \, V_i + 0.188 \, \rho_i \, g \, h}{\lambda}} \tag{2}$$

Safety factor with no cavitation =

$$\frac{\dfrac{\tau_c + (\rho_i \, g \, h - 10 \, \eta \, V_i) \tan \phi}{\lambda}}{\dfrac{3.13 \, \eta \, V_i}{\lambda}} \tag{3}$$

where τ_c = cohesive strength of bedrock (MPa); η = dynamic viscosity of ice taken as 3.15×10^{12} Pa s (Lliboutry, 1987); V_i = velocity of ice (m s^{-1}); ρ_i = density of ice taken as 900 kg m^{-2}; g = acceleration due to gravity taken as 10 m s^{-2}; h = ice thickness (m); λ = wavelength of bed undulations taken as 10 m (taken to be in the ratio amplitude to wavelength of 1:4); ϕ = angle of internal friction.

Failure will occur when the ratio is less than one. As stated previously, the existence of cavities is important when analysing the quarrying process (see equation (2)). Here the tendency towards failure is simply the ratio of rock strength to induced stress. However, when cavities are absent the cryostatic load mobilises the frictional component of the rock strength (the cryostatic load $\rho_i \, g \, h$ is a positive factor in equation (3)). No values for the cohesive strength of gabbroic rock could be found so values for granites were used instead (Farmer, 1968). The cohesive strength of the rock was taken to lie between 14 MPa and 50 MPa.

RESULTS

The reconstructed and measured ice dynamic values were then substituted into equation (2) and the safety factors found. From Figure 27.4, it can be seen that it would not have been possible for ice associated with the LIA advance to have fractured the bed, because the flow-induced stresses do not exceed the cohesive strength of the bed, even when it is taken at its minimum value of 14 MPa.

An investigation of the blockfield area outside the foreland showed the joint pattern repeated that of the foreland. This indicates that the joints were present before the LIA advance. From Jaeger (1959 and 1962), the shear strength of the jointed bedrock can be taken to be between one half and one third the value of the intact rock. Thus the maximum value for τ_c will be approximately 25 MPa and the minimum value

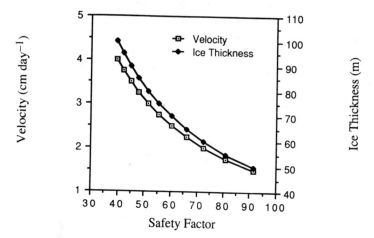

Figure 27.4 Relationship of ice thickness and velocity to the safety factors for a jointed bedrock step of wavelength 10 m and cohesive strength 14 MPa (for all combinations of ice velocity and thicknesses the inequality $\delta P/(\rho_i\, g\, h) > 1$ existed)

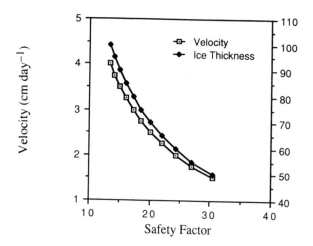

Figure 27.5 Relationship of ice thickness and velocity to the safety factors for a jointed bedrock step of wavelength 10 m and cohesive strength 4.7 MPa (for all combinations of ice velocity and thicknesses the inequality $\delta P/(\rho_i\, g\, h) > 1$ existed)

will be approximately 4.7 MPa. Figure 27.5 indicates that the flow-induced stresses in a joint-weakened bed are again insufficient to allow quarrying.

It would, therefore, appear highly unlikely that the LIA advance could have formed the rock steps shown in Figure 27.6.

PLUCKING

The three joint sets can be seen to produce joint-bounded blocks (Figure 27.7). If these appear at the free face of the cavity, then they can be removed by the glacier

Figure 27.6   Sub-vertical lee side face of a small rock step (tape = 2 m)

simply by means of the basal shear stress produced as the ice flows across the bed, without a need to fracture any of the intact bed. If we take the glacier basal shear stress to be 0.1 MPa (Nye, 1952; Boulton, 1974), then it is possible to calculate the size of a block which can be removed.

If the coefficient of friction (μ) for the joint is taken to be 1.4 (a maximum approximation based on values for the internal angle of friction for granite (Farmer,

Figure 27.7   Joint intersections on the stoss surface of a small rock step showing the production of joint-bounded blocks (tape = 2 m)

481

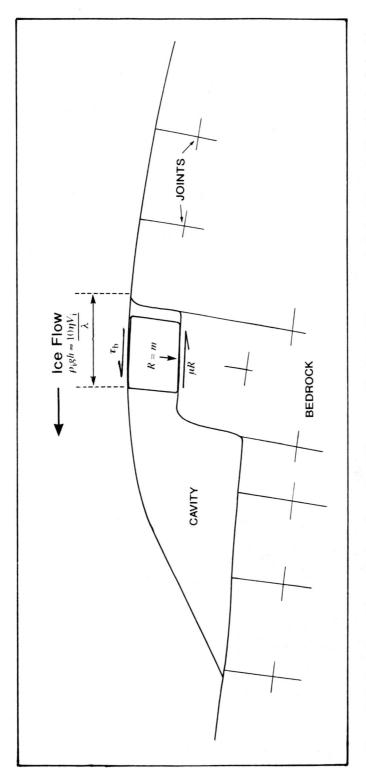

Figure 27.8 Relative position of block, cavity and bed. When $\tau_b > \mu R$ the block will be removed. (It should be noted that, ideally, R should be resolved into components parallel and normal to the bed slope. However, the slope angle is too small to be of significance)

1968)), then the force (F) required to remove a block can be given by equation (4):

$$F = \mu R \tag{4}$$

where R = the normal force acting on the block.

If the block is situated at approximately the point of cavity origin (crest of rock step), then the cryostatic pressure ($\rho_i\, g\, h$) is approximately equal to the normal pressure fluctuation across the bed ($10\, \eta\, V_i/\lambda$), and can be taken as zero (Boulton, 1974), and R is provided solely by the mass of the block, (Figure 27.8). Taking the basal shear stress (τ_b) as 0.1 MPa = 10^5 Nm^{-2} and substituting into equation (4), the normal force of the block is calculated:

$$10^5 = 1.4\, R$$
$$R\ = 7.14 \times 10^4 \text{ Nm}^{-1}$$

From this the volume of a joint-bounded block which can be removed from the bed by the shear stress produced as the ice flows across it can be estimated from equation (5):

$$F = m\, g \tag{5}$$

(where $m = R$)

$$R = 7.14 \times 10^4/10 \text{ kg}$$
$$R = 7.14 \times 10^3 \text{ kg}$$

The density of gabbro (ρ) is taken as 3000 kg m^{-3}, then from equation (5) the volume (V) of the block can be deduced:

$$V = R/\rho$$
$$V = 2.38 \text{ m}^3$$

This is the volume of a block which can be removed if it was just contacting the bed on the bottom face. However, the two vertical sides of the block will also provide an

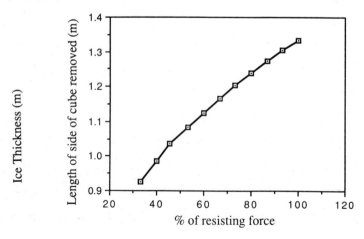

Figure 27.9 Length of side of cube which can be removed in relation to the percentage of the resisting force to removal which is derived from the block mass

additional resistive force which has to be added to the frictional force. The total resistance which this will contribute to prevent removal is unknown. Thus a number of values ranging from 0% to 66.66% (i.e. where each side exerts a resistive force equal to that provided by the bottom face) and their effect upon the size of block which can be removed is shown in Figure 27.9. Even when the value is taken as 66.66%, the ice can still remove blocks greater than 0.729 m^3.

SUBGLACIAL OBSERVATIONS AND IMPLICATIONS

Regelation and plastic deformation of the basal ice are two further mechanisms which contribute to dislodgement of joint-bounded blocks from the bed. Observations in one of the chambers beneath the glacier during the summer of 1992 revealed the extent to which these processes are active in the removal of blocks from the bed and their entrainment into the basal ice. A joint-bounded block shown in Figure 27.10 (of frontal area 0.1 m^2 and third dimension believed to be approximately 0.5 m) appeared to be undergoing removal and ultimately entrainment. The block is bounded on both vertical joints by ice and upon close examination ice could be seen being squeezed out along the horizontal 'bedding' joint. The ice is not just meltwater which has percolated down into the joints and then frozen, but is still flowing, as indicated by the 'ice spike' emanating from the joint on the left side of Figure 27.10. This process is assumed to be similar to a situation described from beneath Grinell Glacier during the years 1976–1979 by Anderson et al. (1982). The ice spike mentioned above is assumed to be a similar feature to the ice horns described in the same paper.

Figure 27.10 Joint-bounded block beneath Camp Glacier showing squeezed ice being extruded from all joints

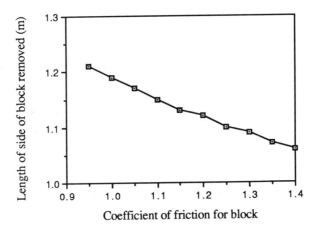

Figure 27.11 Graph showing how a reduction in the coefficient of friction due to squeezed ice affects the size of block removable. (50% of the resisting force is supplied by the normal force of the block)

It is then apparent that when this situation has been reached (i.e. the block is bounded on all sides by ice) the coefficient of friction (μ) from equation (3) will be reduced and the block can be detached either by a lower basal shear stress or a larger block can be dislodged by the same shear stress. Unfortunately, no value for the coefficient of static friction for rock on ice could be found although Barnes *et al.* (1971) and Hobbs (1974) do have some values for the coefficient of kinetic friction of ice on granite. The highest value they found for this was approximately 0.95 so this is the lowest value used in Figure 27.11.

Examination of the foreland shows that the observation in the subglacial chamber is probably not a rare occurrence, and that the glacier has indeed been responsible for the removal and entrainment of blocks. Many of the vertical lee side faces associated with the small rock steps of the LIA advance were found to have striations indicating that ice had moved over them. As other evidence showed that there had been cavities present beneath the glacier, these striations were originally proposed to have been formed by ice obeying the laws of fluid hydrodynamics as originally described by Demorest (1938) from the foreland of Clements Glacier, Montana. In places these faces were over 4 m high and thus substantially higher than those described by Demorest which were approximately 60 cm high. The presence of the squeezed ice observed in the cavity provides an acceptable answer for the above observations. Assuming blocks are being slowly plucked, they will be moved slightly forward, and the gap along the vertical joint at the rear of the block will open up. Squeezed ice containing debris can then be forced down the gap to produce the shallow striations found on these lee side faces.

Examination of the moraine showed that there were indeed a large number of blocks of the size range shown in Figure 27.9 and Figure 27.11. Some were still angular and obviously joint controlled, while others were sub-rounded. This is expected because blocks removed further up-glacier will be gradually rounded as they are used to abrade the bed.

DISCUSSION

The presence of joints in the bed beneath a glacier will play a significant role in the formation of rock steps either through bed weakening allowing stress-induced fracture or by providing blocks for plucking. Boulton (1974) has shown that beneath thick temperate ice masses stress fracture of bedrock with a cohesive strength of 12.5 MPa and wavelength of 10 m can take place. Since this value is less than the cohesive strength of granite and gabbro then stress-induced fracture would not be expected to proceed without jointing. In glacial areas with lithologies of high shear strength the role of joints, therefore, becomes very important. As the study area is a plateau ice field with very few nunataks, the tools which the ice requires for its abrasional work are found at the bed. After any unconsolidated or pre-weathered material is incorporated it is only from the subglacial quarrying process that these tools can be derived. Without the presence of joints the effectiveness of this process would be severely diminished. Joints also play a major role for all lithologies for relatively thin, slow-moving glaciers, as the joints facilitate erosion of material beneath ice thicknesses as little as 5–8 m (Anderson et al., 1982).

Variations in the sliding velocity of the glacier may have an effect upon the rock step formation. The velocity was found to vary between 1.5 and 4 cm day^{-1}, decreasing to lower values during a prolonged period of cold weather, an observation similar to the results reported by Theakstone (1967) at Østerdalsisen Norway, and Anderson et al. (1982) at Grinnell Glacier, USA. During this cold period the meltwater stream discharge was reduced substantially, and much less meltwater was observed in the cavity.

Thus the sliding velocity appears, not surprisingly, to be related to the amount of meltwater present. Associated with these velocity changes there will be changes in the stress pattern produced in the bedrock, and this in time may lead to fatigue failure of the bed, or, more likely, a loosening of the joints. During the winter months the velocity is thus assumed to reduce markedly, and large folds in the ice at the snout end of the cavity suggest that the margins may indeed freeze to the bed as proposed by Anderson et al. (1982). This process will cause a reduction in cavity size as the ice will begin to deform into the cavity. This will allow contact with other parts of the bed, and will also provide a means of incorporation of loosened material and expelled boulders (Vivian and Bocquet, 1973). This process will cause major changes in the stress pattern of the bed as the highest differences are found where the ice separates from and rejoins the bed.

CONCLUSIONS

From the modelling of the flow-induced stresses for the inferred ice dynamics, it has been shown that it is highly unlikely that the small rock steps were formed by lee side fracturing mechanisms. The joints present on the foreland of Camp Glacier have exerted a strong structural control on the formation of the rock steps, shown by the fact that many lee side faces are formed along joints. This is similar to results reported by Sharp et al. (1989) from Snowdon in Wales. Indeed, without the weakening effect of the joints on the bed the glacier could not have produced the small rock steps being exposed at the present time. Thus, joints in bedrock in the subglacial environment play a very important role in the ability of the glacier to erode its bed, especially under thin, relatively slow, sliding ice. At a larger scale, joints under thicker ice masses will probably exert some structural control on quarrying, and indeed may be responsible for the production of landforms such as riegels.

ACKNOWLEDGEMENTS

The author would like to thank W.B. Whalley and J.P. McGreevy for constructive comments during the writing of this paper. Also thanks go to Boots The Chemists PLC, Colman's of Norwich, RHM Foods Ltd, The North Face, Vango Scotland Ltd, Wild Country Ltd, The Royal Geographical Society, The 20th IGC, Dudley Stamp Memorial Fund, Gilchrist Educational Trust, who all provided assistance to the expedition, during which much of the field data was collected. During the production of this paper the author was in receipt of a DENI Postgraduate Studentship.

REFERENCES

Anderson, R. S., Hallet, B. Walder, J. and Aubry, B. F. (1982). Observations in a cavity beneath Grinnell Glacier. *Earth Surface Processes and Landforms*. **7**, 63–70.

Barnes, P., Tabor, D. and Walker, J. C. F. (1971). The friction and creep of polycrystalline ice. *Proceedings Royal Society London*, A, **324**, 127–155.

Boulton, G. S. (1974). Processes and patterns of glacier erosion. In Coates, D. R. (ed.), *Glacial Geomorphology*, State University, New York, pp. 41–87.

Brown, E. T. (ed.) (1986). *Rock Characterisation, Testing and Monitoring*. Pergamon Press, Oxford.

Demorest, M. (1938). Ice flowage as revealed by glacial striae. *Journal of Geology*, **46**, 700–725.

Drewry, D. W. (1986). *Glacial Geologic Processes*. Arnold, London.

Farmer, I. W. (1968). *Engineering Properties of Rocks*. SPON, London.

Hobbs, P. V. (1974). *Ice Physics*. Clarendon Press, Oxford.

Iverson, N. R. (1991). Potential effects of subglacial water-pressure fluctuations on quarrying. *Journal of Glaciology*, 37(125), 27–36.

Jaeger, J. C. (1959). The frictional properties of joints in rock. *Geophysique Pure et Appliqué*, **43**, 148–158.

Jaeger, J. C. (1962). *Elasticity, Fracture and Flow*. Methuen, London.

Krauskopf, K. B. (1954). Igneous and metamorphic rocks of the Øksfjord area, West Finnmark. *Norge Geologiske Undersøgelse Arbok*, **188**, 29–50.

Lliboutry, L. (1987). *Very Slow Flows of Solids. Basics of Modelling in Geodynamics and Glaciology*. Kluwer.

Matthes, F. E. (1930). Geologic History of the Yosemite Valley. *US Geological Survey Professional Paper*, **160**.

Meier, K. D. (1987). *Studien zur periglaziären Landschaftsformung in Finnmark (nordnorwegen)*. Privately published, Selbstverlag der Geographischen Gesellschaft, Hannover.

Nye, J. F. (1952). The mechanics of glacier flow. *Journal of Glaciology*, 31(18), 82–93.

Röthlisberger, H. and Iken, A. (1981). Plucking as an effect of water pressure variations at the glacier bed. *Annals of Glaciology*, **2**, 57–62.

Sharp, M., Dowdswell, J. A. and Gemmell, J. C. (1989). Reconstructing past glacier dynamics and erosion from glacial geomorphic evidence: Snowdon, North Wales. *Journal of Quaternary Science*, 4(2), 115–130.

Theakstone, W. K. (1967). Basal sliding and movement near the margin of the glacier Østerdalsisen, Norway. *Journal of Glaciology*, 6(48), 805–816.

Vivian, R. and Bocquet, G. (1973). Subglacial cavitation phenomena under the Glacier D'Argentière, Mont Blanc, France. *Journal of Glaciology*, 12(66), 439–451.

28 Silt Production from Weathering of Metamorphic Rocks in the Southern Himalaya

RITA A.M. GARDNER
Queen Mary and Westfield College, London, UK

ABSTRACT

Deeply weathered gneisses and mica-schists occur widely in the Middle Hills of Nepal, and similar lithologies exist throughout the Middle Mountains of the Himalayan range. Detailed analysis of three representative weathering profiles in the Likhu Khola basin, taken from elevations not affected by frost weathering, indicate that silts comprise more than 20% by weight throughout most of the weathering profile. The silts are thought to derive from a combination of granular disintegration and chemical decomposition. The depth of the profiles that have not suffered surface truncation through erosion typically exceed 5 metres, and may reach 7 metres or more.

These results highlight the potential of chemical weathering as a producer of large volumes of silt in the southern Himalaya, and provide a major mechanism to account for the large quantities of silt that have been produced in the Himalaya during the Late Tertiary and Quaternary and trapped on the Indo-Gangetic Plains and in the Siwalik molasse of the foreland basins. Moreover, the results suggest that cold processes are not the only mechanisms capable of forming large volumes of loessic silts on a regional scale and in mountain environments.

INTRODUCTION

The Himalayan mountains have been one of the world's major sources of silt since the Later Tertiary. Evidence of past silt production is found marginal to the southern Himalayas in the Siwalik molasse sequences of the foreland basins, which comprise fan and fluvial sediments up to 5 kilometres in thickness along the whole southern flank (approximately 2000 km) of the Himalayas, and in the extensive fluvial deposits of the Indo-Gangetic Plains and deltas. The alluvium in the Terai of Nepal typically contains 50% or more silt and clay sizes (LRMP, 1986). Locally, lacustrine sequences in the intermontane basins of Kashmir (Singh, 1982) and Kathmandu (Dongol, 1988) contain abundant silt-sized particles in sequences that exceed 1300 m thickness in Kashmir.

Climatic conditions in the Southern Himalayas and on the Plains do not generally favour the reworking, accumulation and preservation of the silts into 'loess' deposits (Gardner and Rendell, in press). Minor exceptions are on terraces in intermontane

Rock Weathering and Landform Evolution. Edited by D. A. Robinson and R. B. G. Williams
© 1994 John Wiley & Sons Ltd

basins in the rain shadow of the monsoon (e.g. Kashmir Basin; Gardner, 1989); or in topographical depressions during the drier phases of the Pleistocene (e.g. Potwar Plateau and Peshawar Basin, Pakistan; Rendell, 1989).

In the drier areas to the north and east of the High Himalaya and the Tibetan Uplands, in contrast, the silts are concentrated by aeolian reworking of fluvial sediments into loess deposits (Derbyshire, 1983; Smalley and Smalley, 1983). Most notable and extensive are those in northwest China and the North China Plains, but other substantial deposits occur in the Central Asian Republics from Turkmenistan to Kirghiz (Dodonov, 1984). Thus, Pleistocene age silts are found widely in fluvial and loess sediments marginal to the mountains. They are now believed to derive almost entirely from the Himalayan mountain and upland region (Smalley and Smalley, 1983).

Detailed data are not available from which to estimate current *silt* yields in rivers draining the Himalayas. Even though very high suspended sediment yields and denudation rates have been reported (see Narayana, 1987; Ives and Messerli, 1989; Maskey and Joshi, 1991), there are few details of the proportion of silt, as opposed to sand and clay, sizes within the materials. However, the presence of abundant silts within the Quaternary and older sequences would suggest that silt is an important component in a modern setting.

The continual availability of silt within the Himalayan system during the Quaternary raises the question as to how the silt is produced. This paper reports a case study of the generation of silt-sized particles from weathering *in situ* of high-grade metamorphic rocks in the Middle Hills of Nepal, and examines the wider implications of the findings for loess studies.

THE SOUTHERN HIMALAYAN SETTING AND POTENTIAL MECHANISMS OF SILT PRODUCTION

The southern Himalaya experience a subtropical monsoonal climate, which is modified by both altitude and the effect of the Himalayan range blocking the penetration of cold winter air from Central Asia. The intense seasonal precipitation exceeds 11 000 mm yr^{-1} in the extreme east, but declines westwards to less than 2500 mm over much of Nepal and India. (In the far west, winter precipitation, often in the form of snow, predominates.) Precipitation is also generally greatest on the southern flanks of the Himalayas, and decreases northwards. Hot summer temperatures (*c.* 30°–40°C maximum) and warm winter temperatures (*c.* 15–20°C maximum) characterise the lower and middle mountain altitudes, and at elevations below 2500 m sharp frosts are relatively rare. The regional snowline approximates 5000 m.

Lithologically, the southern Himalaya can be subdivided into three broad zones for the purpose of this paper. The youngest comprises the Siwalik (Miocene–Lower Pleistocene) and alluvial (Quaternary) sequences, whose sediments also underlie the Plains, and which consist mainly of unconsolidated gravels, sands and finer sediments. The boundary between these and the Midland Zone, which equates to the Middle Mountains (or Hills) physiographic region, is marked by the Main Boundary Fault. The Midland Series comprises the Meta-sediments (also termed the Nawakot Group)—a huge succession of phyllite, quartzite, mica and other schists, and limestones—and the Kathmandu Group of schists, phyllites, quartzites, gneisses and limestones. The Midland Series is found the length of the Himalayas, and is separated from the High Himalayan zone by the Main Central Thrust. Precambrian basement gneisses and migmatites, overlain in places by younger metamorphic and sedimentary rocks, dominate the High Himalaya (Ohta and Akiba, 1973; LRMP, 1986).

In the light of the prevailing environmental conditions, four potential mechanisms of silt production could be active in the southern Himalayas; glacial grinding, frost weathering, fluvial abrasion and chemical weathering. To date the contribution of none of these mechanisms has been empirically investigated in the Himalayan region. In particular very little is known about comminution in high energy fluvial environments, although silts have been produced experimentally by grain impacts in water (Moss et al., 1973).

Valley glaciers occupy less than 5% of the area of the southern Himalaya, despite the high elevations, owing to the low latitude. Furthermore, the glacial phases of the Pleistocene led to only a very modest increase in the extent of valley glaciation (Holmes and Street-Perrott, 1989). Taken together with the uncertainty over the ability of glacial environments to produce large amounts of silt (Nahon and Trompette, 1982; Haldorsen, 1981; Sharp and Gomez, 1985), this suggests that glacial action is not likely to be a substantial supplier of the Himalayan silts.

Frost weathering is a highly favoured mechanism for the generation of Himalayan silt. This is best exemplified by Smalley and Smalley (1983) who state 'all across India and into Bangladesh the major rivers carry loess material made by frost weathering in the Himalayan Highlands' (1983, p. 62), and who distinguish 'mountain loess' produced by frost weathering as one of the two major mechanisms responsible for most of the world's loess (the other being glacial action), including desert loesses. Their work suggests that the loesses and loess material in China, Central Asia and the Indian Subcontinent derive from frost weathering in the Himalaya. Certainly, the area of the Himalaya subject to regular and intense frost action, which coincides approximately with the High Himalaya, is far greater than that under ice, but it does not include large areas of the Middle Hills. Furthermore, the little experimental work available indicates that the impact of frost weathering on metamorphic rocks is less significant than on more porous rocks such as limestones. Experiments on schist yielded less than 10% silt sizes (by weight) after approximately 1000 frost cycles (Lautridou and Ozouf, 1982).

Environmental conditions in the non-glaciated southern Himalaya, and the Middle Hills and lower areas in particular, are conducive to chemical weathering. While there have been no studies in the Himalaya, research elsewhere has shown that chemical weathering can form silts, especially in the seasonally humid tropics (Nahon and Trompette, 1982). For example, in Kenya (Pye et al., 1985) a combination of granular disintegration and chemical decomposition is believed to be responsible for some, if not all, of the 10–60% silt-sized grains in regolith developed in high-grade metamorphic and basic igneous rocks.

STUDY AREA AND METHODOLOGY

The detailed study area comprises the Bore catchment of the Likhu Khola drainage basin, which lies immediately north of the Kathmandu intermontane basin in the Middle Hills of Nepal (Figure 28.1). Bedrock comprises gneisses and mica-schists

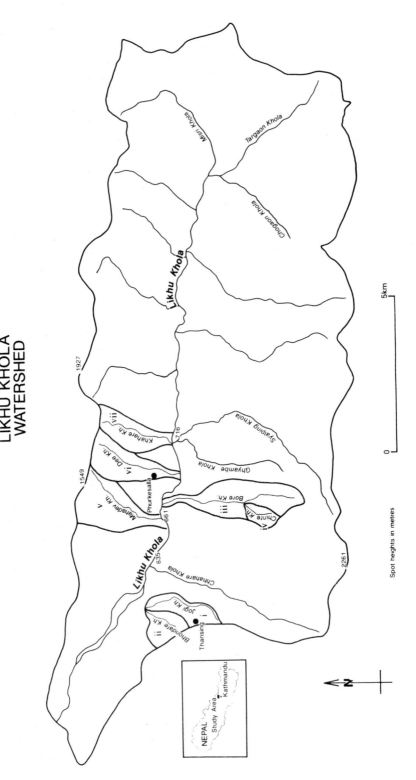

Figure 28.1 Likhu Khola watershed showing the Bore catchment

of the Kathmandu Group. There are no meterological records for the catchment; the nearest station at Kakani is approximately 4 km distant at an elevation of approximately 2500 m on the southern drainage divide of the Likhu basin. Average annual rainfall (1962–1991) recorded at Kakani is 2800 mm, over 80% of which falls between the beginning of June and the end of September. Mean monthly temperature lies between 18 and 20°C for the summer months, and reaches a minimum of 6–8°C in winter. Approximately 16 frost days per year are recorded. Meterological conditions within the Bore catchment differ slightly with, at the lower elevations, both higher temperatures and very few, if any, frost days.

Soil mapping and examination of weathering profiles over a wider range of adjacent subcatchments (as shown on Figure 28.1) indicates that the study area is representative of weathering under these litho-climatological conditions. Within the Bore Khola, a combination of linear transects along deeply gullied footpaths, and auguring was employed to identify areas exhibiting the most complete *in situ* weathering profiles, i.e. where truncation as a result of surface erosion was low; where there was no evidence of colluvial input from upslope; and where the surface soil horizons had not been altered by land management practices.

This preliminary survey showed deep weathering profiles on all land surfaces, except the youngest alluvial terraces, at elevations below 1400 m. At higher elevations deep profiles are less commonly seen, probably owing to higher surface erosion losses. The most complete profiles are to be found on the gentler slopes which comprise (a) former erosional surfaces of the Likhu river cut into bedrock and preserved largely along the east and west drainage divides of the Bore catchment (see Figure 28.2 for a profile of the western divide) and (b) the older alluvial terraces.

Three representative weathering profiles, unaffected by colluviation, were selected for detailed analysis in the Bore catchment: two east-facing and one west-facing. All are at an elevation of between 800 and 900 m and are therefore only rarely, if at all, subject to frost activity. The profiles are located on erosional terraces cut in gneisses on the drainage divides, and exist presently under Sal (*Shorea robusta*) bushes and shrub vegetation. The precise age of the Quaternary terraces and the period of weathering is unknown. An analysis of the bedrock beneath each individual section was not possible as the profiles did not extend far enough to reach unweathered bedrock. Analyses of other samples from the area show quartz, micas and feldspars to be common.

Profiles were sampled at regular intervals of 0.3 or 0.5 m, samples being taken at least 10 cm back from the exposed surface. Samples were dried and split using a riffle box prior to analysis, and subsamples extracted for the following analyses.

1. Particle size determination (dispersion in a solution of 0.05% calgon; separation by wet sieving followed by dry sieving for the >63 μm fraction; and sedimentation analysis by means of a SediGraph for the fine fraction; Gale and Hoare, 1991). (Owing to the presence of substantial amounts of platy mica flakes in the silt fraction, the sedimentation analyses have probably resulted in a systematic error showing size to be slightly finer than is actually the case.)
2. Mineralogical analysis (by means of a petrological microscope; and X-ray diffraction analyses on bulk powder samples, on separated and powdered silt fractions, and on separated <4 μm fractions in the form of pretreated oriented clay mounts; Bridley and Brown, 1984).

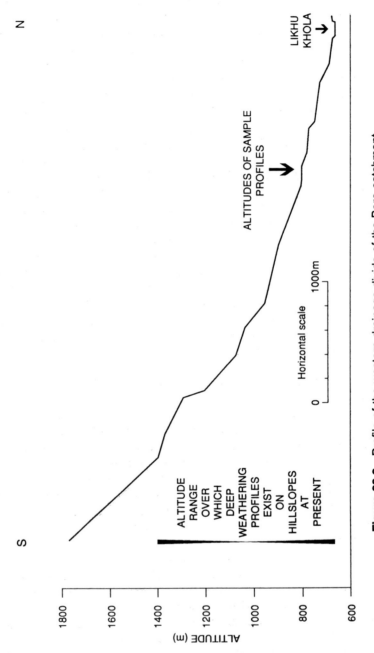

Figure 28.2 Profile of the western drainage divide of the Bore catchment

WEATHERING PRODUCTS

The three profiles show broad similarities (Figure 28.3; Table 28.1). The upper 2 to 3 m comprise a thick, rubified, pedogenic B horizon which can be subdivided into an upper silty clay loam B_1 horizon (dry colour reddish yellow) and a lower, silty loam B_2 horizon (dry colour yellowish red). The pedogenic A horizon, if present, is thin (20 cm) and reddish brown in colour. The downward transitions to the sandier C horizon (c. 1 m thick) and thence to a thick zone of granular, disintegrated bedrock (grus or saprolite) are gradual. The original structure of the gneiss can often be observed within the saprolite and relatively unweathered layers and rock 'cores' of gneiss are typically found towards the base of the profile. Weathering, generally accompanied by rubification, often proceeds preferentially along the schistose layers, at the expense of the more granular layers within the gneiss. The deepest profile examined extended to 7.5 m, and still did not reach to unweathered bedrock.

The fine fraction (<63 μm) increases from 50 to 55%, at the surface, downwards through the A and B_1 horizons, reaching a maximum of between 64 and 73% (by weight) within the lower part of the B_1 horizon at a depth of 1.0–1.5 m (Figures 28.3 and 28.4). This increase reflects a greater proportion of clays and is probably the result of translocation. The fine fraction then declines gradually down to the base of the B horizon (40–60%). Profile 4 consistently has a higher proportion of fines at all depths in the B horizon than either of the other two profiles. The C horizon and the saprolite exhibit, overall, a decline in fine content to a recorded minimum of less than 10% at depths exceeding 7.5 m in the base of profile 4. However, the fine content is highly variable, between 20 and 50%, reflecting variations in crystal sizes in the original bedrock and the differential weathering of certain layers.

Within the fines the clay size fraction (<2 μm) is variable (Table 28.1) and exhibits a strong inverse relationship with the sand fraction (Figure 28.5). In contrast, the silt size fraction lies between 20 and 30% throughout most of the profiles, with a maximum of 34%. This suggests the clay is either derived directly from chemical weathering of the sand fraction independently of the silts, which is unlikely given the particle size distributions within the silt fraction discussed below, or that silt is an early weathering or disintegration product that is subsequently weathered into clay sizes at a rate that is broadly similar to its release from the original rock.

The distribution of sizes within the silt fraction varies according to position within the profile (Table 28.1). The complete range of silt sizes is found within all samples. However, finer silts (<16 μm) dominate the B horizon, whereas coarser silts assume higher proportions lower in the weathering profile. In the B horizon the modal class is generally either 16–8 or 8–4 μm; whereas it is 32–16 μm lower down. The median size within the fine fraction reflects the changing relative proportions of both silt and clay and coarse to fine silt. Values range from 1.4 to 15.7 μm in profile 2; 1.57 to 11.89 in profile 1; and 1.45 to 13.94 in profile 4.

The common minerals present in the silt size range are quartz, micas (biotite and muscovite) and feldspars (especially orthoclase) as shown in Figure 28.6. The clay fraction is dominated by mica and kaolinite, with smaller amounts of gibbsite, hydrobiotite and chlorite. Decomposition of some remaining mica and feldspar grains along cleavages can be seen in thin section.

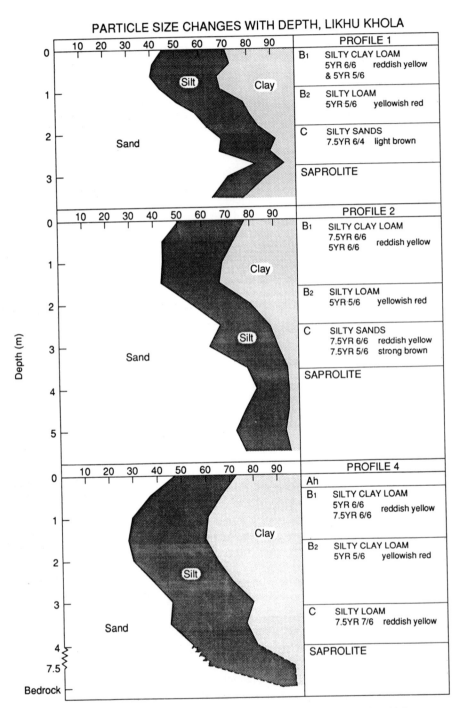

Figure 28.3 Particle size changes with depth, profiles 1, 2 and 4

Depth (m)	Sand %	Fines %	Silt %	Clay %	Ratio (silt/clay)	Median (<63 μm)	%4–5 phi	%5–6 phi	%6–7 phi	%7–8 phi	%8–9 phi
								(as of % of total silt fraction)			
Profile 1											
0.00	44.86	54.88	26.75	28.13	0.9509	1.87	5.98	16.71	26.39	26.39	24.52
0.30	41.29	48.18	31.94	16.24	1.9667	2.29	5.51	11.58	21.51	24.08	37.32
0.60	40.06	59.85	28.22	31.63	0.8922	1.57	5.95	21.65	26.75	24.84	20.80
0.90	44.25	55.02	25.03	29.99	0.8346	1.57	5.79	14.94	22.49	28.73	28.05
1.20	51.50	48.35	27.16	21.19	1.2817	3.01	9.83	20.38	24.82	23.93	20.54
1.50	58.66	40.66	22.59	18.07	1.2501	2.96	4.96	19.03	29.08	24.79	22.13
1.80	63.08	37.06	21.93	15.13	1.4494	3.86	8.25	23.57	28.77	23.07	16.32
2.10	69.70	30.73	22.49	8.24	2.7294	5.99	8.94	24.01	25.66	22.68	18.72
2.40	68.06	32.16	22.45	9.71	2.3120	7.09	14.79	29.18	24.32	19.20	12.52
2.70	82.91	17.38	11.90	5.48	2.1715	7.39	18.82	29.58	23.45	16.81	11.34
3.00	72.06	27.92	23.73	4.19	5.6635	11.89	13.65	32.62	27.10	16.73	9.90
3.50	65.71	34.06	22.06	12.00	1.8383	4.13	20.40	23.62	28.47	23.48	22.39
Profile 2											
0.00	50.80	49.44	27.74	21.70	1.2783	2.65	10.09	16.15	21.99	26.06	25.70
0.50	44.13	54.82	29.45	25.37	1.1608	2.32	3.43	12.16	23.90	27.13	33.38
1.00	43.86	56.19	25.04	31.15	0.8039	1.40	6.07	18.61	26.00	25.32	24.00
1.50	43.46	56.22	24.04	32.18	0.7470	1.38	1.41	13.19	26.12	31.74	27.54
2.00	55.73	44.34	24.91	19.43	1.2820	3.21	11.20	23.28	24.33	23.08	18.11
2.50	68.33	31.95	20.11	11.84	1.6985	4.92	14.32	27.40	23.17	20.93	14.17
3.00	62.75	36.70	29.18	7.52	3.8803	5.80	6.89	22.62	23.00	20.29	27.21
3.50	78.83	21.31	17.00	4.31	3.9443	12.19	19.29	33.88	22.41	15.18	9.24
4.00	82.75	17.27	13.88	3.39	4.0944	19.01	32.71	36.31	16.57	9.58	4.83
4.50	78.16	21.30	18.66	2.64	7.0682	12.89	14.63	33.01	24.01	15.70	12.65
5.00	74.16	26.06	20.71	5.35	3.8710	13.10	21.83	34.09	22.21	13.62	8.26
5.50	78.04	21.85	19.13	2.72	7.0331	15.71	21.33	34.81	23.37	13.85	6.64
Profile 4											
0.00	47.70	52.55	25.57	26.98	0.9477	1.60	12.67	22.29	26.98	22.72	15.33
0.50	37.02	63.92	29.60	34.32	0.8625	1.45	12.57	17.43	21.49	23.61	24.90
1.50	27.73	71.54	32.30	39.24	0.8231	1.53	1.11	13.62	23.65	30.59	31.02
2.00	30.27	68.94	34.43	34.51	0.9977	2.06	0.20	08.71	24.69	31.57	34.82
3.00	46.27	65.04	33.39	31.65	1.0550	2.61	2.43	13.45	26.12	30.97	27.04
3.50	44.85	54.46	31.68	22.78	1.3907	3.30	6.63	20.90	25.76	24.62	22.10
4.00	54.38	45.83	27.02	18.81	1.4365	4.67	12.18	29.72	26.87	20.43	10.81
7.50	62.24	37.57	34.26	3.31	10.3505	13.94	16.40	32.52	24.02	18.36	8.70
8.00	92.11	7.87	5.42	2.45	2.2122	9.19	24.54	30.63	21.96	14.02	8.86

PARTICLE SIZE DISTRIBUTION, PROFILE 2, LIKHU KHOLA

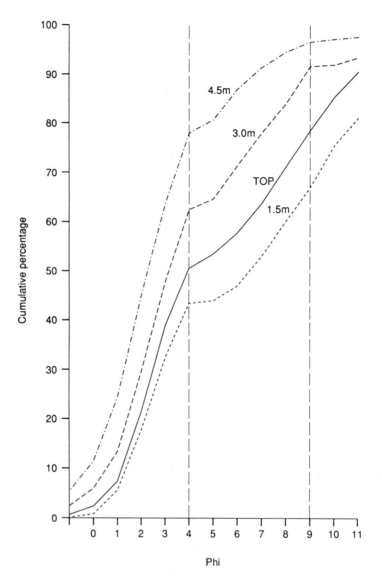

Figure 28.4 Cumulative frequency distributions of selected samples from profile 2

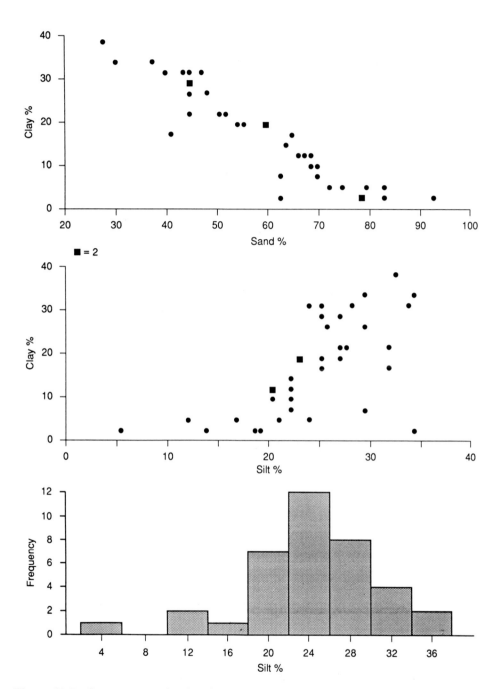

Figure 28.5 Scattergrams showing the relationships between different particle size components and histogram of silt content for all samples from profiles 1, 2 and 4

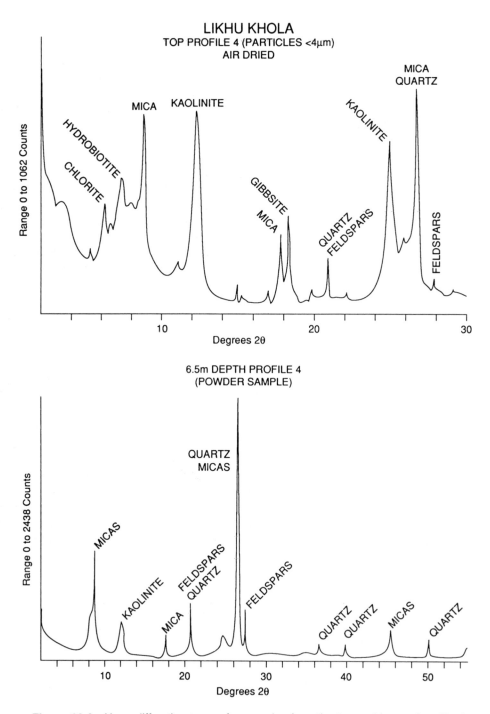

Figure 28.6   X-ray diffraction traces for samples from the top and base of profile 4

DISCUSSION

The presence of relatively uniform silt concentrations with depth in the profiles and the existence of a continuum of silt sizes within each sample suggests that the silts are derived by weathering *in situ* and have not been introduced detritally into the profiles. The elevation of the sample profiles was chosen so as to exclude frost weathering, both under present conditions and during potentially cooler glacial phases. In addition, there is no field evidence for substantial glacial-phase cooling at these elevations in the Himalayan range.

The importance of granular disintegration as an early weathering process is attested by evidence in the base of the weathering profiles. The presence of inter-crystal cracks in the less weathered 'cores'; the high silt/clay ratio; the preservation of ghost rock structures; and the disintegration of samples into individual crystals when crushed in the hand all support this interpretation. The cause of the granular disintegration remains to be investigated (see Pye, 1985 vs. Isherwood and Street, 1976). At a later stage chemical decomposition dominates, as shown by the formation of authigenic clay minerals and the increase in proportion of clay-sized particles higher in the weathering profiles.

The scale of the silt production in the weathering profiles is substantial, and it further illustrates the importance of weathering processes in the seasonally humid tropics in generating silts (Nahon and Trompette, 1982). The percentage of silt sizes compares favourably with silt contents reported to have resulted from chemical weathering of high-grade metamorphic rocks and granites under humid tropical conditions (Boulet, 1973), and from granites under temperate conditions (Norberg, 1980; Table 28.2). The silt content is, however, less than that reported for some of the thin (<1 m thick) soils in semi-arid Kenya (Pye *et al.*, 1985). It is thus the relatively high percentage of silt combined with the vertical distance over which such amounts are formed that makes these profiles so distinctive, and their potential for silt supply so convincing.

The wider implications of these results concern the importance of chemical weathering and granular disintegration in the formation of silts in the southern, humid, Himalayan region as a whole. Production resulting from these processes is likely to be greatest in, but not exclusive to, the Middle Hills zone owing to the particular combination of elevation and temperature, seasonally high rainfall and humidity, and

Table 28.2 Previous studies of the formation of silts by chemical weathering

Location	% silts	Process	Author
Queensland: dune sands	<15%	Dissolution of silica along microcracks	Pye, 1983
Upper Volta: granite	15–20%	Unknown	Boulet, 1973
Kenya: granulite; gneiss; gabbro	10–60%	Granular disintegration + chemical weathering	Pye *et al.* 1985; Pye, 1985
Denmark: granite	10–20%	Weathered quartz and feldspars	Norberg, 1980

metamorphic lithologies containing abundant easily weathered minerals. Given the enormous area over which these conditions exist, the potential for silt supply from chemical weathering is immense. Moreover, the high natural erosion rates in the Himalayan setting have the potential to remove regolith and exhume fresh rock surfaces which, in turn, suffer further weathering. Further research is needed to examine the importance of chemical weathering compared with frost weathering in the higher, cold temperate and subalpine environments at elevations over 2500 m.

Within the southern Himalaya, and on the plains to the south, most of the silt is transported and deposited by the fluvial systems; little of it subsequently accumulates to form true loess deposits owing to the prevailing climatic and tectonic conditions (Gardner and Rendell, in press). Last Glacial loesses in Pakistan and thicker deposits in the drier Kashmir intermontane basin are the exceptions. However, a comparison of the particle size distributions within the weathered regolith with those of loesses marginal to the Himalaya reveals a broad similarity (Figure 28.7).

Typical loess samples from the Himalayan margins in Tajikistan (Goudie et al., 1984), Kashmir (Gardner, 1989), Pakistan (Rendell, 1989), and further away in Lhanzou, China (Lui et al., 1966, as quoted in Derbyshire and Mellors, 1988), lie either within or very close to the envelope of particle size distributions found in the Likhu basin profiles (Figure 28.7; Pakistan details from H. Rendell). Furthermore,

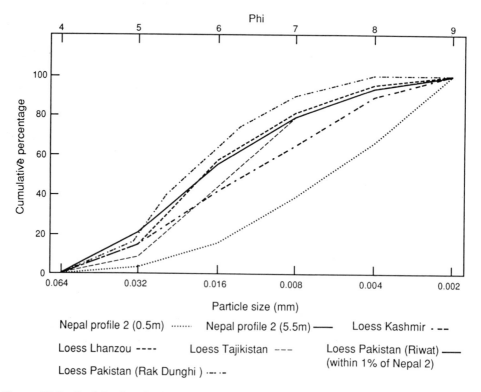

Figure 28.7 Particle size distributions within the silt size range: a comparison of South East Asian loesses and weathered regolith. (Note: different techniques have been used for the determination of the fine fraction)

the loess silts from this small sample appear to be noticeably more similar in terms of the distribution as a whole to the silts from lower in the weathering profile; that is, where chemical weathering and authigenic clay formation is less intense and where, as a result, the concentrations of medium and coarse silts are higher. The median values of the fine fraction (<4.0 phi) in the profiles also reflect the similarities. The medians range between 5.7 and 7.4 phi. This range covers most the median values recorded for loesses in Tajikistan (5.7–6.9 phi; Goudie *et al.*, 1984), China (5.1–7.2 phi; Derbyshire, 1983) and Kashmir (6.0–7.2 phi; Gardner, 1989), as well as Pakistan (5.8–6.0 phi; Rendell, 1989).

In view of the observed similarities in particle size distributions, it is suggested that the silts produced by chemical weathering and granular disintegration of metamorphic rocks in the southern Himalaya could potentially furnish a source for loess. Whether loess deposits resulted, either in the mountains or in marginal areas, would depend upon the extent to which favourable conditions existed for aeolian entrainment, transport and deposition, and for subsequent preservation of the loess.

CONCLUSION

Substantial quantities of quartz, mica and feldspar silt-sized particles are found to occur as a product of deep chemical weathering of gneisses in the Middle Hills of Nepal. Similar profiles are observed on schists in the same area. The particular combination of high- and medium-grade metamorphic lithologies, and a temperate to subtropical monsoonal climate, together with an active geomorphological environment occurs widely in the Middle Hills of the southern Himalayan Range. Overall, the potential for continuous silt production in this area is immense and chemical weathering has probably been a major contributor to the formation of the silts within the Quaternary deposits marginal to the mountains.

The potential for silt production from chemical weathering has not been adequately recognised in a literature that: (a) has tended to apply a cold environment model to all mountain regions; and (b) still sees 'cold' processes—glacial grinding and frost weathering—as the two major mechanisms producing most of the world's loess (Smalley, 1990), and silt production thereby being especially dominant during glacial phases.

ACKNOWLEDGEMENTS

This work was undertaken as part of the Royal Geographical Society's contribution to the joint IH/RGS soil erosion project in Nepal, 1991–1994. Funding by the ODA, and field support from Land Rover, is gratefully acknowledged. Thanks are extended to John Gerrard for assistance with sample collection and for helpful comments on the manuscript, and to Aysen Brown, Nick Cline, Kevin Mawdesley and Kevin Shrapel for assistance with analyses.

REFERENCES

Boulet, R. (1973). Toposequence de sols tropicaux en Haute Volta. Equilibre et disequilibre pedobioclimatique. *Memoir ORSTOM*, Paris, **85**, 1–272.

CHESTER COLLEGE LIBRARY

Brindley, G. W. and Brown, G. (eds) (1984). *Crystal Structures of Minerals and their X-ray Identification*. Mineralogical Society, London.

Derbyshire, E. (1983). Origin and characteristics of some Chinese loess at two locations in China. In Brookfield, M. E. and Ahlbrandt, T. S. (eds), *Eolian Sediments and Processes*. Elsevier, Amsterdam, pp. 69–90.

Derbyshire, E. and Mellors, T. W. (1988). Geological and geotechnical characteristics of some loess and loessic soils from China and Britain: a comparison. *Engineering Geology*, **25**, 135–175.

Dodonov, A. E. (1984). Stratigraphy and correlation of Upper Pliocene–Quaternary deposits of Central Asia. In Pecsi, M. (ed.), *Lithology and Stratigraphy of Loess and Palaeosols*. Hungarian Academy of Sciences, Budapest, pp. 201–211.

Dongol, G. M. S. (1988). Quaternary palaeoenvironments and palaeosols of the Kathmandu Basin, Nepal. Unpublished M.Sc. Thesis, University of Guelph, Canada.

Gale, S. J. and Hoare, P. G. (1991). *Quaternary Sediments*. London, Belhaven.

Gardner, R. A. M. (1989). Late Quaternary loess and palaeosols, Kashmir Valley, India. *Zeitschrift für Geomorphologie NF Supplementband*, **76**, 225–245.

Gardner, R. A. M. and Rendell, H. M. (in press). Loess, climate and orogenesis. *Zeitschrift für Geomorphologie*.

Goudie, A. S., Rendell, H. M. and Bull, P. (1984). The loess of Tajik SSR. In Miller, K. (ed.), *Proceedings of the International Karakoram Project*, Vol. 1, Cambridge University Press, Cambridge, pp. 399–412.

Haldorsen, S. (1981). Grain size distribution of subglacial till and its relation to glacial crushing and abrasion. *Boreas*, **10**, 91–105.

Holmes, J. A. and Street-Perrott, F. A. (1989). The Quaternary glacial history of Kashmir, North-West Himalaya: a revision of De Terra and Paterson's sequence. *Zeitschrift für Geomorphologie, NF Supplementband*, **76**, 195–212.

Isherwood, D. and Street, F. A. (1976). Biotite-induced grussification of the Boulder Creek granodiorite, Boulder Country, Colorado. *Bulletin Geological Society America*, **87**, 366–370.

Ives, J. D. and Messerli, B. (1989). *The Himalayan Dilemma*. Routledge, London.

Lautridou, J. P. and Ozouf, J. C. (1982). Experimental frost shattering: fifteen years of research at the Centre de Géomorphologie du CNRS. *Progress in Physical Geography*, **6**, 215–235.

LRMP. (1986). Land Resource Mapping Project, Geology Report. Kenting Earth Sciences.

Maskey, R. B. and Joshy, D. (1991). Soil and nutrient losses under different soil management practices in the Middle Mountains of central Nepal. In Shah, B. P. *et al.* (eds), *Soil Fertility and Erosion Issues in the Middle Mountains of Nepal*, Workshop Proceedings, Jikhu Khola Watershed, April 1991, pp. 105–120.

Moss, A. J., Walker, P. H. and Hutka, J. (1973). Fragmentation of granitic quartz in water. *Sedimentology*, **20**, 489–511.

Nahon, D. and Trompette, R. (1982). Origin of siltstones: glacial grinding versus weathering. *Sedimentology*, **29**, 15–35.

Narayana, D. V. V. (1987). Downstream impacts of soil conservation in the Himalayan region. *Mountain Research and Development*, **7**, 287–298.

Norberg, P. (1980). Mineralogy of a podzol formed in sandy materials in northern Denmark. *Geoderma*, **24**, 25–43.

Ohta, Y. and Akiba, C. (1973). *Geology of the Nepal Himalayas*. Saikon Publishing Co., Tokyo.

Pye, K. (1983). Formation of quartz silt during humid tropical weathering of dune sands. *Sedimentary Geology*, **34**, 267–282.

Pye, K. (1985). Granular disintegration of gneiss and migmatites. *Catena*, **12**, 191–199.

Pye, K., Goudie, A. S. and Watson, A. (1985). An introduction to the physical geography of the Kora area of central Kenya. *Geographical Journal*, **151**, 168–181.

Rendell, H. M. (1989). Loess deposition during the Late Pleistocene in northern Pakistan. *Zeitschrift für Geomorphologie NF Supplementband*, **76**, 247–255.

Sharp, M. and Gomez, B. (1985). Processes of debris comminution in the glacial environment and implications for quartz sand grain micromorphology. *Sedimentary Geology*, **46**, 33–47.

Singh, I. B. (1982). Sedimentation patterns in the Karewa Basin, Kashmir Valley, India and its geological significance. *Journal Palaeontological Society India*, **27**, 71–110.

Smalley, I. J. (1990). Possible formation mechanisms for the modal coarse-silt quartz particles in loess deposits. *Quaternary International*, **7/8**, 23–27.

Smalley, I. J. and Smalley, V. (1983). Loess material and loess deposits: formation, distribution and consequences. In Brookfield, M. E. and Ahlbrandt, T. S. (eds), *Eolian Sediments and Processes*, Elsevier, Amsterdam, pp. 51–68.

CHESTER COLLEGE LIBRARY

Geographical Index

Subject Index